Lecture Notes in Computer Science   3100

Commenced Publication in 1973
Founding and Former Series Editors:
Gerhard Goos, Juris Hartmanis, and Jan van Leeuwen

Editorial Board

Takeo Kanade
  Carnegie Mellon University, Pittsburgh, PA, USA
Josef Kittler
  University of Surrey, Guildford, UK
Jon M. Kleinberg
  Cornell University, Ithaca, NY, USA
Friedemann Mattern
  ETH Zurich, Switzerland
John C. Mitchell
  Stanford University, CA, USA
Moni Naor
  Weizmann Institute of Science, Rehovot, Israel
Oscar Nierstrasz
  University of Bern, Switzerland
C. Pandu Rangan
  Indian Institute of Technology, Madras, India
Bernhard Steffen
  University of Dortmund, Germany
Madhu Sudan
  Massachusetts Institute of Technology, MA, USA
Demetri Terzopoulos
  New York University, NY, USA
Doug Tygar
  University of California, Berkeley, CA, USA
Moshe Y. Vardi
  Rice University, Houston, TX, USA
Gerhard Weikum
  Max-Planck Institute of Computer Science, Saarbruecken, Germany

James F. Peters   Andrzej Skowron
Jerzy W. Grzymała-Busse   Bożena Kostek
Roman W. Świniarski   Marcin S. Szczuka (Eds.)

Transactions on Rough Sets I

 Springer

Editors-in-Chief

James F. Peters
University of Manitoba, Department of Electrical and Computer Engineering
Manitoba, Winnipeg, Manitoba R3T 5V6 Canada
E-mail: jfpeters@ee.umanitoba.ca

Andrzej Skowron
University of Warsaw, Institute of Mathematics
Banacha 2, 02-097 Warsaw, Poland
E-mail: skowron@mimuw.edu.pl

Volume Editors

Jerzy W. Grzymała-Busse
University of Kansas, Department of Electrical Engineering and Computer Science
3014 Eaton Hall 1520 W. 15th St., #2001 Lawrence, KS 66045-7621, USA
E-mail: Jerzy@ku.edu

Bożena Kostek
Gdansk University of Technology
Faculty of Electronics, Telecommunications and Informatics
Multimedia Systems Department, Narutowicza 11/12, 80-952 Gdansk, Poland
E-mail: bozenka@sound.eti.pg.gda.pl

Roman W. Świniarski
San Diego State University, Department of Computer Science
5500 Campanile Drive, San Diego, CA 92182-7720, USA
E-mail: rswiniar@sciences.sdsu.edu

Marcin S. Szczuka
Warsaw University, Institute of Mathematics
Banacha 2, 02-097 Warsaw, Poland
E-mail: szczuka@mimuw.edu.pl

Library of Congress Control Number: 2004108444

CR Subject Classification (1998): F.4.1, F.1, I.2, H.2.8, I.5.1, I.4

ISSN 0302-9743
ISBN 3-540-22374-6 Springer-Verlag Berlin Heidelberg New York

Springer-Verlag is a part of Springer Science+Business Media

springeronline.com

© Springer-Verlag Berlin Heidelberg 2004
Printed in Germany

Typesetting: Camera-ready by author, data conversion by Olgun Computergrafik
Printed on acid-free paper   SPIN: 11011873   06/3142   5 4 3 2 1 0

Preface

We would like to present, with great pleasure, the first volume of a new journal, Transactions on Rough Sets. This journal, part of the new journal subline in the Springer-Verlag series Lecture Notes in Computer Science, is devoted to the entire spectrum of rough set related issues, starting from logical and mathematical foundations of rough sets, through all aspects of rough set theory and its applications, data mining, knowledge discovery and intelligent information processing, to relations between rough sets and other approaches to uncertainty, vagueness, and incompleteness, such as fuzzy sets, theory of evidence, etc.

The first, pioneering papers on rough sets, written by the originator of the idea, Professor Zdzisław Pawlak, were published in the early 1980s. We are proud to dedicate this volume to our mentor, Professor Zdzisław Pawlak, who kindly enriched this volume with his contribution on philosophical, logical, and mathematical foundations of rough set theory. In his paper Professor Pawlak shows all over again the underlying ideas of rough set theory as well as its relations with Bayes' theorem, conflict analysis, flow graphs, decision networks, and decision rules.

After an overview and introductory article written by Professor Pawlak, the ten following papers represent and focus on rough set theory-related areas. Some papers provide an extension of rough set theory towards analysis of very large data sets, real data tables, data sets with missing values and rough non-deterministic information. Other theory-based papers deal with variable precision fuzzy-rough sets, consistency measure conflict profiles, and layered learning for concept synthesis. In addition, a paper on generalization of rough sets and rule extraction provides two different interpretations of rough sets. The last paper of this group addresses a partition model of granular computing.

Other topics with a more application-oriented view are covered by the following eight articles of this first volume of Transactions on Rough Sets. They can be categorized into the following groups:

- music processing,
- rough set theory applied to software design models and inductive learning programming,
- environmental engineering models,
- medical data processing,
- pattern recognition and classification.

These papers exemplify analysis and exploration of complex data sets from various domains. They provide useful insight into analyzed problems, showing for example how to compute decision rules from incomplete data. We believe that readers of this volume will better appreciate rough set theory-related trends after reading the case studies.

Many scientists and institutions have contributed to the creation and the success of the rough set community. We are very thankful to everybody within the International Rough Set Society who supported the idea of creating a new LNCS journal subline – the Transactions on Rough Sets. It would not have been possible without Professors Peters' and Skowron's invaluable initiative, thus we are especially grateful to them. We believe that this very first issue will be followed by many others, reporting new developments in the rough set domain. This issue would not have been possible without the great efforts of many anonymously acting reviewers. Here, we would like to express our sincere thanks to all of them.

Finally, we would like to express our gratitude to the LNCS editorial staff of Springer-Verlag, in particular Alfred Hofmann, Ursula Barth and Christine Günther, who supported us in a very professional way.

Throughout preparation of this volume the Editors have been supported by various research programs and funds; Jerzy Grzymała-Busse has been supported by NSF award 9972843, Bożena Kostek has been supported by the grant 4T11D01422 from the Polish Ministry for Scientific Research and Information Technology, Roman Świniarski has received support from the *"Adaptive Data Mining and Knowledge Discovery Methods for Distributed Data"* grant, awarded by Lockheed-Martin, and Marcin Szczuka and Roman Świniarski have been supported by the grant 3T11C00226 from the Polish Ministry for Scientific Research and Information Technology.

April 2004

Jerzy W. Grzymała-Busse
Bożena Kostek
Roman Świniarski
Marcin Szczuka

LNCS Transactions on Rough Sets

This journal subline has as its principal aim the fostering of professional exchanges between scientists and practitioners who are interested in the foundations and applications of rough sets. Topics include foundations and applications of rough sets as well as foundations and applications of hybrid methods combining rough sets with other approaches important for the development of intelligent systems.

The journal includes high-quality research articles accepted for publication on the basis of thorough peer reviews. Dissertations and monographs up to 250 pages that include new research results can also be considered as regular papers. Extended and revised versions of selected papers from conferences can also be included in regular or special issues of the journal.

Table of Contents

Rough Sets – Introduction

Rough Sets – Theory

Rough Sets – Applications

Some Issues on Rough Sets

Zdzisław Pawlak[1,2]

[1] Institute for Theoretical and Applied Informatics
Polish Academy of Sciences
ul. Bałtycka 5, 44-100 Gliwice, Poland
[2] Warsaw School of Information Technology*
ul. Newelska 6, 01-447 Warsaw, Poland
zpw@ii.pw.edu.pl

1 Introduction

The aim of this paper is to give rudiments of rough set theory and present some recent research directions proposed by the author.

Rough set theory is a new mathematical approach to imperfect knowledge.

The problem of imperfect knowledge has been tackled for a long time by philosophers, logicians and mathematicians. Recently it became also a crucial issue for computer scientists, particularly in the area of artificial intelligence. There are many approaches to the problem of how to understand and manipulate imperfect knowledge. The most successful one is, no doubt, the fuzzy set theory proposed by Lotfi Zadeh [1].

Rough set theory proposed by the author in [2] presents still another attempt to this problem. This theory has attracted attention of many researchers and practitioners all over the world, who have contributed essentially to its development and applications. Rough set theory overlaps with many other theories. However we will refrain to discuss these connections here. Despite this, rough set theory may be considered as an independent discipline in its own right.

Rough set theory has found many interesting applications. The rough set approach seems to be of fundamental importance to AI and cognitive sciences, especially in the areas of machine learning, knowledge acquisition, decision analysis, knowledge discovery from databases, expert systems, inductive reasoning and pattern recognition.

The main advantage of rough set theory in data analysis is that it does not need any preliminary or additional information about data – like probability in statistics, or basic probability assignment in Dempster-Shafer theory, grade of membership or the value of possibility in fuzzy set theory.

One can observe the following about the rough set approach:

- introduction of efficient algorithms for finding hidden patterns in data,
- determination of minimal sets of data (data reduction),
- evaluation of the significance of data,
- generation of sets of decision rules from data,

---

* Former University of Information Technology and Management.

J.F. Peters et al. (Eds.): Transactions on Rough Sets I, LNCS 3100, pp. 1–58, 2004.

- easy-to-understand formulation,
- straightforward interpretation of obtained results,
- suitability of many of its algorithms for parallel processing.

Rough set theory has been extended in many ways (see, e.g., [3–17]) but we will not discuss these issues in this paper.

Basic ideas of rough set theory and its extensions, as well as many interesting applications can be found in books (see, e.g., [18–27, 12, 28–30]), special issues of journals (see, e.g., [31–34, 34–38]), proceedings of international conferences (see, e.g., [39–49]), tutorials (e.g., [50–53]), and on the internet (see, e.g., www.roughsets.org, logic.mimuw.edu.pl,rsds.wsiz.rzeszow.pl).

The paper is organized as follows:

Section 2 (Basic Concepts) contains general formulation of basic ideas of rough set theory together with brief discussion of its place in classical set theory.

Section 3 (Rough Sets and Reasoning from Data) presents the application of rough set concept to reason from data (data mining).

Section 4 (Rough Sets and Bayes' Theorem) gives a new look on Bayes' theorem and shows that Bayes' rule can be used differently to that offered by classical Bayesian reasoning methodology.

Section 5 (Rough Sets and Conflict Analysis) discuses the application of rough set concept to study conflict.

In Section 6 (Data Analysis and Flow Graphs) we show that many problems in data analysis can be boiled down to flow analysis in a flow network.

This paper is a modified version of lectures delivered at the Taragona University seminar on Formal Languages and Rough Sets in August 2003.

2 Rough Sets – Basic Concepts

2.1 Introduction

In this section we give some general remarks on a concept of a set and the place of rough sets in set theory.

The concept of a set is fundamental for the whole mathematics. Modern set theory was formulated by George Cantor [54].

Bertrand Russell discovered that the intuitive notion of a set proposed by Cantor leads to antinomies [55]. Two kinds of remedy for this discontent have been proposed: axiomatization of Cantorian set theory and alternative set theories.

Another issue discussed in connection with the notion of a set or a concept is vagueness (see, e.g., [56–61]). Mathematics requires that all mathematical notions (including set) must be exact (Gottlob Frege [62]). However, philosophers and recently computer scientists have become interested in vague concepts.

In fuzzy set theory vagueness is defined by graduated membership.

Rough set theory expresses vagueness, not by means of membership, but employing a boundary region of a set. If the boundary region of a set is empty it means that the set is crisp, otherwise the set is rough (inexact). Nonempty boundary region of a set means that our knowledge about the set is not sufficient to define the set precisely.

The detailed analysis of sorities paradoxes for vague concepts using rough sets and fuzzy sets is presented in [63].

In this section the relationship between sets, fuzzy sets and rough sets will be outlined and briefly discussed.

2.2   Sets

The notion of a set is not only basic for mathematics but it also plays an important role in natural language. We often speak about sets (collections) of various objects of interest, e.g., collection of books, paintings, people etc. Intuitive meaning of a set according to some dictionaries is the following:

"A number of things of the same kind that belong or are used together."

Webster's Dictionary

"Number of things of the same kind, that belong together because they are similar or complementary to each other."

The Oxford English Dictionary

Thus a set is a collection of things which are somehow related to each other but the nature of this relationship is not specified in these definitions.

In fact these definitions are due to Cantor [54], which reads as follows:

"Unter einer Mannigfaltigkeit oder Menge verstehe ich nämlich allgenein jedes Viele, welches sich als Eines denken lässt, d.h. jeden Inbegriff bestimmter Elemente, welcher durch ein Gesetz zu einem Ganzen verbunden werden kann."

Thus according to Cantor a set is a collection of any objects, which according to some law can be considered as a whole.

All mathematical objects, e.g., relations, functions, numbers, etc., are some kind of sets. In fact set theory is needed in mathematics to provide rigor.

Russell discovered that the Cantorian notion of a set leads to antinomies (contradictions). One of the best known antinomies called the powerset antinomy goes as follows: consider (infinite) set X of all sets. Thus X is the greatest set. Let Y denote the set of all subsets of X. Obviously Y is greater then X, because the number of subsets of a set is always greater the number of its elements. Hence X is not the greatest set as assumed and we arrived at contradiction.

Thus the basic concept of mathematics, the concept of a set, is contradictory. This means that a set cannot be a collection of arbitrary elements as was stipulated by Cantor.

As a remedy for this defect several improvements of set theory have been proposed. For example,

- Axiomatic set theory (Zermello and Fraenkel, 1904).
- Theory of types (Whitehead and Russell, 1910).
- Theory of classes (v. Neumann, 1920).

All these improvements consist in restrictions, put on objects which can form a set. The restrictions are expressed by properly chosen axioms, which say how the set can be build. They are called, in contrast to Cantors' intuitive set theory, axiomatic set theories.

Instead of improvements of Cantors' set theory by its axiomatization, some mathematicians proposed escape from classical set theory by creating completely new idea of a set, which would free the theory from antinomies. Some of them are listed below.

- Mereology (Leśniewski, 1915).
- Alternative set theory (Vopenka, 1970).
- "Penumbral" set theory (Apostoli and Kanada, 1999).

No doubt the most interesting proposal was given by Stanisaw Leśniewski [64], who proposed instead of membership relation between elements and sets, employed in classical set theory, the relation of "being a part". In his set theory, called *mereology*, this relation is a fundamental one.

None of the three mentioned above "new" set theories were accepted by mathematicians, however Leśniewski's mereology attracted some attention of philosophers and recently also computer scientists, (e.g., Lech Polkowski and Andrzej Skowron [6]).

In classical set theory a set is uniquely determined by its elements. In other words, this means that every element must be uniquely classified as belonging to the set or not. In contrast, the notion of a beautiful painting is vague, because we are unable to classify uniquely all paintings into two classes: beautiful and not beautiful. Thus *beauty* is not a precise but a vague concept. That is to say the notion of a set is a *crisp* (precise) one. For example, the set of odd numbers is crisp because every number is either odd or even. In mathematics we have to use crisp notions, otherwise precise reasoning would be impossible. However philosophers for many years were interested also in *vague* (imprecise) notions.

Almost all concepts we are using in natural language are vague. Therefore common sense reasoning based on natural language must be based on vague concepts and not on classical logic. This is why vagueness is important for philosophers and recently also for computer scientists.

Vagueness is usually associated with the boundary region approach (i.e., existence of objects which cannot be uniquely classified to the set or its complement) which was first formulated in 1893 by the father of modern logic Gottlob Frege [62], who wrote:

"Der Begriff muss scharf begrenzt sein. Einem unscharf begrenzten Begriffe würde ein Bezirk entsprechen, der nicht überall eine scharfe Grenzlinie hätte, sondern stellenweise ganz verschwimmend in die Umgebung überginge. Das wäre eigentlich gar kein Bezirk; und so wird ein unscharf definirter Begriff mit Unrecht Begriff genannt. Solche begriffsartige Bildungen kann die Logik nicht als Begriffe anerkennen; es ist unmöglich, von ihnen genaue Gesetze aufzustellen. Das Gesetz des ausgeschlossenen Dritten ist ja eigentlich nur in anderer Form die Forderung, dass der Begriff scharf begrenzt sei. Ein beliebiger Gegenstand x fällt entweder unter der Begriff y, oder er fällt nicht unter ihn: *tertium non datur*."

Thus according to Frege

"The concept must have a sharp boundary. To the concept without a sharp boundary there would correspond an area that had not a sharp boundary-line all around."

That is, mathematics must use crisp, not vague concepts, otherwise it would be impossible to reason precisely.

Summing up, vagueness is

– Not allowed in mathematics.
– Interesting for philosophy.
– Necessary for computer science.

2.3   Fuzzy Sets

Zadeh proposed completely new, elegant approach to vagueness called *fuzzy set theory* [1]. In his approach an element can belong to a set to a degree $k (0 \leq k \leq 1)$, in contrast to classical set theory where an element must definitely belong or not to a set. For example, in classical set theory language we can state that one is definitely ill or healthy, whereas in fuzzy set theory we can say that someone is ill (or healthy) in 60 percent (i.e., in the degree 0.6). Of course, at once the question arises where we get the value of degree from. This issue raised a lot of discussion, but we will refrain from considering this problem here.

Thus fuzzy membership function can be presented as

$$\mu_X(x) \in < 0, 1 >,$$

where, X is a set and x is an element.

Let us observe that the definition of fuzzy set involves more advanced mathematical concepts, real numbers and functions, whereas in classical set theory the notion of a set is used as a fundamental notion of whole mathematics and is used to derive any other mathematical concepts, e.g., numbers and functions. Consequently fuzzy set theory cannot replace classical set theory, because, in fact, the theory is needed to define fuzzy sets.

Fuzzy membership function has the following properties:

$$\mu_{U-X}(x) = 1 - \mu_X(x) \text{ for any } x \in U, \tag{1}$$
$$\mu_{X \cup Y}(x) = max(\mu_X(x), \mu_Y(x)) \text{ for any } x \in U,$$
$$\mu_{X \cap Y}(x) = min(\mu_X(x), \mu_Y(x)) \text{ for any } x \in U.$$

This means that the membership of an element to the union and intersection of sets is uniquely determined by its membership to constituent sets. This is a very nice property and allows very simple operations on fuzzy sets, which is a very important feature both theoretically and practically.

Fuzzy set theory and its applications developed very extensively over recent years and attracted attention of practitioners, logicians and philosophers worldwide.

2.4 Rough Sets

Rough set theory [2, 18] is still another approach to vagueness. Similarly to fuzzy set theory it is not an alternative to classical set theory but it is embedded in it. Rough set theory can be viewed as a specific implementation of Frege's idea of vagueness, i.e., imprecision in this approach is expressed by a boundary region of a set, and not by a partial membership, like in fuzzy set theory.

Rough set concept can be defined quite generally by means of topological operations, *interior* and *closure*, called *approximations*.

Let us describe this problem more precisely. Suppose we are given a set of objects U called the *universe* and an indiscernibility relation $R \subseteq U \times U$, representing our lack of knowledge about elements of U. For the sake of simplicity we assume that R is an equivalence relation. Let X be a subset of U. We want to characterize the set X with respect to R. To this end we will need the basic concepts of rough set theory given below.

- The *lower approximation* of a set X (with respect to R) is the set of all objects, which can be for *certain* classified as X with respect to R (are *certainly* X with respect to R).
- The *upper approximation* of a set X (with respect to R) is the set of all objects which can be *possibly* classified as X with respect to R (are *possibly* X with respect to R).
- The *boundary region* of a set X (with respect to R) is the set of all objects, which can be classified neither as X nor as not-X with respect to R.

Now we are ready to give the definition of rough sets.

- Set X is *crisp* (exact with respect to R), if the boundary region of X is empty.
- Set X is *rough* (inexact with respect to R), if the boundary region of X is nonempty.

Thus a set is *rough* (imprecise) if it has nonempty boundary region; otherwise the set is *crisp* (precise). This is exactly the idea of vagueness proposed by Frege.

The approximations and the boundary region can be defined more precisely. To this end we need some additional notation.

The equivalence class of R determined by element x will be denoted by $R(x)$. The indiscernibility relation in certain sense describes our lack of knowledge about the universe. Equivalence classes of the indiscernibility relation, called *granules* generated by R, represent elementary portion of knowledge we are able to perceive due to R. Thus in view of the indiscernibility relation, in general, we are unable to observe individual objects but we are forced to reason only about the accessible granules of knowledge.

Formal definitions of approximations and the boundary region are as follows:

R-lower approximation of X

$$R_*(X) = \bigcup_{x \in U} \{R(x) : R(x) \subseteq X\}, \qquad (2)$$

R-upper approximation of X

$$R^*(X) = \bigcup_{x \in U} \{R(x) : R(x) \cap X \neq \emptyset\}, \tag{3}$$

R-boundary region of X

$$BN_R(X) = R^*(X) - R_*(X). \tag{4}$$

As we can see from the definition approximations are expressed in terms of granules of knowledge. The lower approximation of a set is union of all granules which are entirely included in the set; the upper approximation – is union of all granules which have non-empty intersection with the set; the boundary region of set is the difference between the upper and the lower approximation.

In other words, due to the granularity of knowledge, rough sets cannot be characterized by using available knowledge. Therefore with every rough set we associate two *crisp* sets, called its *lower* and *upper approximation*. Intuitively, the lower approximation of a set consists of all elements that *surely* belong to the set, whereas the upper approximation of the set constitutes of all elements that *possibly* belong to the set, and the *boundary region* of the set consists of all elements that cannot be classified uniquely to the set or its complement, by employing available knowledge. Thus any rough set, in contrast to a crisp set, has a non-empty boundary region. The approximation definition is clearly depicted in Figure 1.

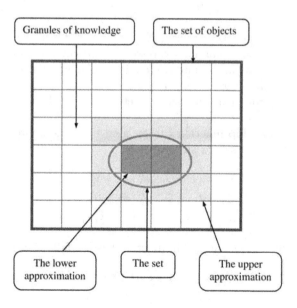

Fig. 1. A rough set

Approximations have the following properties:

$$R_*(X) \subseteq X \subseteq R^*(X), \tag{5}$$
$$R_*(\emptyset) = R^*(\emptyset) = \emptyset; R_*(U) = R^*(U) = U,$$
$$R^*(X \cup Y) = R^*(X) \cup R^*(Y),$$
$$R_*(X \cap Y) = R_*(X) \cap R_*(Y),$$
$$R_*(X \cup Y) \supseteq R_*(X) \cup R_*(Y),$$
$$R^*(X \cap Y) \subseteq R^*(X) \cap R^*(Y),$$
$$X \subseteq Y \rightarrow R_*(X) \subseteq R_*(Y) \& R^*(X) \subseteq R^*(Y),$$
$$R_*(-X) = -R^*(X),$$
$$R^*(-X) = -R_*(X),$$
$$R_*R_*(X) = R^*R_*(X) = R_*(X),$$
$$R^*R^*(X) = R_*R^*(X) = R^*(X).$$

It is easily seen that approximations are in fact interior and closure operations in a topology generated by the indiscernibility relation. Thus fuzzy set theory and rough set theory require completely different mathematical setting.

Rough sets can be also defined employing, instead of approximation, rough membership function [65]

$$\mu_X^R : U \rightarrow < 0, 1 >, \tag{6}$$

where

$$\mu_X^R(x) = \frac{card(X \cap R(x))}{card(R(x))}, \tag{7}$$

and $card(X)$ denotes the cardinality of X.

The rough membership function expresses conditional probability that x belongs to X given R and can be interpreted as a degree that x belongs to X in view of information about x expressed by R.

The meaning of rough membership function can be depicted as shown in Figure 2.

The rough membership function can be used to define approximations and the boundary region of a set, as shown below:

$$R_*(X) = \{x \in U : \mu_X^R(x) = 1\}, \tag{8}$$
$$R^*(X) = \{x \in U : \mu_X^R(x) > 0\},$$
$$BN_R(X) = \{x \in U : 0 < \mu_X^R(x) < 1\}.$$

It can be shown that the membership function has the following properties [65]:

$$\mu_X^R(x) = 1 \text{ iff } x \in R_*(X), \tag{9}$$
$$\mu_X^R(x) = 0 \text{ iff } x \in U - R^*(X),$$
$$0 < \mu_X^R(x) < 1 \text{ iff } x \in BN_R(X),$$

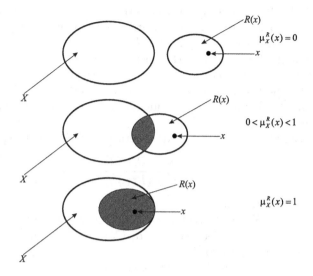

Fig. 2. Rough membership function

$$\mu_{U-X}^{R}(x) = 1 - \mu_{X}^{R}(x) \text{ for any } x \in U,$$
$$\mu_{X \cup Y}^{R}(x) \geq max(\mu_{X}^{R}(x), \mu_{Y}^{R}(x)) \text{ for any } x \in U,$$
$$\mu_{X \cap Y}^{R}(x) \leq min(\mu_{X}^{R}(x), \mu_{Y}^{R}(x)) \text{ for any } x \in U.$$

From the properties it follows that the rough membership differs essentially from the fuzzy membership, because the membership for union and intersection of sets, in general, cannot be computed as in the case of fuzzy sets from their constituents membership. Thus formally the rough membership is a generalization of fuzzy membership. Besides, the rough membership function, in contrast to fuzzy membership function, has a probabilistic flavour.

Now we can give two definitions of rough sets.

Set X is *rough* with respect to R if $R_*(X) \neq R^*(X)$.

Set X *rough* with respect to R if for some x, $0 < \mu_{X}^{R}(x) < 1$.

It is interesting to observe that the above definitions are not equivalent [65], but we will not discuss this issue here.

One can define the following four basic classes of rough sets, i.e., four categories of vagueness:

$$R_*(X) \neq \emptyset \text{ and } R^*(X) \neq U, \text{ iff } X \text{ is } roughly \text{ } R\text{-}definable, \qquad (10)$$
$$R_*(X) = \emptyset \text{ and } R^*(X) \neq U, \text{ iff } X \text{ is } internally \text{ } R\text{-}indefinable,$$
$$R_*(X) \neq \emptyset \text{ and } R^*(X) = U, \text{ iff } X \text{ is } externally \text{ } R\text{-}definable,$$
$$R_*(X) = \emptyset \text{ and } R^*(X) = U, \text{ iff } X \text{ is } totally \text{ } R\text{-}indefinable.$$

The intuitive meaning of this classification is the following.

If X is *roughly R-definable*, this means that we are able to decide for some elements of U whether they belong to X or $-X$, using R.

If X is internally R-indefinable, this means that we are able to decide whether some elements of U belong to $-X$, but we are unable to decide for any element of U, whether it belongs to X or not, using R.

If X is externally R-indefinable, this means that we are able to decide for some elements of U whether they belong to X, but we are unable to decide, for any element of U whether it belongs to $-X$ or not, using R.

If X is totally R-indefinable, we are unable to decide for any element of U whether it belongs to X or $-X$, using R.

A rough set can also be characterized numerically by the following coefficient

$$\alpha_R(X) = \frac{card(R_*(X))}{card(R^*(X))},\qquad(11)$$

called *accuracy of approximation.*

Obviously, $0 \le \alpha_R(X) \le 1$. If $\alpha_R(X) = 1$, X is *crisp* with respect to R (X is *precise* with respect to R), and otherwise, if $\alpha_R(X) < 1$, X is *rough* with respect to R (X is *vague* with respect to R).

It is interesting to compare definitions of classical sets, fuzzy sets and rough sets. Classical set is a primitive notion and is defined intuitively or axiomatically. Fuzzy sets are defined by employing the fuzzy membership function, which involves advanced mathematical structures, numbers and functions. Rough sets are defined by approximations. Thus this definition also requires advanced mathematical concepts.

Let us also mention that rough set theory clearly distinguishes two very important concepts, vagueness and uncertainty, very often confused in the AI literature. Vagueness is the property of sets and can be described by approximations, whereas uncertainty is the property of elements of a set and can expressed by the rough membership function.

3   Rough Sets and Reasoning from Data

3.1   Introduction

In this section we define basic concepts of rough set theory in terms of data, in contrast to general formulation presented in Section 2. This is necessary if we want to apply rough sets to reason from data.

In what follows we assume that, in contrast to classical set theory, we have some additional data (information, knowledge) about elements of a *universe of discourse*. Elements that exhibit the same features are indiscernible (similar) and form blocks that can be understood as *elementary granules* (*concepts*) of knowledge about the universe. For example, patients suffering from a certain disease, displaying the same symptoms are indiscernible and may be thought of as representing a granule (disease unit) of medical knowledge. These granules can be considered as elementary building blocks of knowledge. Elementary concepts can be combined into compound concepts, i.e., concepts that are uniquely determined in terms of elementary concepts. Any union of elementary sets is called a *crisp* set, and any other sets are referred to as *rough* (*vague, imprecise*).

3.2   An Example

Before we will formulate the above ideas more precisely let us consider a simple tutorial example.

Data are often presented as a table, columns of which are labeled by *attributes*, rows by *objects* of interest and entries of the table are *attribute values*. For example, in a table containing information about patients suffering from a certain disease objects are *patients* (strictly speaking their ID's), attributes can be, for example, *blood pressure, body temperature* etc., whereas the entry corresponding to object *Smith* and the attribute *blood preasure* can be *normal*. Such tables are known as *information tables, attribute-value* tables or *information system*. We will use here the term *information system*.

Below an example of information system is presented.

Suppose we are given data about 6 patients, as shown in Table 1.

Table 1. Exemplary information system

Patient	Headache	Muscle-pain	Temperature	Flu
p1	no	yes	high	yes
p2	yes	no	high	yes
p3	yes	yes	very high	yes
p4	no	yes	normal	no
p5	yes	no	high	no
p6	no	yes	very high	yes

Columns of the table are labeled by attributes (symptoms) and rows – by objects (patients), whereas entries of the table are attribute values. Thus each row of the table can be seen as information about specific patient.

For example, patient $p2$ is characterized in the table by the following attribute-value set

(*Headache, yes*), (*Muscle-pain, no*), (*Temperature, high*), (*Flu, yes*),

which form the information about the patient.

In the table patients $p2$, $p3$ and $p5$ are indiscernible with respect to the attribute *Headache*, patients $p3$ and $p6$ are indiscernible with respect to attributes *Muscle-pain* and *Flu*, and patients $p2$ and $p5$ are indiscernible with respect to attributes *Headache, Muscle-pain* and *Temperature*. Hence, for example, the attribute *Headache* generates two elementary sets $\{p2, p3, p5\}$ and $\{p1, p4, p6\}$, whereas the attributes *Headache* and *Muscle-pain* form the following elementary sets: $\{p1, p4, p6\}$, $\{p2, p5\}$ and $\{p3\}$. Similarly one can define elementary sets generated by any subset of attributes.

Patient $p2$ has flu, whereas patient $p5$ does not, and they are indiscernible with respect to the attributes *Headache, Muscle-pain* and *Temperature*, hence flu cannot be characterized in terms of attributes *Headache, Muscle-pain* and

Temperature. Hence $p2$ and $p5$ are the boundary-line cases, which cannot be properly classified in view of the available knowledge. The remaining patients $p1$, $p3$ and $p6$ display symptoms which enable us to classify them with certainty as having flu, patients $p2$ and $p5$ cannot be excluded as having flu and patient $p4$ for sure does not have flu, in view of the displayed symptoms. Thus the lower approximation of the set of patients having flu is the set $\{p1, p3, p6\}$ and the upper approximation of this set is the set $\{p1, p2, p3, p5, p6\}$, whereas the boundary-line cases are patients $p2$ and $p5$. Similarly $p4$ does not have flu and $p2$, $p5$ cannot be excluded as having flu, thus the lower approximation of this concept is the set $\{p4\}$ whereas – the upper approximation – is the set $\{p2, p4, p5\}$ and the boundary region of the concept "not flu" is the set $\{p2, p5\}$, the same as in the previous case.

3.3   Information Systems

Now, we are ready to formulate basic concepts of rough set theory using data.

Suppose we are given two finite, non-empty sets U and A, where U is the *universe*, and A – a set of *attributes*. The pair $S = (U, A)$ will be called an *information system*. With every attribute $a \in A$ we associate a set V_a, of its *values*, called the *domain* of a. Any subset B of A determines a binary relation $I(B)$ on U, which will be called an *indiscernibility relation*, and is defined as follows:

$$xI(B)y \text{ if and only if } a(x) = a(y) \text{ for every } a \in A, \tag{12}$$

where $a(x)$ denotes the value of attribute a for element x.

Obviously $I(B)$ is an equivalence relation. The family of all equivalence classes of $I(B)$, i.e., partition determined by B, will be denoted by $U/I(B)$, or simple U/B; an equivalence class of $I(B)$, i.e., block of the partition U/B, containing x will be denoted by $B(x)$.

If (x, y) belongs to $I(B)$ we will say that x and y are *B-indiscernible*. Equivalence classes of the relation $I(B)$ (or blocks of the partition U/B) are referred to as *B-elementary sets*. In the rough set approach the elementary sets are the basic building blocks (concepts) of our knowledge about reality.

Now approximations can be defined as follows:

$$B_*(X) = \{x \in U : B(x) \subseteq X\}, \tag{13}$$

$$B^*(X) = \{x \in U : B(x) \cap X \neq \emptyset\}, \tag{14}$$

called the *B-lower* and the *B-upper approximation* of X, respectively. The set

$$BN_B(X) = B^*(X) - B_*(X), \tag{15}$$

will be referred to as the *B-boundary region* of X.

If the boundary region of X is the empty set, i.e., $BN_B(X) = \emptyset$, then the set X is *crisp* (*exact*) with respect to B; in the opposite case, i.e., if $BN_B(X) \neq \emptyset$, the set X is referred to as *rough* (*inexact*) with respect to B.

The properties of approximations can be presented now as:

$$B_*(X) \subseteq X \subseteq B^*(X), \tag{16}$$
$$B_*(\emptyset) = B^*(\emptyset) = \emptyset, B_*(U) = B^*(U) = U,$$
$$B^*(X \cup Y) = B^*(X) \cup B^*(Y),$$
$$B_*(X \cap Y) = B_*(X) \cap B_*(Y),$$
$$X \subseteq Y \text{ implies } B_*(X) \subseteq B_*(Y) \text{ and } B^*(X) \subseteq B^*(Y),$$
$$B_*(X \cup Y) \supseteq B_*(X) \cup B_*(Y),$$
$$B^*(X \cap Y) \subseteq B^*(X) \cap B^*(Y),$$
$$B_*(-X) = -B^*(X),$$
$$B^*(-X) = -B_*(X),$$
$$B_*(B_*(X)) = B^*(B_*(X)) = B_*(X),$$
$$B^*(B^*(X)) = B_*(B^*(X)) = B^*(X).$$

3.4 Decision Tables

An information system in which we distinguish two classes of attributes, called *condition* and *decision* (*action*) attributes are called *decision tables*.

The condition and decision attributes define partitions of the decision table universe. We aim at approximation of the partition defined by the decision attributes by means of the partition defined by the condition attributes.

For example, in Table 1 attributes *Headache, Muscle-pain* and *Temperature* can be considered as condition attributes, whereas the attribute *Flu* – as a decision attribute. A decision table with condition attributes C and decision attributes D will be denoted by $S = (U, C, D)$.

Each row of a decision table determines a *decision rule*, which specifies decisions (*actions*) that should be taken when conditions pointed out by *condition* attributes are satisfied. For example, in Table 1 the condition (*Headache, no*), (*Muscle-pain, yes*), (*Temperature, high*) determines uniquely the decision (*Flu, yes*). Objects in a decision table are used as labels of decision rules.

Decision rules 2) and 5) in Table 1 have the same conditions but different decisions. Such rules are called *inconsistent* (*nondeterministic, conflicting*); otherwise the rules are referred to as *consistent* (*certain, deterministic, non-conflicting*). Sometimes consistent decision rules are called *sure* rules, and inconsistent rules are called *possible* rules. Decision tables containing inconsistent decision rules are called *inconsistent* (*nondeterministic, conflicting*); otherwise the table is *consistent* (*deterministic, non-conflicting*).

The number of consistent rules to all rules in a decision table can be used as *consistency factor* of the decision table, and will be denoted by $\gamma(C, D)$, where C and D are condition and decision attributes respectively. Thus if $\gamma(C, D) = 1$ the decision table is consistent and if $\gamma(C, D) \neq 1$ the decision table is inconsistent. For example, for Table 1, we have $\gamma(C, D) = 4/6$.

Decision rules are often presented in a form called *if... then...* rules. For example, rule 1) in Table 1 can be presented as follows

if (Headache,no) and (Muscle-pain,yes) and (Temperature,high) then (Flu,yes).

A set of decision rules is called a *decision algorithm*. Thus with each decision table we can associate a decision algorithm consisting of all decision rules occurring in the decision table.

We must however, make distinction between decision tables and decision algorithms. A decision table is a collection of data, whereas a decision algorithm is a collection of rules, e.g., logical expressions. To deal with data we use various mathematical methods, e.g., statistics but to analyze rules we must employ logical tools. Thus these two approaches are not equivalent, however for simplicity we will often present here decision rules in form of implications, without referring deeper to their logical nature, as it is often practiced in AI.

3.5 Dependency of Attributes

Another important issue in data analysis is discovering *dependencies* between attributes. Intuitively, a set of attributes D *depends totally* on a set of attributes C, denoted $C \Rightarrow D$, if all values of attributes from D are uniquely determined by values of attributes from C. In other words, D depends totally on C, if there exists a functional dependency between values of D and C. For example, in Table 1 there are no total dependencies whatsoever. If in Table 1, the value of the attribute *Temperature* for patient $p5$ were "*no*" instead of "*high*", there would be a total dependency $\{Temperature\} \Rightarrow \{Flu\}$, because to each value of the attribute *Temperature* there would correspond unique value of the attribute *Flu*.

We would need also a more general concept of dependency of attributes, called a *partial dependency* of attributes.

Let us depict the idea by example, referring to Table 1. In this table, for example, the attribute *Temperature* determines uniquely only some values of the attribute *Flu*. That is, (*Temperature, very high*) implies (*Flu, yes*), similarly (*Temperature, normal*) implies (*Flu, no*), but (*Temperature, high*) does not imply always (*Flu, yes*). Thus the partial dependency means that only some values of D are determined by values of C.

Formally dependency can be defined in the following way. Let D and C be subsets of A.

We will say that D *depends on* C in a *degree* k $(0 \leq k \leq 1)$, denoted $C \Rightarrow_k D$, if $k = \gamma(C, D)$.

If $k = 1$ we say that D *depends totally* on C, and if $k < 1$, we say that D *depends partially* (in a *degree* k) on C.

The coefficient k expresses the ratio of all elements of the universe, which can be properly classified to blocks of the partition U/D, employing attributes C.

Thus the concept of dependency of attributes is strictly connected with that of consistency of the decision table.

For example, for dependency {*Headache, Muscle-pain, Temperature*} ⇒ {Flu} we get $k = 4/6 = 2/3$, because four out of six patients can be uniquely classified as having flu or not, employing attributes *Headache, Muscle-pain* and *Temperature*.

If we were interested in how exactly patients can be diagnosed using only the attribute *Temperature*, that is – in the degree of the dependence {*Temperature*} ⇒ {*Flu*}, we would get $k = 3/6 = 1/2$, since in this case only three patients $p3, p4$ and $p6$ out of six can be uniquely classified as having flu. In contrast to the previous case patient $p4$ cannot be classified now as having flu or not. Hence the single attribute *Temperature* offers worse classification than the whole set of attributes *Headache, Muscle-pain* and *Temperature*. It is interesting to observe that neither *Headache* nor *Muscle-pain* can be used to recognize flu, because for both dependencies {*Headache*} ⇒ {*Flu*} and {*Muscle-pain*} ⇒ {*Flu*} we have $k = 0$.

It can be easily seen that if D depends totally on C then $I(C) \subseteq I(D)$. That means that the partition generated by C is finer than the partition generated by D. Observe, that the concept of dependency discussed above corresponds to that considered in relational databases.

If D *depends in degree* $k, 0 \leq k \leq 1$, on C, then

$$\gamma(C, D) = \frac{card(POS_C(D))}{card(U)}, \tag{17}$$

where

$$POS_C(D) = \bigcup_{X \in U/I(D)} C_*(X). \tag{18}$$

The expression $POS_C(D)$, called a *positive region* of the partition U/D with respect to C, is the set of all elements of U that can be uniquely classified to blocks of the partition U/D, by means of C.

Summing up: D is *totally (partially)* dependent on C, if *all (some)* elements of the universe U can be uniquely classified to blocks of the partition U/D, employing C.

3.6   Reduction of Attributes

We often face a question whether we can remove some data from a data table preserving its basic properties, that is – whether a table contains some superfluous data.

For example, it is easily seen that if we drop in Table 1 either the attribute *Headache* or *Muscle-pain* we get the data set which is equivalent to the original one, in regard to approximations and dependencies. That is we get in this case the same accuracy of approximation and degree of dependencies as in the original table, however using smaller set of attributes.

In order to express the above idea more precisely we need some auxiliary notions. Let B be a subset of A and let a belong to B.

- We say that a is *dispensable* in B if $I(B) = I(B - \{a\})$; otherwise a is *indispensable* in B.
- Set B is *independent* if all its attributes are indispensable.
- Subset B' of B is a *reduct* of B if B' is independent and $I(B') = I(B)$.

Thus a reduct is a set of attributes that preserves partition. This means that a reduct is the minimal subset of attributes that enables the same classification of elements of the universe as the whole set of attributes. In other words, attributes that do not belong to a reduct are superfluous with regard to classification of elements of the universe.

Reducts have several important properties. In what follows we will present two of them.

First, we define a notion of a *core of attributes.*

Let B be a subset of A. The *core* of B is the set off all indispensable attributes of B.

The following is an important property, connecting the notion of the core and reducts

$$Core(B) = \bigcap Red(B), \qquad (19)$$

where $Red(B)$ is the set off all reducts of B.

Because the core is the intersection of all reducts, it is included in every reduct, i.e., each element of the core belongs to some reduct. Thus, in a sense, the core is the most important subset of attributes, for none of its elements can be removed without affecting the classification power of attributes.

To further simplification of an information table we can eliminate some values of attribute from the table in such a way that we are still able to discern objects in the table as the original one. To this end we can apply similar procedure as to eliminate superfluous attributes, which is defined next.

- We will say that the value of attribute $a \in B$, is *dispensable* for x, if $B(x) = B^a(x)$, where $B^a = B - \{a\}$; otherwise the value of attribute a is *indispensable* for x.
- If for every attribute $a \in B$ the value of a is indispensable for x, then B will be called *orthogonal* for x.
- Subset $B' \subseteq B$ is a *value reduct* of B for x, iff B' is orthogonal for x and $B(x) = B'(x)$.

The set of all indispensable values of attributes in B for x will be called the *value core* of B for x, and will be denoted $CORE^x(B)$.

Also in this case we have

$$CORE^x(B) = \bigcap Red^x(B), \qquad (20)$$

where $Red^x(B)$ is the family of all reducts of B for x.

Suppose we are given a dependency $C \Rightarrow D$. It may happen that the set D depends not on the whole set C but on its subset C' and therefore we might be interested to find this subset. In order to solve this problem we need the notion of a *relative reduct*, which will be defined and discussed next.

Let $C, D \subseteq A$. Obviously if $C' \subseteq C$ is a D-reduct of C, then C' is a minimal subset of C such that

$$\gamma(C, D) = \gamma(C', D). \tag{21}$$

- We will say that attribute $a \in C$ is D-*dispensable* in C, if $POS_C(D) = POS_{(C-\{a\})}(D)$; otherwise the attribute a is D-*indispensable* in C.
- If all attributes $a \in C$ are C-indispensable in C, then C will be called D-*independent*.
- Subset $C' \subseteq C$ is a D-*reduct* of C, iff C' is D-independent and $POS_C(D) = POS_{C'}(D)$.

The set of all D-indispensable attributes in C will be called $D - core$ of C, and will be denoted by $CORE_D(C)$. In this case we have also the property

$$CORE_D(C) = \bigcap Red_D(C), \tag{22}$$

where $Red_D(C)$ is the family of all D-reducts of C.

If $D = C$ we will get the previous definitions.

For example, in Table 1 there are two relative reducts with respect to *Flu*, {*Headache, Temperature*} and {*Muscle-pain, Temperature*} of the set of condition attributes *Headache, Muscle-pain, Temperature*. That means that either the attribute *Headache* or *Muscle-pain* can be eliminated from the table and consequently instead of Table 1 we can use either Table 2 or Table 3. For Table 1 the relative core of with respect to the set {*Headache, Muscle-pain, Temperature*} is the *Temperature*. This confirms our previous considerations showing that Temperature is the only symptom that enables, at least, partial diagnosis of patients.

Table 2. Data table obtained from Table 1 by drooping the attribute Muscle-pain

Patient	Headache	Temperature	Flu
p1	no	high	yes
p2	yes	high	yes
p3	yes	very high	yes
p4	no	normal	no
p5	yes	high	no
p6	no	very high	yes

Table 3. Data table obtained from Table 1 by drooping the attribute Headache

Patient	Muscle-pain	Temperature	Flu
p1	yes	high	yes
p2	no	high	yes
p3	yes	very high	yes
p4	yes	normal	no
p5	no	high	no
p6	yes	very high	yes

We will need also a concept of a *value reduct* and *value core*. Suppose we are given a dependency $C \Rightarrow D$ where C is relative D-reduct of C. To further investigation of the dependency we might be interested to know exactly how values of attributes from D depend on values of attributes from C. To this end we need a procedure eliminating values of attributes form C which does not influence on values of attributes from D.

- We say that value of attribute $a \in C$, is D-*dispensable* for $x \in U$, if

$$C(x) \subseteq D(x) \text{ implies } C^a(x) \subseteq D(x),$$

  otherwise the value of attribute a is D-*indispensable* for x.
- If for every attribute $a \in C$ value of a is D-indispensable for x, then C will be called D-*independent (orthogonal)* for x.
- Subset $C' \subseteq C$ is a D-*reduct* of C for x (a value reduct), iff C' is D-independent for x and

$$C(x) \subseteq D(x) \text{ implies } C'(x) \subseteq D(x).$$

The set of all D-indispensable for x values of attributes in C will be called the $D - core$ of C for x (the value core), and will be denoted $CORE_D^x(C)$.

We have also the following property

$$CORE_D^x(C) = \bigcap Red_D^x(C), \tag{23}$$

where $Red_D^x(C)$ is the family of all D-reducts of C for x.

Using the concept of a value reduct, Table 2 and Table 3 can be simplified and we obtain Table 4 and Table 5, respectively.

For Table 4 we get its representation by means of rules

if (Headache, no) and (Temperature, high) then (Flu, yes),
if (Headache, yes) and (Temperature, high) then (Flu, yes),
if (Temperature, very high) then (Flu, yes),
if (Temperature, normal) then (Flu, no),
if (Headache, yes) and (Temperature, high) then (Flu, no),
if (Temperature, very high) then (Flu, yes).

For Table 5 we have

if (Muscle-pain, yes) and (Temperature, high) then (Flu, yes),
if (Muscle-pain, no) and (Temperature, high) then (Flu, yes),
if (Temperature, very high) then (Flu, yes),
if (Temperature, normal) then (Flu, no),
if (Muscle-pain, no) and (Temperature, high) then (Flu, no),
if (Temperature, very high) then (Flu, yes).

Table 4. Simplified Table 2

Patient	Headache	Temperature	Flu
p1	no	high	yes
p2	yes	high	yes
p3	–	very high	yes
p4	–	normal	no
p5	yes	high	no
p6	–	very high	yes

Table 5. Simplified Table 3

Patient	Muscle-pain	Temperature	Flu
p1	yes	high	yes
p2	no	high	yes
p3	–	very high	yes
p4	–	normal	no
p5	no	high	no
p6	–	very high	yes

The following important property

a) $B' \Rightarrow B - B'$, where B' is a reduct of B,

connects reducts and dependency.

Besides, we have:

b) If $B \Rightarrow C$, then $B \Rightarrow C'$, for every $C' \subseteq C$,

in particular

c) If $B \Rightarrow C$, then $B \Rightarrow \{a\}$, for every $a \in C$.

Moreover, we have:

d) If B' is a reduct of B, then neither $\{a\} \Rightarrow \{b\}$ nor $\{b\} \Rightarrow \{a\}$ holds, for every $a, b \in B'$, i.e., all attributes in a reduct are pairwise independent.

3.7   Indiscernibility Matrices and Functions

To compute easily reducts and the core we will use discernibility matrix [66], which is defined next.

By an discernibility matrix of $B \subseteq A$ denoted $M(B)$ we will mean $n \times n$ matrix with entries defined by:

$$c_{ij} = \{a \in B : a(x_i) \neq a(x_j)\} \text{ for } i, j = 1, 2, \ldots, n. \qquad (24)$$

Thus entry c_{ij} is the set of all attributes which discern objects x_i and x_j.

The discernibility matrix $M(B)$ assigns to each pair of objects x and y a subset of attributes $\delta(x, y) \subseteq B$, with the following properties:

$$\delta(x, x) = \emptyset, \qquad (25)$$
$$\delta(x, y) = \delta(y, x),$$
$$\delta(x, z) \subseteq \delta(x, y) \cup \delta(y, z).$$

These properties resemble properties of semi-distance, and therefore the function δ may be regarded as *qualitative semi-matrix* and $\delta(x, y)$ – *qualitative semi-distance*. Thus the discernibility matrix can be seen as a *semi-distance* (*qualitative*) matrix.

Let us also note that for every $x, y, z \in U$ we have

$$card(\delta(x, x)) = 0, \qquad (26)$$
$$card(\delta(x, y)) = card(\delta(y, x)),$$
$$card(\delta(x, z)) \leq card(\delta(x, y)) + card(\delta(y, z)).$$

It is easily seen that the core is the set of all single element entries of the discernibility matrix $M(B)$, i.e.,

$$CORE(B) = \{a \in B : c_{ij} = \{a\}, \text{ for some } i, j\}. \qquad (27)$$

Obviously $B' \subseteq B$ is a reduct of B, if B' is the minimal (with respect to inclusion) subset of B such that

$$B' \cap c \neq \emptyset \text{ for any nonempty entry } c \ (c \neq \emptyset) \text{ in } M(B). \qquad (28)$$

In other words reduct is the minimal subset of attributes that discerns all objects discernible by the whole set of attributes.

Every discernibility matrix $M(B)$ defines uniquely a *discernibility* (*boolean*) function $f(B)$ defined as follows.

Let us assign to each attribute $a \in B$ a binary boolean variable $\bar{a}$, and let $\Sigma\delta(x, y)$ denote Boolean sum of all Boolean variables assigned to the set of attributes $\delta(x, y)$. Then the discernibility function can be defined by the formula

$$f(B) = \prod_{(x,y) \in U^2} \{\Sigma\delta(x, y) : (x, y) \in U^2 \text{ and } \delta(x, y) \neq \emptyset\}. \qquad (29)$$

The following property establishes the relationship between disjunctive normal form of the function $f(B)$ and the set of all reducts of B.

All constituents in the minimal disjunctive normal form of the function $f(B)$ are all reducts of B.

In order to compute the value core and value reducts for x we can also use the discernibility matrix as defined before and the discernibility function, which must be slightly modified:

$$f^x(B) = \prod_{y \in U} \{\Sigma\delta(x, y) : y \in U \text{ and } \delta(x, y) \neq \emptyset\}. \qquad (30)$$

Relative reducts and core can be computed also using discernibility matrix, which needs slight modification

$$c_{ij} = \{a \in C : a(x_i) \neq a(x_j) \text{ and } w(x_i, x_j)\}, \tag{31}$$

where $w(x_i, x_j) \equiv x_i \in POS_C(D) \text{ and } x_j \notin POS_C(D) \text{ or}$
$$x_i \notin POS_C(D) \text{ and } x_j \in POS_C(D) \text{ or}$$
$$x_i, x_j \in POS_C(D) \text{ and } (x_j, x_j) \notin I(D),$$

for $i, j = 1, 2, \ldots, n$.

If the partition defined by D is definable by C then the condition $w(x_i, x_j)$ in the above definition can be reduced to $(x_i, x_j) \notin I(D)$.

Thus entry c_{ij} is the set of all attributes which discern objects x_i and x_j that do not belong to the same equivalence class of the relation $I(D)$.

The remaining definitions need little changes.

The D-core is the set of all single element entries of the discernibility matrix $M_D(C)$, i.e.,

$$CORE_D(C) = \{a \in C : c_{ij} = (a), \text{ for some } i, j\}. \tag{32}$$

Set $C' \subseteq C$ is the D-reduct of C, if C' is the minimal (with respect to inclusion) subset of C such that

$$C' \cap c \neq \emptyset \text{ for any nonempty entry } c, (c \neq \emptyset) \text{ in } M_D(C). \tag{33}$$

Thus D-reduct is the minimal subset of attributes that discerns all equivalence classes of the relation $I(D)$.

Every discernibility matrix $M_D(C)$ defines uniquely a *discernibility (Boolean) function* $f_D(C)$ which is defined as before. We have also the following property:

All constituents in the disjunctive normal form of the function $f_D(C)$ are all D-reducts of C.

For computing value reducts and the value core for relative reducts we use as a starting point the discernibility matrix $M_D(C)$ and discernibility function will have the form:

$$f_D^x(C) = \prod_{y \in U} \{\Sigma \delta(x, y) : y \in U \text{ and } \delta(x, y) \neq \emptyset\}. \tag{34}$$

Let us illustrate the above considerations by computing relative reducts for the set of attributes {*Headache, Muscle-pain, Temperature*} with respect to *Flu*.

The corresponding discernibility matrix is shown in Table 6.

In Table 6 H, M, T denote *Headache, Muscle-pain* and *Temperature*, respectively.

The discernibility function for this table is

$$T(H + M)(H + M + T)(M + T),$$

Table 6. Discernibility matrix

	1	2	3	4	5	6
1						
2						
3						
4	T	H, M, T				
5	H, M		M, T			
6				T	H, M, T	

where $+$ denotes the boolean sum and the boolean multiplication is omitted in the formula.

After simplication the discernibility function using laws of Boolean algebra we obtain the following expression

$$TH + TH,$$

which says that there are two reducts TH and TM in the data table and T is the core.

3.8 Significance of Attributes and Approximate Reducts

As it follows from considerations concerning reduction of attributes, they cannot be equally important, and some of them can be eliminated from an information table without losing information contained in the table. The idea of attribute reduction can be generalized by introducing a concept of *significance of attributes*, which enables us evaluation of attributes not only by two-valued scale, *dispensable – indispensable*, but by assigning to an attribute a real number from the closed interval [0,1], expressing how important is an attribute in an information table.

Significance of an attribute can be evaluated by measuring effect of removing the attribute from an information table on classification defined by the table. Let us first start our consideration with decision tables.

Let C and D be sets of condition and decision attributes respectively and let a be a condition attribute, i.e., $a \in A$. As shown previously the number $\gamma(C, D)$ expresses a degree of consistency of the decision table, or the degree of dependency between attributes C and D, or accuracy of approximation of U/D by C. We can ask how the coefficient $\gamma(C, D)$ changes when removing the attribute a, i.e., what is the difference between $\gamma(C, D)$ and $\gamma(C - \{a\}, D)$. We can normalize the difference and define the significance of the attribute a as

$$\sigma_{(C,D)}(a) = \frac{(\gamma(C, D) - \gamma(C - \{a\}, D))}{\gamma(C, D)} = 1 - \frac{\gamma(C - \{a\}, D)}{\gamma(C, D)}, \qquad (35)$$

and denoted simple by $\sigma(a)$, when C and D are understood.

Obviously $0 \leq \sigma(a) \leq 1$. The more important is the attribute a the greater is the number $\sigma(a)$. For example for condition attributes in Table 1 we have the following results:

$$\sigma(Headache) = 0,$$
$$\sigma(Muscle\text{-}pain) = 0,$$
$$\sigma(Temperature) = 0.75.$$

Because the significance of the attribute *Temperature* or *Muscle-pain* is zero, removing either of the attributes from condition attributes does not effect the set of consistent decision rules, whatsoever. Hence the attribute *Temperature* is the most significant one in the table. That means that by removing the attribute *Temperature*, 75% (three out of four) of consistent decision rules will disappear from the table, thus lack of the attribute essentially effects the "decisive power" of the decision table.

For a reduct of condition attributes, e.g., {*Headache, Temperature*}, we get

$$\sigma(Headache) = 0.25,$$
$$\sigma(Temperature) = 1.00.$$

In this case, removing the attribute *Headache* from the reduct, i.e., using only the attribute *Temperature*, 25% (one out of four) of consistent decision rules will be lost, and dropping the attribute *Temperature*, i.e., using only the attribute *Headache* 100% (all) consistent decision rules will be lost. That means that in this case making decisions is impossible at all, whereas by employing only the attribute *Temperature* some decision can be made.

Thus the coefficient $\sigma(a)$ can be understood as an error which occurs when attribute a is dropped. The significance coefficient can be extended to set of attributes as follows:

$$\varepsilon_{(C,D)}(B) = \frac{(\gamma(C,D) - \gamma(C-B,D))}{\gamma(C,D)} = 1 - \frac{\gamma(C-B,D)}{\gamma(C,D)}, \qquad (36)$$

denoted by $\varepsilon(B)$, if C and D are understood, where B is a subset of C.

If B is a reduct of C, then $\varepsilon(B) = 1$, i.e., removing any reduct from a set of decision rules unables to make sure decisions, whatsoever.

Any subset B of C will be called an *approximate reduct* of C, and the number

$$\varepsilon_{(C,D)}(B) = \frac{(\gamma(C,D) - \gamma(B,D))}{\gamma(C,D)} = 1 - \frac{\gamma(B,D)}{\gamma(C,D)}, \qquad (37)$$

denoted simple as $\varepsilon(B)$, will be called an *error of reduct approximation*. It expresses how exactly the set of attributes B approximates the set of condition attributes C. Obviously $\varepsilon(B) = 1 - \sigma(B)$ and $\varepsilon(B) = 1 - \varepsilon(C - B)$. For any subset B of C we have $\varepsilon(B) \leq \varepsilon(C)$. If B is a reduct of C, then $\varepsilon(B) = 0$.

For example, either of attributes *Headache* and *Temperature* can be considered as approximate reducts of {*Headache, Temperature*}, and

$$\varepsilon(Headache) = 1,$$
$$\varepsilon(Temperature) = 0.25.$$

But for the whole set of condition attributes {*Headache, Muscle-pain, Temperature*} we have also the following approximate reduct

$$\varepsilon(\text{Headache, Muscle-pain}) = 0.75.$$

The concept of an approximate reduct is a generalization of the concept of a reduct considered previously. The minimal subset B of condition attributes C, such that $\gamma(C, D) = \gamma(B, D)$, or $\varepsilon_{(C,D)}(B) = 0$ is a reduct in the previous sense. The idea of an approximate reduct can be useful in cases when a smaller number of condition attributes is preferred over accuracy of classification.

4   Rough Sets and Bayes' Theorem

4.1   Introduction

Bayes' theorem is the essence of statistical inference.

"The result of the Bayesian data analysis process is the posterior distribution that represents a revision of the prior distribution on the light of the evidence provided by the data" [67].

"Opinion as to the values of Bayes' theorem as a basic for statistical inference has swung between acceptance and rejection since its publication on 1763" [68].

Rough set theory offers new insight into Bayes' theorem [69–71]. The look on Bayes' theorem presented here is completely different to that studied so far using the rough set approach (see, e.g., [72–85]) and in the Bayesian data analysis philosophy (see, e.g., [67, 86, 68, 87]). It does not refer either to prior or posterior probabilities, inherently associated with Bayesian reasoning, but it reveals some probabilistic structure of the data being analyzed. It states that any data set (decision table) satisfies total probability theorem and Bayes' theorem. This property can be used directly to draw conclusions from data without referring to prior knowledge and its revision if new evidence is available. Thus in the presented approach the only source of knowledge is the data and there is no need to assume that there is any prior knowledge besides the data. We simple look what the data are telling us. Consequently we do not refer to any prior knowledge which is updated after receiving some data.

Moreover, the presented approach to Bayes' theorem shows close relationship between logic of implications and probability, which was first studied by Jan Łukasiewicz [88] (see also [89]). Bayes' theorem in this context can be used to "invert" implications, i.e., to give reasons for decisions. This is a very important feature of utmost importance to data mining and decision analysis, for it extends the class of problem which can be considered in this domains.

Besides, we propose a new form of Bayes' theorem where basic role plays strength of decision rules (implications) derived from the data. The strength of decision rules is computed from the data or it can be also a subjective assessment. This formulation gives new look on Bayesian method of inference and also simplifies essentially computations.

4.2   Bayes' Theorem

"In its simplest form, if H denotes an hypothesis and D denotes data, the theorem says that

$$P(H \mid D) = P(D \mid H) \times P(H)/P(D). \qquad (38)$$

With $P(H)$ regarded as a probabilistic statement of belief about H before obtaining data D, the left-hand side $P(H \mid D)$ becomes an probabilistic statement of belief about H after obtaining D. Having specified $P(D \mid H)$ and $P(D)$, the mechanism of the theorem provides a solution to the problem of how to learn from data.

In this expression, $P(H)$, which tells us what is known about H without knowing of the data, is called the *prior* distribution of H, or the distribution of H *priori*. Correspondingly, $P(H \mid D)$, which tells us what is known about H given knowledge of the data, is called the *posterior* distribution of H given D, or the distribution of H a *posteriori* [87].

"A prior distribution, which is supposed to represent what is known about unknown parameters before the data is available, plays an important role in Bayesian analysis. Such a distribution can be used to represent prior knowledge or relative ignorance" [68].

4.3   Decision Tables and Bayes' Theorem

In this section we will show that decision tables satisfy Bayes' theorem but the meaning of this theorem differs essentially from the classical Bayesian methodology.

Every decision table describes decisions (actions, results etc.) determined, when some conditions are satisfied. In other words each row of the decision table specifies a decision rule which determines decisions in terms of conditions.

In what follows we will describe decision rules more exactly.

Let $S = (U, C, D)$ be a decision table. Every $x \in U$ determines a sequence $c_1(x), \ldots, c_n(x), d_1(x), \ldots, d_m(x)$ where $\{c_1, \ldots, c_n\} = C$ and $\{d_1, \ldots, d_m\} = D$

The sequence will be called a *decision rule induced by* x (in S) and denoted by $c_1(x), \ldots, c_n(x) \rightarrow d_1(x), \ldots, d_m(x)$ or in short $C \rightarrow_x D$.

The number $supp_x(C, D) = card(C(x) \cap D(x))$ will be called a *support* of the decision rule $C \rightarrow_x D$ and the number

$$\sigma_x(C, D) = \frac{supp_x(C, D)}{card(U)}, \qquad (39)$$

will be referred to as the *strength* of the decision rule $C \rightarrow_x D$. With every decision rule $C \rightarrow_x D$ we associate a *certainty factor* of the decision rule, denoted $cer_x(C, D)$ and defined as follows:

$$cer_x(C, D) = \frac{card(C(x) \cap D(x))}{card(C(x))} = \frac{supp_x(C, D)}{card(C(x))} = \frac{\sigma_x(C, D)}{\pi(C(x))}, \qquad (40)$$

where $\pi(C(X)) = \frac{card(C(x))}{card(U)}$.

The certainty factor may be interpreted as a conditional probability that y belongs to $D(x)$ given y belongs to $C(x)$, symbolically $\pi_x(D \mid C)$.

If $cer_x(C, D) = 1$, then $C \to_x D$ will be called a *certain decision rule* in S; if $0 < cer_x(C, D) < 1$ the decision rule will be referred to as an *uncertain decision rule* in S.

Besides, we will also use a *coverage factor* of the decision rule, denoted $cov_x(C, D)$ defined as

$$cov_x(C, D) = \frac{card(C(x) \cap D(x))}{card(D(x))} = \frac{supp_x(C, D)}{card(D(x))} = \frac{\sigma_x(C, D)}{\pi(D(x))}, \quad (41)$$

where $\pi(D(X)) = \frac{card(D(x))}{card(U)}$.

Similarly

$$cov_x(C, D) = \pi_x(C \mid D). \quad (42)$$

The certainty and coverage coefficients have been widely used for years by data mining and rough set communities. However, Łukasiewicz [88] (see also [89]) was first who used this idea to estimate the probability of implications.

If $C \to_x D$ is a decision rule then $C \to_x D$ will be called an *inverse decision rule*. The inverse decision rules can be used to give explanations (*reasons*) for a decision.

Let us observe that

$$cer_x(C, D) = \pi^C_{D(x)}(x) \text{ and } cov_x(C, D). \quad (43)$$

That means that the certainty factor expresses the degree of membership of x to the decision class $D(x)$, given C, whereas the coverage factor expresses the degree of membership of x to condition class $C(x)$, given D.

Decision tables have important probabilistic properties which are discussed next.

Let $C \to_x D$ be a decision rule in S and let $\Gamma = C(x)$ and $\Delta = D(x)$. Then the following properties are valid:

$$\sum_{y \in \Gamma} cer_y(C, D) = 1, \quad (44)$$

$$\sum_{y \in \Gamma} cov_y(C, D) = 1, \quad (45)$$

$$\pi(D(x)) = \sum_{y \in \Gamma} cer_y(C, D) \cdot \pi(C(y)) = \sum_{y \in \Gamma} \sigma_y(C, D), \quad (46)$$

$$\pi(C(x)) = \sum_{y \in \Delta} cov_y(C, D) \cdot \pi(D(y)) = \sum_{y \in \Delta} \sigma_y(C, D), \quad (47)$$

$$cer_x(C, D) = \frac{cov_x(C, D) \cdot \pi(D(x))}{\sum_{y \in \Gamma} cov_y(C, D) \cdot \pi(D(y))} = \frac{\sigma_x(C, D)}{\sum_{y \in \Delta} \sigma_y(C, D)} = \frac{\sigma_x(C, D)}{\pi(C(x))}, \quad (48)$$

$$cov_x(C, D) = \frac{cer_x(C, D) \cdot \pi(C(x))}{\sum\limits_{y \in \Gamma} cer_y(C, D) \cdot \pi(C(y))} = \frac{\sigma_x(C, D)}{\sum\limits_{y \in \Gamma} \sigma_x(C, D)} = \frac{\sigma_x(C, D)}{\pi(D(x))}. \quad (49)$$

That is, any decision table, satisfies (44)-(49). Observe that (46) and (47) refer to the well known *total probability theorem*, whereas (48) and (49) refer to *Bayes' theorem*.

Thus in order to compute the certainty and coverage factors of decision rules according to formula (48) and (49) it is enough to know the strength (support) of all decision rules only. The strength of decision rules can be computed from data or can be a subjective assessment.

4.4   Decision Language and Decision Algorithms

It is often useful to describe decision tables in logical terms. To this end we define a formal language called a *decision language*.

Let $S = (U, A)$ be an information system. With every $B \subseteq A$ we associate a formal language, i.e., a set of formulas $For(B)$. Formulas of $For(B)$ are built up from attribute-value pairs (a, v) where $a \in B$ and $v \in V_a$ by means of logical connectives $\wedge(and)$, $\vee(or)$, $\sim (not)$ in the standard way.

For any $\Phi \in For(B)$ by $\| \Phi \|_S$ we denote the set of all objects $x \in U$ satisfying Φ in S and refer to as the *meaning* of Φ in S.

The meaning $\| \Phi \|_S$ of Φ in S is defined inductively as follows:
$\| (a, v) \|_S = \{x \in U : a(v) = x\}$ for all $a \in B$ and $v \in V_a$, $\| \Phi \wedge \Psi \|_S = \| \Phi \|_S \cup \| \Psi \|_S$, $\| \Phi \wedge \Psi \|_S = \| \Phi \|_S \cap \| \Psi \|_S$, $\| \sim \Phi \|_S = U - \| \Phi \|_S$.

If $S = (U, C, D)$ is a decision table then with every row of the decision table we associate a decision rule, which is defined next.

A *decision rule* in S is an expression $\Phi \rightarrow_S \Psi$ or simply $\Phi \rightarrow \Psi$ if S is understood, read *if Φ then Ψ*, where $\Phi \in For(C)$, $\Psi \in For(D)$ and C, D are condition and decision attributes, respectively; Φ and Ψ are referred to as *conditions* part and *decisions* part of the rule, respectively.

The number $supp_S(\Phi, \Psi) = card((\| \Phi \wedge \Psi \|_S))$ will be called the *support* of the rule $\Phi \rightarrow \Psi$ in S. We consider a probability distribution $p_U(x) = 1/card(U)$ for $x \in U$ where U is the (non-empty) universe of objects of S; we have $p_U(X) = card(X)/card(U)$ for $X \subseteq U$. For any formula Φ we associate its probability in S defined by

$$\pi_S(\Phi) = p_U(\| \Phi \|_S). \quad (50)$$

With every decision rule $\Phi \rightarrow \Psi$ we associate a conditional probability

$$\pi_S(\Psi \mid \Phi) = p_U(\| \Psi \|_S \mid \| \Phi \|_S) \quad (51)$$

called the *certainty factor* of the decision rule, denoted $cer_S(\Phi, \Psi)$. We have

$$cer_S(\Phi, \Psi) = \pi_S(\Psi \mid \Phi) = \frac{card(\| \Phi \wedge \Psi \|_S)}{card(\| \Phi \|_S)}, \quad (52)$$

where $\| \Phi \|_S \neq \emptyset$.

If $\pi_S(\Psi \mid \Phi) = 1$, then $\Phi \to \Psi$ will be called a *certain decision* rule; if $0 < \pi_S(\Psi \mid \Phi) < 1$ the decision rule will be referred to as a *uncertain decision* rule.

There is an interesting relationship between decision rules and their approximations: certain decision rules correspond to the lower approximation, whereas the uncertain decision rules correspond to the boundary region.

Besides, we will also use a *coverage factor* of the decision rule, denoted $cov_S(\Phi, \Psi)$ defined by

$$\pi_S(\Phi \mid \Psi) = p_U(\| \Phi \|_S \mid \| \Psi \|_S). \tag{53}$$

Obviously we have

$$cov_S(\Phi, \Psi) = \pi_S(\Phi \mid \Psi) = \frac{card(\| \Phi \wedge \Psi \|_S)}{card(\| \Psi \|_S)}. \tag{54}$$

There are three possibilities to interpret the certainty and the coverage factors: statistical (frequency), logical (degree of truth) and mereological (degree of inclusion).

We will use here mainly the statistical interpretation, i.e., the certainty factors will be interpreted as the frequency of objects having the property Ψ in the set of objects having the property Φ and the coverage factor – as the frequency of objects having the property Φ in the set of objects having the property Ψ.

Let us observe that the factors are not assumed arbitrarily but are computed from the data.

The number

$$\sigma_S(\Phi, \Psi) = \frac{supp_S(\Phi, \Psi)}{card(U)} = \pi_S(\Psi \mid \Phi) \cdot \pi_S(\Phi), \tag{55}$$

will be called the *strength* of the decision rule $\Phi \to \Psi$ in S.

We will need also the notion of an equivalence of formulas.

Let Φ, Ψ be formulas in $For(A)$ where A is the set of attributes in $S = (U, A)$.

We say that Φ and Ψ are equivalent in S, or simply, equivalent if S is understood, in symbols $\Phi \equiv \Psi$, if and only if $\Phi \to \Psi$ and $\Psi \to \Phi$. This means that $\Phi \equiv$ if and only if $\| \Phi \|_S = \| \Psi \|_S$.

We need also approximate equivalence of formulas which is defined as follows:

$$\Phi \equiv_S \Psi \quad \text{if and only if} \quad cer(\Phi, \Psi) = cov(\Phi, \Psi) = k. \tag{56}$$

Besides, we define also approximate equivalence of formulas with the accuracy ε ($0 \le \varepsilon \le 1$, which is defined as follows:

$$\Phi \equiv_{k,\varepsilon} \Psi \quad \text{if and only if} \quad k = min\{(cer(\Phi, \Psi), cov(\Phi, \Psi)\} \tag{57}$$
$$\text{and } |cer(\Phi, \Psi) - cov(\Phi, \Psi)| \le \varepsilon.$$

Now, we define the notion of a decision algorithm, which is a logical counterpart of a decision table.

Let $Dec(S) = \{\Phi_i \to \Psi\}_{i=1}^m$, $m \ge 2$, be a set of decision rules in a decision table $S = (U, C, D)$.

1) If for every $\Phi \to \Psi$, $\Phi' \to \Psi' \in Dec(S)$ we have $\Phi = \Phi'$ or $\| \Phi \wedge \Phi' \|_S = \emptyset$, and $\Psi = \Psi'$ or $\| \Psi \wedge \Psi' \|_S = \emptyset$, then we will say that $Dec(S)$ is the set of pairwise *mutually exclusive (independent)* decision rules in S.

2) If $\| \bigwedge\limits_{i=1}^{m} \Phi_i \|_S = U$ and $\| \bigwedge\limits_{i=1}^{m} \Psi_i \|_S = U$ we will say that the set of decision rules $Dec(S)$ *covers* U.

3) If $\Phi \to \Psi \in Dec(S)$ and $supp_S(\Phi, \Psi) \neq 0$ we will say that the decision rule $\Phi \to \Psi$ is *admissible* in S.

4) If $\bigcup\limits_{X \in U/D} C_*(X) = \bigwedge\limits_{\Phi \to \Psi \in Dec^+(S)} \| \Phi \|_S$, where $Dec^+(S)$ is the set of all certain decision rules from $Dec(S)$, we will say that the set of decision rules $Dec(S)$ preserves the *consistency* part of the decision table $S = (U, C, D)$.

The set of decision rules $Dec(S)$ that satisfies 1), 2) 3) and 4), i.e., is independent, covers U, preserves the consistency of S and all decision rules $\Phi \to \Psi \in Dec(S)$ are admissible in S – will be called a *decision algorithm* in S. Hence, if $Dec(S)$ is a decision algorithm in S then the conditions of rules from $Dec(S)$ define in S a partition of U. Moreover, the *positive region of D with respect to C*, i.e., the set

$$\bigcup_{X \in U/D} C_*(X), \qquad (58)$$

is partitioned by the conditions of some of these rules, which are certain in S.

If $\Phi \to \Psi$ is a decision rule then the decision rule $\Psi \to \Psi$ will be called an *inverse* decision rule of $\Phi \to \Psi$.

Let $Dec^*(S)$ denote the set of all inverse decision rules of $Dec(S)$.

It can be shown that $Dec^*(S)$ satisfies 1), 2), 3) and 4), i.e., it is a decision algorithm in S.

If $Dec(S)$ is a decision algorithm then $Dec^*(S)$ will be called an *inverse* decision algorithm of $Dec(S)$.

The inverse decision algorithm gives *reasons* (explanations) for decisions pointed out by the decision algorithms.

A decision algorithm is a description of a decision table in the decision language.

Generation of decision algorithms from decision tables is a complex task and we will not discuss this issue here, for it does not lie in the scope of this paper. The interested reader is advised to consult the references (see, e.g., [18, 66, 90–97, 50, 98–104] and the bibliography in these articles).

4.5 An Example

Let us now consider an example of decision table, shown in Table 7.

Attributes *Disease, Age* and *Sex* are condition attributes, whereas *test* is the decision attribute.

We want to explain the test result in terms of patients state, i.e., to describe attribute *Test* in terms of attributes *Disease, Age* and *Sex*.

Table 7. Exemplary decision table

Fact	Disease	Age	Sex	Test	Support
1	yes	old	man	+	400
2	yes	middle	woman	+	80
3	no	old	man	−	100
4	yes	old	man	−	40
5	no	young	woman	−	220
6	yes	middle	woman	−	60

Table 8. Certainty and coverage factors for decision table shown in Table 7

Fact	Strength	Certaint	Coverage
1	0.44	0.92	0.83
2	0.09	0.56	0.17
3	0.11	1.00	0.24
4	0.04	0.08	0.10
5	0.24	1.00	0.52
6	0.07	0.44	0.14

The strength, certainty and coverage factors for decision table are shown in Table 8.

Below a decision algorithm associated with Table 7 is presented.

1) *if (Disease, yes) and (Age, old) then (Test, +);*
2) *if (Disease, yes) and (Age, middle) then (Test, +);*
3) *if (Disease, no) then (Test, −);*
4) *if (Disease, yes) and (Age, old) then (Test, −);*
5) *if (Disease, yes) and (Age, middle) then (Test, −).*

The certainty and coverage factors for the above algorithm are given in Table 9.

Table 9. Certainty and coverage factors for the decision algorithm

Rule	Strength	Certaint	Coverage
1	0.44	0.92	0.83
2	0.09	0.56	0.17
3	0.36	1.00	0.76
4	0.04	0.08	0.10
5	0.24	0.44	0.14

The certainty factors of the decision rules lead the following conclusions:

− 92% ill and old patients have positive test result,
− 56% ill and middle age patients more positive test result,
− all healthy patients have negative test result,
− 8% ill and old patients have negative test result,
− 44% ill and old patients have negative test result.

In other words:

- ill and old patients most probably have positive test result (probability = 0.92),
- ill and middle age patients most probably have positive test result (probability = 0.56),
- healthy patients have certainly negative test result (probability = 1.00).

Now let us examine the inverse decision algorithm, which is given below:

1') *if (Test, +) then (Disease, yes) and (Age, old)*;
2') *if (Test, +) then (Disease, yes) and (Age, middle)*;
3') *if (Test, −) then (Disease, no)*;
4') *if (Test, −) then (Disease, yes) and (Age, old)*;
5') *if (Test, −) then (Disease, yes) and (Age, middle)*.

Employing the inverse decision algorithm and the coverage factor we get the following explanation of test results:

- reason for positive test results are most probably patients disease and old age (probability = 0.83),
- reason for negative test result is most probably lack of the disease (probability = 0.76).

It follows from Table 7 that there are two interesting approximate equivalences of test results and the disease.

According to rule 1) the disease and old age are approximately equivalent to positive test result ($k = 0.83$, $\varepsilon = 0.11$), and lack of the disease according to rule 3) is approximately equivalent to negative test result ($k = 0.76$, $\varepsilon = 0.24$).

5 Rough Sets and Conflict Analysis

5.1 Introduction

Knowledge discovery in databases considered in the previous sections boiled down to searching for functional dependencies in the data set.

In this section we will discuss another kind of relationship in the data – not dependencies, but conflicts.

Formally, the conflict relation can be seen as a negation (not necessarily, classical) of indiscernibility relation which was used as a basis of rough set theory. Thus dependencies and conflict are closely related from logical point of view.

It turns out that the conflict relation can be used to the conflict analysis study.

Conflict analysis and resolution play an important role in business, governmental, political and lawsuits disputes, labor-management negotiations, military operations and others. To this end many mathematical formal models of conflict situations have been proposed and studied, e.g., [105–110].

Various mathematical tools, e.g., graph theory, topology, differential equations and others, have been used to that purpose.

Needless to say that game theory can be also considered as a mathematical model of conflict situations.

In fact there is no, as yet, "universal" theory of conflicts and mathematical models of conflict situations are strongly domain dependent.

We are going to present in this paper still another approach to conflict analysis, based on some ideas of rough set theory – along the lines proposed in [110]. We will illustrate the proposed approach by means of a simple tutorial example of voting analysis in conflict situations.

The considered model is simple enough for easy computer implementation and seems adequate for many real life applications but to this end more research is needed.

5.2   Basic Concepts of Conflict Theory

In this section we give after [110] definitions of basic concepts of the proposed approach.

Let us assume that we are given a finite, non-empty set U called the *universe*. Elements of U will be referred to as *agents*. Let a function $v : U \rightarrow \{-1, 0, 1\}$, or in short $\{-, 0, +\}$, be given assigning to every agent the number $-1, 0$ or 1, representing his opinion, view, voting result, etc. about some discussed issue, and meaning *against, neutral* and *favorable*, respectively.

The pair $S = (U, v)$ will be called a *conflict* situation. In order to express relations between agents we define three basic binary relations on the universe: *conflict, neutrality* and *alliance*. To this end we first define the following auxiliary function:

$$\phi_v(x, y) = \begin{cases} 1, & \text{if } v(x)v(y) = 1 \text{ or } x = y \\ 0, & \text{if } v(x)v(y) = 0 \text{ and } x \neq y \\ -1, & \text{if } v(x)v(y) = -1. \end{cases} \tag{59}$$

This means that, if $\phi_v(x, y) = 1$, agents x and y have the same opinion about issue v (are *allied*) on v); if $\phi_v(x, y) = 0$ means that at least one agent x or y has neutral approach to issue a (is *neutral* on a), and if $\phi_v(x, y) = -1$, means that both agents have different opinions about issue v (are in *conflict* on v).

In what follows we will define three basic relations R_v^+, R_v^0 and R_v^- on U^2 called *alliance, neutrality* and *conflict* relations respectively, and defined as follows:

$$R_v^+(x, y) \text{ iff } \phi_v(x, y) = 1, \tag{60}$$
$$R_v^0(x, y) \text{ iff } \phi_v(x, y) = 0,$$
$$R_v^-(x, y) \text{ iff } \phi_v(x, y) = -1.$$

It is easily seen that the alliance relation has the following properties:

$$R_v^+(x, x), \tag{61}$$
$$R_v^+(x, y) \text{ implies } R_v^+(y, x),$$
$$R_v^+(x, y) \text{ and } R_v^+(y, z) \text{ implies } R_v^+(x, z),$$

i.e., R_v^+ is an *equivalence* relation. Each equivalence class of alliance relation will be called *coalition* with respect to v. Let us note that the last condition in (61) can be expressed as "a friend of my friend is my friend".

For the conflict relation we have the following properties:

$$\text{not } R_v^-(x, x), \tag{62}$$
$$R_v^-(x, y) \text{ implies } R_v^-(y, x),$$
$$R_v^-(x, y) \text{ and } R_v^-(y, z) \text{ implies } R_v^+(x, z),$$
$$R_v^-(x, y) \text{ and } R_v^+(y, z) \text{ implies } R_v^-(x, z).$$

The last two conditions in (62) refer to well known sayings "an enemy of my enemy is my friend" and "a friend of my enemy is my enemy".

For the neutrality relation we have:

$$\text{not } R_v^0(x, x), \tag{63}$$
$$R_v^0(x, y) = R_v^0(y, x).$$

Let us observe that in the conflict and neutrality relations there are no coalitions.

The following property holds: $R_v^+ \cup R_v^0 \cup R_v^- = U^2$ because if $(x, y) \in U^2$ then $\Phi_v(x, y) = 1$ or $\Phi_v(x, y) = 0$ or $\Phi_v(x, y) = -1$ so $(x, y) \in R_v^+$ or $(x, y) \in R_v^-$ or $(x, y) \in R_v^-$. All the three relations R_v^+, R_v^0, R_v^- are pairwise disjoint, i.e., every pair of objects (x, y) belongs to exactly one of the above defined relations (is in conflict, is allied or is neutral).

With every conflict situation we will associate a *conflict graph*

$$G_S = (R_v^+, R_v^0, R_v^-). \tag{64}$$

An example of a conflict graph is shown in Figure 3. Solid lines are denoting conflicts, doted line – alliance, and neutrality, for simplicity, is not shown explicitly in the graph. Of course, $B, C,$ and D form a coalition.

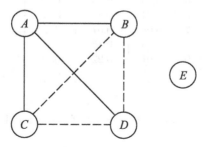

Fig. 3. Exemplary conflict graph

5.3   An Example

In this section we will illustrate the above presented ideas by means of a very simple tutorial example using concepts presented in the previous.

Table 10 presents a decision table in which the only condition attribute is *Party*, whereas the decision attribute is *Voting*. The table describes voting results in a parliament containing 500 members grouped in four political parties denoted A, B, C and D. Suppose the parliament discussed certain issue (e.g., membership of the country in European Union) and the voting result is presented in column *Voting*, where $+$, 0 and $-$ denoted *yes*, *abstention* and *no* respectively. The column *support* contains the number of voters for each option.

Table 10. Decision table with one condition attribute *Party* and the decision *Voting*

Fact	Party	Voting	Support
1	A	+	200
2	A	0	30
3	A	−	10
4	B	+	15
5	B	−	25
6	C	0	20
7	C	−	40
8	D	+	25
9	D	0	35
10	D	−	100

Table 11. Certainty and the coverage factors for Table 10

Fact	Strength	Certainty	Coverage
1	0.40	0.83	0.83
2	0.06	0.13	0.35
3	0.02	0.04	0.06
4	0.03	0.36	0.06
5	0.05	0.63	0.14
6	0.04	0.33	0.23
7	0.08	0.67	0.23
8	0.05	0.16	0.10
9	0.07	0.22	0.41
10	0.20	0.63	0.57

The strength, certainty and the coverage factors for Table 10 are given in Table 11.

From the certainty factors we can conclude, for example, that:

- 83.3% of party A voted *yes*,
- 12.5% of party A *abstained*,
- 4.2% of party A voted *no*.

From the coverage factors we can get, for example, the following explanation of voting results:

- 83.3% *yes* votes came from party A,
- 6.3% *yes* votes came from party B,
- 10.4% *yes* votes came from party C.

6   Data Analysis and Flow Graphs

6.1   Introduction

Pursuit for data patterns considered so far referred to data tables. In this section we will consider data represented not in a form of data table but by means of graphs. We will show that this method od data representation leads to a new look on knowledge discovery, new efficient algorithms, and vide spectrum of novel applications.

The idea presented here are based on some concepts given by Łukasiewicz [88].

In [88] Łukasiewicz proposed to use logic as mathematical foundation of probability. He claims that probability is "purely logical concept" and that his approach frees probability from its obscure philosophical connotation. He recommends to replace the concept of *probability* by *truth values* of *indefinite propositions*, which are in fact propositional functions.

Let us explain this idea more closely. Let U be a non empty finite set, and let $\Phi(x)$ be a propositional function. The meaning of $\Phi(x)$ in U, denoted by $\|\Phi(x)\|$, is the set of all elements of U, that satisfies $\Phi(x)$ in U. The truth value of $\Phi(x)$ is defined by $card(\|\Phi(x)\|)/card(U)$. For example, if $U = \{1, 2, 3, 4, 5, 6\}$ and $\Phi(x)$ is the propositional function $x > 4$, then the truth value of $\Phi(x) = 2/6 = 1/3$. If the truth value of $\Phi(x)$ is 1, then the propositional function is *true*, and if it is 0, then the function is *false*. Thus the truth value of any propositional function is a number between 0 and 1. Further, it is shown that the truth values can be treated as probability and that all laws of probability can be obtained by means of logical calculus.

In this paper we show that the idea of Łukasiewicz can be also expressed differently. Instead of using truth values in place of probability, stipulated by Łukasiewicz, we propose, in this paper, using of deterministic flow analysis in flow networks (graphs). In the proposed setting, flow is governed by some probabilistic rules (e.g., Bayes' rule), or by the corresponding logical calculus proposed by Łukasiewicz, though, the formulas have entirely deterministic meaning, and need neither probabilistic nor logical interpretation. They simply describe flow distribution in flow graphs. However, flow graphs introduced here are different from those proposed by Ford and Fulkerson [111] for optimal flow analysis, because they model rather, e.g., flow distribution in a plumbing network, than the optimal flow.

The flow graphs considered in this paper are basically meant not to physical media (e.g., water) flow analysis, but to information flow examination in decision algorithms. To this end branches of a flow graph are interpreted as decision rules. With every decision rule (i.e. branch) three coefficients are associated, the *strength, certainty* and *coverage factors*. In classical decision algorithms language they have probabilistic interpretation. Using Łukasiewicz's approach we can understand them as truth values. However, in the proposed setting they can be interpreted simply as flow distribution ratios between branches of the flow graph, without referring to their probabilistic or logical nature.

This interpretation, in particular, leads to a new look on Bayes' theorem, which in this setting, has entirely deterministic explanation (see also [86]).

The presented idea can be used, among others, as a new tool for data analysis, and knowledge representation.

We start our considerations giving fundamental definitions of a flow graph and related notions. Next, basic properties of flow graphs are defined and investigated. Further, the relationship between flow graphs and decision algorithms is discussed. Finally, a simple tutorial example is used to illustrate the consideration.

6.2   Flow Graphs

A flow graph is a *directed, acyclic*, finite graph $G = (N, \mathcal{B}, \phi)$, where N is a set of *nodes*, $\mathcal{B} \subseteq N \times N$ is a set of *directed branches*, $\phi : \mathcal{B} \to R^+$ is a *flow function* and R^+ is the set of non-negative reals.

If $(x, y) \in \mathcal{B}$ then x is an *input* of y and y is an *output* of x.

If $x \in N$ then $I(x)$ is the set of all inputs of x and $O(x)$ is the set of all outputs of x.

Input and *output* of a graph G are defined $I(G) = \{x \in N : I(x) = \emptyset\}$, $O(G) = \{x \in N : O(x) = \emptyset\}$.

Inputs and outputs of G are *external nodes* of G; other nodes are *internal nodes* of G.

If $(x, y) \in \mathcal{B}$ then $\phi(x, y)$ is a *troughflow* from x to y. We will assume in what follows that $\phi(x, y) \neq 0$ for every $(x, y) \in \mathcal{B}$.

With every node x of a flow graph G we associate its *inflow*

$$\phi_+(x) = \sum_{y \in I(x)} \phi(y, x), \tag{65}$$

and *outflow*

$$\phi_-(x) = \sum_{y \in O(x)} \phi(x, y). \tag{66}$$

Similarly, we define an inflow and an outflow for the whole flow graph G, which are defined as

$$\phi_+(G) = \sum_{x \in I(G)} \phi_-(x), \tag{67}$$

$$\phi_-(G) = \sum_{x \in O(G)} \phi_+(x). \tag{68}$$

We assume that for any internal node x, $\phi_+(x) = \phi_-(x) = \phi(x)$, where is a *troughflow* of node x.

Obviously, $\phi_+(G) = \phi_-(G) = \phi(G)$, where $\phi(G)$ is a *troughflow* of graph G. The above formulas can be considered as *flow conservation equations* [111]. We will define now a *normalized flow graph*.

A normalized flow graph is a *directed, acyclic, finite* graph $G = (N, \mathcal{B}, \sigma)$, where N is a set of *nodes*, $\mathcal{B} \subseteq N \times N$ is a set of *directed branches* and $\sigma : \mathcal{B} \to < 0, 1 >$ is a *normalized flow* of (x, y) and

$$\sigma(x, y) = \frac{\sigma(x, y)}{\sigma(G)}, \tag{69}$$

is *strength* of (x, y). Obviously, $0 \leq \sigma(x, y) \leq 1$. The strength of the branch expresses simply the percentage of a total flow through the branch.

In what follows we will use normalized flow graphs only, therefore by a flow graphs we will understand normalized flow graphs, unless stated otherwise.

With every node x of a flow graph G we associate its normalized *inflow* and *outflow* defined as

$$\sigma_+(x) = \frac{\phi_+(x)}{\phi(G)} = \sum_{y \in I(x)} \sigma(y, x), \tag{70}$$

$$\sigma_-(x) = \frac{\phi_-(x)}{\phi(G)} = \sum_{y \in O(x)} \sigma(y, x). \tag{71}$$

Obviously for any internal node x, we have $\sigma_+(X) = \sigma_- = \sigma(x)$, where $\sigma(x)$ is a *normalized troughflow* of x.

Moreover, let

$$\sigma_+(G) = \frac{\phi_+(G)}{\phi(G)} = \sum_{x \in I(G)} \sigma_-(x), \tag{72}$$

$$\sigma_-(G) = \frac{\phi_-(G)}{\phi(G)} = \sum_{x \in O(G)} \sigma_+(x). \tag{73}$$

Obviously, $\sigma_+(G) = \sigma_-(G) = \sigma(G) = 1$.

6.3   Certainty and Coverage Factors

With every branch (x, y) of a flow graph G we associate the *certainty* and the *coverage factors*.

The *certainty* and the *coverage* of are defined as

$$cer(x, y) = \frac{\sigma(x, y)}{\sigma(x)}, \tag{74}$$

and

$$cov(x, y) = \frac{\sigma(x, y)}{\sigma(y)}. \tag{75}$$

respectively, where $\sigma(x) \neq 0$ and $\sigma(y) \neq 0$. Below some properties, which are immediate consequences of definitions given above are presented:

$$\sum_{y \in O(x)} cer(x, y) = 1, \tag{76}$$

$$\sum_{y \in I(y)} cov(x, y) = 1, \tag{77}$$

$$\sigma(x) = \sum_{y \in O(x)} cer(x, y)\sigma(x) = \sum_{y \in O(x)} \sigma(x, y), \tag{78}$$

$$\sigma(y) = \sum_{x \in I(y)} cov(x, y)\sigma(y) = \sum_{x \in I(y)} \sigma(x, y), \tag{79}$$

$$cer(x, y) = \frac{cov(x, y)\sigma(y)}{\sigma(x)}, \tag{80}$$

$$cov(x, y) = \frac{cer(x, y)\sigma(x)}{\sigma(y)}. \tag{81}$$

Obviously the above properties have a probabilistic flavor, e.g., equations (78) and (79) have a form of total probability theorem, whereas formulas (80) and (81) are Bayes' rules. However, these properties in our approach are interpreted in a deterministic way and they describe flow distribution among branches in the network.

A (*directed*) *path* from x to y, $x \neq y$ in G is a sequence of nodes $x_1, \ldots, x_n$ such that $x_1 = x$, $x_n = y$ and $(x_i, x_{i+1}) \in B$ for every $i, 1 \leq i \leq n-1$. A path from x to y is denoted by $[x \ldots y]$.

The *certainty*, the *coverage* and the *strength* of the path $[x_1 \ldots x_n]$ are defined as

$$cer[x_1 \ldots x_n] = \prod_{i=1}^{n-1} cer(x_i, x_{i+1}), \tag{82}$$

$$cov[x_1 \ldots x_n] = \prod_{i=1}^{n-1} cov(x_i, x_{i+1}), \tag{83}$$

$$\sigma[x \ldots y] = \sigma(x)cer[x \ldots y] = \sigma(y)cov[x \ldots y], \tag{84}$$

respectively. The set of all paths from x to $y(x \neq y)$ in G denoted $< x, y >$, will be called a *connection* from x to y in G. In other words, connection $< x, y >$ is a sub-graph of G determined by nodes x and y.

For every connection $< x, y >$ we define its certainty, coverage and strength as shown below:

$$cer < x, y >= \sum_{[x...y] \in <x,y>} cer[x ... y], \qquad (85)$$

the *coverage* of the connection $< x, y >$ is

$$cov < x, y >= \sum_{[x...y] \in <x,y>} cov[x ... y], \qquad (86)$$

and the *strength* of the connection $< x, y >$ is

$$\sigma < x, y >= \sum_{[x...y] \in <x,y>} \sigma[x ... y] = \sigma(x)cer < x, y >= \sigma(y)cov < x, y > . \qquad (87)$$

Let $[x ... y]$ be a path such that x and y are input and output of the graph G, respectively. Such a *path* will be referred to as *complete*.

The set of all complete paths from x to y will be called a *complete connection* from x to y in G. In what follows we will consider complete paths and connections only, unless stated otherwise.

Let x and y be an input and output of a graph G respectively. If we substitute for every complete connection $< x, y >$ in G a single branch (x, y) such $\sigma(x, y) = \sigma < x, y >$, $cer(x, y) = cer < x, y >$, $cov(x, y) = cov < x, y >$ then we obtain a new flow graph G' such that $\sigma(G) = \sigma(G')$. The new flow graph will be called a *combined* flow graph. The combined flow graph for a given flow graph represents a relationship between its inputs and outputs.

6.4 Dependencies in Flow Graphs

Let $(x, y) \in \mathcal{B}$. Nodes x and y are independent on each other if

$$\sigma(x, y) = \sigma(x)\sigma(y). \qquad (88)$$

Consequently

$$\frac{\sigma(x, y)}{\sigma(x)} = cer(x, y) = \sigma(y), \qquad (89)$$

and

$$\frac{\sigma(x, y)}{\sigma(y)} = cov(x, y) = \sigma(x). \qquad (90)$$

This idea refers to some concepts proposed by Łukasiewicz [88] in connection with statistical independence of logical formulas.

If

$$cer(x, y) > \sigma(y), \tag{91}$$

or

$$cov(x, y) > \sigma(x), \tag{92}$$

then x and y *depend positively* on each other. Similarly, if

$$cer(x, y) < \sigma(y), \tag{93}$$

or

$$cov(x, y) < \sigma(x), \tag{94}$$

then x and y *depend negatively* on each other.

Let us observe that relations of independency and dependencies are symmetric ones, and are analogous to that used in statistics.

For every $(x, y) \in \mathcal{B}$ we define a *dependency factor* $\eta(x, y)$ defined as

$$\eta(x, y) = \frac{cer(x, y) - \sigma(y)}{cer(x, y) + \sigma(y)} = \frac{cov(x, y) - \sigma(x)}{cov(x, y) + \sigma(x)}. \tag{95}$$

It is easy to check that if $\eta(x, y) = 0$, then x and y are independent on each other, if $-1 < \eta(x, y) < 0$, then x and y are negatively dependent and if $0 < \eta(x, y) < 1$ then x and y are positively dependent on each other.

Thus the dependency factor expresses a degree of dependency, and can be seen as a counterpart of correlation coefficient used in statistics (see also [112]).

6.5   An Example

Now we will illustrate ideas introduced in the previous sections by means of a simple example concerning votes distribution of various age groups and social classes of voters between political parties.

Consider three disjoint age groups of voters y_1 (*old*), y_2 (*middle aged*) and y_3 (*young*) – belonging to three social classes x_1 (*high*), x_2 (*middle*) and x_3 (*low*). The voters voted for four political parties z_1 (*Conservatives*), z_2 (*Labor*), z_3 (*Liberal Democrats*) and z_4 (*others*).

Social class and age group votes distribution is shown in Figure 4.

First we want to find votes distribution with respect to age group. The result is shown in Figure 5. From the flow graph presented in Figure 5 we can see that, e.g., party z_1 obtained 19% of total votes, all of them from age group y_1; party z_2 – 44% votes, which 82% are from age group y_2 and 18% – from age group y_3, etc.

If we want to know how votes are distributed between parties with respects to social classes we have to eliminate age groups from the flow graph. Employing the algorithm presented in Section 6.3 we get results shown in Figure 6.

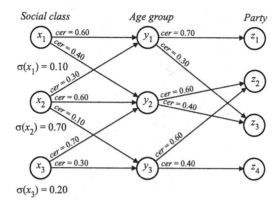

Fig. 4. Social class and age group votes distribution

From the flow graph presented in Figure 6 we can see that party z_1 obtained 22% votes from social class x_1 and 78% – from social class x_2, etc.

We can also present the obtained results employing decision rules. For simplicity we present only some decision rules of the decision algorithm. For example, from Figure 5 we obtain decision rules:

If Party (z_1) then Age group $(y_1)(0.19)$;
If Party (z_2) then Age group $(y_2(0.36)$;
If Party (z_2) then Age group $(y_3)(0.08)$, etc.

The number at the end of each decision rule denotes strength of the rule. Similarly, from Figure 6 we get:

If Party (z_1) then Soc. class $(x_1)(0.04)$;
If Party (z_1) then Soc. class $(x_2)(0.14)$, etc.

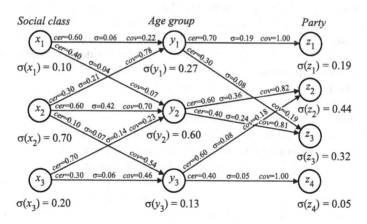

Fig. 5. Votes distribution with respect to the age group

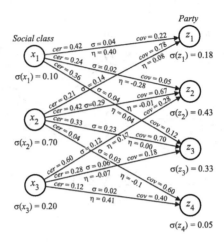

Fig. 6. Votes distribution between parties with respects to the social classes

From Figure 6 we have:

If Soc. class (x_1) then Party $(z_1)(0.04)$;
If Soc. class (x_1) then Party $(z_2)(0.02)$;
If Soc. class (x_1) then Party $(z_3)(0.04)$, etc.

Dependencies between Social class and Parties are shown in Figure 6.

6.6   An Example

In this section we continue the example from Section 5.3. The flow graph associated with Table 11 is shown in Figure 7.

Branches of the flow graph represent decision rules together with their certainty and coverage factors. For example, the decision rule $A \to 0$ has the certainty and coverage factors 0.13 and 0.35, respectively. The flow graph gives a clear insight into the voting structure of all parties. For many applications exact values of certainty of coverage factors of decision rules are not necessary. To this end we introduce "approximate" decision rules, denoted $C \leadsto D$ and read C *mostly implies* D. $C \leadsto D$ if and only if $cer(C, D) > 0.5$. Thus, we can replace flow graph shown in Figure 7 by *approximate* flow graph presented in Figure 8. From this graph we can see that parties B, C and D form a coalition, which is in conflict with party A, i.e., every member of the coalition is in conflict with party A. The corresponding conflict graph is shown in Figure 9.

Moreover, from the flow graph shown in Figure 7 we can obtain an "inverse" approximate flow graph which is shown in Figure 10. This flow graph contains all inverse decision rules with certainty factor greater than 0.5. From this graph we can see that yes votes were obtained *mostly* from party A and *no* votes – *mostly* from party D.

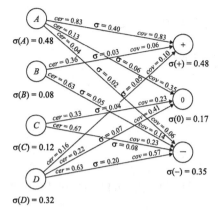

Fig. 7. Flow graph for Table 11

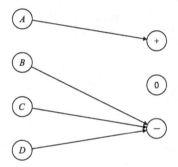

Fig. 8. "Approximate" flow graph

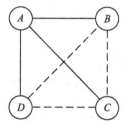

Fig. 9. Conflict graph

We can also compute dependencies between parties and voting results the results are shown in Figure 11.

6.7 Decision Networks

Ideas given in the previous sections can be also presented in logical terms, as shown in what follows.

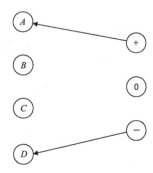

Fig. 10. An "inverse" approximate flow graph

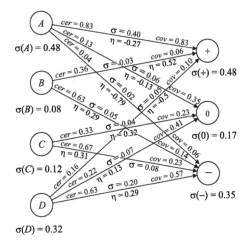

Fig. 11. Dependencies between parties and voting results

The main problem in data mining consists in discovering patterns in data. The patterns are usually expressed in form of decision rules, which are logical expressions in the form *if Φ then Ψ*, where *Φ* and *Ψ* are logical formulas (propositional functions) used to express properties of objects of interest. Any set of decision rules is called a *decision algorithm*. Thus knowledge discovery from data consists in representing hidden relationships between data in a form of decision algorithms. However, for some applications, it is not enough to give only set of decision rules describing relationships in the database. Sometimes also knowledge of relationship between decision rules is necessary in order to understand better data structures. To this end we propose to employ a decision algorithm in which also relationship between decision rules is pointed out, called a *decision network*.

The decision network is a finite, directed acyclic graph, nodes of which represent logical formulas, whereas branches – are interpreted as decision rules. Thus every path in the graph represents a chain of decisions rules, which will be used to describe compound decisions.

Some properties of decision networks will be given and a simple example will be used to illustrate the presented ideas and show possible applications.

Let U be a non empty finite set, called the *universe* and let Φ, Ψ be logical formulas. The meaning of Φ in U, denoted by $\|\Phi\|$, is the set of all elements of U, that satisfies Φ in U. The truth value of Φ denoted $val(\Phi)$ is defined as $card(\|\Phi\|)/card(U)$, where $card(X)$ denotes cardinality of X and $\mathcal{F}$ is a set of formulas.

By *decision network* over $S = (U, \mathcal{F})$ we mean a pair $N = (\mathcal{F}, \mathcal{R})$, where $\mathcal{R} \subseteq \mathcal{F} \times \mathcal{F}$ is a binary relation, called a *consequence relation* and $\mathcal{F}$ is a set of logical formulas.

Any pair $(\Phi, \Psi) \in \mathcal{R}, \Phi \neq \Psi$ is referred to as a decision rule (in N).

We assume that S is known and we will not refer to it in what follows.

A *decision rule* (Φ, Ψ) will be also presented as an expression $\Phi \to \Psi$, read *if Φ then Ψ*, where Φ and Ψ are referred to as *predecessor (conditions)* and *successor (decisions)* of the rule, respectively.

The number $supp(\Phi, \Psi) = card(\|\Phi \wedge \Psi\|)$ will be called a *support* of the rule $\Phi \to \Psi$. We will consider nonvoid decision rules only, i.e., rules such that $supp(\Phi, \Psi) \neq 0$.

With every decision rule $\Phi \to \Psi$ we associate its *strength* defined as

$$str(\Phi, \Psi) = \frac{supp(\Phi, \Psi)}{card(U)}. \tag{96}$$

Moreover, with every decision rule $\Phi \to \Psi$ we associate the *certainty factor* defined as

$$cer(\Phi, \Psi) = \frac{str(\Phi, \Psi)}{val(\Phi)}, \tag{97}$$

and the *coverage factor* of $\Phi \to \Psi$

$$cov(\Phi, \Psi) = \frac{str(\Phi, \Psi)}{val(\Psi)}, \tag{98}$$

where $val(\Phi) \neq 0$ and $val(\Psi) \neq 0$.

The coefficients can be computed from data or can be a subjective assessment.

We assume that

$$val(\Phi) = \sum_{\Psi \in Suc(\Phi)} str(\Phi, \Psi) \tag{99}$$

and

$$val(\Psi) = \sum_{\Phi \in Pre(\Psi)} str(\Phi, \Psi), \tag{100}$$

where $Suc(\Phi)$ and $Pre(\Psi)$ are sets of all successors and predecessors of the corresponding formulas, respectively.

Consequently we have

$$\sum_{Suc(\Phi)} cer(\phi, \Psi) = \sum_{Pre(\Psi)} cov(\Phi, \Psi) = 1. \tag{101}$$

If a decision rule $\Phi \rightarrow \Psi$ uniquely determines decisions in terms of conditions, i.e., if $cer(\Phi, \Psi) = 1$, then the rule is *certain*, otherwise the rule is *uncertain*.

If a decision rule $\Phi \rightarrow \Psi$ covers all decisions, i.e., if $cov(\Phi, \Psi) = 1$ then the decision rule is *total*, otherwise the decision rule is *partial*.

Immediate consequences of (97) and (98) are:

$$cer(\Phi, \Psi) = \frac{cov(\Phi, \Psi)val(\Psi)}{val(\Phi)}, \tag{102}$$

$$cov(\Phi, \Psi) = \frac{cer(\Phi, \Psi)val(\Phi)}{val(\Psi)}. \tag{103}$$

Note, that (102) and (103) are Bayes' formulas. This relationship, as mentioned previously, first was observed by Lukasiewicz [88].

Any sequence of formulas $\Phi_1, \ldots, \Phi_n$, $\Phi_i \in \mathcal{F}$ and for every i, $1 \leq i \leq n-1$, $(\Phi_i, \Phi_{i+1}) \in \mathcal{R}$ will be called a *path* from Φ_1 to Φ_n and will be denoted by $[\Phi_1 \ldots \Phi_n]$.

We define

$$cer[\Phi_1 \ldots \Phi_n] = \prod_{i=1}^{n-1} cer[\Phi_i, \Phi_{i+1}], \tag{104}$$

$$cov[\Phi_1 \ldots \Phi_n] = \prod_{i=1}^{n-1} cov[\Phi_i, \Phi_{i+1}], \tag{105}$$

$$str[\Phi_1 \ldots \Phi_n] = val(\Phi_1)cer[\Phi_1 \ldots \Phi_n] = val(\Phi_n)cov[\Phi_1 \ldots \Phi_n]. \tag{106}$$

The set of all paths form Φ to Ψ, denoted $< \Phi, \Psi >$, will be called a *connection* from Φ to Ψ.

For connection we have

$$cer < \Phi, \Psi >= \sum_{[\Phi \ldots \Psi] \in <\Phi, \Psi>} cer[\Phi \ldots \Psi], \tag{107}$$

$$cov < \Phi, \Psi >= \sum_{[\Phi \ldots \Psi] \in <\Phi, \Psi>} cov[\Phi \ldots \Psi], \tag{108}$$

$$str < \Phi, \Psi > = \sum_{[\Phi \ldots \Psi] \in <\Phi, \Psi>} str[\Phi \ldots \Psi] =$$
$$= val(\Phi)cer < \Phi, \Psi >= val(\Psi)cov < \Phi, \Psi > . \tag{109}$$

With every decision network we can associate a flow graph [70, 71]. Formulas of the network are interpreted as nodes of the graph, and decision rules – as directed branches of the flow graph, whereas strength of a decision rule is interpreted as flow of the corresponding branch.

Let $\Phi \to \Psi$ be a decision rule. Formulas Φ and Ψ are *independent* on each other if

$$str(\Phi, \Psi) = val(\Phi)val(\Psi). \tag{110}$$

Consequently

$$\frac{str(\Phi, \Psi)}{val(\Phi)} = cer(\Phi, \Psi) = val(\Psi), \tag{111}$$

and

$$\frac{str(\Phi, \Psi)}{val(\Psi)} = cov(\Phi, \Psi) = val(\Phi). \tag{112}$$

If

$$cer(\Phi, \Psi) > val(\Psi), \tag{113}$$

or

$$cov(\Phi, \Psi) > val(\Phi), \tag{114}$$

then Φ and Ψ *depend positively* on each other. Similarly, if

$$cer(\Phi, \Psi) < val(\Psi), \tag{115}$$

or

$$cov(\Phi, \Psi) < val(\Phi), \tag{116}$$

then Φ and Ψ *depend negatively* on each other.

For every decision rule $\Phi \to \Psi$ we define a *dependency factor* $\eta(\Phi, \Psi)$ defined as

$$\eta(\Phi, \Psi) = \frac{cer(\Phi, \Psi) - val(\Psi)}{cer(\Phi, \Psi) + val(\Psi)} = \frac{cov(\Phi, \Psi) - val(\Phi)}{cov(\Phi, \Psi) + val(\Phi)}. \tag{117}$$

It is easy to check that if $\eta(\Phi, \Psi) = 0$, then Φ and Ψ are independent on each other, if $-1 < \eta(\Phi, \Psi) < 0$, then Φ and Ψ are negatively dependent and if $0 < \eta(\Phi, \Psi) < 1$ then Φ and Ψ are positively dependent on each other.

6.8   An Example

Flow graphs given in Figures 4–6 can be now presented as shown in Figures 12–14, respectively. These flow graphs show clearly the relational structure between formulas involved in the voting process.

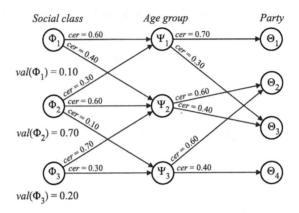

Fig. 12. Decision network for flow graph from Figure 4

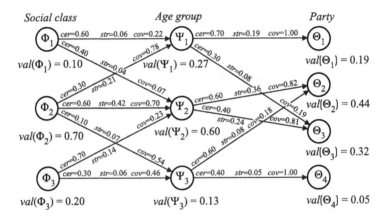

Fig. 13. Decision network for flow graph from Figure 5

6.9   Inference Rules and Decision Rules

In this section we are going to show relationship between previously discussed concepts and reasoning schemes used in logical inference.

Basic rules of inference used in classical logic are *Modus Ponens* (MP) and *Modus Tollens* (MT). These two reasoning patterns start from some general knowledge about reality, expressed by true implication, "*if Φ then Ψ*". Then basing on true *premise* Φ we arrive at true *conclusion* Ψ (MP), or if negation of conclusion Ψ is true we infer that negation of premise Φ is true (MT).

In reasoning from data (data mining) we also use rules *if Φ then Ψ*, called *decision rules*, to express our knowledge about reality, but the meaning of decision rules is different. It does not express general knowledge but refers to partial facts. Therefore decision rules are not true or false but probable (possible) only.

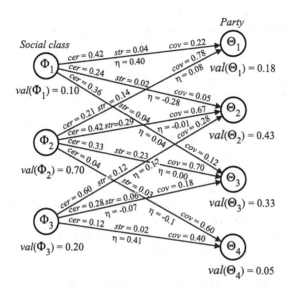

Fig. 14. Decision network for flow graph from Figure 6

In this paper we compare inference rules and decision rules in the context of decision networks, proposed by the author as a new approach to analyze reasoning patterns in data.

Decision network is a set of logical formulas $\mathcal{F}$ together with a binary relation over the set $\mathcal{R} \subseteq \mathcal{F} \times \mathcal{F}$ of formulas, called a *consequence relation*. Elements of the relation are called *decision rules*. The decision network can be perceived as a directed graph, nodes of which are formulas and branches – are decision rules. Thus the decision network can be seen as a knowledge representation system, revealing data structure of a data base.

Discovering patterns in the database represented by a decision network boils down to discovering some patterns in the network. Analogy to the *modus ponens* and *modus tollens* inference rules will be shown and discussed.

Classical rules of inference used in logic are *Modus Ponens* and *Modus Tollens*, which have the form

$$
\begin{array}{ll}
\textit{if} & \varPhi \to \varPsi \ \textit{is true} \\
\textit{and} \ \varPhi & \quad \textit{is true} \\
\hline
\textit{then} & \varPsi \ \textit{is true}
\end{array}
$$

and

$$
\begin{array}{ll}
\textit{if} & \varPhi \to \varPsi \ \textit{is true} \\
\textit{and} & \sim \varPsi \ \textit{is true} \\
\hline
\textit{then} \sim \varPhi & \quad \textit{is true}
\end{array}
$$

respectively.

Modus Ponens allows us to obtain true consequences from true premises, whereas *Modus Tollens* yields true negation of premise from true negation of conclusion.

In reasoning about data (data analysis) the situation is different. Instead of true propositions we consider propositional functions, which are true to a "degree", i.e., they assume truth values which lie between 0 and 1, in other words, they are probable, not true.

Besides, instead of true inference rules we have now decision rules, which are neither true nor false. They are characterized by three coefficients, *strength, certainty* and *coverage factors.* Strength of a decision rule can be understood as a counterpart of truth value of the inference rule, and it represents frequency of the decision rule in a database.

Thus employing decision rules to discovering patterns in data boils down to computation probability of conclusion in terms of probability of the premise and strength of the decision rule, or – the probability of the premise from the probability of the conclusion and strength of the decision rule.

Hence, the role of decision rules in data analysis is somehow similar to classical inference patterns, as shown by the schemes below.

Two basic rules of inference for data analysis are as follows:

$$
\begin{array}{lll}
if & \Phi \to \Psi & has\ cer(\Phi,\Psi)\ and\ cov(\Phi,\Psi) \\
and & \Phi & is\ true\ with\ the\ probability\ val(\Phi) \\
\hline
then & \Psi & is\ true\ with\ the\ probability\ val(\Psi) = \alpha val(\Phi).
\end{array}
$$

Similarly

$$
\begin{array}{lll}
if & \Phi \to \Psi & has\ cer(\Phi,\Psi)\ and\ cov(\Phi,\Psi) \\
and & \Psi & is\ true\ with\ the\ probability\ val(\Psi) \\
\hline
then & \Phi & is\ true\ with\ the\ probability\ val(\Phi) = \alpha^{-1} val(\Phi).
\end{array}
$$

The above inference rules can be considered as counterparts of *Modus Ponens* and *Modus Tollens* for data analysis and will be called Rough *Modus Ponens* (RMP) and Rough *Modus Tollens* (RMT), respectively.

There are however essential differences between MP (MT) and RMP (RMT).

First, instead of truth values associated with inference rules we consider certainly and coverage factors (conditional probabilities) assigned to decision rules.

Second, in the case of decision rules, in contrast to inference rules, truth value of a conclusion (RMP) depends not only on a single premise but in fact depends on truth values of premises of all decision rules having the same conclusions. Similarly, for RMT.

Let us also notice that inference rules are transitive, i.e., *if* $\Phi \to \Psi$ *and* $\Psi \to \Theta$ *then* $\Phi \to \Theta$ and decision rules are not. If $\Phi \to \Psi$ *and* $\Psi \to \Theta$, then we have to compute the certainty, coverage and strength of the rule $\Phi \to \Theta$, employing formulas (104),(105),(107),(108).

This shows clearly the difference between reasoning patterns using classical inference rules in logical reasoning and using decision rules in reasoning about data.

6.10   An Example

Suppose that three models of cars Φ_1, Φ_2 and Φ_3 are sold to three disjoint groups of customers Θ_1, Θ_2 and Θ_3 through four dealers Ψ_1, Ψ_2, Ψ_3 and Ψ_4.

Moreover, let us assume that car models and dealers are distributed as shown in Figure 15. Applying RMP to data shown in Figure 15 we get results shown in Figure 16. In order to find how car models are distributed among customer

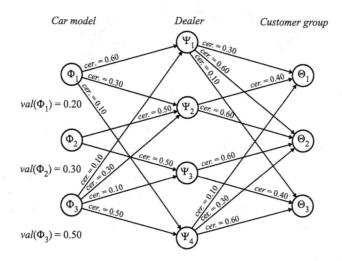

Fig. 15. Distributions of car models and dealers

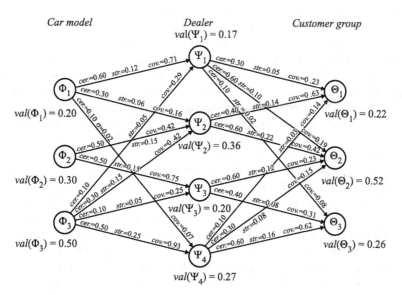

Fig. 16. The result of application of RMP to data from Figure 15

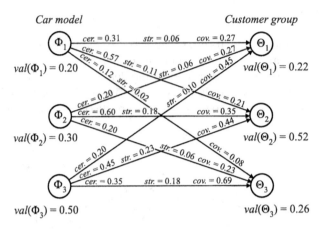

Fig. 17. Distribution of car models among customer groups

groups we have to compute all connections among cars models and consumers groups, i.e., to apply RMP to data given in Figure 16. The results are shown in Figure 17.

For example, we can see from the decision network that consumer group Θ_2 bought 21% of car model Φ_1, 35% of car model Φ_2 and 44% of car model Φ_3. Conversely, for example, car model Φ_1 is distributed among customer groups as follows: 31% cars bought group Θ_1, 57% group Θ_2 and 12% group Θ_3.

7 Summary

Basic concept of mathematics, the set, leads to antinomies, i.e., it is contradictory. This deficiency of sets, has rather philosophical than practical meaning, for sets used in mathematics are free from the above discussed faults. Antinomies are associated with very "artificial" sets constructed in logic but not found in sets used in mathematics. That is why we can use mathematics safely.

Philosophically, fuzzy set theory and rough set theory are two different approaches to vagueness and are not remedy for classical set theory difficulties. Both theories represent two different approaches to vagueness. Fuzzy set theory addresses gradualness of knowledge, expressed by the fuzzy membership whereas rough set theory addresses granularity of knowledge, expressed by the indiscernibility relation.

Practically, rough set theory can be viewed as a new method of intelligent data analysis. Rough set theory has found many applications in medical data analysis, finance, voice recognition, image processing, and others. However the approach presented in this paper is too simple to many real-life applications and was extended in many ways by various authors. The detailed discussion of the above issues can be found in be found in books (see, e.g., [18–27, 12, 28–30]), special issues of journals (see, e.g., [31–34, 34–38]), proceedings of international conferences (see, e.g., [39–49]), tutorials (e.g., [50–53]), and on the internet (see, e.g., www.roughsets.org, logic.mimuw.edu.pl,rsds.wsiz.rzeszow.pl).

Besides, rough set theory inspired new look on Bayes' theorem. Bayesian inference consists in update prior probabilities by means of data to posterior probabilities. In the rough set approach Bayes' theorem reveals data patterns, which are used next to draw conclusions from data, in form of decision rules.

Moreover, we have shown a new mathematical model of flow networks, which can be used to decision algorithm analysis. In particular it has been revealed that the flow in the flow network is governed by Bayes' rule, which has entirely deterministic meaning, and can be used to decision algorithm study.

Also, a new look of dependencies in databases, based on Łukasiewiczs ideas of independencies of logical formulas, is presented.

Acknowledgment

I would like to thank to Prof. Andrzej Skowron for useful discussion and help in preparation of this paper.

References

1. Zadeh, L.A.: Fuzzy sets. Information and Control **8** (1965) 338–353
2. Pawlak, Z.: Rough sets. International Journal of Computer and Information Sciences **11** (1982) 341–356
3. Ziarko, W.: Variable precision rough set model. Journal of Computer and System Sciences **46** (1993) 39–59
4. Polkowski, L., Skowron, A., Żytkow, J.: Rough foundations for rough sets. In [40] 55–58
5. Skowron, A., Stepaniuk, J.: Tolerance approximation spaces. Fundamenta Informaticae **27** (1996) 245–253
6. Polkowski, L., Skowron, A.: Rough mereology: A new paradigm for approximate reasoning. International Journal of Approximate Reasoning **15** (1996) 333–365
7. Słowiński, R., Vanderpooten, D.: Similarity relation as a basis for rough approximations. In Wang, P.P., ed.: Machine Intelligence & Soft-Computing, Vol. IV. Bookwrights, Raleigh, NC (1997) 17–33
8. Słowiński, R., Vanderpooten, D.: A generalized definition of rough approximations based on similarity. IEEE Transactions on Data and Knowledge Engineering **12(2)** (2000) 331–336
9. Stepaniuk, J.: Knowledge discovery by application of rough set models. In [26] 137–233
10. Skowron, A.: Toward intelligent systems: Calculi of information granules. Bulletin of the International Rough Set Society **5** (2001) 9–30
11. Greco, A., Matarazzo, B., Słowiński, R.: Rough approximation by dominance relations. International Journal of Intelligent Systems **17** (2002) 153–171
12. Polkowski, L., ed.: Rough Sets: Mathematical Foundations. Advances in Soft Computing. Physica-Verlag, Heidelberg (2002)
13. Skowron, A., Stepaniuk, J.: Information granules and rough-neural computing. In [30] 43–84
14. Skowron, A.: Approximation spaces in rough neurocomputing. In [29] 13–22
15. Wróblewski, J.: Adaptive aspects of combining approximation spaces. In [30] 139–156

16. Yao, Y.Y.: Informaton granulation and approximation in a decision-theoretical model of rough sets. In [30] 491–520
17. Skowron, A., Swiniarski, R., Synak, P.: Approximation spaces and information granulation (submitted). In: Fourth International Conference on Rough Sets and Current Trends in Computing (RSCTC'04), Uppsala, Sweden, June 1-5, 2004. Lecture Notes in Computer Science. Springer-Verlag, Heidelberg, Germany (2004)
18. Pawlak, Z.: Rough Sets: Theoretical Aspects of Reasoning about Data. Volume 9 of System Theory, Knowledge Engineering and Problem Solving. Kluwer Academic Publishers, Dordrecht, The Netherlands (1991)
19. Słowiński, R., ed.: Intelligent Decision Support - Handbook of Applications and Advances of the Rough Sets Theory. Volume 11 of System Theory, Knowledge Engineering and Problem Solving. Kluwer Academic Publishers, Dordrecht, The Netherlands (1992)
20. Lin, T.Y., Cercone, N., eds.: Rough Sets and Data Mining - Analysis of Imperfect Data. Kluwer Academic Publishers, Boston, USA (1997)
21. Orłowska, E., ed.: Incomplete Information: Rough Set Analysis. Volume 13 of Studies in Fuzziness and Soft Computing. Springer-Verlag/Physica-Verlag, Heidelberg, Germany (1997)
22. Polkowski, L., Skowron, A., eds.: Rough Sets in Knowledge Discovery 1: Methodology and Applications. Volume 18 of Studies in Fuzziness and Soft Computing. Physica-Verlag, Heidelberg, Germany (1998)
23. Polkowski, L., Skowron, A., eds.: Rough Sets in Knowledge Discovery 2: Applications, Case Studies and Software Systems. Volume 19 of Studies in Fuzziness and Soft Computing. Physica-Verlag, Heidelberg, Germany (1998)
24. Pal, S.K., Skowron, A., eds.: Rough Fuzzy Hybridization: A New Trend in Decision-Making. Springer-Verlag, Singapore (1999)
25. Duentsch, I., Gediga, G.: Rough set data analysis: A road to non-invasive knowledge discovery. Methodos Publishers, Bangor, UK (2000)
26. Polkowski, L., Lin, T.Y., Tsumoto, S., eds.: Rough Set Methods and Applications: New Developments in Knowledge Discovery in Information Systems. Volume 56 of Studies in Fuzziness and Soft Computing. Springer-Verlag/Physica-Verlag, Heidelberg, Germany (2000)
27. Lin, T.Y., Yao, Y.Y., Zadeh, L.A., eds.: Rough Sets, Granular Computing and Data Mining. Studies in Fuzziness and Soft Computing. Physica-Verlag, Heidelberg (2001)
28. Demri, S., Orłowska, E., eds.: Incomplete Information: Structure, Inference, Complexity. Monographs in Theoretical Cpmputer Sience. Springer-Verlag, Heidelberg, Germany (2002)
29. Inuiguchi, M., Hirano, S., Tsumoto, S., eds.: Rough Set Theory and Granular Computing. Volume 125 of Studies in Fuzziness and Soft Computing. Springer-Verlag, Heidelberg (2003)
30. Pal, S.K., Polkowski, L., Skowron, A., eds.: Rough-Neural Computing: Techniques for Computing with Words. Cognitive Technologies. Springer-Verlag, Heidelberg, Germany (2003)
31. Słowiński, R., Stefanowski, J., eds.: Special issue: Proceedings of the First International Workshop on Rough Sets: State of the Art and Perspectives, Kiekrz, Poznań, Poland, September 2–4 (1992). Volume 18(3-4) of Foundations of Computing and Decision Sciences. (1993)
32. Ziarko, W., ed.: Special issue. Volume 11(2) of Computational Intelligence: An International Journal. (1995)

33. Ziarko, W., ed.: Special issue. Volume 27(2-3) of Fundamenta Informaticae. (1996)
34. Lin, T.Y., ed.: Special issue. Volume 2(2) of Journal of the Intelligent Automation and Soft Computing. (1996)
35. Peters, J., Skowron, A., eds.: Special issue on a rough set approach to reasoning about data. Volume 16(1) of International Journal of Intelligent Systems. (2001)
36. Cercone, N., Skowron, A., Zhong, N., eds.: (Special issue). Volume 17(3) of Computational Intelligence. (2001)
37. Pal, S.K., Pedrycz, W., Skowron, A., Swiniarski, R., eds.: Special volume: Rough-neuro computing. Volume 36 of Neurocomputing. (2001)
38. Skowron, A., Pal, S.K., eds.: Special volume: Rough sets, pattern recognition and data mining. Volume 24(6) of Pattern Recognition Letters. (2003)
39. Ziarko, W., ed.: Rough Sets, Fuzzy Sets and Knowledge Discovery: Proceedings of the Second International Workshop on Rough Sets and Knowledge Discovery (RSKD'93), Banff, Alberta, Canada, October 12–15 (1993). Workshops in Computing. Springer–Verlag & British Computer Society, London, Berlin (1994)
40. Lin, T.Y., Wildberger, A.M., eds.: Soft Computing: Rough Sets, Fuzzy Logic, Neural Networks, Uncertainty Management, Knowledge Discovery. Simulation Councils, Inc., San Diego, CA, USA (1995)
41. Tsumoto, S., Kobayashi, S., Yokomori, T., Tanaka, H., Nakamura, A., eds.: Proceedings of the The Fourth Internal Workshop on Rough Sets, Fuzzy Sets and Machine Discovery, November 6-8, University of Tokyo , Japan. The University of Tokyo, Tokyo (1996)
42. Polkowski, L., Skowron, A., eds.: First International Conference on Rough Sets and Soft Computing (RSCTC'98), Warsaw, Poland, June 22-26, 1998. Volume 1424 of Lecture Notes in Artificial Intelligence. Springer-Verlag, Heidelberg (1998)
43. Zhong, N., Skowron, A., Ohsuga, S., eds.: Proceedings of the 7-th International Workshop on Rough Sets, Fuzzy Sets, Data Mining, and Granular-Soft Computing (RSFDGrC'99), Yamaguchi, November 9-11, 1999. Volume 1711 of Lecture Notes in Artificial Intelligence. Springer-Verlag, Heidelberg (1999)
44. Ziarko, W., Yao, Y., eds.: Proceedings of the 2-nd International Conference on Rough Sets and Current Trends in Computing (RSCTC'2000), Banff, Canada, October 16-19, 2000. Volume 2005 of Lecture Notes in Artificial Intelligence. Springer-Verlag, Heidelberg (2001)
45. Hirano, S., Inuiguchi, M., Tsumoto, S., eds.: Proceedings of International Workshop on Rough Set Theory and Granular Computing (RSTGC-2001), Matsue, Shimane, Japan, May 20-22, 2001. Volume 5(1-2) of Bulletin of the International Rough Set Society. International Rough Set Society, Matsue, Shimane (2001)
46. Terano, T., Nishida, T., Namatame, A., Tsumoto, S., Ohsawa, Y., Washio, T., eds.: New Frontiers in Artificial Intelligence, Joint JSAI'01 Workshop Post-Proceedings. Volume 2253 of Lecture Notes in Artificial Intelligence. Springer-Verlag, Heidelberg (2001)
47. Alpigini, J.J., Peters, J.F., Skowron, A., Zhong, N., eds.: Third International Conference on Rough Sets and Current Trends in Computing (RSCTC'02), Malvern, PA, October 14-16, 2002. Volume 2475 of Lecture Notes in Artificial Intelligence. Springer-Verlag, Heidelberg (2002)
48. Skowron, A., Szczuka, M., eds.: Proceedings of the Workshop on Rough Sets in Knowledge Discovery and Soft Computing at ETAPS 2003 (RSKD'03), April 12-13, 2003. Volume 82(4) of Electronic Notes in Computer Science. Elsevier, Amsterdam, Netherlands (2003)

49. Wang, G., Liu, Q., Yao, Y., Skowron, A., eds.: Proceedings of the 9-th International Conference on Rough Sets, Fuzzy Sets, Data Mining, and Granular Computing (RSFDGrC'03), Chongqing, China, May 26-29, 2003. Volume 2639 of Lecture Notes in Artificial Intelligence. Springer-Verlag, Heidelberg (2003)

50. Komorowski, J., , Pawlak, Z., Polkowski, L., Skowron, A.: Rough sets: a tutorial. In [24] 3–98

51. Pawlak, Z., Polkowski, L., Skowron, A.: Rough sets and rough logic: A KDD perspective. In [26] 583–646

52. Skowron, A., Pawlak, Z., Komorowski, J., Polkowski, L.: A rough set perspective on data and knowledge. In Kloesgen, W., Żytkow, J., eds.: Handbook of KDD. Oxford University Press, Oxford (2002) 134–149

53. Pawlak, Z., Polkowski, L., Skowron, A.: Rough set theory. In Wah, B., ed.: Encyclopedia Of Computer Science and Engineering. Wiley, New York, USA (2004)

54. Cantor, G.: Grundlagen einer allgemeinen Mannigfaltigkeitslehre, Leipzig, Germany (1883)

55. Russell, B.: The Principles of Mathematics. George Allen & Unwin Ltd., London, Great Britain (1903)

56. Russell, B.: Vagueness. The Australasian Journal of Psychology and Philosophy 1 (1923) 84–92

57. Black, M.: Vagueness: An exercise in logical analysis. Philosophy of Science 4(4) (1937) 427–455

58. Hempel, C.G.: Vagueness and logic. Philosophy of Science 6 (1939) 163–180

59. Fine, K.: Vagueness, truth and logic. Synthese 30 (1975) 265–300

60. Keefe, R., Smith, P.: Vagueness: A Reader. MIT Press, Cambridge, MA (1999)

61. Keefe, R.: Theories of Vagueness. Cambridge University Press, Cambridge, U.K. (2000)

62. Frege, G.: Grundgesetzen der Arithmetik, 2. Verlag von Herman Pohle, Jena, Germany (1903)

63. Read, S.: Thinking about Logic - An Introduction to Philosophy of Logic. Oxford University Press, Oxford (1995)

64. Leśniewski, S.: Grungzüge eines neuen systems der grundlagen der mathematik. Fundamenta Matematicae 14 (1929) 1–81

65. Pawlak, Z., Skowron, A.: Rough membership functions. In Yager, R., Fedrizzi, M., Kacprzyk, J., eds.: Advances in the Dempster-Shafer Theory of Evidence, New York, NY, John Wiley & Sons (1994) 251–271

66. Skowron, A., Rauszer, C.: The discernibility matrices and functions in information systems. In [19] 331–362

67. Berthold, M., Hand, D.J.: Intelligent Data Analysis. An Introduction. Springer-Verlag, Berlin, Heidelberg, New York (1999)

68. Box, G.E.P., Tiao, G.C.: Bayesian Inference in Statistical Analysis. John Wiley and Sons, Inc., New York, Chichester, Brisbane, Toronto, Singapore (1992)

69. Pawlak, Z.: Rough sets and decision algorithms. In [44] 30–45

70. Pawlak, Z.: In pursuit of patterns in data reasoning from data – the rough set way. In [47] 1–9

71. Pawlak, Z.: Probability, truth and flow graphs. In [48] 1–9

72. Wong, S., Ziarko, W.: Algebraic versus probabilistic independence in decision theory. In Ras, Z.W., Zemankova, M., eds.: Proceedings of the ACM SIGART First International Symposium on Methodologies for Intelligent Systems Knoxville (ISMIS'86), Tennessee, USA, October 22-24, 1986. ACM SIGART, USA (1986) 207–212

73. Wong, S., Ziarko, W.: On learning and evaluation of decision rules in the context of rough sets. In Ras, Z.W., Zemankova, M., eds.: Proceedings of the ACM SIGART First International Symposium on Methodologies for Intelligent Systems Knoxville (ISMIS'86), Tennessee, USA, October 22-24, 1986. ACM SIGART, USA (1986) 308-324

74. Pawlak, Z., Wong, S.K.M., Ziarko, W.: Rough sets: Probabilistic versus deterministic approach. International Journal of Man-Machine Studies **29(1)** (1988) 81-95

75. Yamauchi, Y., Mukaidono, M.: Probabilistic inference and bayeasian theorem based on logical implication. In [43] 334-342

76. Intan, R., an Y. Y. Yao, M.M.: Generalization of rough sets with alpha-coverings of the universe induced by conditional probability relations. In [46] 311-315

77. Ślęzak, D.: Approximate decision reducts (in Polish). PhD thesis, Warsaw University, Warsaw, Poland (2002)

78. Ślęzak, D.: Approximate bayesian networks. In Bouchon-Meunier, B., Gutierrez-Rios, J., Magdalena, L., Yager, R., eds.: Technologies for Constructing Intelligent Systems 2: Tools. Volume 90 of Studies in Fuzziness and Soft Computing. Springer-Verlag, Heidelberg, Germany (2002) 313-326

79. Ślęzak, D., Wróblewski, J.: Approximate bayesian network classifiers. In [47] 365-372

80. Yao, Y.Y.: Information granulation and approximation. In [30] 491-516

81. Ślęzak, D.: Approximate markov boundaries and bayesian networks: Rough set approach. In [29] 109-121

82. Ślęzak, D., Ziarko, W.: Attribute reduction in the bayesian version of variable precision rough set model. In [48]

83. Ślęzak, D., Ziarko, W.: Variable precision bayesian rough set model. In [49] 312-315

84. Wong, S.K.M., Wu, D.: A common framework for rough sets, databases, and bayesian networks. In [49] 99-103

85. Ślęzak, D.: The rough bayesian model for distributed decision systems (submitted). In: Fourth International Conference on Rough Sets and Current Trends in Computing (RSCTC'04), Uppsala, Sweden, June 1-5, 2004. Lecture Notes in Computer Science. Springer-Verlag, Heidelberg, Germany (2004)

86. Swinburne, R.: Bayes Theorem. Volume 113 of Proceedings of the British Academy. Oxford University Press, Oxford, UK (2003)

87. Bernardo, J.M., Smith, A.F.M.: Bayesian Theory. Wiley Series in Probability and Mathematical Statistics. John Wiley & Sons, Chichester, New York, Brisbane, Toronto, Singapore (1994)

88. Łukasiewicz, J.: Die logischen grundlagen der wahrscheinilchkeitsrechnung, Kraków 1913. In Borkowski, L., ed.: Jan Łukasiewicz - Selected Works. North Holland Publishing Company, Amstardam, London, Polish Scientific Publishers, Warsaw (1970)

89. Adams, E.W.: The Logic of Conditionals. An Application of Probability to Deductive Logic. D. Reidel Publishing Company, Dordrecht, Boston (1975)

90. Grzymała-Busse, J.W.: LERS - a system for learning from examples based on rough sets. In [19] 3-18

91. Skowron, A.: Boolean reasoning for decision rules generation. In Komorowski, J., Raś, Z.W., eds.: Seventh International Symposium for Methodologies for Intelligent Systems (ISMIS'93), Trondheim, Norway, June 15-18. Volume 689 of Lecture Notes in Artificial Intelligence., Heidelberg, Springer-Verlag (1993) 295-305

92. Pawlak, Z., Skowron, A.: A rough set approach for decision rules generation. In: Thirteenth International Joint Conference on Artificial Intelligence (IJCAI'93), Chambéry, France, Morgan Kaufmann (1993) 114–119

93. Shan, N., Ziarko, W.: An incremental learning algorithm for constructing decision rules. In Ziarko, W., ed.: Rough Sets, Fuzzy Sets and Knowledge Discovery, Berlin, Germany, Springer Verlag (1994) 326–334

94. Nguyen, H.S.: Discretization of Real Value Attributes, Boolean Reasoning Approach. PhD thesis, Warsaw University, Warsaw, Poland (1997)

95. Słowiński, R., Stefanowski, J.: Rough family – software implementation of the rough set theory. In [23] 581–586

96. Nguyen, H.S., Nguyen, S.H.: Pattern extraction from data. Fundamenta Informaticae **34** (1998) 129–144

97. Nguyen, H.S., Nguyen, S.H.: Discretization methods for data mining. In [22] 451–482

98. Skowron, A.: Rough sets in KDD - plenary talk. In Shi, Z., Faltings, B., Musen, M., eds.: 16-th World Computer Congress (IFIP'00): Proceedings of Conference on Intelligent Information Processing (IIP'00). Publishing House of Electronic Industry, Beijing (2002) 1–14

99. Bazan, J., Nguyen, H.S., Nguyen, S.H., Synak, P., Wróblewski, J.: Rough set algorithms in classification problems. In [26] 49–88

100. Grzymala-Busse, J.W., Shah, P.: A comparison of rule matching methods used in aq15 and lers. In: Proceedings of the Twelfth International Symposium on Methodologies for Intelligent Systems (ISMIS'00), Charlotte, NC, October 11-14, 2000. Volume 1932 of Lecture Nites in Artificial Intelligence., Berlin, Germany, Springer-Verlag (2000) 148–156

101. Grzymała-Busse, J., Hu, M.: A comparison of several approaches to missing attribute values in data mining. In [44] 340 – 347

102. Greco, S., Matarazzo, B., Słowiński, R., Stefanowski, J.: An algorithm for induction of decision rules consistent with dominance principle. In [44] 304–313

103. Skowron, A.: Rough sets and boolean reasoning. In Pedrycz, W., ed.: Granular Computing: an Emerging Paradigm. Volume 70 of Studies in Fuzziness and Soft Computing. Springer-Verlag/Physica-Verlag, Heidelberg, Germany (2001) 95–124

104. Greco, S., Matarazzo, B., Słowiński, R.: Rough sets theory for multicriteria decision analysis. European J. of Operational Research **129(1)** (2001) 1–47

105. Casti, J.L.: Alternate Realities: Mathematical Models of Nature and Man. John Wiley and Sons, Inc., New York, Chichester, Brisbane, Toronto, Singapore (1989)

106. Coombs, C.H., Avruin, G.S.: The Structure of Conflicts. Lawrence Erlbaum, London (1988)

107. Deja, R.: Conflict analysis, rough set methods and applications. In [26] 491–520

108. Maeda, Y., Senoo, K., Tanaka, H.: Interval density function in conflict analysis. In [43] 382–389

109. Nakamura, A.: Conflict logic with degrees. In [24] 136–150

110. Pawlak, Z.: An inquiry into anatomy of conflicts. Journal of Information Sciences **109** (1998) 65–68

111. Ford, L.R., Fulkerson, D.R.: Flows in Networks. Princeton University Press, Princeton, New Jersey (1973)

112. Słowiński, R., Greco, S.: A note on dependency factor. (2004) (manuscript).

Learning Rules from Very Large Databases
Using Rough Multisets

Chien-Chung Chan

Department of Computer Science
University of Akron
Akron, OH 44325-4003
chan@cs.uakron.edu

Abstract. This paper presents a mechanism called *LERS-M* for learning production rules from very large databases. It can be implemented using object-
relational database systems, it can be used for distributed data mining, and it
has a structure that matches well with parallel processing. *LERS-M* is based on
rough multisets and it is formulated using relational operations with the objective to be tightly coupled with database systems. The underlying representation
used by LERS-M is multiset decision tables, which are derived from information multisystems. In addition, it is shown that multiset decision tables provide
a simple way to compute Dempster-Shafer's basic probability assignment functions from. data sets.

1 Introduction

The development of computer technologies has provided many useful and efficient
tools to produce, disseminate, store, and retrieve data in electronics forms. As a consequence, ever-increasing streams of data are recorded in all types of databases. For
example, in automated business activities, even simple transactions such as telephone
calls, credit card charges, items in shopping carts, etc. are typically recorded in databases. These data are potentially beneficial to enterprises, because they may be used
for designing effective marketing and sales plans based on consumer's shopping patterns and preferences collectively recorded in the databases. From databases of credit
card charges, some patterns of fraud charges may be detected, hence, preventive actions may be taken.

The raw data stored in databases are potentially lodes of useful information. In order to extract the ore, effective mining tools must be developed. The task of extracting
useful information from data is not a new one. It has been a common interest in research areas such as statistical data analysis, machine learning, and pattern recognition. Traditional techniques developed in these areas are fundamental to the task, but
there are limitations of these methods. For example, these tools usually assume that
the collection of data in the databases is small enough to be fit into the memory of a
computer system so that they can be processed. This condition is no longer true in
very large databases. Another limitation is that these tools are usually applicable to
only static data sets. However, most databases are updated frequently by large streams
of data. It is typical that databases of an enterprise are distributed in different loca-
tions. Issues and techniques related to finding useful information from distributed data
need to be studied and developed.

J.F. Peters et al. (Eds.): Transactions on Rough Sets I, LNCS 3100, pp. 59–77, 2004.
© Springer-Verlag Berlin Heidelberg 2004

There are three classical data mining problems: market basket analysis, clustering, and classification. Traditional machine learning systems are usually developed independent of database technology. One of the recent trends is to develop learning systems that are tightly coupled with relational or object-relational database systems for mining association rules and for mining tree classifiers [1–4]. Due to the maturity of database technology, these systems are more portable and scalable than traditional systems, and they are easier to integrate with *OLAP* (On Line Analytical Processing) and data warehousing systems. Another trend is that more and more data are stored into distributed databases. Some distributed data mining systems have been developed [5]. However, not many have been tightly coupled with database system technology.

In this paper, we introduce a mechanism called *LERS-M* for learning production rules from very large databases. It can be implemented using object-relational database systems, it can be used for distributed data mining, and it has a structure that matches well with parallel processing. *LERS-M* is similar to the *LERS* family of learning programs [6], which is based on rough set theory [7–9]. The main differences are *LERS-M* is based on rough multisets [10] and it is formulated using relational operations with the objective to be tightly coupled with database systems. The underlying representation used by *LERS-M* is multiset decision tables [11], which are derived from information multisystems [10]. In addition to facilitate the learning of rules, multiset decision tables can also be used to compute Dempster-Shafer's belief functions from data [12], [14]. The methodology developed here can be used to design learning systems for knowledge discovery from distributed databases and to develop distributed rule-based expert systems and decision support systems.

The paper is organized as follows. The problem addressed by this paper is formulated in Section 2. In Section 3, we review some related concepts. The concept of multiset decision tables and its properties are presented in Section 4. In Section 5, we present the *LERS-M* learning algorithm with example and discussion. Conclusions are given in Section 6.

2 Problem Statements

In this paper we consider the problem of learning production rules from very large databases. For simplicity, a very large database is considered as a very large data table U defined by a finite nonempty set A of attributes. We assume that a very large data table can be store in one single database or distributed over databases. By distributed databases, we means that the data table U is divided into N smaller tables with sizes manageable by a database management system. In the abstraction, we do not consider communication mechanisms used by a distributed database system. Nor do we consider the costs of transferring data from A to B.

Briefly speaking, the problem of inductive learning of production rules from examples is to generate descriptions or rules to characterize the logical implication $C \rightarrow D$ from a collection U of examples, where C and D are sets of attributes used to describe the examples. The set C is called condition attributes, and the set D is called decision attributes. Usually, set D is a singleton set, and the sets C and D are not overlapped. The objective of learning is to find rules that can be used to predict the logical implication as accurate as possible when applied to new examples.

The objective of this paper is to develop a mechanism for generating production rules by taking into account the following issues: (1) The implication of $C \to D$ may be uncertain, (2) If the set U of examples is divided into N smaller sets, how to determine the implication of $C \to D$, and (3) The result can be implemented using object-relational database technology.

3 Related Concepts

In the following, we will review the concepts of rough sets, information systems, decision tables, rough multisets, information multisystems, and partition of boundary sets.

3.1 Rough Sets, Information Systems, and Decision Tables

The fundamental assumption of the rough set theory is that objects from the domain are perceived only through the accessible information about them, that is, the values of attributes that can be evaluated on these objects. Objects with the same information are indiscernible. Consequently, the classification of objects is based on the accessible information about them, not on objects themselves. The notion of information systems was introduced by Pawlak [8] to represent knowledge about objects in a domain.

In this paper, we use a special case of information systems called *decision tables* or *data tables* to represent data sets. In a decision table there is a designated attribute called *decision attribute* and another set of attributes are called *condition attributes*. A decision attribute can be interpreted as a classification of objects in the domain given by an expert. Given a decision table, values of the decision attribute determine a partition on U. The problem of learning rules from examples is to find a set of classification rules using condition attributes that will produce the partition generated by the decision attribute.

An example of a decision table adapted from [13] is shown in Table 1, where the universe U consists of 28 objects or examples. The set of condition attributes is $\{A, B, C, E, F\}$, and D is the decision attribute with values 1, 2, and 3. The partition on U determined by the decision attribute D is

$$X_1 = [1, 2, 4, 8, 10, 15, 22, 25],$$
$$X_2 = [3, 5, 11, 12, 16, 18, 19, 21, 23, 24, 27],$$
$$X_3 = [6, 7, 9, 13, 14, 17, 20, 26, 28]$$

where X_i is the set of objects whose value of attribute d is i, for $i = 1, 2$, and 3.

Note that Table 1 is an inconsistent decision table. Both objects 8 and 12 have the same condition values $(1, 1, 1, 1, 1)$, but their decision values are different. Object 8 has decision value 1, but object 12 has decision value 2. Inconsistent data sets are also called noisy data sets. This kind of data sets is quite common in real world situations. It is an issue must be addressed by machine learning algorithms. In rough set approach, inconsistency is represented by the concepts of lower and upper approximations. Let $A = (U, R)$ be an *approximation space*, where U is a nonempty set of objects and R is an equivalence relation defined on U. Let X be a nonempty subset of U. Then, the *lower approximation* of X by R *in* A is defined as

Table 1. Example of a decision table.

U	A	B	C	E	F	D
1	0	0	1	0	0	1
2	1	1	1	0	0	1
3	0	1	0	0	0	2
4	1	0	0	0	1	1
5	0	1	0	0	0	2
6	1	0	0	0	1	3
7	0	0	0	1	1	3
8	1	1	1	1	1	1
9	0	0	0	1	1	3
10	0	0	1	0	0	1
11	1	1	1	0	0	2
12	1	1	1	1	1	2
13	1	1	0	1	1	3
14	1	1	0	0	1	3
15	0	0	1	1	1	1
16	1	1	0	1	1	2
17	0	0	0	1	1	3
18	0	0	0	0	0	2
19	0	0	0	0	0	2
20	1	1	1	0	0	3
21	1	1	0	0	1	2
22	0	0	1	0	1	1
23	1	1	1	0	0	2
24	0	0	1	1	1	2
25	1	0	1	0	1	1
26	1	0	1	0	1	3
27	1	0	1	0	1	2
28	1	1	1	1	0	3

$$\underline{R}X = \{\, e \in U \mid [e] \subseteq X\} \text{ and}$$

the *upper approximation* of X by R *in* A is defined as

$$\overline{R}\,X = \{\, e \in U \mid [e] \cap X \neq \varnothing\},$$

where $[e]$ denotes the equivalence class containing e. The difference $\overline{R}\,X - \underline{R}X$ is called the *boundary set* of X in A. A subset X of U is said to be *R-definable in* A if and only if $\underline{R}X = \overline{R}\,X$. The pair $(\underline{R}X, \overline{R}\,X)$ defines a *rough set in* A, which is a family of subsets of U with the same lower and upper approximations as $\underline{R}X$ and $\overline{R}\,X$. In terms of decision tables, the pair (U, A) defines an approximation space. When a decision class $X_i \subseteq U$ is inconsistent, it means that X_i is not A-definable. In this case, we can find classification rules from $\underline{A}X_i$ and $\overline{A}\,X_i$. These rules are called *certain rules* and *possible rules*, respectively [16]. Thus, rough set approach can be used to learn rules from both consistent and inconsistent examples [17], [18].

3.2   Rough Multisets and Information Multisystems

The concepts of rough multisets and information multisystems were introduced by Grzymala-Busse [10]. The basic idea is to represent an information system using *multisets* [15]. Object identifiers represented explicitly in an information system is not represented in an information multisystem. Thus, the resulting data tables are more compact. More precisely, an *information multisystem* is a triple $S = (Q, V, \tilde{Q})$, where Q is a set of attributes, V is the union of domains of attributes in Q, and $\tilde{Q}$ is a *multirelation* on $\underset{q \in Q}{\times} V_q$. In addition, the concepts of lower and upper approximations in rough sets are extended to multisets. Let M be a multiset, and let e be an element of M whose number of occurrences in M is w. The sub-multiset $\{w \cdot e\}$ will be denoted by $[e]_M$. Thus M may be represented as union of all $[e]_M$'s where e is in M. A multiset $[e]_M$ is called an *elementary multiset* in M. The empty multiset is elementary. A finite union of elementary multisets is called a *definable multiset* in M. Let X be a sub-multiset of M. Then, the lower approximation of X in M is the multiset defined as

$$\underline{X} = \{\, e \in M \mid [e]_M \subseteq X\} \text{ and}$$

the upper approximation of X in M is the multiset defined as

$$\overline{X} = \{\, e \in M \mid [e]_M \cap X \neq \varnothing\},$$

where the operations on sets are defined by multisets. Therefore, a *rough multiset* in M is the family of all sub-multisets of M having the same lower and upper approximations in M.

Let P be a subset of Q, a *projection of $\tilde{Q}$ onto P* is defined as the multirelation $\tilde{P}$, obtained by deleting columns corresponding to attributes in $Q - P$. Note that $\tilde{Q}$ and $\tilde{P}$ have same cardinality. Let X be a sub-multiset of $\tilde{P}$. A *P-lower approximation* of X in S is the lower approximation $\underline{X}$ of X in $\tilde{P}$. A *P-upper approximation* of X in S is the upper approximation $\overline{X}$ of X in $\tilde{P}$. A multiset X in $\tilde{P}$ is *P-definable* in S iff $\underline{P}X = \overline{P}X$.

A *multipartition* χ on a multiset X is a multiset $\{X_1, X_2, ..., X_n\}$ of sub-multisets of X such that

$$\sum_{i=1}^{n} X_i = X$$

where the *sum of two multisets X and Y*, denoted $X + Y$, is a multiset of all elements that are members of X or Y with the number of occurrences of each element e in $X + Y$ is the sum of the number of occurrences of e in X and the number of occurrences of e in Y.

Follow from [9], classifications are multipartitions on information multisystems generated with respect to subsets of attributes. Specifically, let $S = (Q, V, \tilde{Q})$ be an information multisystem. Let A and B be subsets of Q with $|A| = i$ and $|B| = j$. Let $\tilde{A}$ be a projection of $\tilde{Q}$ onto A. The subset B generates a multipartition B_A on $\tilde{A}$ defined as follows: each two i-tuples determined by A are in the same multiset X in B_A if and only if their associated j-tuples, determined by B, are equal. The mulitpartition B_A is called a *classification on $\tilde{A}$ generated by B*.

Table 2 shows a multirelation representation of the data table given in Table 1 where the number of occurrences of each row is denoted by integers in the W column. The projection of the multirealtion onto the set P of attributes $\{A, B, C, E, F\}$ is shown in Table 3.

Table 2. An information multisystem *S*.

A	B	C	E	F	D	W
0	0	0	0	0	2	2
0	0	0	1	1	3	3
0	0	1	0	0	1	2
0	0	1	0	1	1	1
0	0	1	1	1	1	1
0	0	1	1	1	2	1
0	1	0	0	0	2	2
1	0	0	0	1	1	1
1	0	0	0	1	3	1
1	0	1	0	1	1	1
1	0	1	0	1	2	1
1	0	1	0	1	3	1
1	1	0	0	1	2	1
1	1	0	0	1	3	1
1	1	0	1	1	2	1
1	1	0	1	1	3	1
1	1	1	0	0	1	1
1	1	1	0	0	2	2
1	1	1	0	0	3	1
1	1	1	1	0	3	1
1	1	1	1	1	1	1
1	1	1	1	1	2	1

Table 3. An information multisystem $\widetilde{P}$.

A	B	C	E	F	W
0	0	0	0	0	2
0	0	0	1	1	3
0	0	1	0	0	2
0	0	1	0	1	1
0	0	1	1	1	2
0	1	0	0	0	2
1	0	0	0	1	2
1	0	1	0	1	3
1	1	0	0	1	2
1	1	0	1	1	2
1	1	1	0	0	4
1	1	1	1	0	1
1	1	1	1	1	2

Let *X* be a sub-multiset of $\widetilde{P}$ with elements shown in Table 4.

Table 4. A sub-multiset *X* of $\widetilde{P}$.

A	B	C	E	F	W
0	0	1	0	0	2
1	1	1	0	0	1
1	0	0	0	1	1
1	1	1	1	1	1
0	0	1	1	1	1
0	0	1	0	1	1
1	0	1	0	1	1

Table 5. *P*-lower approximation of *X*.

A	B	C	E	F	W
0	0	1	0	0	2
0	0	1	0	1	1

Table 6. *P*-upper approximation of *X*.

A	B	C	E	F	W
0	0	1	0	0	2
1	1	1	0	0	4
1	0	0	0	1	2
1	1	1	1	1	2
0	0	1	1	1	2
0	0	1	0	1	1
1	0	1	0	1	3

The *P*-lower and *P*-upper approximations of *X* in $\widetilde{P}$ are shown in Table 5 and 6.

The classification of $\widetilde{P}$ generated by attribute *D* in *S* consists of three sub-multisets which are given in the following Tables 7, 8, and 9 which correspond to the cases where $D = 1$, $D = 2$, and $D = 3$, respectively.

Table 7. Sub-multiset of the multipartition D_p with $D = 1$.

A	B	C	E	F	W
0	0	1	0	0	2
0	0	1	0	1	1
0	0	1	1	1	1
1	0	0	0	1	1
1	0	1	0	1	1
1	1	1	0	0	1
1	1	1	1	1	1

Table 8. Sub-multiset of the m ultipartition D_p with $D = 2$.

A	B	C	E	F	W
0	0	0	0	0	2
0	0	1	1	1	1
0	1	0	0	0	2
1	0	1	0	1	1
1	1	0	0	1	1
1	1	0	1	1	1
1	1	1	0	0	2
1	1	1	1	1	1

Table 9. Sub-multiset of the multipartition D_p with $D = 3$.

A	B	C	E	F	W
0	0	0	1	1	3
1	0	0	0	1	1
1	0	1	0	1	1
1	1	0	0	1	1
1	1	0	1	1	1
1	1	1	0	0	1
1	1	1	1	0	1

3.3 Partition of Boundary Sets

The relationship between rough set theory and Dempster-Shafer's theory of evidence was first shown in [14] and further developed in [13]. The concept of partition of boundary sets was introduced in [13]. The basic idea is to represent an expert's classification on a set of objects in terms of lower approximations and a partition on the boundary set. In information multisystems, the concept of boundary sets is represented by boundary multisets, which is defined as the difference of upper and lower approximations of a multiset. Thus, the partition of a boundary set can be extended as a multipartition on a boundary multiset. The computation of this multipartition will be discussed in next section.

4 Multiset Decision Tables

4.1 Basic Concepts

The idea of multiset decision tables (*MDT*) was first informally introduced in [11]. We will formalize the concept in the following. Let $S = (Q = C \cup D, V, \tilde{Q})$ be an information multisystem, where C are condition attributes and D is a decision attribute. A multiset decision table is an ordered pair $A = (\tilde{C}, C_D)$, where $\tilde{C}$ is a projection of $\tilde{Q}$ onto C and C_D is a multipartition on $\tilde{D}$ generated by C in A. We will call $\tilde{C}$ the *LHS* (Left Hand Side) and C_D the *RHS* (Right Hand Side). Each sub-multiset in C_D is represented by two vectors: a Boolean bit-vector and an integer vector. Similar representational scheme has been used in [19], [20], [21]. The size of each vector is the number of values in the domain V_D of decision attribute D. The Boolean bit-vector labeled by D_i's denotes that a decision value D_i is in a sub-multiset of C_D iff $D_i = 1$ and its number of occurrences is denoted in the integer vector entry labeled by w_i.

The information multisystem of Table 2 is represented as a multiset decision table in Table 10 with $C = \{A, B, C, E, F\}$ and decision attribute D. The Boolean vector is denoted by $[D_1, D_2, D_3]$, and the integer vector is denoted by $[w_1, w_2, w_3]$. Note that $W = w_1 + w_2 + w_3$ on each row.

Table 10. Example of *MDT*.

A	B	C	E	F	W	D₁	D₂	D₃	w₁	w₂	w₃
0	0	0	0	0	2	0	1	0	0	2	0
0	0	0	1	1	3	0	0	1	0	0	3
0	0	1	0	0	2	1	0	0	2	0	0
0	0	1	0	1	1	1	0	0	1	0	0
0	0	1	1	1	2	1	1	0	1	1	0
0	1	0	0	0	2	0	1	0	0	2	0
1	0	0	0	1	2	1	0	1	1	0	1
1	0	1	0	1	3	1	1	1	1	1	1
1	1	0	0	1	2	0	1	1	0	1	1
1	1	0	1	1	2	0	1	1	0	1	1
1	1	1	0	0	4	1	1	1	1	2	1
1	1	1	1	0	1	0	0	1	0	0	1
1	1	1	1	1	2	1	1	0	1	1	0

4.2 Properties of Multiset Decision Tables

Based on multiset decision table representation, we can use relational operations on the table to compute the concepts of rough sets reviewed in Section 3. Let A be a multiset decision table. We will show how to determine the lower and upper approximations of decision classes and partitions of boundary multisets from A. The lower approximation of D_i in terms of the *LHS* columns is defined as the multiset where $D_i = 1$ and $W = w_i$, and the upper approximation of D_i is defined as the multiset where $D_i = 1$ and $W >= w_i$, or simply $D_i = 1$. The boundary multiset of D_i is defined as the multiset where $D_i = 1$ and $W > w_i$. The multipartition of boundary multisets can be identified by an equivalence multirelation defined over the Boolean vector denoted by the decision-value columns D_1, D_2, and D_3. It is clear that one row of a multiset decision table is in some boundary multiset if and only if the sum over D_1, D_2, and D_3 of the row is greater than 1. Therefore, to compute the multipartition of boundary multisets, we will first identify those rows with $D_1 + D_2 + D_3 > 1$, then the rows in the multirelation over D_1, D_2, and D_3 define blocks of the multipartition of the boundary multisets. The above computations are shown in the following example.

Example: Consider the decision class D_1 in Table 10. The C-lower approximation of D_1 is the multiset that satisfies $D_1 = 1$ and $W = w_1$, in table form we have:

Table 11. C-lower approximation of D_1.

A	B	C	E	F	W
0	0	1	0	0	2
0	0	1	0	1	1

The C-upper approximation of D_1 is the multiset that satisfies $D_1 = 1$, in table form we have:

Table 12. C-upper approximation of D_1.

A	B	C	E	F	W
0	0	1	0	0	2
0	0	1	0	1	1
0	0	1	1	1	2
1	0	0	0	1	2
1	0	1	0	1	3
1	1	1	0	0	4
1	1	1	1	1	2

To determine the partition of boundary multisets, we use the following two steps.
Step 1. Identify rows with $D_1 + D_2 + D_3 > 1$, we have the following multiset in table form:

Table 13. Elements in the boundary sets.

A	B	C	E	F	W	D_1	D_2	D_3
0	0	1	1	1	2	1	1	0
1	0	0	0	1	2	1	0	1
1	0	1	0	1	3	1	1	1
1	1	0	0	1	2	0	1	1
1	1	0	1	1	2	0	1	1
1	1	1	0	0	4	1	1	1
1	1	1	1	1	2	1	1	0

Step 2. Grouping the above table in terms of D_1, D_2, and D_3, we have the following blocks in the partition.

Table 14 shows the block where $D_1 = 1$ and $D_2 = 1$ and $D_3 = 0$, i.e., ($1\ 1\ 0$):

Table 14. The block denotes $D = \{1, 2\}$.

A	B	C	E	F	W	D_1	D_2	D_3
0	0	1	1	1	2	1	1	0
1	1	1	1	1	2	1	1	0

Table 15 shows the block where $D_1 = 1$ and $D_2 = 0$ and $D_3 = 1$, i.e., ($1\ 0\ 1$):

Table 15. The block denotes $D = \{1, 3\}$.

A	B	C	E	F	W	D_1	D_2	D_3
1	0	0	0	1	2	1	0	1

Table 16 shows the block where $D_1 = 0$ and $D_2 = 1$ and $D_3 = 1$, i.e., ($0\ 1\ 1$):

Table 16. The block denotes $D = \{2, 3\}$.

A	B	C	E	F	W	D_1	D_2	D_3
1	1	0	0	1	2	0	1	1
1	1	0	1	1	2	0	1	1

Table 17 shows the block where $D_1 = 1$ and $D_2 = 1$ and $D_3 = 1$, i.e., ($1\ 1\ 1$):

Table 17. The block denotes $D = \{1, 2, 3\}$.

A	B	C	E	F	W	D_1	D_2	D_3
1	0	1	0	1	3	1	1	1
1	1	1	0	0	4	1	1	1

From the above example, it is clear that an expert's classification on the decision attribute D can be obtained by grouping similar values over columns D_1, D_2, and D_3 and by taking the sum over the W column in a multiset decision table. Based on this grouping and summing operation, we can derive a basic probability assignment (bpa) function as required in Dempster-Shafer theory for computing belief functions. This is shown in Table 18.

Table 18. Grouping over D_1, D_2, D_3 and sum over W.

D_1	D_2	D_3	W
1	0	0	3
0	1	0	4
0	0	1	4
0	1	1	4
1	0	1	2
1	1	0	4
1	1	1	7

Let $\Theta = \{1, 2, 3\}$. Table 19 shows the basic probability assignment function derived from the information multisystem shown in Table 2. The computation is based on the partition of boundary multisets shown in Table 18.

Table 19. The bpa derived from Table 2.

X	{1}	{2}	{3}	{1, 2}	{1, 3}	{2, 3}	{1, 2, 3}
m(X)	3/28	4/28	4/28	4/28	2/28	4/28	7/28

5 Learning Rules From MDT

5.1 LERS-M (Learning Rules from Examples Using Rough MultiSets)

In this section, we will present an algorithm *LERS-M* for learning production rules from a database table based on multiset decision table. A multiset decision table can be computed directly using typical *SQL* commands from a database table once the condition and decision attributes are specified. For efficiency reason, we will associate entries in an *MDT* with a sequence of integer numbers. This can be accomplished by using extensions to relational database management system such as the *UDF* (User Defined Functions) and *UDT* (User defined Data Type) available on IBM's *DB2* [22]. The emphasis of this paper is more on algorithms, implementation details will be covered somewhere else.

The basic idea of *LERS-M* is to generate a multiset decision table with a sequence of integer numbers. Then, for each value d_i of the decision attribute D, the upper approximation of d_i, *UPPER(d_i)*, is computed, and a set of rules is generated for each *UPPER(d_i)*. The algorithm *LERS-M* is given in the following. The detail for generation of rules is presented in Section 5.2.

> **procedure** *LERS-M*
> **Inputs**: a table S with condition attributes $C_1, C_2, ..., C_n$ and decision attribute D.
> **Outputs**: a set of production rules represented as a multiset data table.
> **begin**
>   Create a Multiset Decision Table (*MDT*) from S with sequence numbers;
>   **for** each decision value d_i of D **do**
>   **begin**
>   find the upper approximation *UPPER(d_i)* of d_i;
>   Generate rules for *UPPER(d_i)*;
>   **end**;
> **end**;

5.2 Rule Generation Strategy

The basic idea of rule generation is to create an *AVT* (Attribute-Value pairs Table) table containing all *a-v* pairs appeared in the set UPPER(d_i). Then, we will partition the a-v pairs into different groups based on a grouping criterion such as degree of relevancy, which is also used to rank the groups. The left hand sides of rules are identified by taking conjunctions of *a-v* pairs within the same group (*intra-group conjuncts*) and by taking natural join over different groups (*inter-group conjuncts*). Strategies for generating and validating candidate conjuncts are encapsulated in a module called *GenerateAndTestConjuncts*. Once a set of valid conjuncts is identified, minimal conjuncts can be generated using the method of dropping conditions. The process of rule generation is an iterative one. It starts with the set *UPPER(d_i)* as an initial *TargetSet*. In each iteration, a set of rules is generated, and the instances covered by the rule-set are removed from the *TargetSet*. It stops when all instances in *UPPER(d_i)* are covered by the generated rules. In *LERS-M*, the stopping condition is guaranteed by the fact that upper approximations are always definable based on the theory of rough sets.

The above strategy is presented in the following procedures *RULE_GEN*, *GroupAVT*, and *GenerateAndTestConjuncts*. A working example will be given in next section. Specifically, we have adopted the following notions. The extension of an *a-v* pair (a, v) denoted by $[(a, v)]$, i.e., the set of instances covered by the *a-v* pair, is a subset of the sequence numbers in the original *MDT*. The extension of an *a-v* pair is encoded by a Boolean bit-vector. A conjunct is a nonempty finite set of *a-v* pairs. The extension of a conjunct is the intersection of extensions of all the *a-v* pairs in the conjunct. Note that the extension of a group of conjunct is the union of extensions of all the conjuncts in the group, and the extension of an empty group of conjuncts is an empty set.

procedure *RULE_GEN*
Inputs: an upper approximation of a decision value d_i, *UPPER(d_i)* and an *MDT*.
Outputs: a set of rules for *UPPER(d_i)* represented as a multiset decision table.
 begin
 TargetSet := UPPER(d_i);
 Ruleset := empty set;
  Select a grouping criteria *G :=* degree of relevance;
  Create an *a-v* pair table *AVT* contains all *a-v* pairs appeared in *UPPER(d_i)*;
 while *TargetSet* is not empty **do**
 begin
   AVT := GroupAVT(G, TargetSet);
   NewRules := GenerateAndTestConjuncts(AVT, UPPER(d_i));
   RuleSet := RuleSet + NewRules;
   TargetSet := TargetSet – [NewRules];
 end;
 minimalCover(*RuleSet*); /* applying dropping condition technique to remove redundant rules from *RuleSet* linearly starting from the first rule to the last rule in the set
                */

 end; // *RULE_GEN*

procedure *GroupAVT*
 Inputs: a grouping criterion such as degree of relevance and
   a subset of the upper approximation of a decision value d_i.
 Outputs: a list of groups of equivalent a-v pairs relevant to the target set.
 begin
   Initialize the *AVT* table to be empty;

Select a subtable *T* from the target set where decision value = d_i;
Create a query to get a vector of condition attributes from the subtable *T*;
for each condition attribute **do** /* Generate distinct values for each condition
attribute */
begin
    Create query string to select distinct values;
    for each distinct value **do**
  begin
        Create a query string to select count of occurrences;
        relevance := count of occurrences;
        if (*relevance* > 0)
            Add the condition-value pair to *AVT* table;
   end;// for each distinct value
  end; // end of for each condition
  Select the list of distinct values of the relevance column;
  Sort the list of distinct values in descending order;
  Use the list of distinct values to generate a list of groups of a-v pairs;
end; // *GroupAVT*

procedure GenerateAndTestConjuncts
  Inputs: a list *AVT* of groups of equivalent a-v pairs and
        the upper approximation of decision value d_i.
  Outputs: a set of rules.
  begin
    RuleList := $\varnothing$;
    CarryOverList := $\varnothing$; // a list of groups of a-v pairs
    CandidateList := $\varnothing$; // a list of
    TargetSet := *UPPER*(d_i);
    // Generate Candidate List
    repeat
     L := getNext(*AVT*); // L is a list of equivalent a-v pairs
     if (*L* is empty)
      then break;
     if ([conjunct(*L*)] $\subseteq$ *TargetSet*)
      then Add conjunct(*L*) to *CandidateList*;
      /* conjunct(*L*) returns a conjunction of all a-v pairs in *L* */
     if (*CarryOverList* is empty)
      then Add all a-v pairs in *L* to *CarryOverList*
      else   begin
            FilterList := $\varnothing$;
            Add join(*CarryOverList*, *L*) to *FilterList*;
            /*join is a function that creates new lists of a-v pairs by taking
            and joining one element each from the *CarryOverList* and *L* */
            CarryOverList := $\varnothing$;
            for each *list* in *FilterList* **do**
            if ([*list*] $\subseteq$ *TargetSet*)
              then Add *list* to *CandidateList*

                                      else Add *list* to *CarryOverList*;
                        end;
              until (*CandidateList* is not empty);
              // Test CandidateList
              for each *list* in *CandidateList* **do**
              begin
                      list := minimalConjunct(*list*); /* applying dropping condition to get minimal
                                            list of a-v pairs */
                      Add *list* to *RuleList*;
              end;
              return *RuleList*;
              end; // *GenerateAndTestConjuncts*

Example

Consider the information multisystem in Table 2 as input to *LERS-M*. The result of
generating an *MDT* with sequence numbers is shown in Table 20.

Table 20. *MDT* with sequence numbers.

Seq	A	B	C	E	F	W	D_1	D_2	D_3	w_1	w_2	w_3
1	0	0	0	0	0	2	0	1	0	0	2	0
2	0	0	0	1	1	3	0	0	1	0	0	3
3	0	0	1	0	0	2	1	0	0	2	0	0
4	0	0	1	0	1	1	1	0	0	1	0	0
5	0	0	1	1	1	2	1	1	0	1	1	0
6	0	1	0	0	0	2	0	1	0	0	2	0
7	1	0	0	0	1	2	1	0	1	1	0	1
8	1	0	1	0	1	3	1	1	1	1	1	1
9	1	1	0	0	1	2	0	1	1	0	1	1
10	1	1	0	1	1	2	0	1	1	0	1	1
11	1	1	1	0	0	4	1	1	1	1	2	1
12	1	1	1	1	0	1	0	0	1	0	0	1
13	1	1	1	1	1	2	1	1	0	1	1	0

The *C*-upper approximation of the class *D = 1* is the sub-*MDT* shown in Table 21.

Table 21. Table of *UPPER(D₁)*.

Seq	A	B	C	E	F	W	D_1	D_2	D_3	w_1	w_2	w_3
3	0	0	1	0	0	2	1	0	0	2	0	0
4	0	0	1	0	1	1	1	0	0	1	0	0
5	0	0	1	1	1	2	1	1	0	1	1	0
7	1	0	0	0	1	2	1	0	1	1	0	1
8	1	0	1	0	1	3	1	1	1	1	1	1
11	1	1	1	0	0	4	1	1	1	1	2	1
13	1	1	1	1	1	2	1	1	0	1	1	0

The following is how *RULE_GEN* will generate rules for *UPPER(D₁)*. Table 22
shows the *AVT* table created by procedure *GroupAVT* before sorting is applied to the

table to generate the final list of groups of equivalent a-v pairs. The grouping criterion used is based on the size of intersection between the extension of an a-v pair and the set $UPPER(D_i)$. Each entry in the Relevance column denotes the number of rows in the $UPPER(D_i)$ table matched with the a-v pair. For example, the relevance of $(A, 0)$ is 3 means that there are three rows in $UPPER(D_i)$ that satisfy $A = 0$. The ranking of a-v pairs is based on maximum degree of relevance, i.e., larger relevance number has higher priority. The ranks are ordered in ascending order, i.e., smaller rank number has higher priority.

The encoding for extensions of a-v pairs in the AVT is shown in Table 23, and the Target set $UPPER(D_i) = \{3, 4, 5, 7, 8, 11, 13\}$ is considered with the encoding $(0, 0, 1, 1, 1, 0, 1, 1, 0, 0, 1, 0, 1)$.

Table 22. AVT table created from $UPPER(D_i)$.

Name	Value	Relevance	Rank
A	0	3	4
A	1	4	3
B	0	5	2
B	1	2	5
C	0	1	6
C	1	6	1
E	0	5	2
E	1	2	5
F	0	2	5
F	1	5	2

Table 23. Extensions of a-v pairs encoded as Boolean bit-vector.

N	V	1	2	3	4	5	6	7	8	9	10	11	12	13
A	0	1	1	1	1	1	1	0	0	0	0	0	0	0
A	1	0	0	0	0	0	0	1	1	1	1	1	1	1
B	0	1	1	1	1	1	0	1	1	0	0	0	0	0
B	1	0	0	0	0	0	1	0	0	1	1	1	1	1
C	0	1	1	0	0	0	1	1	0	1	0	0	0	0
C	1	0	0	1	1	1	0	0	1	0	0	1	1	1
E	0	1	0	1	1	0	1	1	1	1	1	0	0	0
E	1	0	1	0	0	1	0	0	0	0	1	0	1	1
F	0	1	0	1	0	0	1	0	0	0	0	1	1	0
F	1	0	1	0	1	1	0	1	1	1	1	0	0	1

Based on the Rank of the AVT table shown in Table 22, the a-v pairs are grouped into the following six groups listed from higher to lower rank:

$$\{(C, 1)\}$$
$$\{(B, 0), (E, 0), (F, 1)\}$$
$$\{(A, 1)\}$$
$$\{(A, 0)\}$$
$$\{(B, 1), (E, 1), (F, 0)\}$$
$$\{(C, 0)\}$$

Candidate conjuncts are generated and tested by the *GenerateAndTestConjuncts* procedure based on the above list. The basic strategy used here is to generate the intra-group conjuncts first, then followed by generating inter-group conjuncts. The procedure proceeds sequentially starting from the highest ranked group downward. It stops when at least one rule is found. The heuristics employed here is trying to find rules with maximum coverage of instances in $UPPER(d_i)$.

In our example, the first group contains only one *a-v* pair $(C, 1)$; therefore, no need to generate intra-group conjuncts. From Table 21, we can see that $[\{(C, 1)\}]$ is not a subset of the $UPPER(D_j)$. Thus, inter-group join is needed. In addition, the second group $\{(B, 0), (E, 0), (F, 1)\}$ is also included in the candidate list. This results in the following list of candidate conjuncts, which are listed with their corresponding externsions.

$$[\{(C, 1), (B, 0)\}] = \{3, 4, 5, 8\}$$
$$[\{(C, 1), (E, 0)\}] = \{3, 4, 8, 11\}$$
$$[\{(C, 1), (F, 1)\}] = \{4, 5, 8, 13\}$$
$$[\{(B, 0), (E, 0), (F, 1)\}] = \{4, 7, 8\}$$

Following the generating stage, a testing stage is performed to identify valid conjuncts. Because all the conjuncts are valid, i.e., their extensions are subset of $UPPER(d_j)$. Four new rules are found in this iteration. The next step is to find minimal conjuncts by using dropping condition method.

Consider the conjunction of $\{(B, 0), (E, 0), (F, 1)\}$. Dropping the *a-v* pair $(B, 0)$ from the group, we have $[\{(E, 0), (F, 1)\}] = \{1, 4, 7, 8, 9\}$, which is not a subset of TargetSet, $\{3, 4, 5, 7, 8, 11, 13\}$.

Next, try to drop the *a-v* pair $(E, 0)$ from the group, we have
$$[\{(B, 0), (F, 1)\}] = \{1, 4, 5, 7, 8\},$$
which is not a subset of TargetSet, $\{3, 4, 5, 7, 8, 11, 13\}$.

Finally, try to drop the *a-v* pair $(F, 1)$ from the group, we have
$$[\{(B, 0), (E, 0)\}] = \{1, 3, 4, 7, 8\},$$
which is not a subset of TargetSet, $\{3, 4, 5, 7, 8, 11, 13\}$.

We can conclude that the conjunction of $\{(B, 0), (E, 0), (F, 1)\}$ contains no redundant *a-v* pairs, and it is a minimal conjunct. Similarly, it can be verified that the conjuncts $\{(C, 1), (B, 0)\}$, $\{(C, 1), (E, 0)\}$, and $\{(C, 1), (F, 1)\}$ are minimal.

All minimal conjuncts found are added to the new rule set R. Thus, we have the extension $[R]$ of the new rules as

$$[R] = [\{(C, 1), (B, 0)\}] + [\{(C, 1), (E, 0)\}]$$
$$+ [\{(C, 1), (F, 1)\}] + [\{(B, 0), (E, 0), (F, 1)\}]$$
$$= \{3, 4, 5, 7, 8, 11, 13\}.$$

The target set is updated by the following

$$TargetSet = \{3, 4, 5, 7, 8, 11, 13\} - [R] = \text{empty set}.$$

Therefore, we have found the rule set. The last step in procedure *RULE_GEN* is to remove redundant rules from the rule set. The basic idea is similar to finding minimal conjuncts. Here, we try to remove one rule at a time and to test if the remaining rules cover all examples of the target set. More specifically, we try to remove the conjunct $\{(C, 1), (B, 0)\}$ from the collection. Then, we have

$$[R] = [\{(C, 1), (E, 0)\}] + [\{(C, 1), (F, 1)\}] + [\{(B, 0), (E, 0), (F, 1)\}]$$
$$= \{3, 4, 5, 7, 8, 11, 13\}$$
$$= TargetSet = \{3, 4, 5, 7, 8, 11, 13\}.$$

Therefore, the conjunct $\{(C, 1), (B, 0)\}$ is redundant and is removed from the rule set. Next, we try to remove the conjunct $\{(C, 1), (E, 0)\}$ from the rule set, we have

$[R] = [\{(C, 1), (F, 1)\}] + [\{(B, 0), (E, 0), (F, 1)\}]$
$\quad = \{4, 5, 7, 8, 13\}$
$\quad \neq TargetSet.$

Therefore, the conjunct $\{(C, 1), (B, 0)\}$ is not redundant, and it is kept in the rule set. Similarly, it can be verified that both conjuncts $\{(C, 1), (F, 1)\}$ and $\{(B, 0), (E, 0), (F, 1)\}$ are not redundant. The resulting rule set generated is shown in Table 22, where the w_1, w_2, and w_3 are column sums extracted from the table $UPPER(D_1)$ of Table 21.

Table 24. Rules generated for $UPPER(D_1)$.

A	B	C	E	F	D	w_1	w_2	w_3
null	null	1	0	null	1	5	3	2
null	null	1	null	1	1	4	3	1
null	0	null	0	1	1	3	1	2

The *LERS-M* algorithm tries to find only one minimal set of rules, it does not try to find all minimal sets of rules.

5.3 Discussion

There are several advantages of developing *LERS-M* using relational database technology. Relational database systems have been highly optimized and scalable in dealing with large amount of data. They are very portable. They provide smooth integration with *OLAP* or data warehousing systems. However, one typical disadvantage of SQL implementation is extra computational overhead. Experiments are needed to identify impacts of computational overhead to the performance of *LERS-M*.

When a database is very large, we can divide the database into smaller *n* databases and run *LERS-M* on each small database. Similarly, this scheme can be applied to homogeneous distributed databases. To integrate the distributed answers provided by multiple *LERS-M* programs, we can take the sum over the number of occurrences (i.e., w_1, w_2, and w_3 in previous example) provided by local *LERS-M* programs. When single answer is desirable, then the decision value D_i with maximum sum of w_i can be returned, or the entire vector of number of occurrences can be returned as an answer. It is possible to develop other inference mechanisms that will make use of the number of occurrences when performing the task of classification.

Based on our discussion, there are two major parameters of *LERS-M*, namely, grouping criteria and generation of conjuncts. New criteria and heuristics based on numerical measures such as *gini index* and entropy function may be used. In the paper, we have used the minimal length criterion for the generation of candidate conjuncts. The search strategy is not exhaustive, and it stops when at least one candidate conjunct is identified. There are rooms for developing more extensive and efficient strategies for generating candidate conjuncts.

The proposed algorithm is under implementation on IBM's DB2 database system running on Redhat Linux with web-based interface implemented using Java servlets and JSP. Performance evaluation and comparison to systems based on classical rough set methods will need further work.

6 Conclusions

In this paper we have formulated the concept of multiset decision tables based on the concept of information multisystems. The concept is then used to develop an algorithm *LERS-M* for learning rules from databases. Based on the concept of partition of boundary sets, we have shown that it is straightforward to compute basic probability assignment functions of the Dempster-Shafer theory from multiset decision tables. A nice feature of multiset decision tables is that we can use the sum over number of occurrences of decision values as a simple mechanism to integrate distributed answers. Developing *LERS-M* on top of relational database technology will make the system scalable and portable. Our next step is to evaluate the time and space complexities of *LERS-M* over very large data sets. It would be interesting to compare the SQL-based implementation to classical rough set methods for learning rules from very large data sets. In addition, we have considered only homogenous data tables, which may be very large or distributed. Generalization to multiple heterogeneous tables needs further work.

References

1. Sarawagi, S., S. Thomas, and R. Agrawal, "Integrating association rule mining with relational database systems: alternatives and implications," *Data Mining and Knowledge Discovery*, 4, 89–125, (2000).
2. Agrawal, R. and K. Shim, "Developing tightly-coupled data mining applications on a relational database systems," *Proc. of the 2nd Int. Conference on Knowledge Discovery in Databases and Data Mining*, Portland, Oregon, (1996).
3. Wang, M., B. Iyer, and J.S. Vitter, "Scalable mining for classification rules in relational databases," *IDEAS*, 58-67, (1998).
4. Fernández-Baizán, M.C., Menasalvas Ruiz E., Peña Sánchez J.M., Pardo Pastrana B., "Integrating KDD Algorithms and RDBMS Code," *Rough Sets and Current Trends in Computing* (1998), 210-213.
5. Stolfo, S., A. Prodromidis, S. Tselepis, W. Lee, W. Fan, and P. Chan, "JAM: Java agents for meta-learning over distributed databases," *Proc. Third Intl. Conf. Knowledge Discovery and Data Mining*, 74-81, (1997).
6. Grzymala-Busse, J.W., "The LERS family of learning systems based on rough sets," *Proc. of the 3rd Midwest Artificial Intelligence and Cognitive Science Society Conference*, Carbondale, IL, April 12-14, 103-107, (1991).
7. Pawlak, Z., "Rough sets: basic notion," *Int. J. of Computer and Information Science* 11, 344-56, (1982).
8. Pawlak, Z., "Rough sets and decision tables," *Lecture Notes in Computer Science* 208, 186-196, Berlin, Heidelberg, Springer-Verlag, (1985).
9. Pawlak, Z., J. Grzymala-Busse, R. Slowinski, and W. Ziarko, "Rough sets," *Communication of ACM*, Vol. 38, No. 11, November, (1995), 89-95.
10. Grzymala-Busse, J.W., "Learning from examples based on rough multisets," *Proc. of the 2nd Int. Symposium on Methodologies for Intelligent Systems*, Charlotte, North Carolina, October 14-17, 325-332, (1987).
11. Chan, C.-C., "Distributed incremental data mining from very large databases: a rough multiset approach," *Proc. the 5th World Multi-Conference on Systemics, Cybernetics and Informatics, SCI 2001*, Orlando, Florida, July 22-25, (2001), 517-522.

12. Shafer, G., *A Mathematical Theory of Evidence*. Princeton, NJ, Princeton University Press, (1976).
13. Skowron, A. and J. Grzymala-Busse, "From rough set theory to evidence theory." in *Advances in the Dempster-Shafer Theory of Evidence*, edited by R. R. Yager, J. Kacprzyk, and M. Fedrizzi, 193-236, John Wiley & Sons, Inc, New York, (1994).
14. Grzymala-Busse, J.W., "Rough set and Dempster-Shafer approaches to knowledge acquisition under uncertainty - a comparison," *manuscript*, (1987).
15. Knuth, D.E., *The Art of Computer Programming. Vol. III, Sorting and Searching*. Addison-Wesley, (1973).
16. Grzymala-Busse, J.W., "Knowledge acquisition under uncertainty: a rough set approach," *J. of Intelligent and Robotic Systems*, Vol. 1, 3-16, (1988).
17. Chan, C.-C., "Incremental learning of production rules from examples under uncertainty: a rough set approach," *Int. J. of Software Engineering and Knowledge Engineering*, Vol. 1, No. 4, 439 - 461, (1991).
18. Grzymala-Busse, J.W., *Managing Uncertainty in Expert Systems*. Morgan Kaufmann Pub., San Mateo, CA, (1991).
19. Hu, X., T.Y. Lin, E. Louie, "Bitmap techniques for optimizing decision support queries and association rule algorithms," IDEAS, (2003), pp. 34-43.
20. Kryszkiewicz, M., "Rough Set Approach to Rules Generation from Incomplete Information Systems," In *The Encyclopedia of Computer Science and Technology*, Marcel Dekker, Inc., New York, Vol. 44, 319–346, (2001).
21. Ślęzak, D., "Various approaches to reasoning with frequency based decision reducts: a survey," in *Rough Set Methods and Applications*, L. Polkowski, S. Tsumoto, T.Y. Lin (eds.), Physica-Verlag, Heidelberg, New York, (2000).
22. Chamberlin, D. *A Complete Guide to DB2 Universal Database*. Morgan Kaufmann Publishers. (1998).

Data with Missing Attribute Values: Generalization of Indiscernibility Relation and Rule Induction

Jerzy W. Grzymala-Busse[1,2]

[1] Department of Electrical Engineering and Computer Science, University of Kansas
Lawrence, KS 66045, USA
[2] Institute of Computer Science, Polish Academy of Sciences, 01-237 Warsaw, Poland
Jerzy@ku.edu
http://lightning.eecs.ku.edu/index.html

Abstract. Data sets, described by decision tables, are incomplete when for some cases (examples, objects) the corresponding attribute values are missing, e.g., are lost or represent "do not care" conditions. This paper shows an extremely useful technique to work with incomplete decision tables using a block of an attribute-value pair. Incomplete decision tables are described by characteristic relations in the same way complete decision tables are described by indiscernibility relations. These characteristic relations are conveniently determined by blocks of attribute-value pairs. Three different kinds of lower and upper approximations for incomplete decision tables may be easily computed from characteristic relations. All three definitions are reduced to the same definition of the indiscernibility relation when the decision table is complete. This paper shows how to induce certain and possible rules for incomplete decision tables using MLEM2, an outgrow of the rule induction algorithm LEM2, again, using blocks of attribute-value pairs. Additionally, the MLEM2 may induce rules from incomplete decision tables with numerical attributes as well.

1   Introduction

We will assume that data sets are presented as decision tables. In such a table columns are labeled by variables and rows by case names. In the simplest case such case names, also called cases, are numbers. Variables are categorized as either independent, also called attributes, or dependent, called decisions. Usually only one decision is given in a decision table. The set of all cases that correspond to the same decision value is called a concept (or a class).

In most articles on rough set theory it is assumed that for all variables and all cases the corresponding values are specified. For such tables the indiscernibility relation, one of the most fundamental ideas of rough set theory, describes cases that can be distinguished from other cases.

However, in many real-life applications, data sets have missing attribute values, or, in other words, the corresponding decision tables are incompletely spec-

J.F. Peters et al. (Eds.): Transactions on Rough Sets I, LNCS 3100, pp. 78–95, 2004.

ified. For simplicity, incompletely specified decision tables will be called incomplete decision tables.

In this paper we will assume that there are two reasons for decision tables to be incomplete. The first reason is that an attribute value, for a specific case, is lost. For example, originally the attribute value was known, however, due to a variety of reasons, currently the value is not recorded. Maybe it was recorded but is erased. The second possibility is that an attribute value was not relevant – the case was decided to be a member of some concept, i.e., was classified, or diagnosed, in spite of the fact that some attribute values were not known. For example, it was feasible to diagnose a patient regardless of the fact that some test results were not taken (here attributes correspond to tests, so attribute values are test results). Since such missing attribute values do not matter for the final outcome, we will call them "do not care" conditions. The main objective of this paper is to study incomplete decision tables, i.e., incomplete data sets, or, yet in different words, data sets with missing attribute values. We will assume that in the same decision table some attribute values may be lost and some may be "do not care" conditions. The first paper dealing with such decision tables was [6].

For such incomplete decision tables there are two special cases: in the first case, all missing attribute values are lost, in the second case, all missing attribute values are "do not care" conditions. Incomplete decision tables in which all attribute values are lost, from the viewpoint of rough set theory, were studied for the first time in [8], where two algorithms for rule induction, modified to handle lost attribute values, were presented. This approach was studied later in [13–15], where the indiscernibility relation was generalized to describe such incomplete decision tables.

On the other hand, incomplete decision tables in which all missing attribute values are "do not care" conditions, from the view point of rough set theory, were studied for the first time in [3], where a method for rule induction was introduced in which each missing attribute value was replaced by all values from the domain of the attribute. Originally such values were replaced by all values from the entire domain of the attribute, later, by attribute values restricted to the same concept to which a case with a missing attribute value belongs. Such incomplete decision tables, with all missing attribute values being "do not care conditions", were extensively studied in [9], [10], including extending the idea of the indiscernibility relation to describe such incomplete decision tables.

In general, incomplete decision tables are described by characteristic relations, in a similar way as complete decision tables are described by indiscernibility relations [6].

In rough set theory, one of the basic notions is the idea of lower and upper approximations. For complete decision tables, once the indiscernibility relation is fixed and the concept (a set of cases) is given, the lower and upper approximations are unique.

For incomplete decision tables, for a given characteristic relation and concept, there are three different possibilities to define lower and upper approximations, called singleton, subset, and concept approximations [6]. Singleton lower and upper approximations were studied in [9], [10], [13–15]. Note that similar three definitions of lower and upper approximations, though not for incomplete decision tables, were studied in [16–18]. In this paper we further discuss applications to data mining of all three kinds of approximations: singleton, subset and concept. As it was observed in [6], singleton lower and upper approximations are not applicable in data mining.

The next topic of this paper is demonstrating how certain and possible rules may be computed from incomplete decision tables. An extension of the well-known LEM2 algorithm [1], [4], MLEM2, was introduced in [5]. Originally, MLEM2 induced certain rules from incomplete decision tables with missing attribute values interpreted as lost and with numerical attributes. Using the idea of lower and upper approximations for incomplete decision tables, MLEM2 was further extended to induce both certain and possible rules from a decision table with some missing attribute values being lost and some missing attribute values being "do not care" conditions, while some attributes may be numerical.

2 Blocks of Attribute-Value Pairs and Characteristic Relations

Let us reiterate that our basic assumption is that the input data sets are presented in the form of a *decision table*. An example of a decision table is shown in Table 1.

Table 1. A complete decision table

	Attributes			Decision
Case	Temperature	Headache	Nausea	Flu
1	high	yes	no	yes
2	very_high	yes	yes	yes
3	high	no	no	no
4	high	yes	yes	yes
5	high	yes	yes	no
6	normal	yes	no	no
7	normal	no	yes	no
8	normal	yes	no	yes

Rows of the decision table represent *cases*, while columns are labeled by *variables*. The set of all cases will be denoted by U. In Table 1, $U = \{1, 2, ..., 8\}$. Independent variables are called *attributes* and a dependent variable is called a *decision* and is denoted by d. The set of all attributes will be denoted by A. In Table 1, $A = \{Temperature, Headache, Nausea\}$. Any decision table defines a function ρ that maps the direct product of U and A into the set of all values. For example, in Table 1, $\rho(1, Temperature) = high$. Function ρ describing Table 1 is completely specified (total). A decision table with completely specified function ρ will be called *completely specified*, or, for the sake of simplicity, *complete*.

Rough set theory [11], [12] is based on the idea of an indiscernibility relation, defined for complete decision tables. Let B be a nonempty subset of the set A of all attributes. The indiscernibility relation $IND(B)$ is a relation on U defined for $x, y \in U$ as follows

$$(x, y) \in IND(B) \quad if \ and \ only \ if \ \rho(x, a) = \rho(y, a) \ for \ all \ a \in B.$$

The indiscernibility relation $IND(B)$ is an equivalence relation. Equivalence classes of $IND(B)$ are called elementary sets of B and are denoted by $[x]_B$. For example, for Table 1, elementary sets of $IND(A)$ are {1}, {2}, {3}, {4, 5}, {6, 8}, {7}. The indiscernibility relation $IND(B)$ may be computed using the idea of blocks of attribute-value pairs. Let a be an attribute, i.e., $a \in A$ and let v be a value of a for some case. For complete decision tables if $t = (a, v)$ is an attribute-value pair then a block of t, denoted $[t]$, is a set of all cases from U that for attribute a have value v. For Table 1,

[(Temperature, high)] = {1, 3, 4, 5},
[(Temperature, very_high)] = {2},
[(Temperature, normal)] = {6, 7, 8},
[(Headache, yes)] = {1, 2, 4, 5, 6, 8},
[(Headache, no)] = {3, 7},
[(Nausea, no)] = {1, 3, 6},
[(Nausea, yes)] = {2, 4, 5, 7}.

The indiscernibility relation $IND(B)$ is known when all elementary blocks of IND(B) are known. Such elementary blocks of B are intersections of the corresponding attribute-value pairs, i.e., for any case $x \in U$,

$$[x]_B = \cap \{[(a, v)] | a \in B, \rho(x, a) = v\}.$$

We will illustrate the idea how to compute elementary sets of B for Table 1 and $B = A$.

$[1]_A = [(Temperature, high)] \cap [(Headache, yes)] \cap [(Nausea, no)] = \{1\}$,
$[2]_A = [(Temperature, very\_high)] \cap [(Headache, yes)] \cap [(Nausea, yes)] = \{2\}$,
$[3]_A = [(Temperature, high)] \cap [(Headache, no)] \cap [(Nausea, no)] = \{3\}$,
$[4]_A = [5]_A = [(Temperature, high)] \cap [(Headache, yes)] \cap [(Nausea, yes)] = \{4, 5\}$,
$[6]_A = [8]_A = [(Temperature, normal)] \cap [(Headache, yes)] \cap [(Nausea, no] = \{6, 8\}$,
$[7]_A = [(Temperature, normal)] \cap [(Headache, no] \cap [(Nausea, yes)] = \{7\}$.

In practice, input data for data mining are frequently affected by missing attribute values. In other words, the corresponding function ρ is incompletely specified (partial). A decision table with an incompletely specified function ρ will be called *incompletely specified*, or *incomplete*.

For the rest of the paper we will assume that all decision values are specified, i.e., they are not missing. Also, we will assume that all missing attribute values are denoted either by "?" or by "*", lost values will be denoted by "?", "do not

Table 2. An incomplete decision table

Case	Temperature	Headache	Nausea	Flu
1	high	?	no	yes
2	very_high	yes	yes	yes
3	?	no	no	no
4	high	yes	yes	yes
5	high	?	yes	no
6	normal	yes	no	no
7	normal	no	yes	no
8	*	yes	*	yes

care" conditions will be denoted by "*". Additionally, we will assume that for each case at least one attribute value is specified.

Incomplete decision tables are described by characteristic relations instead of indiscernibility relations. Also, elementary blocks are replaced by characteristic sets. An example of an incomplete table is presented in Table 2.

For incomplete decision tables the definition of a block of an attribute-value pair must be modified. If for an attribute a there exists a case x such that $\rho(x, a) = ?$, i.e., the corresponding value is lost, then the case x is not included in the block $[(a, v)]$ for any value v of attribute a. If for an attribute a there exists a case x such that the corresponding value is a "do not care" condition, i.e., $\rho(x, a) = *$, then the corresponding case x should be included in blocks $[(a, v)]$ for all values v of attribute a. This modification of the definition of the block of attribute-value pair is consistent with the interpretation of missing attribute values, lost and "do not care" condition. Thus, for Table 2

$[(\text{Temperature, high})] = \{1, 4, 5, 8\}$,
$[(\text{Temperature, very\_high})] = \{2, 8\}$,
$[(\text{Temperature, normal})] = \{6, 7, 8\}$,
$[(\text{Headache, yes})] = \{2, 4, 6, 8\}$,
$[(\text{Headache, no})] = \{3, 7\}$,
$[(\text{Nausea, no})] = \{1, 3, 6, 8\}$,
$[(\text{Nausea, yes})] = \{2, 4, 5, 7, 8\}$.

The *characteristic set* $K_B(x)$ is the intersection of blocks of attribute-value pairs (a, v) for all attributes a from B for which $\rho(x, a)$ is specified and $\rho(x, a) = v$. For Table 2 and $B = A$,

$K_A(1) = \{1, 4, 5, 8\} \cap \{1, 3, 6, 8\} = \{1, 8\}$,
$K_A(2) = \{2, 8\} \cap \{2, 4, 6, 8\} \cap \{2, 4, 5, 7, 8\} = \{2, 8\}$,
$K_A(3) = \{3, 7\} \cap \{1, 3, 6, 8\} = \{3\}$,
$K_A(4) = \{1, 4, 5, 8\} \cap \{2, 4, 6, 8\} \cap \{2, 4, 5, 7, 8\} = \{4, 8\}$,
$K_A(5) = \{1, 4, 5, 8\} \cap \{2, 4, 5, 7, 8\} = \{4, 5, 8\}$,
$K_A(6) = \{6, 7, 8\} \cap \{2, 4, 6, 8\} \cap \{1, 3, 6, 8\} = \{6, 8\}$,
$K_A(7) = \{6, 7, 8\} \cap \{3, 7\} \cap \{2, 4, 5, 7, 8\} = \{7\}$, and
$K_A(8) = \{2, 4, 6, 8\}$.

Characteristic set $K_B(x)$ may be interpreted as the smallest set of cases that are indistinguishable from x using all attributes from B, using a given interpretation of missing attribute values. Thus, $K_A(x)$ is the set of all cases that cannot be distinguished from x using all attributes.

The characteristic relation $R(B)$ is a relation on U defined for $x, y \in U$ as follows

$$(x, y) \in R(B) \; if \; and \; only \; if \; y \in K_B(x).$$

The characteristic relation $R(B)$ is reflexive but – in general – does not need to be symmetric or transitive. Also, the characteristic relation $R(B)$ is known if we know characteristic sets $K(x)$ for all $x \in U$. In our example, $R(A) = \{(1, 1),$ $(1, 8), (2, 2), (2, 8), (3, 3), (4, 4), (4, 8), (5, 4), (5, 5), (5, 8), (6, 6), (6, 8), (7, 7), (8,$ $2), (8, 4), (8, 6), (8, 8)\}$. The most convenient way is to define the characteristic relation through the characteristic sets. Nevertheless, the characteristic relation $R(B)$ may be defined independently of characteristic sets in the following way:

$$(x, y) \in R(B) \; if \; and \; only \; if \; \rho(x, a) = \rho(y, a) \; or \; \rho(x, a) = * \; or \rho(y, a) = *$$
$$for \; all \; a \in B \; such \; that \; \rho(x, a)?.$$

For decision tables, in which all missing attribute values are lost, a special characteristic relation was defined by J. Stefanowski and A. Tsoukias in [14], see also, e.g., [13], [15]. In this paper that characteristic relation will be denoted by $LV(B)$, where B is a nonempty subset of the set A of all attributes. For $x, y \in U$ characteristic relation $LV(B)$ is defined as follows:

$$(x, y) \in LV(B) \; if \; and \; only \; if \; \rho(x, a) = r(y, a)$$
$$for \; all \; a \in B \; such \; that \; \rho(x, a) \neq ?.$$

For any decision table in which all missing attribute values are lost, the characteristic relation $LV(B)$ is reflexive, but – in general – does not need to be symmetric or transitive.

For decision tables where all missing attribute values are "do not care" conditions a special characteristic relation, in this paper denoted by $DCC(B)$, was defined by M. Kryszkiewicz in [9], see also, e.g., [10]. For $x, y \in U$, the characteristic relation $DCC(B)$ is defined as follows:

$$(x, y) \in DCC(B) \; if \; and \; only \; if \; \rho(x, a) = \rho(y, a) \; or \rho(x, a) = * \; or \; \rho(y, a) = *$$
$$for \; all \; a \in B.$$

Relation $DCC(B)$ is reflexive and symmetric but – in general – not transitive.

Obviously, characteristic relations $LV(B)$ and $DCC(B)$ are special cases of the characteristic relation $R(B)$. For a completely specified decision table, the characteristic relation $R(B)$ is reduced to $IND(B)$.

3   Lower and Upper Approximations

For completely specified decision tables lower and upper approximations are defined on the basis of the indiscernibility relation. Any finite union of elementary sets, associated with B, will be called a *B-definable set*. Let X be any subset of the set U of all cases. The set X is called a *concept* and is usually defined as the set of all cases defined by a specific value of the decision. In general, X is not a B-definable set. However, set X may be approximated by two B-definable sets, the first one is called a *B-lower approximation* of X, denoted by $\underline{B}X$ and defined as follows

$$\{x \in U | [x]_B \subseteq X\}.$$

The second set is called a *B-upper approximation* of X, denoted by $\overline{B}X$ and defined as follows

$$\{x \in U | [x]_B \cap X \neq \emptyset.$$

The above shown way of computing lower and upper approximations, by constructing these approximations from singletons x, will be called the *first method*. The B-lower approximation of X is the greatest B-definable set, contained in X. The B-upper approximation of X is the smallest B-definable set containing X.

As it was observed in [12], for complete decision tables we may use a *second method* to define the B-lower approximation of X, by the following formula

$$\cup\{[x]_B | x \in U, [x]_B \subseteq X\},$$

and the B-upper approximation of x may de defined, using the second method, by

$$\cup\{[x]_B | x \in U, [x]_B \cap X \neq \emptyset).$$

For incompletely specified decision tables lower and upper approximations may be defined in a few different ways. First, the definition of definability should be modified. Any finite union of characteristic sets of B is called a *B-definable set*. In this paper we suggest three different definitions of lower and upper approximations. Again, let X be a concept, let B be a subset of the set A of all attributes, and let $R(B)$ be the characteristic relation of the incomplete decision table with characteristic sets $K(x)$, where $x \in U$. Our first definition uses a similar idea as in the previous articles on incompletely specified decision tables [9], [10], [13], [14], [15], i.e., lower and upper approximations are sets of singletons from the universe U satisfying some properties. Thus, lower and upper approximations are defined by analogy with the above first method, by constructing both sets from singletons. We will call these approximations *singleton*. A singleton B-lower approximation of X is defined as follows:

$$\underline{B}X = \{x \in U | K_B(x) \subseteq X\}.$$

A singleton B-upper approximation of X is

$$\overline{B}X = \{x \in U | K_B(x) \cap X \neq \emptyset\}.$$

In our example of the decision table presented in Table 2 let us say that $B = A$. Then the singleton A-lower and A-upper approximations of the two concepts: $\{1, 2, 4, 8\}$ and $\{3, 5, 6, 7\}$ are:

$$\underline{A}\{1, 2, 4, 8\} = \{1, 2, 4\},$$

$$\underline{A}\{3, 5, 6, 7\} = \{3, 7\},$$

$$\overline{A}\{1, 2, 4, 8\} = \{1, 2, 4, 5, 6, 8\},$$

$$\overline{A}\{3, 5, 6, 7\} = \{3, 5, 6, 7, 8\}.$$

The second method of defining lower and upper approximations for complete decision tables uses another idea: lower and upper approximations are unions of elementary sets, subsets of U. Therefore we may define lower and upper approximations for incomplete decision tables by analogy with the second method, using characteristic sets instead of elementary sets. There are two ways to do this. Using the first way, a *subset* B-lower approximation of X is defined as follows:

$$\underline{B}X = \cup\{K_B(x)|x \in U, K_B(x) \subseteq X\}.$$

A *subset* B-upper approximation of X is

$$\overline{B}X = \cup\{K_B(x)|x \in U, K_B(x) \cap X \neq \emptyset\}.$$

Since any characteristic relation $R(B)$ is reflexive, for any concept X, singleton B-lower and B-upper approximations of X are subsets of the subset B-lower and B-upper approximations of X, respectively. For the same decision table, presented in Table 2, the subset A-lower and A-upper approximations are

$$\underline{A}\{1, 2, 4, 8\} = \{1, 2, 4, 8\},$$

$$\underline{A}\{3, 5, 6, 7\} = \{3, 7\},$$

$$\overline{A}\{1, 2, 4, 8\} = \{1, 2, 4, 5, 6, 8\},$$

$$\overline{A}\{3, 5, 6, 7\} = \{2, 3, 4, 5, 6, 7, 8\}.$$

The second possibility is to modify the subset definition of lower and upper approximation by replacing the universe U from the subset definition by a concept X. A *concept* B-lower approximation of the concept X is defined as follows:

$$\underline{B}X = \cup\{K_B(x)|x \in X, K_B(x) \subseteq X\}.$$

Obviously, the subset B-lower approximation of X is the same set as the concept B-lower approximation of X. A concept B-upper approximation of the concept X is defined as follows:

$$\overline{B}X = \cup\{K_B(x)|x \in X, K_B(x) \cap X \neq \emptyset\} = \cup\{K_B(x)|x \in X\}.$$

The concept B-upper approximation of X is a subset of the subset B-upper approximation of X. Besides, the concept B-upper approximations are truly the smallest B-definable sets containing X. For the decision table presented in Table 2, the concept A-lower and A-upper approximations are

$$\underline{A}\{1, 2, 4, 8\} = \{1, 2, 4, 8\},$$

$$\underline{A}\{3, 5, 6, 7\} = \{3, 7\},$$

$$\overline{A}\{1, 2, 4, 8\} = \{1, 2, 4, 6, 8\},$$

$$\overline{A}\{3, 5, 6, 7\} = \{3, 4, 5, 6, 7, 8\}.$$

Note that for complete decision tables, all three definitions of lower approximations, singleton, subset and concept, coalesce to the same definition. Also, for complete decision tables, all three definitions of upper approximations coalesce to the same definition. This is not true for incomplete decision tables, as our example shows.

4   Rule Induction

In the first step of processing the input data file, the data mining system LERS (Learning from Examples based on Rough Sets) checks if the input data file is *consistent* (i.e., if the file does not contain conflicting examples). Table 1 is inconsistent because the fourth and the fifth examples are conflicting. For these examples, the values of all three attributes are the same (*high, yes, yes*), but the decision values are different, *yes* for the fourth example and *no* for the fifth example. If the input data file is inconsistent, LERS computes lower and upper approximations of all concepts. Rules induced from the lower approximation of the concept *certainly* describe the concept, so they are called *certain*. On the other hand, rules induced from the upper approximation of the concept describe the concept only *possibly* (or *plausibly*), so they are called *possible* [2].

The same idea of blocks of attribute-value pairs is used in a rule induction algorithm LEM2 (Learning from Examples Module, version 2), a component of LERS. LEM2 learns *discriminant description*, i.e., the smallest set of minimal rules, describing the concept. The option LEM2 of LERS is most frequently used since – in most cases – it gives best results. LEM2 explores the search space of attribute-value pairs. Its input data file is a lower or upper approximation of a concept, so its input data file is always consistent. In general, LEM2 computes a local covering and then converts it into a rule set. We will quote a few definitions to describe the LEM2 algorithm.

Let B be a nonempty lower or upper approximation of a concept represented by a decision-value pair (d, w). Set B *depends* on a set T of attribute-value pairs $t = (a, v)$ if and only if

$$\emptyset \neq [T] = \bigcap_{t \in T} [t] \subseteq B.$$

Set T is a *minimal complex* of B if and only if B depends on T and no proper subset T' of T exists such that B depends on T'. Let $\mathcal{T}$ be a nonempty collection of nonempty sets of attribute-value pairs. Then $\mathcal{T}$ is a *local covering* of B if and only if the following conditions are satisfied:

(1) each member T of $\mathcal{T}$ is a minimal complex of B,

(2) $\bigcap_{t\in\mathcal{T}}[T] = B$, and

$\mathcal{T}$ is minimal, i.e., $\mathcal{T}$ has the smallest possible number of members.

The procedure LEM2 is presented below.

Procedure LEM2
(**input**: a set B,
output: a single local covering $\mathcal{T}$ of set B);
begin

    $G := B$;
    $\mathcal{T} := \emptyset$;
    while $G \neq \emptyset$
        begin
        $T := \emptyset$;
        $T(G) := \{t|[t] \cap G \neq \emptyset\}$;
        while $T = \emptyset$ **or** $[T] \not\subseteq B$
            begin
                select a pair $t \in T(G)$ such that $|[t] \cap G|$
                is maximum; if a tie occurs, select a pair $t \in T(G)$
                with the smallest cardinality of $[t]$;
                if another tie occurs, select first pair;
                $T := T \cup \{t\}$;
                $G := [t] \cap G$;
                $T(G) := \{t|[t] \cap G \neq \emptyset\}$;
                $T(G) := T(G) - T$;
            end {while}
            for each $t \in T$ **do**
                if $[T - \{t\}] \subseteq B$ **then** $T := T - \{t\}$;
        $\mathcal{T} := \mathcal{T} \cup \{T\}$;
        $G := B - \bigcup_{T\in\mathcal{T}}[T]$;
    end {while};
    for each $T \in \mathcal{T}$ **do**
        if $\bigcup_{S\in\mathcal{T}-\{T\}}[S] = B$ **then** $\mathcal{T} := \mathcal{T} - \{T\}$;
end {procedure}.

MLEM2 is a modified version of the algorithm LEM2. The original algorithm LEM2 needs discretization, a preprocessing, to deal with numerical attributes. The MLEM2 algorithm can induce rules from incomplete decision tables with numerical attributes. Its previous version induced certain rules from incomplete decision tables with missing attribute values interpreted as lost and with numerical attributes. Recently, MLEM2 was further extended to induce both certain and possible rules from a decision table with some missing attribute values being lost and some missing attribute values being "do not care" conditions, while some attributes may be numerical. Rule induction from decision tables with numerical attributes will be described in the next section. In this section we will describe a new way in which MLEM2 handles incomplete decision tables.

Since all characteristic sets $K_B(x)$, where $x \in U$, are intersections of attribute-value pair blocks for attributes from B, and for subset and concept definitions of B–lower and B–upper approximations are unions of sets of the type $K_B(x)$, it is most natural to use an algorithm based on blocks of attribute-value pairs, such as LEM2 [1], [4] for rule induction.

First of all let us examine rule induction usefulness for the three different definition of lower and upper approximations: singleton, subset and concept. The first observation is that singleton lower and upper approximations should not be used for rule induction. Let us explain that on the basis of our example of the decision table from Table 2. The singleton A-lower approximation of the concept $\{1, 2, 4, 8\}$ is the set $\{1, 2, 4\}$. Our expectation is that we should be able to describe the set $\{1, 2, 4\}$ using given interpretation of missing attribute values, while in the rules we are allowed to use conditions being attribute-value pairs. However, this is impossible, because, as follows from the list of all sets $K_A(x)$ there is no way to describe case 1 not describing at the same time case 8, but $\{1, 8\} \not\subseteq \{1, 2, 4\}$. Similarly, there is no way to describe the singleton A-upper approximation of the concept $\{3, 5, 6, 7\}$, i.e., the set $\{3, 5, 6, 7, 8\}$, since there is no way to describe case 5 not describing, at the same time, cases 4 and 8, however, $\{4, 5, 8\} \not\subseteq \{3, 5, 6, 7, 8\}$. On the other hand, both subset and concept A-lower and A-upper approximations are unions of the characteristic sets of the type $K_A(x)$, therefore, it is always possible to induce certain rules from subset and concept A-lower approximations and possible rules from concept and subset A-upper approximations. Subset A-lower approximations are identical with concept A-lower approximations so it does not matter which approximations we are going to use. Since concept A-upper approximations are subsets of the corresponding subset A-upper approximations, it is more feasible to use concept A-upper approximations, since they are closer to the concept X, and rules will more precisely describe the concept X. Moreover, it better fits into the idea that the upper approximation should be the smallest set containing the concept. Therefore, we will use for rule induction only concept lower and upper approximations.

In order to induce certain rules for our example of the decision table presented in Table 2, we have to compute concept A-lower approximations for both concepts, $\{1, 2, 4, 8\}$ and $\{3, 5, 6, 7\}$. The concept lower approximation of $\{1, 2, 4, 8\}$ is the same set $\{1, 2, 4, 8\}$, so we are going to pass to the procedure LEM2 as the set B. Initially $G = B$. The set $T(G)$ is the following set $\{$(Temperature, high), (Temperature, very_high), (Temperature, normal), (Headache, yes), (Nausea, no), Nausea, yes)$\}$.

For three attribute-value pairs from $T(G)$, namely, (Temperature, high), (Headache, yes) and (Nausea, yes), the following value

$$[(attribute, value)] \cap G$$

is maximum. The second criterion, the smallest cardinality of [(attribute, value)], indicates (Temperature, high), (Headache, yes) (in both cases that cardinality is equal to four). The last criterion, "first pair", selects (Temperature, high). Thus $T = \{(\text{Temperature, high})\}$, $G = \{1, 4, 8\}$, and the new $T(G)$ is equal to $\{(\text{Temperature, very\_high}), (\text{Temperature, normal}), (\text{Headache, yes}), (\text{Nausea, no}), \text{Nausea, yes})\}$.

Since $[(Temperature, high)] \not\subseteq B$, we have to perform the next iteration of the inner WHILE loop. This time (Headache, yes) will be selected, the new T $= \{(\text{Temperature, high}), (\text{Headache, yes})\}$ and new G is equal to $\{4, 8\}$. Since $[T] = [(Temperature, high)] \cap [(Headache, yes)] = \{4, 8\} \subseteq B$, the first minimal complex is computed.

It is not difficult to see that we cannot drop any of these two attribute-value pairs, so $T = \{T\}$, and the new G is equal to $B - \{4, 8\} = \{1, 2\}$.

During the second iteration of the outer WHILE loop, the next minimal complex T is identified as $\{(\text{Temperature, very\_high})\}$, so $T = \{\{(\text{Temperature, high}), (\text{Headache, yes})\}, \{(\text{Temperature, very\_high})\}\}$ and $G = \{1\}$.

We need one additional iteration of the outer WHILE loop, the next minimal complex T is computed as $\{(\text{Temperature, high}), (\text{Nausea, no})\}$, and $T = \{\{(\text{Temperature, high}), (\text{Headache, yes})\}, \{(\text{Temperature, very\_high})\}, \{(\text{Temperature, high}), (\text{Nausea, no})\}\}$ becomes the first local covering, since we cannot drop any of minimal complexes from T. The set of certain rules, corresponding to T and describing the concept $\{1, 2, 4, 8\}$, is

(Temperature, high) & (Headache, yes) -> (Flu, yes),
(Temperature, very_high) -> (Flu, yes),
(Temperature, high) & (Nausea, no) -> (Flu, yes).

Remaining rule sets, certain for the second concept equal to $\{3, 5, 6, 7\}$, and both sets of possible rules are compute in a similar manner. Eventually, rules in the LERS format (every rule is equipped with three numbers, the total number of attribute-value pairs on the left-hand side of the rule, the total number of examples correctly classified by the rule during training, and the total number of training cases matching the left-hand side of the rule) are:
certain rule set:

2, 2, 2
(Temperature, high) & (Headache, yes) -> (Flu, yes)
1, 2, 2
(Temperature, very_high) -> (Flu, yes)
2, 2, 2
(Temperature, high) & (Nausea, no) -> (Flu, yes)
1, 2, 2
(Headache, no) -> (Flu, no)

and possible rule set:

1, 3, 4
(Headache, yes) -> (Flu, yes)
2, 2, 2
(Temperature, high) & (Nausea, no) -> (Flu, yes)
2, 1, 3
(Nausea, yes) & (Temperature, high) -> (Flu, no)
1, 2, 2
(Headache, no) -> (Flu, no)
1, 2, 3
(Temperature, normal) -> (Flu, no)

5   Other Approaches to Missing Attribute Values

So far we have used two approaches to missing attribute values, in the first one a missing attribute value was interpreted as lost, in the second as a "do not care" condition. There are many other possible approaches to missing attribute values, for some discussion on this topic see [7]. Our belief is that for any possible interpretation of a missing attribute vale, blocks of attribute-value pairs may be re-defined, a new characteristic relation may be computed, corresponding lower and upper approximations computed as well, and eventually, corresponding certain and possible rules induced.

As an example we may consider another interpretation for "do not care" conditions. So far, in computing the block for an attribute-value pair (a, v) we added all cases with value "*" to such block $[(a, v)]$. Following [7], we may consider another interpretation of "do not care conditions": If for an attribute a there exists a case x such that the corresponding value is a "do not care" condition, i.e., $\rho(x, a) = *$, then the corresponding case x should be included in blocks $[(a, v)]$ for all values v of attribute a with the same decision value as for x (i.e., we will add x only to members of the same concept to which x belongs). With this new interpretation of "*"s, blocks of attribute-value pairs for Table 2 are:

[(Temperature, high)] = {1, 4, 5, 8},
[(Temperature, very_high)] = {2, 8},
[(Temperature, normal)] = {6, 7},
[(Headache, yes)] = {2, 4, 6, 8},
[(Headache, no)] = {3, 7},
[(Nausea, no)] = {1, 3, 6},
[(Nausea, yes)] = {2, 4, 5, 7, 8}.

The characteristic set $K_B(x)$ for Table 2, a new interpretation of "*"s, and $B = A$, are:

$K_A(1) = \{1, 4, 5, 8\} \cap \{1, 3, 6\} = \{1, 8\}$,
$K_A(2) = \{2, 8\} \cap \{2, 4, 6, 8\} \cap \{2, 4, 5, 7, 8\} = \{2, 8\}$,
$K_A(3) = \{3, 7\} \cap \{1, 3, 6\} = \{3\}$,
$K_A(4) = \{1, 4, 5, 8\} \cap \{2, 4, 6, 8\} \cap \{2, 4, 5, 7, 8\} = \{4, 8\}$,

$K_A(5) = \{1, 4, 5, 8\} \cap \{2, 4, 5, 7, 8\} = \{4, 5, 8\},$
$K_A(6) = \{6, 7\} \cap \{2, 4, 6, 8\} \cap \{1, 3, 6\} = \{6\},$
$K_A(7) = \{6, 7\} \cap \{3, 7\} \cap \{2, 4, 5, 7, 8\} = \{7\},$ and
$K_A(8) = \{2, 4, 6, 8\}.$

The characteristic relation $R(B)$ is $\{(1, 1), (1, 8), (2, 2), (2, 8), (3, 3), (4, 4), (4, 8), (5, 4), (5, 5), (5, 8), (6, 6), (7, 7), (8, 2), (8, 4), (8, 6), (8, 8)\}$. Then we may define lower and upper approximations and induce rules using a similar technique as in the previous section.

6   Incomplete Decision Tables with Numerical Attributes

An example of an incomplete decision table with a numerical attribute is presented in Table 3.

Table 3. An incomplete decision table with a numerical attribute

	Attributes			Decision
Case	Temperature	Headache	Nausea	Flu
1	98	?	no	yes
2	101	yes	yes	yes
3	?	no	no	no
4	99	yes	yes	yes
5	99	?	yes	no
6	96	yes	no	no
7	96	no	yes	no
8	*	yes	*	yes

Numerical attributes should be treated in a little bit different way as symbolic attributes. First, for computing characteristic sets, numerical attributes should be considered as symbolic. For example, for Table 3 the blocks of the numerical attribute Temperature are:

[(Temperature, 96)] = {6, 7, 8},
[(Temperature, 98)] = {1, 8},
[(Temperature, 99)] = {4, 5, 8},
[(Temperature, 101)] = {2, 8}.

Remaining blocks of attribute-value pairs, for attributes Headache and Nausea, are the same as for Table 2. The characteristic sets $K_B(x)$ for Table 3 and $B = A$ are:

$K_A(1) = \{1, 8\} \cap \{1, 3, 6, 8\} = \{1, 8\},$
$K_A(2) = \{2, 8\} \cap \{2, 4, 6, 8\} \cap \{2, 4, 5, 7, 8\} = \{2, 8\},$
$K_A(3) = \{3, 7\} \cap \{1, 3, 6, 8\} = \{3\},$
$K_A(4) = \{4, 5, 8\} \cap \{2, 4, 6, 8\} \cap \{2, 4, 5, 7, 8\} = \{4, 8\},$
$K_A(5) = \{4, 5, 8\} \cap \{2, 4, 5, 7, 8\} = \{4, 5, 8\},$

$$K_A(6) = \{6,7,8\} \cap \{2,4,6,8\} \cap \{1,3,6,8\} = \{6,8\},$$
$$K_A(7) = \{6,7,8\} \cap \{3,7\} \cap \{2,4,5,7,8\} = \{7\}, \text{ and}$$
$$K_A(8) = \{2,4,6,8\}.$$

The characteristic relation $R(B)$ is $\{(1, 1), (1, 8), (2, 2), (2, 8), (3, 3), (4, 4), (4, 8), (5, 4), (5, 5), (5, 8), (6, 6), (6, 8), (7, 7), (8, 2), (8, 4), (8, 6), (8, 8)\}$. For the decision presented in Table 3, the concept A-lower and A-upper approximations are

$$\underline{A}\{1,2,4,8\} = \{1,2,4,8\},$$

$$\underline{A}\{3,5,6,7\} = \{3,7\},$$

$$\overline{A}\{1,2,4,8\} = \{1,2,4,6,8\},$$

$$\overline{A}\{3,5,6,7\} = \{3,4,5,6,7,8\}.$$

For inducing rules, blocks of attribute-value pairs are defined differently than in computing characteristic sets. MLEM2 has an ability to recognize integer and real numbers as values of attributes, and labels such attributes as numerical. For numerical attributes MLEM2 computes blocks in a different way than for symbolic attributes. First, it sorts all values of a numerical attribute, ignoring missing attribute values. Then it computes cutpoints as averages for any two consecutive values of the sorted list. For each cutpoint c MLEM2 creates two blocks, the first block contains all cases for which values of the numerical attribute are smaller than c, the second block contains remaining cases, i.e., all cases for which values of the numerical attribute are larger than c. The search space of MLEM2 is the set of all blocks computed this way, together with blocks defined by symbolic attributes. Starting from that point, rule induction in MLEM2 is conducted the same way as in LEM2. Note that if in a rule there are two attribute value pairs with overlapping intervals, a new condition is computed with the intersection of both intervals. Thus, the corresponding blocks for Temperature are:

[(Temperature, 96..97)] = {6, 7, 8},
[(Temperature, 97..101)] = {1, 2, 4, 5, 8},
[(Temperature, 96..98.5)] = {1, 6, 7, 8},
[(Temperature, 98.5..101)] = {2, 4, 5, 8},
[(Temperature, 96..100)] = {1, 4, 5, 6, 7, 8},
[(Temperature, 100..101)] = {2, 8}.

Remaining blocks of attribute-value pairs, for attributes Headache and Nausea, are the same as for Table 2. Using the MLEM2 algorithm, the following rules are induced from the concept approximations:

certain rule set:

2, 3, 3
(Temperature, 98.5..101) & (Headache, yes) -> (Flu, yes)
1, 2, 2
(Temperature, 97..98.5) -> (Flu, yes)
1, 2, 2
(Headache, no) -> (Flu, no)

possible rule set:

    1, 3, 4
    (Headache, yes) -> (Flu, yes)
    2, 2, 3
    (Temperature, 96..98.5) & (Nausea, no) -> (Flu, yes)
    2, 2, 4
    (Temperature, 96..100) & (Nausea, yes) -> (Flu, no)
    1, 2, 3
    (Temperature, 96..97) -> (Flu, no)
    1, 2, 2
    (Headache, no) -> (Flu, no)

7 Conclusions

It was shown in the paper that the idea of attribute-value pair blocks is an extremely useful tool. That idea may be used for computing characteristic relations for incomplete decision tables; in turn, characteristic sets are used for determining lower and upper approximations. Furthermore, the same idea of

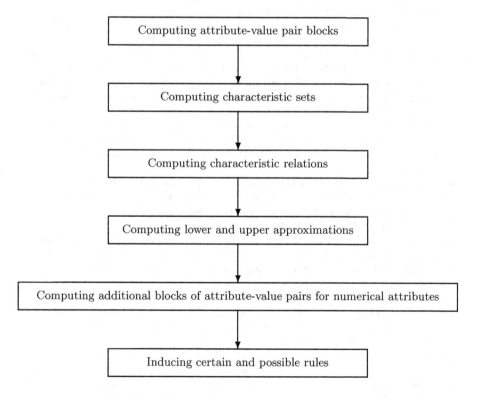

Fig. 1. Using attribute-value pair blocks for rule induction from incomplete decision tables

attribute-value pair blocks may be used for rule induction, for example, using the MLEM2 algorithm. The process is depicted in Figure 1.

Note that it is much more convenient to define the characteristic relations through the two-stage process of determining blocks of attribute-value pairs and then computing characteristic sets than to define characteristic relations, for every interpretation of missing attribute values, separately.

For completely specified decision tables any characteristic relation is reduced to an indiscernibility relation. Also, it is shown that the most useful way of defining lower and upper approximations for incomplete decision tables is a new idea of concept lower and upper approximations. Two new ways to define lower and upper approximations for incomplete decision tables, called subset and concept, and the third way, defined previously in a number of papers [9], [10], [13], [14], [15] and called here singleton lower and upper approximations, are all reduced to respective well-known definitions of lower and upper approximations for complete decision tables.

References

1. Chan, C.C. and Grzymala-Busse, J.W.: On the attribute redundancy and the learning programs ID3, PRISM, and LEM2. Department of Computer Science, University of Kansas, TR-91-14, December 1991, 20 pp.

2. Grzymala-Busse, J.W.: Knowledge acquisition under uncertainty – A rough set approach. *Journal of Intelligent & Robotic Systems* **1** (1988), 3–16.

3. Grzymala-Busse, J.W.: On the unknown attribute values in learning from examples. Proc. of the ISMIS-91, 6th International Symposium on Methodologies for Intelligent Systems, Charlotte, North Carolina, October 16–19, 1991. Lecture Notes in Artificial Intelligence, vol. 542, Springer-Verlag, Berlin, Heidelberg, New York (1991) 368–377.

4. Grzymala-Busse, J.W.: LERS – A system for learning from examples based on rough sets. In Intelligent Decision Support. Handbook of Applications and Advances of the Rough Sets Theory, ed. by R. Slowinski, Kluwer Academic Publishers, Dordrecht, Boston, London (1992) 3–18.

5. Grzymala-Busse., J.W.: MLEM2: A new algorithm for rule induction from imperfect data. Proceedings of the 9th International Conference on Information Processing and Management of Uncertainty in Knowledge-Based Systems, IPMU 2002, July 1–5, Annecy, France, 243–250.

6. Grzymala-Busse, J.W.: Rough set strategies to data with missing attribute values. Workshop Notes, Foundations and New Directions of Data Mining, the 3-rd International Conference on Data Mining, Melbourne, FL, USA, November 19–22, 2003, 56–63.

7. Grzymala-Busse, J.W. and Hu, M.: A comparison of several approaches to missing attribute values in data mining. Proceedings of the Second International Conference on Rough Sets and Current Trends in Computing RSCTC'2000, Banff, Canada, October 16–19, 2000, 340–347.

8. Grzymala-Busse, J.W. and A. Y. Wang A.Y.: Modified algorithms LEM1 and LEM2 for rule induction from data with missing attribute values. Proc. of the Fifth International Workshop on Rough Sets and Soft Computing (RSSC'97) at the Third Joint Conference on Information Sciences (JCIS'97), Research Triangle Park, NC, March 2–5, 1997, 69–72.

9. Kryszkiewicz, M.: Rough set approach to incomplete information systems. Proceedings of the Second Annual Joint Conference on Information Sciences, Wrightsville Beach, NC, September 28–October 1, 1995, 194–197.

10. Kryszkiewicz, M.: Rules in incomplete information systems. *Information Sciences* **113** (1999) 271–292.

11. Pawlak, Z.: Rough Sets. *International Journal of Computer and Information Sciences* **11** (1982) 341–356.

12. Pawlak, Z.: Rough Sets. Theoretical Aspects of Reasoning about Data. Kluwer Academic Publishers, Dordrecht, Boston, London (1991).

13. Stefanowski, J.: Algorithms of Decision Rule Induction in Data Mining. Poznan University of Technology Press, Poznan, Poland (2001).

14. Stefanowski, J. and Tsoukias, A.: On the extension of rough sets under incomplete information. Proceedings of the 7th International Workshop on New Directions in Rough Sets, Data Mining, and Granular-Soft Computing, RSFDGrC'1999, Ube, Yamaguchi, Japan, November 8–10, 1999, 73–81.

15. Stefanowski, J. and Tsoukias, A.: Incomplete information tables and rough classification. *Computational Intelligence* **17** (2001) 545–566.

16. Yao, Y.Y.: Two views of the theory of rough sets in finite universes. *International J. of Approximate Reasoning* **15** (1996) 291–317.

17. Yao, Y.Y.: Relational interpretations of neighborhood operators and rough set approximation operators. *Information Sciences* **111** (1998) 239–259.

18. Yao, Y.Y.: On the generalizing rough set theory. Proc. of the 9th Int. Conference on Rough Sets, Fuzzy Sets, Data Mining and Granular Computing (RSFDGrC'2003), Chongqing, China, October 19–22, 2003, 44–51.

Generalizations of Rough Sets
and Rule Extraction

Masahiro Inuiguchi

Division of Mathematical Science for Social Systems
Department of Systems Innovation
Graduate School of Engineering Science, Osaka University
1-3, Machikaneyama, Toyonaka, Osaka 560-8531, Japan
inuiguti@sys.es.osaka-u.ac.jp
http://www-inulab.sys.es.osaka-u.ac.jp/~inuiguti/

Abstract. In this paper, two kinds of generalizations of rough sets are proposed based on two different interpretations of rough sets: one is an interpretation of rough sets as approximation of a set by means of elementary sets and the other is an interpretation of rough sets as classification of objects into three different classes, i.e., positive objects, negative objects and boundary objects. Under each interpretation, two different definitions of rough sets are given depending on the problem setting. The fundamental properties are shown. The relations between generalized rough sets are given. Moreover, rule extraction underlying each rough set is discussed. It is shown that rules are extracted based on modified decision matrices. A simple example is given to show the differences in the extracted rules by underlying rough sets.

1 Introduction

Rough sets [7] are useful in applications to data mining, knowledge discovery, decision making, conflict analysis, and so on. Rough set approaches [7] have been developed under equivalence relations. The equivalence relation implies that attributes are all nominal. Because of this weak assumption, unreasonable results for human intuition have been exemplified when some attributes are ordinal [3]. To overcome such unreasonableness, the dominance-based rough set approach has been proposed by Greco et al. [3]. On the other hand, the generalization of rough sets is an interesting topic not only in mathematical point of view but also in practical point of view. Along this direction, rough sets have been generalized under similarity relations [5, 10], covers [1, 5] and general relations [6, 11–13]. Those results demonstrate a diversity of generalizations. Moreover, recently, the introduction of fuzziness into rough set approaches attracts researchers in order to obtain more realistic and useful tools (see, for example, [2]).

Considering applications of rough sets in the generalized setting, the interpretation of rough sets plays an important role. This is because any mathematical model cannot be properly applied without its interpretation. In other words, the interpretation should be proper for the aim of application. The importance of the

J.F. Peters et al. (Eds.): Transactions on Rough Sets I, LNCS 3100, pp. 96–119, 2004.
© Springer-Verlag Berlin Heidelberg 2004

interpretation increases as the problem setting becomes more generalized such as a fuzzy setting. This is because the diversity of definitions and treatments which are same in the original setting is increased by the generalization.

Two major interpretations have been traditionally given to rough sets. One is an interpretation of rough sets as approximation of a set by means of elementary sets. The other is an interpretation of rough sets as classification of objects into three different classes, i.e., positive objects, negative objects and boundary objects. Those interpretations can be found in the terminologies, 'lower approximation' (resp. 'upper approximation') and 'positive region' (resp. 'possible region') in the classical rough sets. The lower approximation of a set equals to the positive region of the set in the classical rough sets, i.e., rough sets under equivalence relations. However, they can be different in a general setting. For example, Inuiguchi and Tanino [5] showed the difference under a similarity relation. They described the difference under a more generalized setting (see [6]). However, fundamental properties have not been considerably investigated yet. Moreover, from definitions of rough sets in the previous papers, we may have some other definitions of rough sets under generalized settings.

When generalized rough sets are given, we may have a question how we can extract decision rules based on them. The type of extracted decision rules would be different depending on the underlying generalized rough set. To this question, Inuiguchi and Tanino [6] demonstrated the difference in rule extraction based on generalized rough sets.

In this paper, we discuss the generalized rough sets in the two different interpretations restricting ourselves into crisp setting as extensions of a previous paper [6]. Such investigations are necessary and important also for proper definitions and applications of fuzzy rough sets. We introduce some new definitions of generalized rough sets. The fundamental properties of those generalized rough sets are newly given. The relations between rough sets under two different interpretations are discussed. In order to see the differences of those generalized rough sets in applications, we discuss rule extraction based on the generalized rough sets. We demonstrate the difference in the types of decision rules depending on underlying generalized rough sets. Moreover, we show that decision rules with minimal conditions can be extracted by modifying the decision matrix.

This paper is organized as follows. The classical rough sets are briefly reviewed in the next section. In Section 3, interpreting rough sets as classification of objects, we define rough sets under general relations. The fundamental properties of the generalized rough sets are investigated. In Section 4, using the interpretation of rough sets as approximation by means of elementary sets, we define rough sets under a family of sets. The fundamental properties of this generalized rough sets are also investigated. Section 5 is devoted to relations between those two kinds of rough sets. In Section 6, we discuss decision rule extraction based on generalized rough sets. Extraction methods using modified decision matrices are proposed. In Section 7, a few numerical examples are given to demonstrate the differences among the extracted decision rules based on different generalized rough sets. Some concluding remarks are given in Section 8.

2   Classical Rough Sets

2.1   Definitions, Interpretations and Fundamental Properties

Let R be an equivalence relation in the finite universe U, i.e., $R \subseteq U \times U$. In rough set literature, R is referred to as an indiscernibility relation and a pair (U, R) is called an approximation space. By the equivalence relation R, U can be partitioned into a collection of equivalence classes or elementary sets, $U|R = \{E_1, E_2, \ldots, E_p\}$. Define $R(x) = \{y \in U \mid (y, x) \in R\}$. Then we have $x \in E_i$ if and only if $E_i = R(x)$. Note that $U|R = \{R(x) \mid x \in U\}$.

Let X be a subset of U. Using $R(x)$, a rough set of X is defined by a pair of the following lower and upper approximations;

$$R_*(X) = \{x \in X \mid R(x) \subseteq X\} = U - \bigcup\{R(y) \mid y \in U - X\}$$

$$= \bigcup\{E_i \mid E_i \subseteq X, \ i = 1, 2, \ldots, p\},$$

$$= \bigcup\left\{\bigcap_{i \in I}(U - E_i) \ \middle| \ \bigcap_{i \in I}(U - E_i) \subseteq X, \ I \subseteq \{1, 2, \ldots .p\}\right\}, \qquad (1)$$

$$R^*(X) = \bigcup\{R(x) \mid x \in X\} = U - \{y \in U - X \mid R(y) \subseteq U - X\}$$

$$= \bigcap\left\{\bigcup_{i \in I} E_i \ \middle| \ \bigcup_{i \in I} E_i \supseteq X, \ I \subseteq \{1, 2, \ldots, p\}\right\}$$

$$= \bigcap\{U - E_i \mid U - E_i \supseteq X\}. \qquad (2)$$

Let us interpret $R(x)$ as a set of objects we intuitively identify as members of X from the fact $x \in X$. Then, from the first expression of $R_*(X)$ in (1), $R_*(X)$ is interpreted as a set of objects which are consistent with the intuition that $R(x) \subseteq X$ if $x \in X$. Under the same interpretation of $R(x)$, $R^*(X)$ is interpreted as a set of objects which can be intuitively inferred as members of X from the first expression of $R^*(X)$ in (2). In other words, $R_*(X)$ and $R^*(X)$ show positive (consistent) and possible members of X. Moreover, $R^*(X) - R_*(X)$ and $U - R^*(X)$ show ambiguous (boundary) and negative members of X. In this way, a rough set classifies objects of U into three classes, i.e., positive, negative and boundary regions.

On the contrary let us interpret $R(x)$ as a set of objects we intuitively identify as members of $U - X$ from the fact $x \in U - X$. In the same way as previous discussion, $\bigcup\{R(y) \mid y \in U - X\}$ and $\{y \in U - X \mid R(y) \subseteq U - X\}$ show possible and positive members of $U - X$, respectively. From the second expression of $R_*(X)$ in (1), $R_*(X)$ can be regarded as a set of impossible members of $U - X$. In other words, $R_*(X)$ show certain members of X. Similarly, from the second expression of $R^*(X)$ in (2), $R^*(X)$ can be regarded as a set of non-positive members of $U - X$. Namely, $R^*(X)$ show conceivable members of X. $R^*(X) - R_*(X)$ and $U - R^*(X)$ show border and inconceivable members of X. In this case, a rough set again classifies objects of U into three classes, i.e., certain, inconceivable and border regions.

Table 1. Fundamental properties of rough sets

(i) $R_*(X) \subseteq X \subseteq R^*(X)$.
(ii) $R_*(\emptyset) = R^*(\emptyset) = \emptyset$, $R_*(U) = R^*(U) = U$.
(iii) $R_*(X \cap Y) = R_*(X) \cap R_*(Y)$, $R^*(X \cup Y) = R^*(X) \cup R^*(Y)$.
(iv) $X \subseteq Y$ implies $R_*(X) \subseteq R_*(Y)$, $X \subseteq Y$ implies $R^*(X) \subseteq R^*(Y)$.
(v) $R_*(X \cup Y) \supseteq R_*(X) \cup R_*(Y)$, $R^*(X \cap Y) \subseteq R^*(X) \cap R^*(Y)$.
(vi) $R_*(U - X) = U - R^*(X)$, $R^*(U - X) = U - R_*(X)$.
(vii) $R_*(R_*(X)) = R^*(R_*(X)) = R_*(X)$,
$\quad\quad R^*(R^*(X)) = R_*(R^*(X)) = R^*(X)$.

From the third expression of $R_*(X)$ in (1), $R_*(X)$ is the best approximation of X by means of the union of elementary sets E_i such that $E_i \subseteq X$. On the other hand, from the third expression of $R^*(X)$ in (2), $R^*(X)$ is the minimal superset of X by means of the union of elementary sets E_i.

Finally, from the fourth expression of $R_*(X)$ in (1), $R_*(X)$ is the maximal subset of X by means of the intersection of complements of elementary sets $U - E_i$. From the fourth expression of $R^*(X)$ in (2), $R^*(X)$ is the best approximation of X by means of the intersection of complements of elementary sets $U - E_i$ such that $U - E_i \supseteq X$.

We introduced only four kinds of expressions of lower and upper approximations but there are other many expressions [5, 10–13]. The interpretation of rough sets depends on the expression of lower and upper approximations. Thus we may have more interpretations by adopting the other expressions. However the interpretations described above seem appropriate for applications of rough sets. Those interpretations can be divided into two categories: interpretation of rough sets as classification of objects and interpretation of rough sets as approximation of a set.

The fundamental properties listed in Table 1 are satisfied with the lower and upper approximations of classical rough sets.

3   Classification-Oriented Generalization

3.1   Proposed Definitions

We generalize classical rough sets under interpretation of rough sets as classification of objects. As described in the previous section, there are two expressions in this interpretation, i.e., the first and second expressions of (1) and (2).

First we describe the generalization based on the second expressions of (1) and (2). In this case, we assume that there exists a relation $P \subseteq U \times U$ such that $P(x) = \{y \in U \mid (y, x) \in P\}$ means a set of objects we intuitively identify as members of X from the fact $x \in X$. Then if $P(x) \subseteq X$ for an object $x \in X$ then there is no objection against $x \in X$. In this case, $x \in X$ is consistent with the intuitive knowledge based on the relation P. Such an object $x \in X$ can be considered as a positive member of X. Hence the positive region of X can be defined as

$$P_*(X) = \{x \in X \mid P(x) \subseteq X\}. \tag{3}$$

On the other hand, by the intuition from the relation P, an object $y \in P(x)$ for $x \in X$ can be a member of X. Such an object $y \in U$ is a possible member of X. Moreover, every object $x \in X$ is evidently a possible member of X. Hence the possible region of X can be defined as

$$P^*(X) = X \cup \bigcup \{P(x) \mid x \in X\}. \tag{4}$$

Using the positive region $P_*(X)$ and the possible region $P^*(X)$, we can define a rough set of X as a pair $(P_*(X), P^*(X))$. We can call such rough sets as classification-oriented rough sets under a positively extensive relation P of X (for short CP-rough sets).

The relation P depends on the meaning of X whose positive and possible regions we are interested in. Thus, we cannot always define the CP-rough set of $U - X$ by using the same relation P. To define a CP-rough set of $U - X$, we should introduce another relation $Q \subseteq U \times U$ such that $Q(x) = \{y \in U \mid (y, x) \in Q\}$ means a set of objects we intuitively identify as members of $U - X$ from the fact $x \in U - X$. Using Q we have positive and possible regions of $U - X$ by

$$Q_*(U - X) = \{x \in U - X \mid Q(x) \subseteq U - X\}, \tag{5}$$

$$Q^*(U - X) = (U - X) \cup \bigcup \{Q(x) \mid x \in U - X\}. \tag{6}$$

Using those, we can define certain and conceivable regions of X by

$$\bar{Q}_*(X) = U - Q^*(U - X) = X \cap \left(U - \bigcup \{Q(x) \mid x \in U - X\}\right), \tag{7}$$

$$\bar{Q}^*(X) = U - Q_*(U - X) = U - \{x \in U - X \mid Q(x) \subseteq U - X\}. \tag{8}$$

Those definitions correspond to the second expressions of (1) and (2).

We can define another rough set of X as a pair $(\bar{Q}_*(X), \bar{Q}^*(X))$ with the certain region $\bar{Q}_*(X)$ and the conceivable region $\bar{Q}^*(X)$. We can call this type of rough sets as classification-oriented rough sets under a negatively extensive relation Q of X (for short CN-rough sets).

Let $Q^{-1}(x) = \{y \in U \mid (x, y) \in Q\}$. As is shown in [10], we have

$$\bigcup \{Q(x) \mid x \in U - X\} = \{x \in U \mid Q^{-1}(x) \cap (U - X) \neq \emptyset\}. \tag{9}$$

Therefore, we have

$$\bar{Q}_*(X) = X \cap \{x \in U \mid Q^{-1}(x) \cap (U - X) = \emptyset\}$$
$$= \{x \in X \mid Q^{-1}(x) \subseteq X\} = Q_*^{\mathrm{T}}(X) \tag{10}$$

$$\bar{Q}^*(X) = U - \{x \in U - X \mid Q(x) \cap X = \emptyset\} = X \cup \{x \in U \mid Q(x) \cap X \neq \emptyset\}$$
$$= X \cup \bigcup \{Q^{-1}(x) \mid x \in X\} = Q^{\mathrm{T}^*}(X), \tag{11}$$

where Q^{T} is the converse relation of Q, i.e., $Q^{\mathrm{T}} = \{(x, y) \mid (y, x) \in Q\}$. Note that we have $Q^{\mathrm{T}}(x) = \{x \in U \mid (x, y) \in Q^{\mathrm{T}}\} = \{x \in U \mid (y, x) \in Q\} = Q^{-1}(x)$.

From (10) and (11), the classification-oriented rough sets under a negatively extensive relation Q can be seen as the classification-oriented rough sets under a positively extensive relation Q^T. By the same discussion, the classification-oriented rough sets under a positively extensive relation P can be also seen as the classification-oriented rough sets under a negatively extensive relation P^T. Moreover, when $P = Q^T$, we have the classification-oriented rough sets under a positively extensive relation P coincides with the classification-oriented rough sets under a negatively extensive relation Q.

3.2   Relationships to Previous Definitions

Rough sets were previously defined under a general relation. We discuss the relationships of the proposed generalized rough sets with previous ones.

First of all, let us review the previous generalized rough sets briefly. In analogy to Kripke model in modal logic, Yao and Lin [13] and Yao [11, 12] proposed a generalized rough set with the following lower and upper approximations:

$$T_*(X) = \{x \in U \mid T^{-1}(x) \subseteq X\}, \tag{12}$$
$$T^*(X) = \{x \mid T^{-1}(x) \cap X \neq \emptyset\}, \tag{13}$$

where T is a general binary relation and $T^{-1}(x) = \{y \in U \mid (x, y) \in T\}$. In Yao [12], $T^{-1}(x)$ is replaced with a neighborhood $n(x)$ of $x \in U$.

Słowiński and Vanderpooten [10] proposed rough sets under a similarity relation S. They assume the reflexivity of S $((x, x) \in S$, for each $x \in U)$. They classify all objects in U into the following four categories under the intuition that $y \in U$ to which $x \in U$ is similar must be in the same set containing x: (i) positive objects, i.e., objects $x \in U$ such that $x \in X$ and $S^{-1}(x) \subseteq X$, (ii) ambiguous objects of type I, i.e., objects $x \in U$ such that $x \in X$ but $S^{-1}(x) \cap (U - X) \neq \emptyset$, (iii) ambiguous objects of type 2, i.e., objects $x \in U$ such that $x \in U - x$ but $S^{-1}(x) \cap X \neq \emptyset$, and (iv) negative objects, i.e., $x \in U - X$ and $S^{-1}(x) \subseteq U - X$. Based on this classification, lower and upper approximations are defined by (12) and (13) with substitution of S for T. Namely, the lower approximation is a collection of positive objects and the upper approximation is a collection of positive and ambiguous objects. Note that they expressed the upper approximation as $S^*(X) = \bigcup \{S(x) \mid x \in X\}$ which is equivalent to (13) with the substitution of S for T (see Słowiński and Vanderpooten [10]), where $S(x) = \{y \in U \mid (y, x) \in S\}$.

Greco, Matarazzo and Słowiński [3] proposed rough sets under a dominance relation D. They assume the reflexivity of D $((x, x) \in D$, for each $x \in U)$. Let X be a set of objects better than x. Under the intuition that $y \in U$ by which $x \in U$ is dominated must be better than x, i.e., y must be at least in X, they defined lower and upper approximations as

$$D_*(X) = \{x \in U \mid D(x) \subseteq X\}, \tag{14}$$
$$D^*(X) = \bigcup \{D(x) \mid x \in X\}, \tag{15}$$

where $D(x) = \{y \in U \mid (y, x) \in D\}$. It can be shown that $D^*(X) = \{x \mid D^{-1}(x) \cap X \neq \emptyset\}$, where $D^{-1}(x) = \{y \in U \mid (x, y) \in D\}$.

Finally, Inuiguchi and Tanino [5] assume that a set X corresponds to an ambiguous concept so that we may have a set $\underline{X}$ composed of objects that everyone agrees their membership and a set $\overline{X}$ composed of objects that only someone agrees their membership. A given X can be considered a set of objects whose memberships are evaluated by a certain person. Thus, we assume that $\underline{X} \subseteq X \subseteq \overline{X}$. Let S be a reflexive similarity relation. Assume that only objects which are similar to a member of $\underline{X}$ are possible candidates for members of $\overline{X}$ for any set X. Then we have

$$\overline{X} \subseteq \bigcup \{S(x) \mid x \in \underline{X}\} = \{x \in U \mid S^{-1}(x) \cap \underline{X} \neq \emptyset\}. \tag{16}$$

From the definitions of $\underline{X}$ and $\overline{X}$, we can have $U - \underline{X} = \overline{U - X}$ and $U - \overline{X} = \underline{U - \overline{X}}$. Hence we also have

$$\underline{X} \subseteq \{x \in U \mid S^{-1}(x) \subseteq \overline{X}\}. \tag{17}$$

We do not know $\underline{X}$ and $\overline{X}$ but X. With substitution of S for T, we obtain the lower approximation of $\underline{X}$ by (12) and the upper approximation of $\overline{X}$ by (13).

Now, let us discuss relationships between the previous definitions and the proposed definitions. The previous definitions are formally agreed in the definition of upper approximation by (13) with substitution of a certain relation for T. However, the proposed definition (4) is similar but different since they take a union with X. By this union, $X \subseteq P^*(X)$ is guaranteed. In order to have this property of the upper approximation, Słowiński and Vanderpooten [10], Greco, Matarazzo and Słowiński [3] and Inuiguchi and Tanino [5] assumed the reflexivity of the binary relations S and D.

The idea of the proposed CP-rough set follows that of rough sets under a dominance relation proposed by Greco, Matarazzo and Słowiński [3]. On the other hand the idea of the proposed CN-rough set is similar to those of Słowiński and Vanderpooten [10] and Inuiguchi and Tanino [5] since we may regard S as a negatively extensive relation, i.e., $S(x)$ means a set of objects we intuitively identify as members of $U - X$ from the fact $x \in U - X$. However, the differences are found in the restrictions, i.e., $x \in U$ in (14) versus $x \in X$ in (3). In other words, we take an intersection with X, i.e., $P_*(X) = X \cap \{x \in U \mid P(x) \subseteq X\}$ and $\bar{Q}_*(X) = X \cap \{x \in U \mid Q^{-1}(x) \subseteq X\}$. This intersection guarantees $X \subseteq P_*(X)$ and $X \subseteq \bar{Q}_*(X)$. In order to guarantee those relations, the reflexivity of the relation is assumed in Słowiński and Vanderpooten [10], Greco, Matarazzo and Słowiński [3] and Inuiguchi and Tanino [5].

Finally, we remark that, in definitions by Słowiński and Vanderpooten [10] and Inuiguchi and Tanino [5], S acts as a positively extensive relation P and a negatively extensive relation Q at the same time.

3.3   Fundamental Properties

The fundamental properties of the CP- and CN-rough sets can be obtained as in Table 2. In property (vii), we assume that P can be regarded as positively

Table 2. Fundamental properties of CP- and CN-rough sets

(i) $P_*(X) \subseteq X \subseteq P^*(X)$, $\bar{Q}_*(X) \subseteq X \subseteq \bar{Q}^*(X)$.

(ii) $P_*(\emptyset) = P^*(\emptyset) = \emptyset$, $P_*(U) = P^*(U) = U$,
$\bar{Q}_*(\emptyset) = \bar{Q}^*(\emptyset) = \emptyset$, $\bar{Q}_*(U) = \bar{Q}^*(U) = U$.

(iii) $P_*(X \cap Y) = P_*(X) \cap P_*(Y)$, $P^*(X \cup Y) = P^*(X) \cup P^*(Y)$,
$\bar{Q}_*(X \cap Y) = \bar{Q}_*(X) \cap \bar{Q}_*(Y)$, $\bar{Q}^*(X \cup Y) = \bar{Q}^*(X) \cup \bar{Q}^*(Y)$.

(iv) $X \subseteq Y$ implies $P_*(X) \subseteq P_*(Y)$, $P^*(X) \subseteq P^*(Y)$,
$X \subseteq Y$ implies $\bar{Q}_*(X) \subseteq \bar{Q}_*(Y)$, $\bar{Q}^*(X) \subseteq \bar{Q}^*(Y)$.

(v) $P_*(X \cup Y) \supseteq P_*(X) \cup P_*(Y)$, $P^*(X \cap Y) \subseteq P^*(X) \cap P^*(Y)$,
$\bar{Q}_*(X \cup Y) \supseteq \bar{Q}_*(X) \cup \bar{Q}_*(Y)$, $\bar{Q}^*(X \cap Y) \subseteq \bar{Q}^*(X) \cap \bar{Q}^*(Y)$.

(vi) When Q is the converse of P, i.e., $(x, y) \in P$ if and only if $(y, x) \in Q$,
$P^*(X) = U - Q_*(U - X) = \bar{Q}^*(X)$, $P_*(X) = U - Q^*(U - X) = \bar{Q}_*(X)$.

(vii) $X \supseteq P^*(P_*(X)) \supseteq P_*(X) \supseteq P_*(P_*(X))$,
$X \subseteq P_*(P^*(X)) \subseteq P^*(X) \subseteq P^*(P^*(X))$,
$X \supseteq \bar{Q}^*(\bar{Q}_*(X)) \supseteq \bar{Q}_*(X) \supseteq \bar{Q}_*(\bar{Q}_*(X))$,
$X \subseteq \bar{Q}_*(\bar{Q}^*(X)) \subseteq \bar{Q}^*(X) \subseteq \bar{Q}^*(\bar{Q}^*(X))$.
When P is transitive, $P_*(P_*(X)) = P_*(X)$, $P^*(P^*(X)) = P^*(X)$.
When Q is transitive, $\bar{Q}_*(\bar{Q}_*(X)) = \bar{Q}_*(X)$, $\bar{Q}^*(\bar{Q}^*(X)) = \bar{Q}^*(X)$.
When P is reflexive and transitive,
$P^*(P_*(X)) = P_*(X) = P_*(P_*(X))$, $P_*(P^*(X)) = P^*(X) = P^*(P^*(X))$.
When Q is reflexive and transitive,
$\bar{Q}^*(\bar{Q}_*(X)) = \bar{Q}_*(X) = \bar{Q}_*(\bar{Q}_*(X))$, $\bar{Q}_*(\bar{Q}^*(X)) = \bar{Q}^*(X) = \bar{Q}^*(\bar{Q}^*(X))$.

---

extensive relations of $P_*(X)$, $P^*(X)$, $P_*(P_*(X))$, $P^*(P_*(X))$, $P_*(P^*(X))$ and $P^*(P^*(X))$. Similarly, we assume also that Q can be regarded as negatively extensive relations of $Q_*(X)$, $Q^*(X)$, $Q_*(Q_*(X))$, $Q^*(Q_*(X))$, $Q_*(Q^*(X))$ and $Q^*(Q^*(X))$. Properties (i)–(v) are obvious. The proofs of (vi) and (vii) are given in Appendix.

As shown in Table 2, (i)–(v) in Table 1 are preserved by classification-oriented generalization. However (vi) and (vii) in Table 1 are conditionally preserved. A part of (vii) in Table 1 is unconditionally preserved. However the other part is satisfied totally when P is reflexive and transitive. When P is transitive, we have $P^*(\cdots (P^*(P_*(X))) \cdots) = P^*(P_*(X)) \subseteq X$ and $P_*(\cdots (P_*(P^*(X))) \cdots) = P_*(P^*(X)) \supseteq X$. Similarly, we have $\bar{Q}^*(\cdots (\bar{Q}^*(\bar{Q}_*(X))) \cdots) = \bar{Q}^*(\bar{Q}_*(X)) \subseteq X$ and $\bar{Q}_*(\cdots (\bar{Q}_*(\bar{Q}^*(X))) \cdots) = \bar{Q}_*(\bar{Q}^*(X)) \supseteq X$ when Q is transitive. Those facts mean that the first operation governs the relations with the original set when the relation is transitive. When relations P and Q represent the similarity between objects, P and Q can be equal each other. In such case, the condition for (vi) implies that P, or equivalently, Q is symmetric.

4   Approximation-Oriented Generalization

4.1   Proposed Definitions

In order to generalize classical rough sets under the interpretation of rough sets as approximation of a set by means of elementary sets, we introduce a family with a finite number of elementary sets on U, $\mathcal{F} = \{F_1, F_2, \ldots, F_p\}$, as a generalization of a partition $U|R = \{E_1, E_2, \ldots, E_p\}$. Each F_i is a group of objects collected according to some specific meaning.

There are two ways to define lower and upper approximations of a set X under a family $\mathcal{F}$: one is approximations by means of the union of elementary sets F_i and the other is approximations by means of the intersection of complements of elementary sets $U - F_i$. Namely, from the third and fourth expressions of lower and upper approximations in (1) and (2), lower and upper approximations of a set X under $\mathcal{F}$ are defined straightforwardly in the following two ways:

$$\mathcal{F}_*^{\cup}(X) = \bigcup \{F_i \mid F_i \subseteq X, \ i = 0, 1, \ldots, p\}, \tag{18}$$

$$\mathcal{F}_*^{\cap}(X) = \bigcup \left\{ \bigcap_{i \in I}(U - F_i) \ \middle| \ \bigcap_{i \in I}(U - F_i) \subseteq X, \ I \subseteq \{1, 2, \ldots, p+1\} \right\}, \tag{19}$$

$$\mathcal{F}_{\cup}^*(X) = \bigcap \left\{ \bigcup_{i \in I} F_i \ \middle| \ \bigcup_{i \in I} F_i \supseteq X, \ I \subseteq \{1, 2, \ldots, p+1\} \right\}, \tag{20}$$

$$\mathcal{F}_{\cap}^*(X) = \bigcap \{U - F_i \mid U - F_i \supseteq X, \ i = 0, 1, \ldots, p\}, \tag{21}$$

where, for convenience, we define $F_0 = \emptyset$ and $F_{p+1} = U$. Because $F_i \cap F_j \neq \emptyset$ for $i \neq j$ does not always hold, $\mathcal{F}_*^{\cap}(X)$ (resp. $\mathcal{F}_{\cup}^*(X)$) is not always an intersection of complements of elementary sets $U - F_i$ (resp. a union of elementary sets F_i) but a union of several maximal intersections $\mathcal{F}_j^{\cap}(X)$, $j = 1, 2, \ldots, t_1$ of complements of elementary sets $U - F_i$ (resp. an intersection of several minimal unions $\mathcal{F}_j^{\cup}(X)$, $j = 1, 2, \ldots, t_2$ of elementary sets F_i) if $\bigcap_{i=1,2,\ldots,p}(U - F_i) \subseteq X$ (resp. $\bigcup_{i=1,2,\ldots,p} F_i \supseteq X$) is satisfied. Namely, we have $\mathcal{F}_*^{\cap}(X) = \bigcup_{j=1,2,\ldots,t_1} \mathcal{F}_j^{\cap}(X)$ and $\mathcal{F}_{\cup}^*(X) = \bigcap_{j=1,2,\ldots,t_2} \mathcal{F}_j^{\cup}(X)$. We can call a pair $(\mathcal{F}_*^{\cup}(X), \mathcal{F}_{\cup}^*(X))$ an approximation-oriented rough set by means of the union of elementary sets F_i under a family $\mathcal{F}$ (for short, an AU-rough set) and a pair $(\mathcal{F}_*^{\cap}(X), \mathcal{F}_{\cap}^*(X))$ an approximation-oriented rough set by means of the intersection of complements of elementary sets $U - F_i$ under a family $\mathcal{F}$ (for short, an AI-rough set).

4.2   Relationships to Previous Definitions

So far, rough sets have been generalized also under a finite cover and neighborhoods. In this subsection, we discuss relationships of AU- and AI-rough sets with the previous rough sets.

First, we describe previous definitions. Bonikowski, Bryniarski and Wybraniec-Skardowska [1] proposed rough sets under a finite cover $\mathcal{C} = \{C_1, C_2 \ldots, C_p\}$ such that $\bigcup_{i=1,2,\ldots,p} C_i = U$. They defined the lower approximation of $X \subseteq U$ by

$$\mathcal{C}_*'(X) = \bigcup \{C_i \in \mathcal{C} \mid C_i \subseteq X\}. \tag{22}$$

In order to define the upper approximation, we should define the minimal description of an object $x \in U$ and the boundary of X. The minimal description of an object $x \in U$ is a family defined by

$$Md(x) = \{C_i \in \mathcal{C} \mid x \in C_i, \; \forall C_j \in \mathcal{C}(x \in C_j \wedge C_j \subseteq C_i \rightarrow C_i = C_j)\}. \quad (23)$$

Then the boundary of X is a family defined by $Bn(X) = \bigcup\{Md(x) \mid x \in X, x \notin \mathcal{C}'_*(X)\}$. The upper approximation of X is defined by

$$\mathcal{C}'^*(X) = \bigcup Bn(X) \cup \mathcal{C}'_*(X). \quad (24)$$

Owing to $Md(x)$, we have

$$\mathcal{C}'^*(X) \subseteq \mathcal{C}'_*(X) \cup \bigcup\{C_i \mid C_i \cap (X - \mathcal{C}'_*(X)) \neq \emptyset\} \subseteq \{C_i \mid C_i \cap X \neq \emptyset\}. (25)$$

Yao [12] proposed rough sets under neighborhoods $\{n(x) \mid x \in U\}$ where $n : U \rightarrow 2^U$ and $n(x)$ is interpreted as the neighbothood of $x \in U$. Three kinds of rough sets were proposed. One of them has been described in subsection 3.2. The lower and upper approximations in the second kind of rough sets are

$$\nu_*(X) = \bigcup\{n(x) \mid x \in U, \; n(x) \subseteq X\}$$
$$= \{x \in U \mid \exists y(x \in n(y) \wedge n(y) \subseteq X)\}, \quad (26)$$
$$\nu^*(X) = U - \nu_*(U - X) = \{x \in U \mid \forall y(x \in n(y) \rightarrow n(y) \cap X \neq \emptyset)\}. \quad (27)$$

As shown above, those lower and upper approximations are closely related with interior and closure operations in topology.

The upper and lower approximations in the third kind of rough sets are defined as follows:

$$\nu'^*(X) = \bigcup\{n(x) \mid x \in U, \; n(x) \cap X \neq \emptyset\}$$
$$= \{x \in U \mid \exists y(x \in n(y) \wedge n(y) \cap X \neq \emptyset)\}, \quad (28)$$
$$\nu'_*(X) = U - \nu'^*(U - X) = \{x \in U \mid \forall y(x \in n(y) \rightarrow n(y) \subseteq X)\}. \quad (29)$$

Inuiguchi and Tanino [5] also proposed rough sets under a cover $\mathcal{C}$. They defined upper and lower approximations as

$$\mathcal{C}''_*(X) = \bigcup\{C_i \mid C_i \subseteq X, \; i = 1, 2, \ldots, p\}, \quad (30)$$
$$\mathcal{C}''^*(X) = U - \mathcal{C}''_*(U - X) = \bigcap\{U - C_i \mid U - C_i \supseteq X, \; i = 1, 2, \ldots, p\}. (31)$$

Now let us discuss the relationships with AU- and AI-rough sets. When $\mathcal{F} = \mathcal{C}$, we have $\mathcal{F}^{\cup}_*(X) = \mathcal{C}'_*(X) = \mathcal{C}''_*(X)$, $\mathcal{F}^*_*(X) \subseteq \mathcal{C}'^*(X)$ and $\mathcal{F}^*_{\cap}(X) = \mathcal{C}''^*(X)$. We also have $\mathcal{C}'^*(X) \supseteq \bigcup_{j=1,2,\ldots,t_2} \mathcal{F}^{\cup}_j(X)$. The equality does not hold always because we have the possibility of $\mathcal{F}^{\cup}_j(X) \subset \mathcal{C}'_*(X) \cup \bigcup_{x \in X - \mathcal{C}'_*(X)} C(x)$, where $C(x)$ is an arbitrary $C_i \in Md(x)$. When $\mathcal{F} = \{n(x) \mid x \in U\}$, we have $\mathcal{F}^{\cup}_*(X) = \nu_*(X) \supseteq \nu'_*(X)$ and $\mathcal{F}^*_{\cap}(X) = \nu^*(X) \subseteq \nu'^*(X)$. Generally, $\mathcal{F}^*_{\cap}(X)$ (resp. $\mathcal{F}^*_{\cup}(X)$) has no relation with $\nu_*(X)$ and $\nu'_*(X)$ (resp. $\nu^*(X)$ and $\nu'^*(X)$). From those relations, we know that $F^{\cup}_*(X)$ and $F^{\cap}_*(X)$ are maximal approximations among six lower approximations while $F^*_{\cup}(X)$ and $F^*_{\cap}(X)$ are minimal approximations among six upper approximations. This implies that the proposed lower and upper

Table 3. Fundamental properties of AU- and AI-rough sets

(i) $\mathcal{F}_*^\cup(X) \subseteq X \subseteq \mathcal{F}_\cup^*(X)$, $\mathcal{F}_*^\cap(X) \subseteq X \subseteq \mathcal{F}_\cap^*(X)$.

(ii) $\mathcal{F}_*^\cup(\emptyset) = \mathcal{F}_*^\cap(\emptyset) = \emptyset$, $\mathcal{F}_\cup^*(U) = \mathcal{F}_\cap^*(U) = U$.

$\qquad$ When $\bigcup \mathcal{F} = \bigcup_{i=1,\ldots,p} F_i = U$, $\mathcal{F}_\cup^*(\emptyset) = \emptyset$, $\mathcal{F}_*^\cup(U) = U$.

$\qquad$ When $\bigcap \mathcal{F} = \bigcap_{i=1,\ldots,p} F_i = \emptyset$, $\mathcal{F}_\cap^*(\emptyset) = \emptyset$, $\mathcal{F}_*^\cap(U) = U$.

(iii) $\mathcal{F}_*^\cup(X \cap Y) \subseteq \mathcal{F}_*^\cup(X) \cap \mathcal{F}_*^\cup(Y)$, $\mathcal{F}_*^\cap(X \cap Y) = \mathcal{F}_*^\cap(X) \cap \mathcal{F}_*^\cap(Y)$,

$\qquad \mathcal{F}_\cup^*(X \cup Y) = \mathcal{F}_\cup^*(X) \cup \mathcal{F}_\cup^*(Y)$, $\mathcal{F}_\cap^*(X \cup Y) \supseteq \mathcal{F}_\cap^*(X) \cup \mathcal{F}_\cap^*(Y)$.

$\qquad$ When $F_i \cap F_j = \emptyset$, for any $i \neq j$,

$\qquad\qquad \mathcal{F}_*^\cup(X \cap Y) = \mathcal{F}_*^\cup(X) \cap \mathcal{F}_*^\cup(Y)$, $\mathcal{F}_\cap^*(X \cup Y) = \mathcal{F}_\cap^*(X) \cup \mathcal{F}_\cap^*(Y)$.

(iv) $X \subseteq Y$ implies $\mathcal{F}_*^\cup(X) \subseteq \mathcal{F}_*^\cup(Y)$, $\mathcal{F}_*^\cap(X) \subseteq \mathcal{F}_*^\cap(Y)$,

$\qquad X \subseteq Y$ implies $\mathcal{F}_\cup^*(X) \subseteq \mathcal{F}_\cup^*(Y)$, $\mathcal{F}_\cap^*(X) \subseteq \mathcal{F}_\cap^*(Y)$.

(v) $\mathcal{F}_*^\cup(X \cup Y) \supseteq \mathcal{F}_*^\cup(X) \cup \mathcal{F}_*^\cup(Y)$, $\mathcal{F}_*^\cap(X \cup Y) \supseteq \mathcal{F}_*^\cap(X) \cup \mathcal{F}_*^\cap(Y)$,

$\qquad \mathcal{F}_\cup^*(X \cap Y) \subseteq \mathcal{F}_\cup^*(X) \cap \mathcal{F}_\cup^*(Y)$, $\mathcal{F}_\cap^*(X \cap Y) \subseteq \mathcal{F}_\cap^*(X) \cap \mathcal{F}_\cap^*(Y)$.

(vi) $\mathcal{F}_*^\cup(U - X) = U - \mathcal{F}_\cup^*(X)$, $\mathcal{F}_*^\cap(U - X) = U - \mathcal{F}_\cap^*(X)$,

$\qquad \mathcal{F}_\cup^*(U - X) = U - \mathcal{F}_*^\cap(X)$, $\mathcal{F}_\cap^*(U - X) = U - \mathcal{F}_*^\cup(X)$.

(vii) $\mathcal{F}_*^\cup(\mathcal{F}_*^\cup(X)) = \mathcal{F}_*^\cup(X)$, $\mathcal{F}_\cap^*(\mathcal{F}_\cap^*(X)) = \mathcal{F}_\cap^*(X)$,

$\qquad \mathcal{F}_\cup^*(\mathcal{F}_\cup^*(X)) = \mathcal{F}_\cup^*(X)$, $\mathcal{F}_\cap^*(\mathcal{F}_\cap^*(X)) = \mathcal{F}_\cap^*(X)$,

$\qquad \mathcal{F}_\cup^*(\mathcal{F}_*^\cup(X)) = \mathcal{F}_*^\cup(X)$, $\mathcal{F}_*^\cap(\mathcal{F}_\cap^*(X)) = \mathcal{F}_\cap^*(X)$,

$\qquad \mathcal{F}_\cap^*(\mathcal{F}_*^\cap(X)) \supseteq \mathcal{F}_*^\cap(X)$, $\mathcal{F}_\cap^*(\mathcal{F}_j^\cap(X)) = \mathcal{F}_j^\cap(X)$, $j = 1, 2, \ldots, t_1$,

$\qquad \mathcal{F}_*^\cup(\mathcal{F}_\cup^*(X)) \subseteq \mathcal{F}_\cup^*(X)$, $\mathcal{F}_*^\cup(\mathcal{F}_j^\cup(X)) = \mathcal{F}_j^\cup(X)$, $j = 1, 2, \ldots, t_2$,

$\qquad$ When $F_i \cap F_j = \emptyset$, for any $i \neq j$, $\mathcal{F}_\cap^*(\mathcal{F}_*^\cap(X)) = \mathcal{F}_*^\cap(X)$, $\mathcal{F}_*^\cup(\mathcal{F}_\cup^*(X)) = \mathcal{F}_\cup^*(X)$.

approximations are better approximations of X so that they are suitable for our interpretation of rough sets. Moreover, the proposed definitions are applicable under a more general setting since we neither assume that $\mathcal{F}$ is a cover nor that $p = \mathrm{Card}(\mathcal{F}) \leq n = \mathrm{Card}(U)$, i.e., the number of elementary sets F_i is not less than the number of objects.

4.3   Fundamental Properties

The fundamental properties of AU- and AI-rough sets are shown in Table 3. Properties (i), (iv) and (v) in Table 1 are preserved for both of AU- and AI-rough sets. Parts of (ii) and (iii) in Table 1 are preserved, however, some conditions are necessary for full preservation. The duality, i.e., property (vi) in Table 1 is preserved between upper (resp. lower) approximations of AU-rough sets (resp. AI-rough sets) and lower (resp. upper) approximations of AI-rough sets (resp. AU-rough sets). Property (vii) in Table 1 is almost preserved. $\mathcal{F}_\cap^*(\mathcal{F}_*^\cap(X)) = \mathcal{F}_*^\cap(X)$ (resp. $\mathcal{F}_*^\cup(\mathcal{F}_\cup^*(X)) = \mathcal{F}_\cup^*(X)$) is not always preserved because $\mathcal{F}_*^\cap(X)$ (resp. $\mathcal{F}_\cup^*(X)$) is not always a union of elementary sets F_i (resp. an intersection of complements of elementary sets $U - F_i$). However, for the minimal union $\mathcal{F}_j^\cup$ (resp. $\mathcal{F}_j^\cap$), the property corresponding to (vii) holds. The proof of property (iii) is given in Appendix. The other properties can be proved easily.

Table 4. Relationships between two kinds of rough sets

(a) When P is reflexive, $P_*(X) \subseteq \mathcal{P}_*^\cup(X) = P^*(P_*(X)) \subseteq X \subseteq \mathcal{P}_\cup^*(X) \subseteq P^*(X)$.
When Q is reflexive, $\bar{Q}^*(X) \supseteq \mathcal{Q}_\cap^*(X) = \bar{Q}_*(\bar{Q}^*(X)) \supseteq X \supseteq \mathcal{Q}_*^\cap(X) \supseteq \bar{Q}_*(X)$.
(b) When P is transitive, $\mathcal{P}_\cup^*(X) \supseteq P^*(X) \supseteq X \supseteq P_*(X) \supseteq \mathcal{P}_*^\cup(X)$.
When Q is transitive, $\mathcal{Q}_*^\cap(X) \subseteq \bar{Q}_*(X) \subseteq X \subseteq \bar{Q}^*(X) \subseteq \mathcal{Q}_\cap^*(X)$.
(c) When P is reflexive and transitive,
$\qquad P_*(X) = \mathcal{P}_*^\cup(X) = P^*(P_*(X)) \subseteq X \subseteq \mathcal{P}_\cup^*(X) = P^*(X)$.
When Q is reflexive and transitive,
$\qquad \bar{Q}^*(X) = \mathcal{Q}_\cap^*(X) = \bar{Q}_*(\bar{Q}^*(X)) \supseteq X \supseteq \mathcal{Q}_*^\cap(X) = \bar{Q}_*(X)$.

5 Relationships between Two Kinds of Rough Sets

Given a relation P, we may define a family by

$$\mathcal{P} = \{P(x) \mid x \in U\}. \tag{32}$$

Therefore, under a positively extensive relation P is given, we obtain not only
CP-rough sets but also AU- and AI-rough sets. This is the same for a negatively
extensive relation Q. Namely, by a family $\mathcal{Q} = \{Q(x) \mid x \in U\}$, we obtain AU-
and AI-rough sets.

The relationships between CP-/CN-rough sets and AU/AI-rough sets are
listed in Table 4. In Table 4 we recognize a strong relation between CP- and
AU-rough sets as well as a strong relation between CN- and AI-rough sets. The
proofs of (a) and (b) in Table 4 are given in Appendix.

6 Rule Extraction

6.1 Decision Table and Problem Setting

In this section, we discuss rule extraction from decision tables based on gen-
eralized rough sets. Consider a decision table $\mathcal{I} = \langle U, C \cup \{d\}, V, \rho \rangle$, where
$U = \{x_1, x_2, \ldots, x_n\}$ is a universe of objects, C is a set of all condition attributes, d is a unique decision attribute, $V = \bigcup_{a \in C \cup \{d\}} V_a$, V_a is a finite set of
attribute values of attribute a, and $\rho : U \times C \cup \{d\} \to V$ is the information function such that $\rho(x, a) \in V_a$ for all $a \in C \cup \{d\}$. By decision attribute value $\rho(x, d)$,
we assume that we can group objects into several classes D_k, $k = 1, 2, \ldots, m$.
D_k, $k = 1, 2, \ldots, m$ do not necessary form a partition but a cover. Namely,
$D_k \cap D_j = \emptyset$ does not always hold but $\bigcup_{k=1, 2, \ldots, m} D_k = U$.

Corresponding to D_k, $k = 1, 2, \ldots, m$, we assume that there is a relation
$P_a \in V_a^2$ is given to each condition attribute $a \in C$ so that if $x \in D_k$ and
$(y, x) \in P_a$ then we intuitively conclude $y \in D_k$ from the viewpoint of attribute
a. For each $A \subseteq C$, we define a positively extensive relation by

$$P_A = \{(x, y) \mid (\rho(x, a), \rho(y, a)) \in P_a, \ \forall a \in A\}. \tag{33}$$

Moreover, we also assume that there is a relation $Q_a \in V_a \times V_a$ is given to
each condition attribute $a \in C$ so that if $x \in U - D_k$ and $(y, x) \in Q_a$ then

we intuitively conclude $y \in U - D_k$ from the viewpoint of attribute a. For each $A \subseteq C$, we define a negatively extensive relation by

$$Q_A = \{(x, y) \mid (\rho(x, a), \rho(y, a)) \in Q_a, \; \forall a \in A\}. \tag{34}$$

For the purpose of the comparison, we may built finite families based on relations P_a and Q_a as described below. We can build families using P_A and Q_A as

$$\mathcal{P} = \{P_A(x) \mid x \in U, \; A \subseteq C\}, \tag{35}$$
$$\mathcal{Q} = \{Q_A(x) \mid x \in U, \; A \subseteq C\}, \tag{36}$$

where $P_A(x) = \{y \in U \mid (y, x) \in P_A\}$ and $Q_A(x) = \{y \in U \mid (y, x) \in Q_A\}$. For $A = \{a_1, a_2, \ldots, a_s\}$ and $\boldsymbol{v} = (v_1, v_2, \ldots, v_s) \in V_{a_1} \times V_{a_2} \times \cdots \times V_{a_s}$, let us define

$$Z_A(\boldsymbol{v}) = \{x \in U \mid (\rho(x, a_i), v_i) \in P_{a_i}, \; i = 1, 2, \ldots, s\}, \tag{37}$$
$$W_A(\boldsymbol{v}) = \{x \in U \mid (\rho(x, a_i), v_i) \in Q_{a_i}, \; i = 1, 2, \ldots, s\}. \tag{38}$$

Using those sets, we may build the following families defined by

$$\mathcal{Z} = \{Z_A(\boldsymbol{v}) \mid \boldsymbol{v} \in V_{a_1} \times V_{a_2} \times \cdots \times V_{a_s}, \; A = \{a_1, a_2, \ldots, a_s\} \subseteq U\}, \tag{39}$$
$$\mathcal{W} = \{W_A(\boldsymbol{v}) \mid \boldsymbol{v} \in V_{a_1} \times V_{a_2} \times \cdots \times V_{a_s}, \; A = \{a_1, a_2, \ldots, a_s\} \subseteq U\}. \tag{40}$$

6.2   Rule Extraction Based on Positive and Certain Regions

As shown in (10), a positive region $P_*(X)$ and a certain region $\bar{Q}_*(X)$ has the same representation. The difference is the adoption of the relation, i.e., P versus Q^{T}. Therefore the rule extraction method is the same. In this subsection, we describe the rule extraction method based on a positive region $P_*(X)$. The rule extraction method based on a certain region $\bar{Q}_*(X)$ is obtained by replacing a relation P with a relation Q^{T}.

We discuss the extraction of decision rules from the decision table $\mathcal{I} = \langle U, C \cup \{d\}, V, \rho \rangle$ described in the previous subsection. First, let us discuss the type of decision rule corresponding to the positive region (3). For any object $y \in U$ satisfying the condition of the decision rules, $y \in D_k$ and $P_C(y) \subseteq D_k$. D_k should not be in the condition part since we would like to infer the members of D_k. Considering those requirements, we should explore suitable conditions of the decision rules. When we confirm $y = x$ for an object $x \in P_{C*}(D_k)$, we may obviously conclude $y \in P_{C*}(D_k)$. Since each object is characterized by conditional attributes $a \in C$, $y = x$ can be conjectured from $\rho(y, a) = \rho(x, a)$, $\forall a \in C$. However, it is possible that there exists $z \in U$ such that $\rho(z, a) = \rho(x, a)$, $\forall a \in C$ but $z \notin D_k$. When P_C is reflexive, we always have $x \notin P_{C*}(D_k)$ if such an object $z \in U$ exists. Since we do not assume the reflexivity, $x \in P_{C*}(D_k)$ is possible even in the case such an object $z \in U$ exists. From these observations, we obtain the following type of decision rule based on $x \in P_{C*}(D_k)$ only when there is no object $z \in U$ such that $\rho(z, a) = \rho(x, a)$, $\forall a \in C$ but $z \notin D_k$:

$$\text{if } \rho(y, a_1) = v_1 \text{ and } \cdots \text{ and } \rho(y, a_l) = v_l \text{ then } y \in D_k,$$

where $v_j = \rho(x, a_i)$, $i = 1, 2, \ldots, l$ and we assume $C = \{a_1, a_2, \ldots, a_l\}$. Let us call this type of the decision rule, an identity if-then rule (for short, id-rule).

When P_C is transitive, we may conclude $y \in P_{C*}(D_k)$ from the fact that $(y, x) \in P_C$ and $x \in P_{C*}(D_k)$. This is because we have $P_C(y) \subseteq P_C(x) \subseteq D_k$ and $y \in P_C(x) \subseteq X$ from transitivity and the fact $x \in P_{C*}(D_k)$. In this case, we may have the following type of decision rule,

$$\text{if } (\rho(y, a_1), v_1) \in P_{a_1} \text{ and } \cdots \text{ and } (\rho(y, a_l), v_l) \in P_{a_l} \text{ then } y \in D_k.$$

This type of if-then rule is called a relational if-then rule (for short, R-rule). When the relation P_C is reflexive and transitive, an R-rule includes the corresponding id-rule.

As discussed above, based on an object $x \in P_{C*}(D_k)$, we can extract id-rules, and R-rules if P_C is transitive. We prefer to obtain decision rules with minimum length conditions. To this end, we should calculate the minimal condition attribute set $A \subseteq C$ such that $x \in P_{A*}(D_k)$. Let $A = \{a'_1, a'_2, \ldots, a'_q\}$ be such a minimal condition attribute set. Then we obtain the following id-rule when there is no object $z \in U - D_k$ such that $\rho(z, a'_i) = v'_i$, $i = 1, 2, \ldots, q$,

$$\text{if } \rho(y, a'_1) = v'_1 \text{ and } \cdots \text{ and } \rho(y, a'_q) = v'_q \text{ then } y \in D_k,$$

where $v'_i = \rho(x, a'_i)$, $i = 1, 2, \ldots, q$. When P_C is transitive, we obtain an R-rule,

$$\text{if } (\rho(y, a'_1), v'_1) \in P_{a'_1} \text{ and } \cdots \text{ and } (\rho(y, a'_q), v'_q) \in P_{a'_q} \text{ then } y \in D_k.$$

Note that P_C is transitive if and only if P_A is transitive for each $A \subseteq C$. Moreover, the minimal condition attribute set is not always unique and, for each minimal condition attribute set, we obtain id- and R-rules.

Through the above procedure, we will obtain many decision rules. The decision rules are not always independent. Namely, we may have two decision rules 'if $Cond_1$ then Dec' and 'if $Cond_2$ then Dec' such that $Cond_1$ implies $Cond_2$. Eventually, the decision rule 'if $Cond_1$ then Dec' is superfluous and then omitted. This is different from the rule extraction based on the classical rough set.

For extracting all decision rules with minimal length conditions, we can utilize a decision matrix [8] with modifications. Consider the extraction of decision rules concluding $y \in D_k$. We begin with the calculation of $P_{C*}(D_k)$. Based on the obtained $P_{C*}(D_k)$, we define two disjoint index sets $K^+ = \{i \mid x_i \in P_{C*}(D_k)\}$ and $K^- = \{i \mid x_i \notin D_k\}$. The decision matrix $M^{id}(\mathcal{I}) = (M_{ij}^{id})$ is defined by

$$M_{ij}^{id} = \{(a, \tilde{v}_i) \mid \tilde{v}_i = \rho(x_i, a), \ (\rho(x_j, a), \rho(x_i, a)) \notin P_a,$$
$$\rho(x_j, a) \neq \rho(x_i, a), \ a \in C\}, \ i \in K^+, \ j \in K^-. \tag{41}$$

Note that the size of the decision matrix $M^{id}(\mathcal{I})$ is $\text{Card}(K^+) \times \text{Card}(K^-)$. An element $(a, \tilde{v}_i)$ of M_{ij}^{id} corresponds to a condition '$\rho(y, a) = \tilde{v}_i$' which is not satisfied with $y = x_j$ but with $y = x_i$. Moreover $M_{ij}^{id}(\mathcal{I})$ can be empty and in this case, we cannot obtain any id-rule from $x_i \in P_{C*}(D_k)$.

When P_C is transitive, we should consider another decision matrix for R-rules. The decision matrix $M^R(\mathcal{I}) = (M_{ij}^R)$ is defined by

$$M_{ij}^R = \{(a, \tilde{v}_i) \mid \tilde{v}_i = \rho(x_i, a), \ (\rho(x_j, a), \rho(x_i, a)) \notin P_a, \ a \in C\},$$
$$i \in K^+, \ j \in K^-. \tag{42}$$

Note that the size of the decision matrix $M^{\mathrm{R}}(\mathcal{I})$ is $\mathrm{Card}(K^+) \times \mathrm{Card}(K^-)$. An element $(a, \tilde{v}_i)$ of M^{R}_{ij} shows a condition '$(\rho(y, a), \tilde{v}_i) \in P_a$' which is not satisfied with $y = x_j$ but with $y = x_i$.

Let $Id((a, v))$ be a statement '$\rho(x, a) = v$' and $\tilde{P}((a, v))$ a statement '$(\rho(x, a), v) \in P_a$'. Then all minimal conditions in all possible decision rules with respect to D_k are obtained as conjunctive terms in the disjunctive normal form of the following logical function:

$$B_k = \begin{cases} \displaystyle\bigvee_{i \in K^+} \bigwedge_{j \in K^-} \bigvee Id(M^{\mathrm{id}}_{ij}), & \text{if } P_C \text{ is not transitive,} \\[2em] \left(\displaystyle\bigvee_{i \in K^+} \bigwedge_{j \in K^-} \bigvee Id(M^{\mathrm{id}}_{ij}) \right) \vee \left(\displaystyle\bigvee_{i \in K^+} \bigwedge_{j \in K^-} \bigvee \tilde{P}(M^{\mathrm{R}}_{ij}) \right), \\ \hfill \text{if } P_C \text{ is transitive.} \end{cases} \quad (43)$$

By the construction, it is obvious that $z \notin D_k$ does not satisfy the conditions of decision rules and that $x \in P_{C*}(D_k)$ satisfies them. Moreover we can prove that $z \in D_k - P_{C*}(D_k)$ does not satisfy the conditions. The proof is as follows.

Let $z \in D_k - P_{C*}(D_k)$ and let $y \notin D_k$ such that $(\rho(y, a), \rho(z, a)) \in P_a$ for all $a \in C$. The existence of y is guaranteed by the definition of z. First consider the condition of an arbitrary id-rule, '$\rho(w, a'_1) = v'_1$, $\rho(w, a'_2) = v'_2$ and $\cdots$ and $\rho(w, a'_q) = v'_q$', where $A = \{a'_1, a'_2, \ldots, a'_q\} \subseteq C$. Suppose z satisfies this condition, i.e., '$\rho(z, a'_1) = v'_1$, $\rho(z, a'_2) = v'_2$ and $\cdots$ and $\rho(z, a'_q) = v'_q$'. Since $v'_i = \rho(x, a'_i)$, $i = 1, 2, \ldots, q$, the fact $z \in D_k - P_{C*}(D_k)$ implies that $(\rho(y, a), \rho(x, a)) \in P_a$ for all $a \in A$, i.e., $(y, x) \in P_A$ $(y \in P_A(x))$. From $y \notin D_k$, we have $P_A(x) \not\subseteq D_k$. On the other hand, by the construction of M^{id}_{ij}, for each $y \notin D_k$, there exists $a \in A$ such that $(\rho(y, a), \rho(x, a)) \notin P_a$. This implies $P_A(x) \subseteq D_k$. A contradiction. Thus, for each id-rule, there is no $z \in D_k - P_{C*}(D_k)$ satisfying the condition. Next, assuming that P_C is transitive, we consider the condition of an arbitrary R-rule, '$(\rho(w, a'_1), v'_1) \in P_{a'_1}$, and $\cdots$ and $(\rho(w, a'_q), v'_q) \in P_{a'_q}$', where $\{a'_1, a'_2, \ldots, a'_q\} \subseteq C$. Suppose z satisfies this condition, i.e., '$(\rho(z, a'_1), v'_1) \in P_{a'_1}$, and $\cdots$ and $(\rho(z, a'_q), v'_q) \in P_{a'_q}$'. From the transitivity and the fact $(\rho(y, a), \rho(z, a)) \in P_a$ for all $a \in C$, we have '$(\rho(y, a'_1), v'_1) \in P_{a'_1}$, and $\cdots$ and $(\rho(y, a'_q), v'_q) \in P_{a'_q}$'. This contradicts the construction of the condition of R-rule. Therefore, for each R-rule, there is no $z \in D_k - P_{C*}(D_k)$ satisfying the condition.

The rule extraction method based on certain region is obtained by replacing P_C, P_A and P_a of the above discussion with Q^{T}_C, Q^{T}_A and Q^{T}_a, respectively.

6.3   Rule Extraction Based on Lower Approximations of AU-Rough Sets

As in the previous subsection, we discuss the extraction of decision rules from the decision table $\mathcal{I} = \langle U, C \cup \{d\}, V, \rho \rangle$. First, let us discuss the type of decision rule corresponding to the lower approximation of AU-rough set (18). For any object $y \in U$ satisfying the condition of the decision rules, we should have $y \in F_i$ and

$F_i \subseteq D_k$. When we confirm $y \in F_i$ for an elementary set $F_i \in \mathcal{F}$ such that $F_i \subseteq D_k$, we may obviously conclude $y \in D_k$. From this fact, when $F_i \subseteq D_k$ we have the following type of decision rule;

$$\text{if } y \in F_i \text{ then } y \in D_k.$$

For the decision table $\mathcal{I} = \langle U, C \cup \{d\}, V, \rho \rangle$, we consider two cases; (a) a case when $\mathcal{F} = \mathcal{P}$ and (b) a case when $\mathcal{F} = \mathcal{Z}$. In those cases the corresponding decision rules from the facts $P_A(x) \subseteq X$ and $Z_A(v) \subseteq X$ become

Case (a): if $(\rho(y, a_1), \bar{v}_1) \in P_{a_1}$ and $\cdots$ and $(\rho(y, a_s), \bar{v}_s) \in P_{a_s}$ then $y \in D_k$,

Case (b): if $(\rho(y, a_1), v_1) \in P_{a_1}$ and $\cdots$ and $(\rho(y, a_s), v_s) \in P_{a_s}$ then $y \in D_k$,

where $A = \{a_1, a_2, \ldots, a_s\}$, $\bar{v}_i = \rho(x, a_i)$, $i = 1, 2, \ldots, s$ and $v = (v_1, v_2, \ldots, v_s)$.

By the construction of $\mathcal{P}$ and $\mathcal{Z}$, we have $P_{A'}(x) \supseteq P_A(x)$ and $Z_{A'}(v') \supseteq Z_A(v)$ for $A' \subseteq A$, where $A' = \{a_{k_1}, a_{k_2}, \ldots, a_{k_t}\} \subseteq A$ and $v' = (v_{k_1}, v_{k_2}, \ldots, v_{k_t})$ is a sub-vector of v. Therefore, the decision rules with respect to minimal attribute sets A' are sufficient since they cover all decision rules with larger attribute sets $A \supseteq A'$. By this observation, we enumerate all decision rules with respect to minimal attribute sets.

The enumeration can be done by a modification of the decision matrix [8]. We describe the method in Case (a). Consider an enumeration of all decision rules with respect to a decision class D_k. To apply the decision matrix method, we first obtain a CP-rough set $P_*(D_k) = \{x \in U \mid P_C(x) \subseteq D_k\}$. Using $K^+ = \{i \mid x_i \in P_*(D_k)\}$ and $K^- = \{i \mid x_i \notin D_k\}$, we define the decision matrix $\tilde{M}(\mathcal{I}) = (\tilde{M}_{ij})$ by

$$\tilde{M}_{ij} = \{(a, v) \mid v = \rho(x_i, a), \ (\rho(x_j, a), \rho(x_i, a)) \notin P_a, \ a \in C\}, \quad i \in K^+, \ j \in K^-. \tag{44}$$

Then all minimal conditions in all possible decision rules with respect to D_k are obtained as conjunctive terms in the disjunctive normal form of the following logical function:

$$\tilde{B}_k = \bigvee_{i \in K^+} \bigwedge_{j \in K^-} \tilde{P}(\tilde{M}_{ij}). \tag{45}$$

In Case (b), we calculate $K(D_k) = \{v \in V_1 \times V_2 \times \cdots V_l \mid Z(v) \subseteq D_k\}$ instead of $P_*(D_k)$. Number elements of $K(D_k)$ such that $K(D_k) = \{v^1, v^2, \ldots, v^r\}$, where $r = \text{Card}(K(D_k))$. Then we define the decision matrix $\tilde{M}'(\mathcal{I}) = (\tilde{M}'_{ij})$ by

$$\tilde{M}'_{ij} = \{(a_d, v^i_d) \mid (\rho(x_j, a_d), v^i_d) \notin P_{a_d}, \ a_d \in C\}, \ i \in \{1, \ldots, r\}, \ j \in K^-, \tag{46}$$

where v^i_d is the d-th component of v^i, i.e., $v^i = (v^i_1, v^i_2, \ldots, v^i_l)$. All minimal conditions in all possible decision rules with respect to D_k are obtained as conjunctive terms in the disjunctive normal form of the following logical function:

$$\tilde{B}'_k = \bigvee_{i \in \{1, 2, \ldots, r\}} \bigwedge_{j \in K^-} \tilde{P}(\tilde{M}'_{ij}). \tag{47}$$

6.4   Rule Extraction Based on Lower Approximations of AI-Rough Sets

Let us discuss the type of decision rule corresponding to the lower approximation of AI-rough set (19). For any object $y \in U$ satisfying the condition of the decision rules, $y \in \bigcup_{j=1,2,\ldots,t_1} \mathcal{F}_j^\cap$. Therefore, for each $\mathcal{F}_j^\cap$, we have the following type of decision rule:

$$\text{if } y \in \mathcal{F}_j^\cap \text{ then } y \in D_k.$$

By the definition, $\mathcal{F}_j^\cap$ is represented by an intersection of a number of complementary sets of elementary sets, i.e., $\bigcap_{i \in I}(U - F_i)$ for a certain $I \subseteq \{1, 2, \ldots, p\}$. Therefore the condition part of the decision rule can be represented by '$y \notin F_{i_1}$, $y \notin F_{i_2}, \ldots$, and $y \notin F_{i_{\mathrm{Card}(I)}}$', where $I = \{i_1, i_2, \ldots, i_{\mathrm{Card}(I)}\}$. Since each F_{i_z} is a conjunction of sentences $(\rho(y, a_c), v_c) \in Q_{a_c}$, $a_c \in C$ in our problem setting, at first glance, the condition of the above decision rule seems to be very long. Note that we should use the relation Q_a, $a \in C$. Otherwise, it is not suitable for the meaning of the relation P_a because we approximate D_k by monotonone set operations of $U - P_a(v_a)$, $a \in C$, $v_a \in V_a$. Accordingly, we consider two cases; (c) $\mathcal{F} = \mathcal{Q}$ and (d) $\mathcal{F} = \mathcal{W}$.

By the construction of $\mathcal{Q}$ and $\mathcal{W}$, the condition part of the decision rule becomes simpler. This relies on the following fact. Suppose $\mathcal{F}_i^\cap = (U - (Q_{a_1}(x) \cap Q_{a_2}(x))) \cap (U - (Q_{a_3}(y) \cap Q_{a_4}(y)))$, where $x, y \in U$, $\{a_1, a_2\}$, $\{a_3, a_4\} \subseteq C$ and it is possible that $x = y$ and $\{a_1, a_2\} \cap \{a_3, a_4\} \neq \emptyset$. Then we have $\mathcal{F}_i^\cap = ((U - Q_{a_1}(x)) \cap (U - Q_{a_3}(y))) \cup ((U - Q_{a_1}(x)) \cap (U - Q_{a_4}(y))) \cup ((U - Q_{a_2}(x)) \cap (U - Q_{a_3}(y))) \cup ((U - Q_{a_2}(x)) \cap (U - Q_{a_4}(y)))$. Let $\mathcal{F}_{i1}^{\mathrm{sub}} = (U - Q_{a_1}(x)) \cap (U - Q_{a_3}(y))$, $\mathcal{F}_{i2}^{\mathrm{sub}} = (U - Q_{a_1}(x)) \cap (U - Q_{a_4}(y))$, $\mathcal{F}_{i3}^{\mathrm{sub}} = (U - Q_{a_2}(x)) \cap (U - Q_{a_3}(y))$ and $\mathcal{F}_{i4}^{\mathrm{sub}} = (U - Q_{a_2}(x)) \cap (U - Q_{a_4}(y))$. We have $\mathcal{F}_i^\cap = \mathcal{F}_{i1}^{\mathrm{sub}} \cup \mathcal{F}_{i2}^{\mathrm{sub}} \cup \mathcal{F}_{i3}^{\mathrm{sub}} \cup \mathcal{F}_{i4}^{\mathrm{sub}}$. This implies that the decision rule 'if $y \in \mathcal{F}_i^\cap$ then $y \in D_k$' can be decomposed to 'if $y \in \mathcal{F}_{ij}^{\mathrm{sub}}$ then $y \in D_k$', $j = 1, 2, 3, 4$.

From this observation, for $\mathcal{F}_j^\cap$, $j = 1, 2, \ldots, t_1$, we have the following body of if-then rules:

$$\text{if } y \in \mathcal{F}_{ji}^{\mathrm{sub}} \text{ then } y \in D_k, \; i = 1, 2, \ldots, i(j), \; j = 1, 2, \ldots, t_1,$$

where $\mathcal{F}_j^\cap = \bigcup_{i=1,2,\ldots,i(j)} \mathcal{F}_{ji}^{\mathrm{sub}}$. It can be seen that $\mathcal{F}_{ji}^{\mathrm{sub}}$, $i = 1, 2, \ldots, i(j)$, $j = 1, 2, \ldots, t_1$ include all maximal sets of the form $\bigcap_{a \subseteq C, \, x \in U}(U - Q_a(x))$ such that $\bigcap_{a \in A \subseteq C, \, x \in I \subseteq U}(U - Q_a(x)) \subseteq D_k$. This can be proved as follows. Suppose that $\mathcal{G}^{\mathrm{sub}}$ is one of the maximal sets which does not included in $\mathcal{F}_{ji}^{\mathrm{sub}}$, $i = 1, 2, \ldots, i(j)$, $j = 1, 2, \ldots, t_1$. By the construction of $\mathcal{Q}$, $\mathcal{G}^{\mathrm{sub}}$ is a member of $\mathcal{Q}$. This implies that there is a set $\bigcap_{A \subseteq C, \, x \in I \subseteq U}(U - Q_A(x))$ such that $\mathcal{G}^{\mathrm{sub}} \subseteq \bigcap_{A \subseteq C, \, x \in I \subseteq U}(U - Q_A(x)) \subseteq D_k$. This contradicts to the fact that $\mathcal{F}_j^\cap$, $j = 1, 2, \ldots, t_1$ are maximal. Hence, $\mathcal{F}_{ji}^{\mathrm{sub}}$, $i = 1, 2, \ldots, i(j)$, $j = 1, 2, \ldots, t_1$ include all maximal sets of the form $\bigcap_{a \subseteq C, \, x \in U}(U - Q_a(x))$ such that $\bigcap_{a \in A \subseteq C, \, x \in I \subseteq U}(U - Q_a(x)) \subseteq D_k$. The same discussion is valid in Case (d), i.e., $\mathcal{F} = \mathcal{W}$.

Therefore we consider the type of decision rule,

Case (c): if $(\rho(y, a_1), \bar{v}_1) \notin Q_{a_1}$ and $\cdots$ $(\rho(y, a_s), \bar{v}_s) \notin Q_{a_s}$ then $y \in D_k$,

Case (d): if $(\rho(y, a_1), v_1) \notin Q_{a_1}$ and $\cdots$ $(\rho(y, a_s), v_s) \notin Q_{a_s}$ then $y \in D_k$, where $\bar{v}_i = \rho(x_i, a_i)$, $x_i \in U$, $a_i \in C$, $i = 1, 2, \ldots, s$ and $v_i \in V_{a_i}$, $i = 1, 2, \ldots, s$. We should enumerate all minimal conditions of the decision rules above. This can be done also by a decision matrix method with modifications described below. In Case (c), let $K^+ = \{i \mid x_i \in \underline{Q}^{\cap}(D_k)\}$ and $K^- = \{i \mid x_i \notin D_k\}$. We define a decision matrix $\mathcal{M}^{\mathcal{Q}} = (M_{ij}^{\mathcal{Q}})$ by

$$M_{ij}^{\mathcal{Q}} = \{(a, v) \mid (\rho(x_j, a), v) \in Q_a, \ (\rho(x_i, a), v) \notin Q_a, \\ v = \rho(a, x), \ x \in U, \ a \in C\}, \ i \in K^+, \ j \in K^-. \quad (48)$$

Let $\neg\tilde{Q}((a, v))$ be a statement '$(\rho(y, a), v) \notin Q$'. Then the all minimal conditions are obtained as conjunctive terms in the disjunctive normal form of the following logical function:

$$B_k^{\mathcal{Q}} = \bigvee_{i \in K^+} \bigwedge_{j \in K^-} \bigvee \neg\tilde{Q}(M_{ij}^{\mathcal{Q}}).$$

In Case (d), let $K^+ = \{i \mid x_i \in \underline{W}_*^{\cap}(D_k)\}$ and $K^- = \{i \mid x_i \notin D_k\}$. We define a decision matrix $\mathcal{M}^{\mathcal{W}} = (M_{ij}^{\mathcal{W}})$ by

$$M_{ij}^{\mathcal{W}} = \{(a, v) \mid (\rho(x_j, a), v) \in Q_a, \ (\rho(x_i, a), v) \notin Q_a, \ v \in V_a, \ a \in C\}, \\ i \in K^+, \ j \in K^-. \quad (49)$$

All minimal conditions in all possible decision rules with respect to D_k are obtained as conjunctive terms in the disjunctive normal form of the following logical function:

$$B_k^{\mathcal{W}} = \bigvee_{i \in K^+} \bigwedge_{j \in K^-} \bigvee \neg\tilde{Q}(M_{ij}^{\mathcal{W}}). \quad (50)$$

6.5   Comparison and Correspondence between Definitions and Rules

As shown in the previous sections, the extracted decision rules are different by the underlying generalized rough sets. The correspondences between underlying generalized rough sets and types of decision rules are arranged in Table 5.

Table 5. Correspondence between generalized rough sets and types of decision rules

definition of rough set	type of decision rule
CP-rough set: $\{x \in X \mid P(x) \subseteq X\}$	if $\rho(x, a_1) = v_1$ and $\cdots$ and $\rho(x, a_p) = v_p$ then $x \in D_k$, if $(\rho(x, a_1), v_1) \in P_{a_1}$ and $\cdots$ and $(\rho(x, a_p), v_p) \in P_{a_p}$ then $x \in D_k$. (when P_a is transitive)
CN-rough set: $X \cap \left(U - \bigcup\{Q(x) \mid x \in U - X\}\right)$	if $\rho(x, a_1) = v_1$ and $\cdots$ and $\rho(x, a_p) = v_p$ then $x \in D_k$, if $(v_1, \rho(x, a_1)) \in Q_{a_1}$ and $\cdots$ and $(v_p, \rho(x, a_p)) \in Q_{a_p}$ then $x \in D_k$. (when Q_a is transitive)
AU-rough set: $\bigcup\{F_i \in \mathcal{F} \mid F_i \subseteq X\}$	if $(\rho(x, a_1), v_1) \in P_{a_1}$ and $\cdots$ and $(\rho(x, a_p), v_p) \in P_{a_p}$ then $x \in D_k$. (when $\mathcal{F} = \mathcal{P}$ of (35) or $\mathcal{F} = \mathcal{Z}$ of (39))
AI-rough set: (∗1)	if $(\rho(y, a_1), v_1) \notin Q_{a_1}$ and $\cdots$ and $(\rho(y, a_p), v_p) \notin Q_{a_p}$ then $y \in D_k$. (when $\mathcal{F} = \mathcal{Q}$ of (36) or $\mathcal{F} = \mathcal{W}$ of (40))

(∗1) $\bigcup\{\bigcap_{i \in I}(U - F_i) \mid \bigcap_{i \in I}(U - F_i) \subseteq X, \ I \subseteq \{1, \ldots, p + 1\}\}$

When P_a, $a \in C$ are reflexive and transitive, the type of decision rules are the same between CP- and AU-rough sets. However, extracted decision rules are not always the same. More specifically, condition parts of extracted decision rules based on CP-rough sets are the same as those based on AU-rough sets when $\mathcal{F} = \mathcal{P}$ of (35) but usually stronger than those based on AU-rough sets when $\mathcal{F} = \mathcal{Z}$ of (39). This is because we have $M_{ij}^{\mathrm{R}} = \tilde{M}_{ij} \subseteq \tilde{M}'_{ij}$. When P_a, $a \in C$ are only transitive, the extracted R-rules based on CP-rough sets are the same as extracted decision rules based on AU-rough sets with $\mathcal{F} = \mathcal{P}$. In this case, the extracted decision rules include id-rules. Namely, extracted decision rules based on CP-rough sets are more than those based on AU-rough sets.

While converse relations Q_a^{T}, $a \in C$ appear in extracted R-rules based on CN-rough sets when Q is transitive, complementary relations $(U \times U) - Q_a$, $a \in C$ appear in extracted decision rules based on AI-rough sets.

Table 6. Car evaluation

Car	fuel consumption (Fu)	selling price (Pr)	size (Si)	marketability (Ma)
Car1	medium	medium	medium	poor
Car2	high	medium	[medium,large]	poor
Car3	[medium,high]	low	[medium,large]	poor
Car4	low	[low,medium]	large	good
Car5	high	[low,high]	[small,medium]	poor
Car6	[low,medium]	low	[medium,large]	good

7   Simple Examples

Example 1.   Let us consider a decision table with interval attribute values about car evaluation, Table 6. An interval attribute value in this table shows that we do not know the exact value but the possible range within which the exact value exists. Among attribute values, we have orderings, low $\leq$ medium $\leq$ high and small $\leq$ medium $\leq$ large. Let us consider the decision class of good marketability, i.e., $D_1 = \{\text{Car4}, \text{Car6}\}$, and extract conditions of good marketability. A car with low fuel consumption, low selling price and large size is preferable. Therefore, we can define $P_F = \leq^{\mathrm{st}}$, $P_P = \leq^{\mathrm{st}}$ and $P_S = \geq^{\mathrm{st}}$, $Q_F = \geq^{\mathrm{st}}$, $Q_P = \geq^{\mathrm{st}}$ and $Q_S = \leq^{\mathrm{st}}$, where for intervals $E_1 = [\rho_1^{\mathrm{L}}, \rho_1^{\mathrm{R}}]$ and $E_2 = [\rho_2^{\mathrm{L}}, \rho_2^{\mathrm{R}}]$, we define $E_1 \leq^{\mathrm{st}} E_2 \Leftrightarrow \rho_1^{\mathrm{R}} \leq \rho_2^{\mathrm{L}}$ and $E_1 \geq^{\mathrm{st}} E_2 \Leftrightarrow \rho_1^{\mathrm{L}} \geq \rho_2^{\mathrm{R}}$. We consider $\mathcal{P}$ of (35), $\mathcal{Z}$ of (39), $\mathcal{Q}$ of (36) and $\mathcal{W}$ of (40). P and Q are not reflexive but transitive.

We obtain $P_{C*}(D_1) = \mathcal{P}_*^{\cup}(D_1) = \mathcal{Z}_*^{\cup}(D_1) = \mathcal{Q}_*^{\cap}(D_1) = \mathcal{W}_*^{\cap}(D_1) = \{\text{Car4}, \text{Car6}\}$, $\bar{Q}_{C*}(D1) = \{\text{Car4}\}$, where $C = \{Fu, Pr, Si\}$. Applying the proposed methods, we obtain the following decision rules:

$P_{C*}(D_1)$: if Pr=[low,medium] then Ma=good,

  if Fu=[low,medium] then Ma=good,

  if $Fu^{\mathrm{R}} \leq$ low then Ma=good,

  if $Si^{\mathrm{L}} \geq$ large then Ma=good,

  if $Fu^{\mathrm{R}} \leq$ medium and $Pr^{\mathrm{R}} \leq$ low then Ma=good,

$\bar{Q}_{C*}(D_1)$: if Fu=low then Ma=good,

  if Pr=[low,medium] then Ma=good,

  if $Fu^{\mathrm{L}} \leq$ low then Ma=good,

$\mathcal{P}^{\cup}_{*}(D_1)$: if $Fu^{\mathrm{R}} \leq$ low then Ma=good,

  if $Si^{\mathrm{L}} \geq$ large then Ma=good,

  if $Fu^{\mathrm{R}} \leq$ medium and $Pr^{\mathrm{R}} \leq$ low then Ma=good,

$\mathcal{Q}^{\cap}_{*}(D_1)$: if $Fu^{\mathrm{L}} <$ medium then Ma=good,

where we use $Fu = [Fu^{\mathrm{L}}, Fu^{\mathrm{R}}] = \rho(y, Fu)$, $Pr = [Pr^{\mathrm{L}}, Pr^{\mathrm{R}}] = \rho(y, Pr)$, $Si = [Si^{\mathrm{L}}, Si^{\mathrm{R}}] = \rho(y, Si)$ and $Ma = \rho(y, Ma)$ for convenience. Extracted decision rules based on $\mathcal{Z}^{\cup}_{*}(D_1)$ and $\mathcal{W}^{\cap}_{*}(D_1)$ are same as those based on $\mathcal{P}^{\cup}_{*}(D_1)$ and $\mathcal{Q}^{\cap}_{*}(D_1)$. We can observe the similarity between rules based on $P_{C*}(D_1)$ and $\mathcal{P}^{\cup}_{*}(D_1)$ and between rules based on $\bar{Q}_{C*}(D_1)$ and $\mathcal{Q}^{\cap}_{*}(D_1)$, respectively.

Table 7. Survivability of alpinists with respect to foods and tools

	foods (Fo)	tools (To)	survivability (Sur)
Alp1	$\{a\}$	$\{A, B\}$	low
Alp2	$\{a, b, c\}$	$\{A, B\}$	high
Alp3	$\{a, b\}$	$\{A\}$	low
Alp4	$\{b\}$	$\{A\}$	low
Alp5	$\{a, b\}$	$\{A, B\}$	high

Example 2. Consider an alpinist problem. There are three packages a, b and c of foods and two packages A and B of tools. When an alpinist climbs a mountain, he/she should carry foods and tools in order to be back safely. Assume the survivability Sur is determined by foods Fo and tools To packed in his/her knapsack and a set of data is given as in Table 7. Discarding the weight, we think that the more foods and tools, the higher the survivability. In this sense, we consider an inclusion relation $\supseteq$ for both attributes Fo and To. Namely, we adopt $\supseteq$ for the positively extensive relation P and $\subseteq$ for the negatively extensive relation Q. Since $\supseteq$ satisfies the reflexivity and transitivity and $\subseteq$ is the converse of $\supseteq$, all generalized rough sets described in this paper, i.e., CP-rough sets, CN-rough sets, AU-rough sets and AI-rough sets coincide one another. Indeed, for a class D_1 of Sur =high, we have $P_{C*}(D_1) = \bar{Q}_{C*}(D_1) = \mathcal{P}^{\cup}_{*}(D_1) = \mathcal{Z}^{\cup}_{*}(D_1) = \mathcal{Q}^{\cap}_{*}(D_1) = \mathcal{W}^{\cap}_{*}(D_1) = D_1 = \{\text{Alp2, Alp3}\}$, where $C = \{Fo, To, Sur\}$ and $\mathcal{P}$, $\mathcal{Q}$, $\mathcal{Z}$ and $\mathcal{W}$ are defined by (35), (36), (39) and (40).

  Extracting decision rules based on rough sets $P_{C*}(D_1)$, $\bar{Q}_{C*}(D_1)$, $\mathcal{P}^{\cup}_{*}(D_1)$, $\mathcal{Z}^{\cup}_{*}(D_1)$, $\mathcal{Q}^{\cap}_{*}(D_1)$ and $\mathcal{W}^{\cap}_{*}(D_1)$, we have the following decision rules;

$P_{C*}(D_1)$: if $Fo \supseteq \{a, b, c\}$ then $Sur = $ high,
         if $Fo \supseteq \{a, b\}$ and $To \supseteq \{A, B\}$ then $Sur = $ high,

$\mathcal{Z}_*^\cup(D_1)$: if $Fo \supseteq \{c\}$ then $Sur = $ high,
         if $Fo \supseteq \{b\}$ and $To \supseteq \{B\}$ then $Sur = $ high,

$\mathcal{Q}_*^\cap(D_1)$: if $Fo \not\subseteq \{a, b\}$ then $Sur = $ high,
         if $Fo \not\subseteq \{a\}$ and $To \not\subseteq \{A\}$ then $Sur = $ high,

where we use $Fo = \rho(y, Fo)$, $To = \rho(y, To)$ and $Sur = \rho(y, Sur)$ for convenience. Extracted decision rules based on $\bar{Q}_{C*}(D_1)$, $\mathcal{P}_*^\cup(D_1)$ and $\mathcal{W}_*^\cap(D_1)$ are same as those based on $\mathcal{P}_*^\cup(D_1)$, $\mathcal{P}_*^\cup(D_1)$ and $\mathcal{Q}_*^\cap(D_1)$, respectively.

Unlike the previous example, the extracted decision rules based on $\mathcal{Q}_*^\cap(D_1)$ are not very similar to those based on $\bar{Q}_{C*}(D_1)$, i.e., those based on $P_{C*}(D_1)$. This is because an inclusion relation $\subseteq$ is a partial order so that the negation of an inclusion relation is very different from the converse of the inclusion relation. As shown in this example, even if positive region, certain region and lower approximations coincide each other, the extracted if-then rules are different by underlying generalized rough sets.

8   Concluding Remarks

We have proposed four kinds of generalized rough sets based on two different interpretations of rough sets: rough sets as classification of objects into positive, negative and boundary regions and rough sets as approximation by means of elementary sets in a given family. We have described relationships of the proposed rough sets to the previous rough sets in general settings. Fundamental properties of the generalized rough sets have been investigated. Moreover relations among four generalized rough sets have been also discussed. Rule extraction based on the generalized rough sets has been proposed. We have shown the differences in the types of extracted decision rules by underlying rough sets. Rule extraction methods based on modified decision matrices have been proposed. A few numerical examples have been given to illustrate the differences among extracted decision rules. One of the examples has demonstrated that extracted decision rules can be different by underlying generalized rough sets even when positive region, certain region and lower approximations coincide one another.

For rule extraction, we did not utilize possible regions, conceivable regions and upper approximations. It would be possible to extract decision rules corresponding to those sets. The proposed rule extraction methods are all based on decision matrices and require a lot of computational effort. The other extraction methods like LERS [4] should be investigated. In this case, we should abandon to extract all decision rules but extract only useful decision rules or a minimal body of decision rules which covers all objects. In all methods proposed in this paper, we extracted all minimal conditions. This may increase the risk to give wrong conclusions for objects when we apply the obtained decision rules to infer conclusions of new objects. Risk and minimal descriptions of conditions are in a trade-off relation. We should investigate an extraction method of decision rules with moderate risk and sufficiently weak conditions. Those topics and applications to real world problems would be our future work.

References

1. Bonikowski, Z., Bryniarski, E., Wybraniec-Skardowska, U.: Extensions and intensions in the rough set theory. *Information Sciences* **107** (1998) 149–167
2. Dubois, D., Grzymala-Busse, J., Inuiguchi, M., Polkowski, L. (eds.): *Fuzzy Rough Sets: Fuzzy and Rough and Fuzzy along Rough.* Springer-Verlag, Berlin (to appear)
3. Greco, S., Matarazzo, B., Słowiński, R.: The use of rough sets and fuzzy sets in MCDM. in: Gal, T., Stewart, T. J., Hanne, T. (Eds.) *Multicriteria Decision Making: Advances in MCDM Models, Algorithms, Theory, and Applications*, Kluwer Academic Publishers, Boston, MA (1999) 14-1–14-59
4. Grzymala-Busse, J. W.: LERS: A system for learning from examples based on rough sets. in: Słowiński (ed.): *Intelligent Decision Support: Handbook pf Applications and Advances of the Rough Sets Theory*, Kluwer Academic Publishers, Dordrecht, (1992) 3–18.
5. Inuiguchi, M., Tanino, T.: On rough sets under generalized equivalence relations. *Bulletin of International Rough Set Society* **5**(1/2) (2001) 167–171
6. Inuiguchi, M., Tanino, T.: Generalized rough sets and rule extraction. in: Alpigini, J. J., Peters, J. F., Skowron, A., Zhong, N. (eds.): *Rough Sets and Current Trends in Computing*, Springer-Verlag, Berlin (2002) 105–112
7. Pawlak, Z.: *Rough Sets: Theoretical Aspects of Reasoning About Data*, Kluwer Academic Publishers, Boston, MA (1991)
8. Shan, N., Ziarko, W.: Data-based acquisition and incremental modification of classification rules. *Computational Intelligence* **11** (1995) 357–370
9. Skowron, A., Rauszer, C. M.: The discernibility matrix and functions in information systems. in: Słowiński, R. (ed.) *Intelligent Decision Support: Handbook of Applications and Advances of the Rough Sets Theory*, Kluwer Academic Publishers, Dordrecht (1992) 331–362
10. Słowiński, R., Vanderpooten, D.: A generalized definition of rough approximations based on similarity. *IEEE Transactions on Data and Knowledge Engineering* **12**(2) (2000) 331–336
11. Yao, Y.Y.: Two views of the theory of rough sets in finite universes. *International Journal of Approximate Reasoning* **15** (1996) 291–317
12. Yao, Y.Y.: Relational interpretations of neighborhood operators and rough set approximation operators. *Information Sciences* **111** (1998) 239–259
13. Yao, Y.Y., Lin, T.Y.: Generalization of rough sets using modal logics. *Intelligent Automation and Soft Computing* **2**(2) (1996) 103–120

Appendix: Proofs of Fundamental Properties

(a) The proof of (vi) in Table 2

When Q is the converse of P, we have $y \in P(x)$ if and only if $x \in Q(y)$. Then we obtain $Q^*(U - X) = (U - X) \cup \bigcup\{Q(x) \mid x \in U - X\} = (U - X) \cup \{x \in U \mid P(x) \cap (U - X) \neq \emptyset\}$. Hence we have

$$\bar{Q}_*(X) = U - Q^*(U - X) = X \cap \{x \in U \mid P(x) \cap (U - X) = \emptyset\}$$
$$= X \cap \{x \in U \mid P(x) \subseteq X\} = \{x \in X \mid P(x) \subseteq X\} = P_*(X).$$

The other equation can be obtained similarly.

(b) The proof of (vii) in Table 2

$$P^*(P_*(X)) = P_*(X) \cup \bigcup \{P(x) \mid x \in P_*(X)\}$$
$$= P_*(X) \cup \bigcup \{P(x) \mid P(x) \subseteq X, \ x \in X\} \subseteq X,$$
$$P_*(P^*(X)) = \{x \in P^*(X) \mid P(x) \subseteq P^*(X)\}$$
$$= P^*(X) \cap \left\{x \in U \mid P(x) \subseteq X \cup \bigcup \{P(x) \mid x \in X\}\right\} \supseteq X$$

are valid. Thus we have $X \supseteq P^*(P_*(X))$ and $X \subseteq P_*(P^*(X))$. This implies also $X \supseteq \bar{Q}^*(\bar{Q}_*(X))$ and $X \subseteq \bar{Q}_*(\bar{Q}^*(X))$ because we obtain $U - X \supseteq Q^*(Q_*(U - X))$ and $U - X \subseteq Q_*(Q^*(U - X))$. Hence, first four relations are obvious.

When P is transitive, $x \in P(y)$ implies $P(x) \subseteq P(y)$. Let $z \in P_*(X)$, i.e., $z \in X$ and $P(z) \subseteq X$. Suppose $z \notin P_*(P_*(X))$. Then we obtain $P(z) \not\subseteq P_*(X)$. Namely, there exists $y \in P(z)$ such that $y \notin P_*(X)$. Since $P(z) \subseteq X$, $y \in X$. Combining this with $y \notin P_*(X)$, we have $P(y) \not\subseteq X$. From the transitivity of P, $y \in P(z) \subseteq X$ implies $P(y) \subseteq X$. Contradiction. Therefore, we proved $P_*(X) \subseteq P_*(P_*(X))$. The opposite inclusion is obvious. Hence $P_*(P_*(X)) = P_*(X)$.

Now, let us prove $P^*(P^*(X)) = P^*(X)$ when P is transitive. It suffices to prove $P^*(P^*(X)) \subseteq P^*(X)$ since the opposite inclusion is obvious. Let $z \in P^*(P^*(X))$, i.e, (i) $z \in P^*(X)$ or (ii) there exists $y \in P^*(X)$ such that $z \in P(y)$. We prove $z \in P^*(X)$. Thus, in case of (i), it is straightforward. Consider case (ii). Since $y \in P^*(X)$, (iia) $y \in X$ or (iib) there exists $w \in X$ such that $y \in P(w)$. In case of (iia), we obtain $z \in P^*(X)$ from $z \in P(y)$. In case of (iib), from the transitivity of P, we have $P(y) \subseteq P(w)$. Combining this fact with $z \in P(y)$, $z \in P(w)$. Since $w \in X$, we obtain $z \in P^*(X)$. Therefore, in any case, we obtain $z \in P^*(X)$. Hence, $P^*(P^*(X)) = P^*(X)$.

The same properties with respect to a relation Q can be proved similarly.

When P is reflexive and transitive, we can prove $\{x \in U \mid P(x) \subseteq X\} = \bigcup \{P(x) \mid P(x) \subseteq X\}$. This equation can be proved in the following way. Let $y \in \bigcup \{P(x) \mid P(x) \subseteq X\}$. There exists $z \in U$ such that $y \in P(z) \subseteq X$. Because of the transitivity, $P(y) \subseteq P(z) \subseteq X$. This implies that $y \in \{x \in U \mid P(x) \subseteq X\}$. Hence, $\{x \in U \mid P(x) \subseteq X\} \supseteq \bigcup \{P(x) \mid P(x) \subseteq X\}$. The opposite inclusion is obvious from the reflexivity.

From the reflexivity, we have $P_*(X) = \{x \in X \mid P(x) \subseteq X\}$, $P^*(X) = \bigcup \{P(x) \mid x \in X\}$. Using these equations, we obtain

$$P^*(P_*(X)) = \bigcup \{P(x) \mid x \in P_*(x)\} = \bigcup \{P(x) \mid P(x) \subseteq X\} = P_*(X),$$
$$P_*(P^*(X)) = \{x \in X \mid P(x) \subseteq P^*(X)\} = \bigcup \{P(x) \mid P(x) \subseteq P^*(X)\}$$
$$= \bigcup \{P(x) \mid x \in X\} = P^*(X).$$

The properties with respect to the relation Q can be proved in the same way.

(c) The proof of (iii) in Table 3

The first and fourth inclusion relations are obvious. We prove $\mathcal{F}_\cup^*(X \cup Y) = \mathcal{F}_\cup^*(X) \cup \mathcal{F}_\cup^*(Y)$ only. The second equality can be proved by the duality (vi).

$\mathcal{F}_{\cup}^*(X \cup Y) \supseteq \mathcal{F}_{\cup}^*(X) \cup \mathcal{F}_{\cup}^*(Y)$ is straightforward. We prove the opposite inclusion. Let $x \in \mathcal{F}_{\cup}^*(X \cup Y)$. Suppose $x \notin \mathcal{F}_{\cup}^*(X)$ and $x \notin \mathcal{F}_{\cup}^*(Y)$. Then there exist $J, K \subseteq \{1, 2, \ldots, p\}$ such that $x \notin \bigcup_{j \in J} F_j \supseteq X$ and $x \notin \bigcup_{j \in K} F_j \supseteq Y$. This fact implies that $x \notin \bigcup_{j \in J} F_j \cup \bigcup_{j \in K} F_j = \bigcup_{j \in J \cup K} F_j \supseteq X \cup Y$. This contracts with $x \in \mathcal{F}_{\cup}^*(X \cup Y)$. Hence, we have $x \in \mathcal{F}_{\cup}^*(X) \cup \mathcal{F}_{\cup}^*(Y)$.

(d) The proof of (a) in Table 4

We only prove $P_*(X) \subseteq \mathcal{P}_*^{\cup}(X) = P^*(P_*(X)) \subseteq X \subseteq \mathcal{P}_{\cup}^*(X) \subseteq P^*(X)$ when P is reflexive. The other assertion can be proved similarly.

First inclusion and the inequality are obvious from the reflexivity. Relations with a set X is obtained from (i) in Table 3. From the reflexivity, we have

$$P^*(X) = \bigcup_{x \in X} P(x) \in \left\{ \bigcup_{x \in Y} P(x) \;\middle|\; X \subseteq \bigcup_{x \in Y} P(x), \; Y \subseteq U \right\}$$

Then the last inclusion is proved as follows:

$$P^*(X) \supseteq \bigcap \left\{ \bigcup_{x \in Y} P(x) \;\middle|\; X \subseteq \bigcup_{x \in Y} P(x), \; Y \subseteq U \right\} = \mathcal{P}_{\cup}^*(X).$$

(e) The proof of (b) in Table 4

We only prove the first part. The second part can be obtained similarly.

Let $x \in \mathcal{P}_*^{\cup}(X)$. There exists y such that $x \in P(y) \subseteq X$. Because of the transitivity, $P(x) \subseteq P(y)$. Therefore, $P(x) \subseteq X$. This fact together with $x \in X$ implies $x \in P_*(X)$. Hence, $\mathcal{P}_*^{\cup}(X) \subseteq P_*(X)$.

The relation $P_*(X) \subseteq X \subseteq P^*(X)$ has been given as (i) in Table 2.

Finally, we prove $P^*(X) \subseteq \mathcal{P}_{\cup}^*(X)$. Let $z \in X \subseteq \mathcal{P}_{\cup}^*(X)$. Then, for all W_i such that $X \subseteq \bigcup_{w \in W_i} P(w)$, there exists $w_i \in W_i$ such that $z \in P(w_i)$. By transitivity, $P(z) \subseteq P(w_i)$. Therefore,

$$P(z) \subseteq \bigcap \left\{ P(w_i) \;\middle|\; X \subseteq \bigcup_{w \in W_i} P(w) \right\} \subseteq \bigcap \left\{ \bigcup_{w \in W_i} P(w) \;\middle|\; X \subseteq \bigcup_{w \in W_i} P(w) \right\}$$

Hence, we have

$$P^*(X) = \bigcup_{z \in X} P(z) \subseteq \bigcap \left\{ \bigcup_{w \in W_i} P(w) \;\middle|\; X \subseteq \bigcup_{w \in W_i} P(w) \right\} = \mathcal{P}_{\cup}^*(X).$$

Towards Scalable Algorithms
for Discovering Rough Set Reducts

Marzena Kryszkiewicz[1] and Katarzyna Cichoń[1,2]

[1] Institute of Computer Science, Warsaw University of Technology
Nowowiejska 15/19, 00-665 Warsaw, Poland
`mkr@ii.pw.edu.pl`
[2] Institute of Electrical Apparatus, Technical University of Lodz
Stefanowskiego 18/22, 90-924 Lodz, Poland
`cichon@p.lodz.pl`

Abstract. Rough set theory allows one to find reducts from a decision table, which are minimal sets of attributes preserving the required quality of classification. In this article, we propose a number of algorithms for discovering all generalized reducts (preserving generalized decisions), all possible reducts (preserving upper approximations) and certain reducts (preserving lower approximations). The new *RAD* and *CoreRAD* algorithms, we propose, discover exact reducts. They require, however, the determination of all maximal attribute sets that are not supersets of reducts. In the case, when their determination is infeasible, we propose *GRA* and *CoreGRA* algorithms, which search approximate reducts. These two algorithms are well suited to the discovery of supersets of reducts from very large decision tables.

1 Introduction

Rough set theory has been conceived as a non-statistical tool for analysis of imperfect data [17]. Rough set methodology allows one to discover interesting data dependencies, decision rules, repetitive data patterns and to analyse conflict situations [24]. The reasoning in the rough set approach is based solely on available information. Objects are perceived as indiscernible if they have the same description in the system. This may be a reason for uncertainty. Two or more objects identically described in the system may belong to different classes (concepts). Such concepts, though vague, can be defined roughly by means of a pair of crisp sets: lower approximation and upper approximation. Lower approximation of a concept is a set of objects that surely belong to that concept, whereas upper approximation is a set of objects that possibly belong to that concept.

Rough set theory allows one to find reducts from a decision table, which are minimal sets of attributes preserving the required quality of classification. For example, a reduct may preserve lower approximations of decision classes, or upper approximations of decision classes, or both. A number of methods for discovering reducts have already been proposed in the literature [2-8, 11, 15-17, 20-31]. The most popular

J.F. Peters et al. (Eds.): Transactions on Rough Sets I, LNCS 3100, pp. 120–143, 2004.

methods are based on discernibility matrices [20]. Other methods are based, e.g., on the theory of cones and fences [7, 19]. Unfortunately, the existing methods are not capable to discover all reducts from very large decision tables, although research on discovering rough set decision rules in large data sets started a few years ago (see e.g., [9-10, 14]). One may try to overcome this problem either by applying heuristics or data sampling or both, or by restricting search to looking for some reducts instead of all of them.

Recently, we have proposed the *GRA*-like (*GeneralizedReductsApriori*) algorithms for discovering approximate generalized, possible and certain reducts from very large decision tables [13]. This article extends the results obtained in [13]. Here, we propose new algorithms - *RAD* and *CoreRAD* - for discovering exact generalized, possible and certain reducts. *CoreRAD* is a variation of *RAD*, which uses information on the so-called core in order to restrict the number of candidates for reducts and the number of scans of the decision table. The new algorithms require the determination of all maximal sets that are not supersets of reducts ($\mathcal{MNSR}$). The knowledge of $\mathcal{MNSR}$ is sufficient to evaluate candidates for reducts correctly. The method of creating and pruning candidates is very similar to the one proposed in *GRA* [13]. In the case, when the calculation of $\mathcal{MNSR}$ is infeasible, we advocate to search approximate reducts. In the article, we first introduce the theory behind approximate reducts and then present in detail respective algorithms (*GRA* and *CoreGRA*).

The layout of the article is as follows: In Section 2, we remind basic rough set notions and prove some of their properties that will be applied in the proposed algorithms. In Section 3, we propose the *RAD* algorithm for discovering generalized and possible reducts. A number of optimizations of the basic algorithm are discussed as well. The *CoreRAD* algorithm, which calculates both the core and the reducts, is offered in Section 4. In Section 5, we discuss briefly how to adapt *RAD* and *Core-RAD* for the discovery of certain reducts. The notions of approximate reducts are introduced in Section 6. We prove that approximate reducts are supersets of exact reducts. The properties of approximate generalized reducts are used in the construction of the *GRA* algorithm, which is presented in Section 7. In Section 8, we discuss the *CoreGRA* algorithm, which calculates both the approximate generalized reducts and the approximate core. In Section 9, we propose simple modifications of *GRA* and *CoreGRA* that enable the usage of these algorithms for discovering approximate certain reducts. Section 10 concludes the results indicating that the proposed solutions can be applied in the case of incomplete decision tables as well.

2 Basic Notions

2.1 Information Systems

An *information system* (*IS*) is a pair S = (O, *AT*), where O is a non-empty finite set of *objects* and *AT* is a non-empty finite set of *attributes*, such that $a: O \rightarrow V_a$ for any $a \in AT$, where V_a is called *domain* of the attribute a.

An attribute-value pair (a,v), where $a \in AT$ and $v \in V_a$, is called an *atomic descriptor*. An *atomic descriptor* or its conjunction is called a *descriptor* [20]. A conjunction of atomic descriptors for attributes $A \subseteq AT$ is called A-*descriptor*.

Let $S = (O, AT)$. Each subset of attributes $A \subseteq AT$ determines a binary *indiscernibility relation* $IND(A)$, $IND(A) = \{(x,y) \in O \times O| \ \forall a \in A, \ a(x) = a(y)\}$. The relation $IND(A)$, $A \subseteq AT$, is an equivalence relation and constitutes a partition of O. Objects indiscernible with regard to their description on attribute set A in the system will be denoted by $I_A(x)$; that is, $I_A(x) = \{y \in O| \ (x,y) \in IND(A)\}$.

Property 1 [9]. Let $A, B \subseteq AT$.
a) If $A \subseteq B$, then $I_B(x) \subseteq I_A(x)$.
b) $I_{A \cup B}(x) = I_A(x) \cap I_B(x)$.

c) $I_A(x) = \bigcap_{a \in A} I_a(x)$.

Let $X \subseteq O$ and $A \subseteq AT$. $\underline{A}X$ is defined as a *lower approximation* of X iff $\underline{A}X = \{x \in O| \ I_A(x) \subseteq X\} = \{x \in X \ | \ I_A(x) \subseteq X\}$. $\overline{A}X$ is defined as an *upper approximation* of X iff $\overline{A}X = \{x \in O| \ I_A(x) \cap X \neq \emptyset\} = \bigcup\{I_A(x)| \ x \in X\}$. $\underline{A}X$ is the set of objects that belong to X with certainty, while $\overline{A}X$ is the set of objects that possibly belong to X.

2.2   Decision Tables

A *decision table* is an information system $DT = (O, AT \cup \{d\})$, where $d \notin AT$ is a distinguished attribute called the *decision*, and the elements of AT are called *conditions*. The set of all objects whose decision value equals k, $k \in V_d$, will be denoted by X_k. Let us define the function $\partial_A: O \to P(V_d)$, $A \subseteq AT$, as follows [18]:

$$\partial_A(x) = \{d(y)| \ y \in I_A(x)\}.$$

∂_A will be called A-*generalized decision* in DT. For $A = AT$, an A-*generalized decision* will be also called briefly a *generalized decision*.

Table 1. $DT = (O, AT \cup \{f\})$ extended by generalized decision ∂_{AT}

Table 2. $DT' = (O, AT \cup \{\partial_{AT}\})$ – sorted and reduced version of DT from Table 1.

$x \in O$	a	b	c	D	e	f	∂_{AT}
1	1	0	0	1	1	1	{1}
2	1	1	1	1	2	1	{1}
3	0	1		1	0	3	1	{1,2}
4	0	1	1	0	3	2	{1,2}
5	0	1	1	2	2	2	{2}
6	1	1	0	2	2	2	{2,3}
7	1	1	0	2	2	3	{2,3}
8	1	1	0	3	2	3	{3}
9	1	0	0	3	2	3	{3}

$x \in O$ in DT' ($x \in O$ in DT)	a	b	c	d	e	∂_{AT}
1 (3,4)	0	1	1	0	3	{1,2}
2 (5)	0	1	1	2	2	{2}
3 (1)	1	0	0	1	1	{1}
4 (9)	1	0	0	3	2	{3}
5 (6,7)	1	1	0	2	2	{2,3}
6 (8)	1	1	0	3	2	{3}
7 (2)	1	1	1	1	2	{1}

Example 1. Table 1 describes a sample decision table DT. The conditional attributes are as follows: $AT = \{a, b, c, d, e\}$. The decision attribute is f. One may note that objects 3 and 4 are indiscernible with respect to the conditional attributes in AT.

Hence, ∂_{AT} for object 3 contains both the decision 1 for object 3, as well as the decision 2 for object 4. Analogously, ∂_{AT} for object 4 contains both its own decision (2), as well as the decision of object 3 (1). Please see the last column in Table 1 for generalized decision ∂_{AT} for all objects in DT. Let X_1 be the class of objects determined by decision 1; that is, $X_1 = \{1,2,3\}$. The lower and upper approximations of X_1 are as follows: $\underline{A}TX_1 = \{1,2\}$ and $\overline{A}TX_1 = \{1,2,3,4\}$.   □

Property 2 shows that the approximations of decision classes can be expressed by means of an A-generalized decision.

Property 2 [9-11]. Let $X_i \subseteq O$ and $A \subseteq AT$.
a) $I_A(x) \subseteq X_i$ iff $\partial_A(x) = \{i\}$.
b) $I_A(x) \cap X_i \neq \emptyset$ iff $i \in \partial_A(x)$.
c) $\underline{A}X_i = \{x \in O| \partial_A(x) = \{i\}\}$.
d) $\overline{A}X_i = \{x \in O| i \in \partial_A(x)\}$.
e) $\partial_A(x) = \partial_A(y)$ for any $(x,y) \in IND(A)$.

By Property 2e, objects having the same A-descriptor have also the same A-generalized decision value; that is, the A-descriptor uniquely determines the A-generalized decision value for all objects satisfying this descriptor. In the sequel, the A-generalized decision value determined by A-descriptor t, such that t is satisfied by at least one object in the system, will be denoted by ∂_t. Table 2 shows the generalized decision values determined by atomic descriptors that occur in Table 1.

Table 3. Generalized decision values $\partial_{(a,v)}$ determined by atomic descriptors (a,v), where $a \in AT$, $v \in V_a$, supported by DT from Table 1.

(a,v)	$(a,0)$	$(a,1)$	$(b,0)$	$(b,1)$	$(c,0)$	$(c,1)$	$(d,0)$	$(d,1)$	$(d,2)$	$(d,3)$	$(e,1)$	$(e,2)$	$(e,3)$
$\partial_{(a,v)}$	$\{1,2\}$	$\{1,2,3\}$	$\{1,3\}$	$\{1,2,3\}$	$\{1,2,3\}$	$\{1,2\}$	$\{1,2\}$	$\{1\}$	$\{2,3\}$	$\{3\}$	$\{1\}$	$\{1,2,3\}$	$\{1,2\}$

We note that the A- and B-generalized decision values for object x provide an upper bound on the $A \cup B$-generalized decision value for x.

Property 3 [13]. Let $A,B \subseteq AT$, $x \in DT$. $\partial_{A \cup B}(x) \subseteq \partial_A(x) \cap \partial_B(x)$.

Proof: $\partial_{A \cup B}(x) = \{d(y)| y \in I_{A \cup B}(x)\} = $ /* by Property 1b */ $= \{d(y)| y \in (I_A(x) \cap I_B(x))\}$ $\subseteq \{d(y)| y \in I_A(x)\} \cap \{d(y)| y \in I_B(x)\} = \partial_A(x) \cap \partial_B(x)$.   □

We conclude further that the elementary a-generalized decision values for x, $a \in A$, can be used for calculating an upper bound on the A-generalized decision value for x.

Corollary 1. Let $A \subseteq AT$ and $x \in DT$. $\partial_A(x) \subseteq \bigcap_{a \in A} \partial_a(x) = \bigcap_{a \in A} \partial_{(a, a(x))}$.

Example 2. The $\{ce\}$-generalized decision value calculated from DT in Table 1 for object 5 ($\partial_{\{ce\}}(5) = \{1,2\}$) equals its upper bound $\partial_c(5) \cap \partial_e(5) = \partial_{(c,1)} \cap \partial_{(e,2)} = \{1,2\}$ $\cap \{1,2,3\} = \{1,2\}$. On the other hand, the $\{ce\}$-generalized decision value for object 6 ($\partial_{\{ce\}}(6) = \{2,3\}$) is a proper subset of its upper bound $\partial_c(6) \cap \partial_e(6) = \partial_{(c,0)} \cap \partial_{(e,2)}$ $= \{1,2,3\} \cap \{1,2,3\} = \{1,2,3\}$.   □

Corollary 2. Let $A \subseteq B \subseteq AT$, $x \in DT$. $\partial_B(x) \subseteq \partial_A(x)$.

Proof: By Property 3, $\partial_B(x) \subseteq \partial_A(x) \cap \partial_{B \setminus A}(x)$. Hence, $\partial_B(x) \subseteq \partial_A(x)$. □

Finally, we observe that A- and B-generalized decision values for object x, where $A \subseteq B \subseteq AT$, are identical when their cardinalities are identical.

Proposition 1. Let $A \subseteq B \subseteq AT$ and $x \in DT$. $\partial_A(x) = \partial_B(x)$ iff $|\partial_A(x)| = |\partial_B(x)|$.

Proof: ($\Rightarrow$) Straightforward.

($\Leftarrow$) Let $|\partial_A(x)| = |\partial_B(x)|$ (*). Since, $A \subseteq B$, then by Corollary 2, $\partial_A(x) \supseteq \partial_B(x)$. Taking into account (*), we conclude $\partial_A(x) = \partial_B(x)$. □

2.3 Reducts for Decision Tables

Reducts for decision tables are minimal sets of conditional attributes that preserve the required properties of classification. In what follows, we provide definitions of reducts preserving lower and upper approximations of decision classes and objects' generalized decisions, respectively.

Let $\emptyset \neq A \subseteq AT$. A is a *certain reduct (c-reduct)* of DT iff A is a minimal attribute set such that

$$\forall x \in O,\ x \in \underline{ATX}_{d(x)} \Rightarrow I_A(x) \subseteq X_{d(x)} \tag{c}$$

A certain reduct is a set of attributes that allows us to distinguish each object x belonging to the lower approximation of its decision class in DT from the objects that do not belong to this approximation.

A is a *possible reduct (p-reduct)* of DT iff A is a minimal attribute set such that

$$\forall x \in O,\ I_A(x) \subseteq \overline{ATX}_{d(x)} \tag{p}$$

A possible reduct is a set of attributes that allows us to distinguish each object x in DT from objects that do not belong to the upper approximation of its decision class.

A is a *generalized decision reduct (g-reduct)* of DT iff A is a minimal set such that

$$\forall x \in O,\ \partial_A(x) = \partial_{AT}(x) \tag{g}$$

A generalized decision reduct is a set of attributes that preserves the generalized decision value for each object x in DT. In the sequel, a superset of a t-reduct, where $t \in \{c, p, g\}$, will be called a *t-super-reduct*.

Corollary 3. AT is a superset of all c-reducts, p-reducts, and g-reducts for any DT.

Proposition 2. Let $A \subseteq AT$.
a) If A satisfies property (**c**), then all of its supersets satisfy property (**c**).
b) If A does not satisfy property (**c**), then all of its subsets do not satisfy (**c**).
c) If A satisfies property (**p**), then all of its supersets satisfy property (**p**).
d) If A does not satisfy property (**p**), then all of its subsets do not satisfy (**p**).
e) If A satisfies property (**g**), then all of its supersets satisfy property (**g**).
f) If A does not satisfy property (**g**), then all of its subsets do not satisfy (**g**).

Proof: Let $A \subseteq B \subseteq AT$ and $x \in O$.

Ad a) Let A satisfy property (**c**) and $x \in \underline{AT}X_{d(x)}$. We are to prove that $I_B(x) \subseteq X_{d(x)}$. Since A satisfies property (**c**), then $I_A(x) \subseteq X_{d(x)}$ (*). By Property 1a, $I_B(x) \subseteq I_A(x)$ (**). By (*) and (**), $I_B(x) \subseteq X_{d(x)}$.

Ad b) Analogous to a).

Ad c) Let A satisfy property (**g**). We are to prove that $\partial_B(x) = \partial_{AT}(x)$. Since A satisfies property (**g**), then $\partial_A(x) = \partial_{AT}(x)$ (*). By Corollary 2, $\partial_{AT}(x) \subseteq \partial_B(x) \subseteq \partial_A(x)$ (**). By (*) and (**), $\partial_B(x) = \partial_{AT}(x)$.

Ad b, d, f) Follow immediately from Proposition 2a, b, c, respectively.   □

Corollary 4.
a) c-super-reducts are all and the only attribute sets that satisfy property (**c**).
b) p-super-reducts are all and the only attribute sets that satisfy property (**p**).
c) g-super-reducts are all and the only attribute sets that satisfy property (**g**).

Proof: By definition of reducts and Proposition 2.   □

Interestingly, not only g-reducts, but also p-reducts and c-reducts, can be determined by examining generalized decisions.

Theorem 1 [11]. The set of all generalized decision reducts of DT equals the set of all possible reducts of DT.

Lemma 1 [13]. $A \subseteq AT$ is a c-reduct of DT iff A is a minimal set such that $\forall x \in O, \partial_{AT}(x) = \{d(x)\} \Rightarrow \partial_A(x) = \{d(x)\}$.

Proof: By Property 2a,c.   □

Corollary 5 [13]. $A \subseteq AT$ is a c-reduct of DT iff A is a minimal set such that $\forall x \in O, \partial_{AT}(x) = \{d(x)\} \Rightarrow \partial_A(x) = \partial_{AT}(x)$.

2.4   Core

The notion of a *core* is meant to be the greatest set of attributes without which an attribute set does not satisfy the required classification property (i.e. is not a super-reduct). The generic notion of a t-core, $t \in \{c, p, g\}$, corresponding to c-reducts, p-reducts and g-reducts, respectively, is defined as follows:

$$t\text{-}core = \{a \in AT |\; AT\backslash\{a\} \text{ is not a } t\text{-super-reduct}\}.$$

Clearly, the p-core and g-core are the same.

Proposition 3. Let $\mathcal{R}$ be all reducts of the same type t, where $t \in \{c, p, g\}$.

$$t\text{-}core = \cap \mathcal{R}.$$

Proof: Let us consider the case when $\mathcal{R}$ is the set of all c-reducts. Let $b \in c$-core. Hence b is an attribute in AT such that $AT\backslash\{b\}$ is not a superset of c-reduct. By Corollary 4a and Proposition 2b, no attribute set without b satisfies property (**c**). Hence, no attribute set without b is a c-reduct. Thus, all c-reducts contain b; that is, $\cap \mathcal{R} \supseteq \{b\}$.

Generalizing this observation, $\cap \mathcal{R} \supseteq c$-core.

Now, we will prove by contradiction that $\bigcap \mathcal{R} \setminus c$-core is an empty set. Let $d \in \bigcap \mathcal{R}$ and $d \notin c$-core. Since $d \notin c$-core, then, by definition of a core, $AT \setminus \{d\}$ is a superset of some c-reduct, say B. Since B is a subset of $AT \setminus \{d\}$, then B does not contain d either. This means that among c-reducts, there is an attribute set (B), which does not contain d. Therefore, $d \notin \bigcap \mathcal{R}$, which contradicts the assumption.

The cases when $\mathcal{R}$ is the set of all p-reducts or g-reducts can be proved analogously from Corollary 4b,c and Proposition 2d,f, respectively.   □

3 Discovering Generalized Reducts

3.1 Main Algorithm

Notation for *RAD*

- $\mathcal{R}_k$ – candidate k attribute sets (potential g-reducts);
- $\mathcal{A}_k$ – k attribute sets that are not g-super-reducts;
- $\mathcal{MNSR}$ – all maximal conditional attribute sets that are not g-super-reducts;
- $\mathcal{MNSR}_k$ – k attribute sets in $\mathcal{MNSR}$;
- DT' – reduced DT;
- $x.a$ – the value of an attribute a for object x;
- $x.\partial_{AT}$ – the generalized decision value for object x.

Algorithm. *RAD*;

$DT' = GenDecRepresentation\text{-}of\text{-}DT(DT)$;
$\mathcal{MNSR} = MaximalNonSuperReducts(DT')$;
/* search g-reducts - note: g-reducts are all attribute sets that are not subsets of any set in $\mathcal{MNSR}$ */
if $|\mathcal{MNSR}_{|AT|-1}| = |AT|$ **then return** AT;   // optional optimizing step 1
$\mathcal{R}_1 = \{\{a\} \mid a \in AT\}$; $\mathcal{A}_1 = \{\}$;   // initialize 1 attribute candidates for g-reducts
forall $B \in \mathcal{MNSR}$ **do** move subsets of B from $\mathcal{R}_1$ to $\mathcal{A}_1$; // subsets of non-super-reducts are not reducts
for $(k = 1; \mathcal{A}_k \neq \{\}; k++)$ **do begin**
    if $|\mathcal{MNSR}| = 1$ **then return** $\cup_k \mathcal{R}_k$;   // optional optimizing step 2
    $\mathcal{MNSR} = \mathcal{MNSR} \setminus \mathcal{MNSR}_k$;   // $\mathcal{MNSR}_k$ is not useful any more – optional optimizing step 3
    /* create $k+1$ attribute g-reducts $\mathcal{R}_{k+1}$ and non-g-super-reducts $\mathcal{A}_{k+1}$ from $\mathcal{A}_k$ and $\mathcal{MNSR}$ */
    $RADGen(\mathcal{R}_{k+1}, \mathcal{A}_{k+1}, \mathcal{A}_k, \mathcal{MNSR})$;
endfor;
return $\cup_k \mathcal{R}_k$;

The *RAD* (*ReductsAprioriDiscovery*) algorithm we propose starts by determining the reduced decision table DT' that stores only conditional attributes AT and the AT-generalized decision for each object in DT instead of the original decision (see Section 3.2 for the description of the *GenDecRepresentation-of-DT* function). Each class of objects indiscernible w.r.t. $AT \cup \{\partial_{AT}\}$ in DT (see Table 1) is represented by one object in DT' (see Table 2). Next, DT' is examined in order to find all maximal attribute sets $\mathcal{MNSR}$ that are not g-super-reducts (see Section 3.3 for the description of the *MaximalNonSuperReducts* function). The information on $\mathcal{MNSR}$ is sufficient to derive all g-reducts; namely, g-reducts are these sets each of which has no superset in $\mathcal{MNSR}$ (i.e., is a g-super-reduct), but all proper subsets of which have supersets in $\mathcal{MNSR}$ (i.e., are not g-reducts).

Now, *RAD* creates initial candidates for *g*-reducts that are singleton sets and are stored in $\mathcal{R}_1$. The candidates in $\mathcal{R}_1$ that are subsets of $\mathcal{MNSR}$ are moved to 1 attribute non-*g*-super-reducts $\mathcal{A}_1$. The main loop starts. In each *k*-th iteration, $k \geq 1$, *k*+1 attribute candidates $\mathcal{R}_{k+1}$ are created from *k* attribute sets in $\mathcal{A}_k$, which are not *g*-super-reducts (see Section 3.4 for the description of the *RADGen* procedure). The information on non-*g*-super-reducts $\mathcal{MNSR}$ is used to prune candidates in $\mathcal{R}_{k+1}$. Namely, each candidate in $\mathcal{R}_{k+1}$ that has a superset in $\mathcal{MNSR}$ is not a *g*-super-reduct. Therefore it is moved from $\mathcal{R}_{k+1}$ to $\mathcal{A}_{k+1}$. The algorithm stops when $\mathcal{A}_k = \{\}$. Optional optimizing steps in *RAD* are discussed in Section 3.5.

3.2   Determining Generalized Decision Representation of Decision Table

The *GenDecRepresentation-of-DT* function starts with sorting the given decision table *DT* w.r.t. the set of all conditional attributes and (optionally) the decision attribute. The sorting enables fast determination of the generalized decision values for all classes of objects indiscernible w.r.t. *AT*. Each such class will be represented by one object in the new decision table $DT' = (AT, \{\partial_{AT}\})$, where the decision attribute is replaced by the generalized decision.

---

function *GenDecRepresentation-of-DT*(decision table *DT*);

    $DT' = \{\}$;
    sort *DT* with respect to *AT* and *d*;   // apply any ordering of attributes in *AT*, e.g. lexicographical
    x = first object in *DT*;   // or **null** if *DT* is empty
    while *x* is not **null do begin**
        forall $a \in AT$ **do** $x'.a = x.a$; $x'.\partial_{AT} = \{d(y) | y \in I_{AT}(x)\}$; add *x'* to *DT'*;
        x = the first object located just after $I_{AT}(x)$ in *DT*;
    endwhile;
    return *DT'*;

3.3   Calculating Maximal Non-super-reducts

The purpose of the *MaximalNonSuperReducts* function is to determine all maximal conditional attribute sets that are not *g*-super-reducts. To this end, each object in the reduced decision table *DT'* is compared with all other objects from different generalized decision classes. The result of the comparison of two objects, say *x* and *y*, belonging to different classes is the set of all attributes on which *x* and *y* are indiscernible. Clearly, such a resulting set is not a *g*-super-reduct, since it does not discern at least one pair of objects belonging to different generalized decision classes. The comparison results, which are non-*g*-super-reducts, are stored in the $\mathcal{NSR}$ variable. After the comparison of objects is accomplished, $\mathcal{NSR}$ contains a superset of all maximal non-*g*-super-reducts. The function returns MAX($\mathcal{NSR}$), which can be calculated as the final step or on the fly. For *DT'* from Table 2, *MaximalNonSuperReducts* will find $\mathcal{NSR} = \{abc, b, bc, e, bde, be, bce, ac, ace, ae, abce, abe\}$, and eventually will return MAX($\mathcal{NSR}$) = $\{abce, bde\}$.

function *MaximalNonSuperReducts*(reduced decision table *DT'*);

$\mathcal{NSR} = \{\}$;
  forall objects x in *DT'* **do**
    forall objects y following x in *DT'* **do**
    if $x.\partial_{AT} \neq y.\partial_{AT}$ **then**
      /* objects x and y should be distinguishable as they belong to different generalized decision classes; */
      /* the set $\{a \in AT \mid x.a = y.a\}$ is not a g-super-reduct since it does not distinguish between x and y */
      insert in $\{a \in AT \mid x.a = y.a\}$, if non-empty, to $\mathcal{NSR}$;
return MAX($\mathcal{NSR}$);   // note: MAX($\mathcal{NSR}$) contains all maximal non-g-super-reducts

3.4   Generating Candidates for Reducts

The *RADGen* procedure has 4 arguments. Two of them are input ones: k attribute non-g-super-reducts $\mathcal{A}_k$ and the maximal non-g-super-reducts $\mathcal{MNSR}$. The two remaining candidates $\mathcal{R}_{k+1}$ and $\mathcal{A}_{k+1}$ are output ones. After the completion of the function, $\mathcal{R}_{k+1}$ contains $k+1$ attribute g-reducts and $\mathcal{A}_{k+1}$ contains $k+1$ attribute non-g-super-reducts. During the first phase of the procedure, new $k+1$ attribute candidates are created by merging k attribute non-g-super-reducts in $\mathcal{A}_k$ that differ only in their final attributes. The characteristic feature of such a method of creating candidates is that no candidate that is likely to be a solution (here: g-reduct) is missed and that no candidate is generated twice (please, see the detailed description of the *Apriori* algorithm [1] for justification). In the second phase, it is checked for each newly obtained $k+1$ attribute candidate whether all its proper k attribute subsets are contained in non-g-super-reducts $\mathcal{A}_k$. If yes, then a candidate remains in $\mathcal{R}_{k+1}$; otherwise it is pruned as a proper superset of some g-super-reduct. Finally, all candidates in $\mathcal{R}_{k+1}$ that are subsets of maximal non-g-super-reducts $\mathcal{MNSR}$ are found non-g-super-reducts too, and thus are moved to $\mathcal{A}_{k+1}$.

procedure *RADGen*(**var** $\mathcal{R}_{k+1}$, **var** $\mathcal{A}_{k+1}$, **in** $\mathcal{A}_k$, **in** $\mathcal{MNSR}$);

  forall $B, C \in \mathcal{A}_k$ **do**   /* Merging */
    if $B[1] = C[1] \wedge ... \wedge B[k-1] = C[k-1] \wedge B[k] < C[k]$ **then begin**
    $A = B[1] \bullet B[2] \bullet ... \bullet B[k] \bullet C[k]$;   add A to $\mathcal{R}_{k+1}$;
    endif;
  forall $A \in \mathcal{R}_{k+1}$ **do**   /* Pruning */
    forall k attribute sets $B \subset A$ **do**
      if $B \notin \mathcal{A}_k$ **then** delete A from $\mathcal{R}_{k+1}$;   // A is a proper superset of g-super-reduct B
  forall $B \in \mathcal{MNSR}$ **do** move subsets of B from $\mathcal{R}_{k+1}$ to $\mathcal{A}_{k+1}$; /* Removing subsets of non-g-super-reducts */
return;

3.5   Optimizing Steps in *RAD*

In the main algorithm, we offer an optimization that may speed up checking which candidates are not g-reducts (optimizing step 3) and two optimizations for reducing the number of useless iterations (optimizing steps 1 and 2).

In step 3, k attribute sets are deleted from $\mathcal{MNSR}$ since they are useless for identifying non-g-superset-reducts among l attribute candidates, where $l > k$.

Optimizing step 1 is based on the following observation: the condition $|\mathcal{MNSR}_{|AT|-1}|$ = $|AT|$ implies that all $AT\backslash\{a\}$ sets are not g-super-reducts. Hence, AT is the only g-reduct for DT and thus the algorithm can be stopped.

Optimizing step 2 can be applied when $|\mathcal{MNSR}| = 1$. This condition implies that all sets in $\mathcal{A}_k$, which are not g-super-reducts, have exactly one - the same superset, say B, in maximal non-g-super-reducts $\mathcal{MNSR}$. If one continues the creation of $k+1$ attribute candidates $\mathcal{R}_{k+1}$ by merging sets in $\mathcal{A}_k$, then the new $k+1$ attribute candidates would be still subsets of B. Hence, they would be pruned by the $RADGen$ procedure from $\mathcal{R}_{k+1}$ to $\mathcal{A}_{k+1}$. As a result, one would obtain $\mathcal{R}_{k+1} = \{\}$ and $|\mathcal{MNSR}| =$ 1. Such a scenario would continue when creating longer candidates until $\mathcal{A}_l = \{B\}$, $l > k$. Then, $RADGen$ will produce empty $\mathcal{R}_{l+1}$ and empty $\mathcal{A}_{l+1}$; that is, the condition, which stops the RAD algorithm. In conclusion, the condition $|\mathcal{MNSR}| = 1$ implies that no more g-reducts will be discovered, so the algorithm can be stopped.

3.6 Illustration of *RAD*

Let us illustrate now the discovery of g-reducts of DT from Table 1. We assume that maximal non-g-super-reducts $\mathcal{MNSR}$ are already found and are equal to $\{\{abce\},$ $\{bde\}\}$. Table 4 shows how candidates for g-reducts change in each iteration.

Table 4. $\mathcal{R}_k$ and $\mathcal{A}_k$ after verification w.r.t. $\mathcal{MNSR}$ in subsequent iterations of *New*.

k	$\mathcal{A}_k$ (each X in $\mathcal{A}_k$ has a superset in $\mathcal{MNSR}$)	$\mathcal{R}_k$ (each X in $\mathcal{R}_k$ has no superset in $\mathcal{MNSR}$)
1	$\{a\}, \{b\}, \{c\}, \{d\}, \{e\}$	
2	$\{ab\}, \{ac\}, \{ae\}, \{bc\}, \{bd\}, \{be\}, \{ce\}, \{de\}$	$\{ad\}, \{cd\}$
3	$\{abc\}, \{abe\}, \{ace\}, \{bce\}, \{bde\}$	
4	$\{abce\}$	

4 Core-Oriented Discovery of Generalized Reducts

4.1 Main Algorithm

In this section, we offer the *CoreRAD* procedure, which finds not only g-reducts, but also their core. The layout of *CoreRAD* reminds that of *RAD*. *CoreRAD*, however, differs from *RAD* in that it first checks if the set of all maximal non-g-super-reducts $\mathcal{MNSR}$ is empty. If yes, then each single conditional attribute is a g-reduct, so *CoreRAD* returns $\{\{a\}|\ a \in AT\}$ as the set of all g-reducts and $\bigcap_{a \in AT} \{a\} = \emptyset$ as the g-core (by Proposition 3). Otherwise, *CoreRAD* determines the g-core by definition from all maximal $|AT|-1$ non-g-super-reducts in $\mathcal{MNSR}$. All sets in $\mathcal{MNSR}$ that are not supersets of the g-core are deleted, since the only candidates considered in *CoreRAD* will be the g-core and its supersets. If the reduced $\mathcal{MNSR}$ is an empty set, then the g-core does not have subsets in $\mathcal{MNSR}$ and thus it is the only g-reduct. Otherwise, the g-core is not a g-reduct, and the new candidates $\mathcal{R}_{|core|+1}$ are created by merging the g-core with the remaining attributes in AT. Clearly, the new candidates which have supersets in maximal non-g-super-reducts $\mathcal{MNSR}$ are not g-reducts either, and hence are moved from $\mathcal{R}_{|core|+1}$ to $\mathcal{A}_{|core|+1}$. From now on, *CoreRAD* is performed in the same way as *RAD*.

Algorithm. *CoreRAD*;

$DT' = GenDecRepresentation\text{-}of\text{-}DT(DT)$;
$\mathcal{MNSR} = MaximalNonSuperReducts(DT')$;
if $\mathcal{MNSR} = \{ \}$ **then return** $(\varnothing, \{\{a\}|\ a{\in}AT\})$;   // each conditional attribute is a g-reduct
$core = \varnothing$;
forall $A{\in}\mathcal{MNSR}_{|AT|-1}$ **do begin** $\{a\} = AT{\backslash}A$; $core = core \cup \{a\}$ **endfor**;
if $|\mathcal{MNSR}_{|AT|-1}| = |AT|$ **then return** (AT, AT);   // or **if** $core = AT$ **then** - optional optimizing step 1
$\mathcal{MNSR} = \{B \in \mathcal{MNSR}| B \supseteq core\}$;   // g-reducts are supersets of the g-core
if $\mathcal{MNSR} = \{ \}$ **then return** $(core, \{core\})$; // g-core is a g-reduct as there is no its superset in $\mathcal{MNSR}$
$\mathcal{MNSR} = \mathcal{MNSR} \setminus \mathcal{MNSR}_{|core|}$;   // or equivalently $\mathcal{MNSR} = \mathcal{MNSR} \setminus \{core\}$;
/* initialize candidate for reducts as g-core's supersets */
$startLevel = |core| + 1$; $\mathcal{R}_{startLevel} = \{ \}$; $\mathcal{A}_{startLevel} = \{ \}$;
forall $a{\in}AT \setminus core$ **do begin** $A = core \cup \{a\}$; $\mathcal{R}_{startLevel} = \mathcal{R}_{startLevel} \cup \{A\}$ **endfor**;
forall $B \in \mathcal{MNSR}$ **do** move subsets of B from $\mathcal{R}_{startLevel}$ to $\mathcal{A}_{startLevel}$;
for $(k = startLevel;\ \mathcal{A}_k \neq \{ \};\ k{+}{+})$ **do begin**
  if $|\mathcal{MNSR}| = 1$ **then return** $(core, \cup_k \mathcal{R}_k)$;   // optional optimizing step 2
  $\mathcal{MNSR} = \mathcal{MNSR} \setminus \mathcal{MNSR}_k$;   // $\mathcal{MNSR}_k$ is not useful any more – optional optimizing step 3
  /* create $k{+}1$ attribute g-reducts $\mathcal{R}_{k+1}$ and non-g-super-reducts $\mathcal{A}_{k+1}$ from $\mathcal{A}_k$ and $\mathcal{MNSR}$ */
  $GRAGen(\mathcal{R}_{k+1}, \mathcal{A}_{k+1}, \mathcal{A}_k, \mathcal{MNSR})$;
endfor;
return $(core, \cup_k \mathcal{R}_k)$;

4.2 Illustration of *CoreRAD*

We will illustrate now the core-oriented discovery of g-reducts of DT from Table 1. We assume that $\mathcal{MNSR}$ has already been calculated and equals $\{\{abce\}, \{bde\}\}$. Hence, $core = AT / \{abce\} = \{d\}$. Now, we leave only the supersets of the $core$ in $\mathcal{MNSR}$; thus $\mathcal{MNSR}$ becomes equal to $\{\{bde\}\}$. Table 5 shows how candidates for g-reducts change in each iteration (here: only 1 iteration was sufficient).

Table 5. $\mathcal{R}_k$ and $\mathcal{A}_k$ after verification w.r.t. $\mathcal{MNSR}$ in subsequent iterations of *CoreRAD*.

K	$\mathcal{A}_k$ (each X in $\mathcal{A}_k$ has a superset in $\mathcal{MNSR}$)	$\mathcal{R}_k$ (each X in $\mathcal{R}_k$ has no superset in $\mathcal{MNSR}$)
2	$\{bd\}, \{de\}$	$\{ad\}, \{cd\}$

5 Discovering Certain Reducts

RAD and *CoreRAD* can easily be adapted for the discovery of certain reducts. It suffices to modify line 4 of the *MaximalNonSuperReducts* function as follows:

$$\textbf{if } (x.\partial_{AT} \neq y.\partial_{AT}) \textbf{ and } (|x.\partial_{AT}| = 1 \textbf{ or } |y.\partial_{AT}| = 1) \textbf{ then}$$

This modification guarantees that all objects from lower approximations of all decision classes, which have singleton generalized decisions, will be compared with all objects not belonging to the lower approximations of their decision classes.

6 Approximate Attribute Reduction

6.1 Approximate Reducts for Decision Table

The discovery of reducts may be very time consuming. Therefore, one may resign from calculating strict reducts and search more efficiently for approximate reducts, which however, should be supersets of exact reducts and subsets of AT. In this section, we introduce the notion of such approximate reducts based on the observation that for any object x in O: $\bigcap_{a \in A} \partial_a(x) \supseteq \partial_A(x)$ (by Corollary 1).

Let $\varnothing \neq A \subseteq AT$. AT is defined an approximate generalized decision reduct (ag-reduct) of DT iff $\exists x \in O, \bigcap_{a \in AT} \partial_a(x) \supset \partial_{AT}(x)$. Otherwise, A is an approximate generalized decision reduct (g-reduct) of DT iff A is a minimal set such that

$$\forall x \in O, \bigcap_{a \in A} \partial_a(x) = \partial_{AT}(x) \tag{ag}$$

Corollary 5 specifies properties of certain decision reducts in terms of generalized decisions. By analogy to this corollary, we define an *approximate certain decision reduct* as follows:

AT is defined an *approximate certain decision reduct* (ac-reduct) of DT iff $\exists x \in O$, $\partial_{AT}(x) = \{d(x)\} \Rightarrow \bigcap_{a \in AT} \partial_a(x) \supset \partial_{AT}(x)$. Otherwise, A is defined an *approximate certain reduct* (ac-reduct) of DT iff A is a minimal attribute set such that

$$\forall x \in O, \partial_{AT}(x) = \{d(x)\} \Rightarrow \bigcap_{a \in A} \partial_a(x) = \partial_{AT}(x) \tag{ac}$$

In the sequel, a superset of a t-reduct, $t \in \{ac, ag\}$, will be called a t-*super-reduct*.

Corollary 6. AT is a superset of all ac-reducts and ag-reducts for any DT.

Proposition 4. Let $x \in O$ and $A \subseteq AT$. If $\bigcap_{a \in A} \partial_a(x) = \partial_{AT}(x)$, then:

a) $\bigcap_{a \in A} \partial_a(x) = \partial_A(x) = \partial_{AT}(x)$.

b) $\forall B \subseteq AT, B \supset A \Rightarrow \bigcap_{a \in B} \partial_a(x) = \partial_B(x) = \partial_{AT}(x)$.

Proof: Let $\bigcap_{a \in A} \partial_a(x) = \partial_{AT}(x)$ (*).

Ad a) By Corollaries 1-2, $\bigcap_{a \in A} \partial_a(x) \supseteq \partial_A(x) \supseteq \partial_{AT}(x)$. Taking into account (*), $\bigcap_{a \in A} \partial_a(x) = \partial_A(x) = \partial_{AT}(x)$.

Ad b) Let $B \subseteq AT$, $B \supset A$. By Corollary 2, $\partial_A(x) \supseteq \partial_B(x) \supseteq \partial_{AT}(x)$. Taking into account Proposition 4a, $\bigcap_{a \in A} \partial_a(x) = \partial_A(x) = \partial_B(x) = \partial_{AT}(x)$ (**). Clearly, $\bigcap_{a \in A} \partial_a(x) \supseteq \bigcap_{a \in B} \partial_a(x) \supseteq \bigcap_{a \in AT} \partial_a(x)$. Taking into account (**), $\partial_B(x) = \partial_{AT}(x) = \bigcap_{a \in A} \partial_a(x) \supseteq \bigcap_{a \in B} \partial_a(x) \supseteq \bigcap_{a \in AT} \partial_a(x) \supseteq \partial_{AT}(x)$. Hence, $\bigcap_{a \in B} \partial_a(x) = \partial_B(x) = \partial_{AT}(x)$. □

Corollary 7.
a) An ag-reduct is a g-super-reduct.
b) An ag-reduct is a p-super-reduct.
c) An ac-reduct is a c-super-reduct.

Proof: Ad a) Let A be an ag-reduct. If $\exists x \in O$, $\bigcap_{a \in AT} \partial_a(x) \supset \partial_{AT}(x)$, then $A = AT$, which by Corollary 3 is a g-super-reduct. Otherwise, by definition of an ag-reduct and Proposition 4a, $\forall x \in O$, $\bigcap_{a \in A} \partial_a(x) = \partial_A(x) = \partial_{AT}(x)$. Thus A satisfies property (**g**). Hence, by Corollary 4c, A is a g-super-reduct.
Ad b) Follows from Theorem 1 and Corollary 7a.
Ad c) Analogous, to the proof of Corollary 7a. Follows from the definition of an ac-reduct, Corollary 3, Corollary 5, Corollary 4a and Proposition 4a.

Proposition 5. Let $A \subseteq AT$.
a)   If A satisfies property (**ag**), then all of its supersets satisfy property (**ag**).
b)   If A does not satisfy property (**ag**), then all of its subsets do not satisfy (**ag**).
c)   If A satisfies property (**ac**), then all of its supersets satisfy property (**ac**).
d)   If A does not satisfy property (**ac**), then all of its subsets do not satisfy (**ac**).

Proof: Ad a,c) Follow from Proposition 4.
Ad b, d) Follow immediately from Proposition 5a, c, respectively.   □

Corollary 8.
a) ag-super-reducts are all and the only attribute sets that satisfy property (**ag**).
b) ac-super-reducts are all and the only attribute sets that satisfy property (**ac**).

Proof: By definition of respective approximate reducts and Proposition 5.   □

6.2 Approximate Core

An *approximate core* will be defined in usual way; that is,

$$t\text{-}core = \{a \in AT \mid AT \backslash \{a\} \text{ is not a } t\text{-super-reduct}\}, \text{ where } t \in \{ac, ag\}.$$

Proposition 6. Let $\mathcal{R}$ be all approximate reducts of the same type t, $t \in \{ac, ag\}$.

$$t\text{-}core = \bigcap \mathcal{R}.$$

Proof: Follows from Corollary 8 and Proposition 5, and is analogous to the proof of Proposition 3.   □

7 Discovering Approximate Generalized Reducts

7.1 Main Algorithm

The *GRA* (*GeneralizedReductsApriori*) algorithm, we have recently introduced in [13], finds all ag-reducts from the decision table *DT*. Unlike in *RAD*, *GRA*, does not need to store all maximal non-g-super-reducts $\mathcal{MNSR}$. On the other hand, *GRA* requires the candidates for reducts to be evaluated against the decision table. The validation of the candidate solution against the decision table *DT* in our algorithm consists in checking if the candidate satisfies property (**ag**); that is, if the intersection of the elementary generalized decisions of the attributes in the candidate set determines the same generalized decision value as the set of all conditional attributes *AT* does for each object in *DT*. We will use the following properties in the process of searching reducts in order to prune the search space efficiently:

- Proper supersets of ag-reducts are not ag-reducts, and hence such sets shall not be evaluated against the decision table.
- Subsets of attribute sets that are not ag-super-reducts are not ag-reducts, and thus such sets shall not be evaluated against the decision table.
- An attribute set whose all proper subsets are not ag-super-reducts may or may not be an ag-reduct, and hence should be evaluated against the decision table.

Since our algorithm is to work with very large decision tables, we propose to restrict the number of decision table objects against which a candidate should be evaluated. Our proposal is based on the following observation:

- If an attribute set A satisfies property (**ag**) for the first n objects in DT (or reduced DT') and does not satisfy it for object $n+1$, then A is certainly not an ag-reduct and thus evaluating it against the remaining objects in DT (DT') is useless.
- If an attribute set A satisfies property (**ag**) for the first n objects in DT (or DT'), then property (**ag**) will be satisfied for these objects for all supersets of A. Hence, the evaluation of the first n objects should be skipped for a candidate that is a proper superset of A.

The *GRA* algorithm starts with building the reduced version DT' of decision table DT (see Section 3.2 for the description of the *GenDecRepresentation-of-DT* function). DT' stores only the AT-generalized decisions instead of the original decisions. Next, the a-generalized decision value for each atomic descriptor (a,v) occurring in DT (or in DT') is calculated as the set of the decisions (or the union of the AT-generalized decisions) of the objects supporting (a,v) in DT (or in DT'). Each pair: (*atomic descriptor, its generalized decision*) is stored in Γ. Now *GRA* creates initial candidates for ag-reducts. The initial candidates are singleton sets and are stored in $\mathcal{R}_1$. The set of 1 attribute non-ag-super-reducts $\mathcal{A}_1$, as well as known maximal non-ag-super-reducts $\mathcal{NSR}$, are initialized to an empty set. The main loop starts. In each k-th iteration, $k \geq 1$, the k attribute candidates $\mathcal{R}_k$ are evaluated during one pass over DT' (see Section 7.2 for the description of the *EvaluateCandidates* procedure). As a side effect of evaluating of $\mathcal{R}_k$, all k attribute non-ag-super-reducts $\mathcal{A}_k$ are found and known maximal non-ag-super-reducts $\mathcal{NSR}$ are updated. The case when $\mathcal{NSR}_{|AT|} = AT$ indicates that AT does not satisfy property (**ag**) for some object. Hence, by definition AT is the only ag-reduct and the algorithms stops. Otherwise, $k+1$ attribute candidates $\mathcal{R}_{k+1}$ are created from k attribute sets in $\mathcal{A}_k$, which turned out not to be ag-super-reducts (see Section 7.4 for the description of the *GRAGen* procedure). The information on non-ag-super-reducts $\mathcal{NSR}$ is used to prune the candidates in $\mathcal{R}_{k+1}$. Namely, each candidate in $\mathcal{R}_{k+1}$ that has a superset in $\mathcal{NSR}$ is known a priori not to be an ag-reduct. Therefore it is moved from $\mathcal{R}_{k+1}$ to $\mathcal{A}_{k+1}$. The algorithm stops when $\mathcal{R}_k = \mathcal{A}_k = \{\}$. Optimizations steps 1-2 in *GRA* are analogous to steps 1-2 in *RAD*, which were discussed in Section 3.5.

Modified or additional notation for *GRA*

- $\mathcal{R}_k$ – candidate k attribute sets (potential *ag*-reducts);
- $\mathcal{A}_k$ – k attribute sets that are not *ag*-super-reducts;
- *A.id* – the identifier of the object against which attribute set *A* should be evaluated;
- $\mathcal{NSR}$ – quasi maximal attribute sets found not to be *ag*-super-reducts;
- $\mathcal{NSR}_k$ – k attribute sets in $\mathcal{NSR}$;
- *x.identifier* – the identifier of object *x*;
- Γ - the set containing generalized decision values determined by atomic descriptors supported by objects in *DT* (*DT'*); that is: $\Gamma = \cup_{a \in AT, v \in V_a} \{\{(a,v), \partial_{(a,v)}\}\}$.

Algorithm. *GRA*;

DT' = *GenDecRepresentation-of-DT(DT)*;
/* calculate *a*-generalized decision value for each atomic descriptor (a,v) supported by *DT* (or *DT'*) */
for each conditional attribute $a \in AT$ **do**
    for each domain value $v \in V_a$ **do begin** compute $\partial_{(a,v)}$; store $((a,v), \partial_{(a,v)})$ in Γ **endfor**;
/* initialize 1 attribute candidates */
  $\mathcal{R}_1 = \{\{a\} \mid a \in AT\}$; $\mathcal{A}_1 = \{\}$; $\mathcal{NSR} = \{\}$;   // conditional attributes are candidates for *ag*-reducts
  for each *A* in $\mathcal{R}_1$ **do** *A.id* = 1;   // the evaluation of candidate *A* should start from object 1 in *DT'*
/* search reducts */
  for (k = 1; $\mathcal{A}_k \neq \{\} \vee \mathcal{R}_k \neq \{\}$; k++) **do begin**
  if $\mathcal{R}_k \neq \{\}$ **then begin**
    /* find and move non-*ag*-reducts from $\mathcal{R}_k$ to $\mathcal{A}_k$ and determine maximal non-*ag*-super-reducts $\mathcal{NSR}$ */
    EvaluateCandidates($\mathcal{R}_k$, $\mathcal{A}_k$, Γ, $\mathcal{NSR}$);
    if $|\mathcal{NSR}_{|AT|}| = 1$ **then return** *AT*;   // or equivalently, **if** $\mathcal{NSR}_{|AT|} = AT$ **then**
    if $|\mathcal{NSR}_{|AT|-1}| = |AT|$ **then return** *AT*;   // optional optimizing step 1
    elseif $|\mathcal{NSR}| = 1$ **then return** $\cup_k \mathcal{R}_k$;   // optional optimizing step 2
    endif;
    /* create k+1 attribute candidates $\mathcal{R}_{k+1}$ and non-*ag*-super-reducts $\mathcal{A}_{k+1}$ from $\mathcal{A}_k$ and $\mathcal{NSR}$ */
    GRAGen($\mathcal{R}_{k+1}$, $\mathcal{A}_{k+1}$, $\mathcal{A}_k$, $\mathcal{NSR}$);
  endfor;
return $\cup_k \mathcal{R}_k$;

A characteristic feature of our algorithm, which is shared by all *Apriori*-like algorithms (see [1] for the *Apriori* algorithm), is that the evaluation of candidates requires no more than *n* scans of the data set (decision table), where *n* is the length of a longest candidate (here: $n \leq |AT|$).

GRA, however, differs from *Apriori* in several ways. First of all, our candidates are sets of attributes instead of descriptors. Next, we evaluate candidates whether they satisfy property (**ag**), while the evaluation in *Apriori* consists in calculating the number of objects satisfying candidate descriptors. Additionally, our algorithm uses dynamically obtained information on non-*ag*-super-reducts to restrict the search space as quickly as possible. Another distinct optimizing feature of our algorithm is that the majority of candidates is evaluated against a fraction of the decision table instead of the entire decision table (see Section 7.2). Namely, having found that a candidate *A* does not satisfy the required property (**ag**) for some object *x*, the next objects are not considered for evaluating this candidate at all. In addition, the evaluation of candidates that are proper supersets of the invalidated candidate *A* starts from object *x*. These two optimizations may speed up the evaluation process considerably.

7.2 Evaluating Candidates for Approximate Reducts

The *EvaluateCandidates* procedure takes 4 arguments: k attribute candidates for *ag*-reducts $\mathcal{R}_k$, k attribute sets that are known not to be *ag*-super-reducts $\mathcal{A}_k$, the generalized decisions determined by atomic descriptors Γ, and known maximal non- *ag*-approximate-super-reducts $\mathcal{NSR}$. For each object read from DT', the candidates in $\mathcal{R}_k$ that should be evaluated against this object are identified. These are candidates A such that $A.id$ equals the *identifier* of the object. Let x be the object under consideration and A be a candidate such that $A.id = x.identifier$. The upper bound ∂ on $\partial_A(x)$ is calculated from the generalized decisions determined by the atomic descriptors stored in Γ. If ∂ equals $x.\partial_{AT}$, then A satisfies property (**ag**) for object x and still has a chance to be an *ag*-reduct. Hence, $A.id$ is incremented to indicate that A should be evaluated against the next object after x in DT' too. Otherwise, if $\partial \neq x.\partial_{AT}$, then A is certainly not an *ag*-reduct and thus is moved from candidates $\mathcal{R}_k$ to non-*ag*-super-reducts $\mathcal{A}_k$. Additionally, the *MaximalNonAGSuperReduct* procedure (see Section 7.3) is called to determine a quasi maximal superset (*nsr*) of A that does not satisfy property (**ag**) for object x either. If *nsr* obtains the maximal possible length (i.e. $|nsr| = |AT|$), AT is returned as the maximal set the approximate generalized decision of which differs from the real AT-generalized decision, and the procedure stops. Otherwise, the found non-*ag*-super-reduct is stored in $\mathcal{NSR}'$. Since the evaluation of candidates against objects may result in moving all candidates from $\mathcal{R}_k$ to $\mathcal{A}_k$, scanning of DT' is stopped as soon as all candidates turned out false ones.

The last step of the *EvaluateCandidates* procedure consists in updating maximal non-*ag*-super-reducts $\mathcal{NSR}$ with $\mathcal{NSR}'$. Please note that k attribute sets are not stored in the final $\mathcal{NSR}$ since they are useless for identifying non-super-reducts among l attribute candidates, where $l > k$.

---

procedure *EvaluateCandidates*(**var** $\mathcal{R}_k$, **var** $\mathcal{A}_k$, **in** Γ, **var** $\mathcal{NSR}$);

```
/* assert: Γ = ∪ₐ∈AT, v∈Va {{(a,v), ∂₍ₐ,ᵥ₎)} */
    NSR' = {};
    for each object x in DT' do begin
        for each candidate A in Rₖ do
            if A.id = x.identifier then begin
                ∂ = ∩ₐ∈A ∂₍ₐ, ₓ.ₐ₎;                     // note: each ((a, x.a), ∂₍ₐ, ₓ.ₐ₎) ∈ Γ
                if ∂ ≠ x.∂ₐₜ then begin                   // or equivalently: if |∂| ≠ |x.∂ₐₜ| then
                    move A from Rₖ to Aₖ;
                    nsr = MaximalNonAGSuperReduct(A, x, ∂, Γ);  // find a quasi maximal non-ag-super-reduct
                    if nsr = AT then begin NSR = {AT}; return endif; // or equivalently: if |nsr| = |AT| then
                    add nsr to NSR';
                else A.id = x.identifier + 1              // A should be evaluated against the next object
                endif
            endif;
        if Rₖ = {} then break;
    endfor;
        NSR = MAX((NSR' \ NSRₖ') ∪ (NSR \ NSRₖ));
    return;
```

7.3 Calculating Quasi Maximal Non-approximate Generalized Super-reducts

The *MaximalNonAGSuperReduct* function is called whenever a candidate, say A, does not satisfy property (**ag**) for some object x. This function returns a quasi maximal superset of A that does not satisfy property (**ag**) for x. Clearly, there may be many such supersets of A; however the function creates and evaluates supersets of A in a specific order. Namely, *nsr* variable, which initially equals A, is extended in each iteration with one attribute (assigned to variable a) that is next after the one recently added to *nsr*. Please note that the first attribute in AT is assumed to be next to the last attribute in AT. The creation of supersets stops when an evaluated attribute $nsr \cup \{a\}$ satisfies property (**ag**) for object x. Then, *MaximalNonAGSuperReduct* returns *nsr* as a known maximal superset of A, which is not an *ag*-super-reduct.

---

function *MaximalNonAGSuperReduct*(**in** A, **in** x, **in** ∂, **in** Γ);

/* assert: $\partial \neq x.\partial_{AT}$ */
  $nsr = A$; $\partial_{nsr} = \partial$, *previous_a* = last attribute in A;
  for ($i{=}1$; $i <= |AT|$; $i{+}{+}$) **do**
    if *previous_a* = last attribute in AT **then** a = first attribute in AT
    else a = next attribute after *previous_a* in AT;
    previous_a = a; $\partial_{nsr} = \partial_{nsr} \cap \partial_{(a, x.a)}$;   // note: each $((a, x.a), \partial_{(a, x.a)}) \in \Gamma$
    if $\partial_{nsr} = x.\partial_{AT}$ **then** **break** **else** add a to *nsr*;
  endfor;
  return *nsr*;

7.4 Generating Candidates for Reducts

The *GRAGen* procedure differs from *RADGen* only in the pruning phase in that it determines the *id* field of each $k+1$ attribute candidate, say A, as a side effect of checking if A has all its k attribute subsets in $\mathcal{A}_k$. Namely, *A.id* is assigned the maximum of the *id* fields of A's subsets in $\mathcal{A}_k$. Such value of *A.id* field means that there was a subset of A in $\mathcal{A}_k$ that satisfied property (**ag**) for *A.id*-1 objects. Hence, A is known a priori to satisfy this property for *A.id*-1 objects and the first object against which it should be evaluated has identifier equal to *A.id*.

---

procedure *GRAGen*(**var** $\mathcal{R}_{k+1}$, **var** $\mathcal{A}_{k+1}$, **in** $\mathcal{A}_k$, **in** $\mathcal{NSR}$);

  forall $B, C \in \mathcal{A}_k$ **do**   /* Merging */
      if $B[1] = C[1] \wedge ... \wedge B[k{-}1] = C[k{-}1] \wedge B[k] < C[k]$ **then begin**
        $A = B[1] \bullet B[2] \bullet ... \bullet B[k] \bullet C[k]$; add A to $\mathcal{R}_{k+1}$; $A.id = 1$;
      endif;
  forall $A \in \mathcal{R}_{k+1}$ **do**   /* Pruning */
    forall k attribute sets $B \subset A$ **do**
      if $B \in \mathcal{A}_k$ **then** $A.id = \max(A.id, B.id)$
      else delete A from $\mathcal{R}_{k+1}$;   // A is a proper superset of super-reduct B
    forall $B \in \mathcal{NSR}$ **do** move subsets of B from $\mathcal{R}_{k+1}$ to $\mathcal{A}_{k+1}$; /* Removing subsets of non-super-reducts */
  return;

Example 3. We will illustrate *GRAGen* by showing how the candidates of size 3 are created. Let $\mathcal{A}_2 = \{\{ab\}_{[id:2]}, \{ac\}_{[id:3]}, \{ae\}_{[id:2]}, \{bc\}_{[id:3]}, \{bd\}_{[id:2]}, \{be\}_{[id:3]},$

$\{ce\}_{[id:3]}$, $\{de\}_{[id:2]}\}$ (the indices provide information on identifiers of objects recently evaluated for respective attribute sets) and $\mathcal{NSR} = \{\{abce\}, \{bde\}\}$.

The first phase of the procedure consists in creating candidates $\mathcal{R}_3$ from pairs of sets in $\mathcal{A}_2$ that differ only in their final attributes. Thus, we receive the following candidates: $\mathcal{R}_3 = \{\{abc\}_{[id:1]}, \{abe\}_{[id:1]}, \{ace\}_{[id:1]}, \{bcd\}_{[id:1]}, \{bce\}_{[id:1]}, \{bde\}_{[id:1]}\}$.

The pruning phase deletes these candidates from $\mathcal{R}_3$ that do not have at least one of their 2 attribute subsets in $\mathcal{A}_2$. In addition, the field id of each candidate is set to maximum of id values of all proper subsets of the candidates in $\mathcal{A}_2$. The only candidate in $\mathcal{R}_3$ that does not have some of its 2 attribute subsets in $\mathcal{A}_2$ is $\{bcd\}$. Namely, $\{cd\}$ is such a subset of $\{bcd\}$, which does not belong to $\mathcal{A}_2$. The fact that $\{cd\} \notin \mathcal{A}_2$ means that $\{cd\}$ is an ag-super-reduct. Hence, $\{bcd\}$ is known a priori to be a proper superset of an ag-reduct. Thus, this candidate is pruned from candidates $\mathcal{R}_3$. As a result, $\mathcal{R}_3$ becomes equal to $\{\{abc\}_{[id:3]}, \{abe\}_{[id:3]}, \{ace\}_{[id:3]}, \{bce\}_{[id:3]}, \{bde\}_{[id:3]}\}$.

The final phase determines candidates that are certainly not ag-reducts as they are subsets of previously found non-ag-super-reducts $\mathcal{NSR}$. Such candidates are moved from $\mathcal{R}_3$ to $\mathcal{A}_3$. Eventually, $\mathcal{A}_3 = \{\{abc\}_{[id:3]}, \{abe\}_{[id:3]}, \{ace\}_{[id:3]}, \{bce\}_{[id:3]}, \{bde\}_{[id:3]}\}$ and $\mathcal{R}_3 = \{\}$. Hence, no 3 attribute candidates should be evaluated against the decision table. □

7.5 Illustration of *GRA*

In this section, we illustrate the discovery of ag-reducts for DT from Table 1. We assume that the reduced decision table DT' (see Table 2) has already been determined. Table 6 shows how candidates change in each iteration before and after validation (if any) against DT'. In this process, the reduced decision table was scanned twice in order to evaluate the candidate sets. Only 8 candidates were evaluated against DT', although 21 attribute sets were enumerated (that is, occurred in $\mathcal{R}$ or $\mathcal{A}$). As a result, 2 approximate ag-reducts were found; namely, $\{ad\}$ and $\{cd\}$, which are exact g-reducts (see Section 3.6).

Table 6. $\mathcal{R}_k$, $\mathcal{A}_k$, and $\mathcal{NSR}$ in subsequent iterations of *GRA*.

k	$\mathcal{R}_k$ before validation	$\mathcal{A}_k$ before validation	$\mathcal{R}_k$ after validation	$\mathcal{A}_k$ after validation	$\mathcal{NSR}'$	$\mathcal{NSR}$
1	$\{a\}_{[id:1]}, \{b\}_{[id:1]},$ $\{c\}_{[id:1]}, \{d\}_{[id:1]},$ $\{e\}_{[id:1]}$			$\{a\}_{[id:2]}, \{b\}_{[id:1]},$ $\{c\}_{[id:3]}, \{d\}_{[id:2]},$ $\{e\}_{[id:2]}$	$\{abc\}, \{b\},$ $\{c\}, \{de\},$ $\{abce\}$	$\{abce\},$ $\{de\}$
2	$\{ad\}_{[id:2]}, \{bd\}_{[id:2]},$ $\{cd\}_{[id:3]}$	$\{ab\}_{[id:2]}, \{ac\}_{[id:3]},$ $\{ae\}_{[id:2]}, \{bc\}_{[id:3]},$ $\{be\}_{[id:2]}, \{ce\}_{[id:3]},$ $\{de\}_{[id:2]}$	$\{ad\}_{[id:8]},$ $\{cd\}_{[id:8]}$	$\{ab\}_{[id:2]}, \{ac\}_{[id:3]},$ $\{ae\}_{[id:2]}, \{bc\}_{[id:3]},$ $\{bd\}_{[id:2]}, \{be\}_{[id:3]},$ $\{ce\}_{[id:3]}, \{de\}_{[id:2]}$	$\{bde\}$	$\{abce\},$ $\{bde\}$
3		$\{abc\}_{[id:3]}, \{abe\}_{[id:3]},$ $\{ace\}_{[id:3]}, \{bce\}_{[id:3]},$ $\{bde\}_{[id:3]}$				$\{abce\}$
4		$\{abce\}_{[id:3]}$				

8 Core-Oriented Discovery of Approximate Generalized Reducts

8.1 Main Algorithm

Algorithm. *CoreGRA*;

$DT' = GenDecRepresentation\text{-}of\text{-}DT(DT)$;

for each conditional attribute $a \in AT$ **do**

    for each domain value $v \in V_a$ **do begin** compute $\partial_{(a,v)}$; store $((a,v), \partial_{(a,v)})$ in Γ **endfor**;

/*initialize 1 attribute candidates */

$\mathcal{R}_1 = \{\{a\} | a \in AT\}$; $\mathcal{A}_1 = \{\}$; $\mathcal{NSR} = \{\}$;   // conditional attributes are candidates for *ag*-reducts

for each A in $\mathcal{R}_1$ **do begin** $A.id = 1$; $A.nsr = A$ **endfor**;

/* find and move non-reducts from $\mathcal{R}_1$ to $\mathcal{A}_1$; determine maximal non-reducts */

$EvaluateCandidates1(\mathcal{R}_1, \mathcal{A}_1, \Gamma, \mathcal{NSR})$;

if $|\mathcal{NSR}_{|AT|}| = 1$ **then return** (AT, AT);   // or equivalently, **if** $\mathcal{NSR}_{|AT|} = AT$ **then**

/* determine core */

$core = \varnothing$; $core.id = 1$;

forall $A \in \mathcal{NSR}_{|AT|-1}$ **do begin** $\{a\} = AT \backslash A$; $core = core \cup \{a\}$; $core.id = \max(core.id, \{a\}.id)$ **endfor**;

/* create candidate for reducts as core's supersets */

if $core = \varnothing$ **then begin**

  $startLevel = 2$;

  $ReductsAprioriGen(\mathcal{R}_2, \mathcal{A}_2, \mathcal{A}_1, \mathcal{NSR})$; //create 2 attribute candidates from 1 attribute non-*ag*-reducts

else begin

  $\mathcal{NSR} = \{B \in \mathcal{NSR} | B \supseteq core\}$;   // *ag*-reducts are supersets of *ag*-core

  if $|core| > 1$ **then**

    if $\mathcal{NSR} \neq \{\}$ **then**   // *ag*-core is not an *ag*-reduct as there is its superset in $\mathcal{NSR}$

      $\mathcal{NSR} = \mathcal{NSR} \backslash \mathcal{NSR}_{|core|}$ // or equivalently $\mathcal{NSR} = \mathcal{NSR} \backslash \{core\}$;

    else begin

      $\mathcal{R}_{|core|} = \{core\}$; $\mathcal{A}_{|core|} = \{\}$; $EvaluateCandidates(\mathcal{R}_{|core|}, \mathcal{A}_{|core|}, \Gamma, \mathcal{NSR})$;

      if $|\mathcal{NSR}_{|AT|}| = 1$ **then return** (AT, AT);

    endif;

    if $\mathcal{R}_{|core|} = \{core\}$ **then return**$(core, \mathcal{R}_{|core|})$ // or equivalently **if** $|\mathcal{R}_{|core|}| = 1$ **then**

    else begin

      $startLevel = |core| + 1$; $\mathcal{R}_{startLevel} = \{\}$; $\mathcal{A}_{startLevel} = \{\}$;

      forall $\{a\} \in \mathcal{A}_1$ such that $a \notin core$ **do begin**

        $A = core \cup \{a\}$; $A.id = \max(core.id, \{a\}.id)$;   // candidates should contain *ag*-core

        $\mathcal{R}_{startLevel} = \mathcal{R}_{startLevel} \cup \{A\}$

      endfor;

      forall $B \in \mathcal{NSR}$ **do** move subsets of B from $\mathcal{R}_{startLevel}$ to $\mathcal{A}_{startLevel}$;

  endif

endif;

for $(k = startLevel;$ $\mathcal{A}_k \neq \{\} \vee \mathcal{R}_k \neq \{\};$ $k{+}{+})$ **do begin**   /* *ag*-reducts' regular search */

  if $\mathcal{R}_k \neq \{\}$ **then begin**

  /* find and move non-*ag*-reducts from $\mathcal{R}_k$ to $\mathcal{A}_k$ and determine maximal non-*ag*-super-reducts $\mathcal{NSR}$ */

    $EvaluateCandidates(\mathcal{R}_k, \mathcal{A}_k, \Gamma, \mathcal{NSR})$;

    if $|\mathcal{NSR}_{|AT|}| = 1$ **then return** (AT, AT) **endif;**

    elseif $|\mathcal{NSR}| = 1$ **then return** $(core; \cup_k \mathcal{R}_k)$;   // optional optimizing step

  endif;

  $GRAGen(\mathcal{R}_{k+1}, \mathcal{A}_{k+1}, \mathcal{A}_k, \mathcal{NSR})$;   // create $(k{+}1)$-candidates from k attribute non-*ag*-reducts

endfor;

return $(core; \cup_k \mathcal{R}_k)$;

The *CoreGRA* algorithm, we propose, finds not only *ag*-reducts, but also their core. The layout of *CoreGRA* reminds that of *GRA*. *CoreGRA*, however, differs from *GRA*, in that it evaluates 1 attribute candidates in special way that provides sufficient information to determine the *ag*-core, and next creates subsequent candidates only as supersets of the found *ag*-core. *CoreGRA* calls the *EvaluateCandidate1* procedure (see Section 8.2) in order to evaluate 1 attribute candidates. Unlike the *EvaluateCandidate* procedure, *EvaluateCandidate1* guarantees that all maximal $|AT|$-1 non-*ag*-super-reducts will be determined and returned in $\mathcal{NSR}$. Using this information, the *ag*-core will then be calculated according to its definition.

If the *ag*-core is an empty set, then 2 attribute and longer candidates are created and evaluated as in *GRA*. Otherwise, all sets in $\mathcal{NSR}$ that are not supersets of the *ag*-core are deleted, since the only candidates considered in *CoreGRA* will be the *ag*-core and its supersets. If the *ag*-core contains only one attribute, it is not evaluated because singleton attributes were already evaluated. The *ag*-core is not evaluated also in the case, when $\mathcal{NSR}$, already restricted to non-*ag*-super-reducts being the core's supersets, is not empty. In this case, the *ag*-core is also a non-*ag*-super-reduct as a subset of some non-*ag*-super-reduct in $\mathcal{NSR}$. Otherwise, the *ag*-core is evaluated. Provided the *ag*-core is found an *ag*-reduct, it is returned as the only *ag*-reduct. If the *ag*-core is not a reduct, the new candidates $\mathcal{R}_{|core|+1}$ are created by merging the core with the remaining attributes in *AT*. Clearly, the new candidates which have supersets in maximal known non-*ag*-super-reducts $\mathcal{NSR}$, are not *ag*-reducts either, and hence are moved from $\mathcal{R}_{|core|+1}$ to $\mathcal{A}_{|core|+1}$. From now on, *CoreGRA* is performed in the same way as *GRA*.

It is expected that *CoreGRA* should perform better than *GRA*, when the *ag*-core consists of a sufficient number of attributes. Then fewer iterations should be performed and probably fewer candidates will be evaluated. Nevertheless, when the number of attributes in the *ag*-core is small, *CoreGRA* may be less effective than *GRA* because of the more exhaustive evaluation of 1 attribute candidates (their *nsr* fields are likely to be evaluated against the entire decision table in *CoreGRA*).

8.2 Evaluating Singleton Candidates

Below we describe the *EvaluateCandidates1* procedure, which is primarily intended to be applied only to 1 attribute candidates in *CoreGRA*, although it can be applied for evaluating candidates of any length. It is assumed that an additional field *nsr* is associated with each k attribute candidate A in $\mathcal{R}_k$.

The *EvaluateCandidates1* procedure differs from *EvaluateCandidates* in that after discovering that a candidate A is not an *ag*-reduct, it is not removed from $\mathcal{R}_k$ immediately. Nevertheless, *EvaluateCandidates1* stops advancing $A.id$ field as soon as the first object invalidating A is found (like *EvaluateCandidates* does). In such a case, instead of evaluating A, its *nsr* field is extended and evaluated against the remaining objects in the decision table as long as *nsr* obtains the maximal possible length (i.e. $|nsr| = |AT|$) or the end of the decision table is reached. In the former case, *AT* is returned as the maximal set the approximate generalized decision of which differs from the real AT-generalized decision, and the procedure stops. In the latter case, the remaining candidates A in $\mathcal{R}_k$ that turned out not ag-reducts (i.e. such that $A.id \neq |DT|+1$), are moved to $\mathcal{A}_k$ and $\mathcal{NSR}'$ is updated with their nsr fields.

---

procedure $EvaluateCandidates1($**var** $\mathcal{R}_k,$ **var** $\mathcal{A}_k,$ **in** $\Gamma,$ **var** $\mathcal{NSR});$

$\mathcal{NSR}' = \{\};$
for each object x in DT **do begin**
  for each candidate A in $\mathcal{R}_k$ **do begin**
    $\partial = \cap_{a \in A.nsr}\, \partial_{(a,x.a)};$ // note: each $((a,x.a),\, \partial_{(a,x.a)}) \in \Gamma$
    if $\partial \neq x.\partial_{AT}$ **then begin**   // or equivalently: **if** $|\partial| = |x.\partial_{AT}|$ **then**
      $A.nsr = MaximalNonAGSuperReduct(A.nsr, x, \partial, \Gamma);$ // find a maximal non-ag-super-reduct
      if $A.nsr = AT$ **then begin** $\mathcal{NSR} = \{AT\};$ **return endif**   // or equivalently: **if** $|A.nsr| = |AT|$ **then**
    elseif $A.id = x.identifier$ **then** $A.id = x.identifier + 1$ // evaluate A's supersets against the next object
    endif;
  endfor;
  if $\mathcal{R}_k = \{\}$ **then break**;
endfor;
for each candidate A in $\mathcal{R}_k$ **do**
  if $A.id \neq |DT|+1$ **then** move A from $\mathcal{R}_k$ to $\mathcal{A}_k$;   add $A.nsr$ to $\mathcal{NSR}'$ **endif**;   // A is not an ag-reduct
  $\mathcal{NSR} = MAX(\mathcal{NSR}' \setminus \mathcal{NSR}_k');$// $\mathcal{NSR} = MAX((\mathcal{NSR}' \setminus \mathcal{NSR}_k') \cup (\mathcal{NSR} \setminus \mathcal{NSR}_k))$ for $k > 1$
return;

8.3 Illustration of *CoreGRA*

In this section, we illustrate how *CoreGRA* searches ag-reducts in the decision table DT from Table 1. Table 7 shows how candidates change in each iteration before and after validation against the reduced decision table DT' from Table 2.

After 1 attribute candidates were evaluated by *EvaluateCandidates1*, $\mathcal{NSR}$ became equal to $\{\{abce\}, \{de\}\}$. Thus, $\{abce\}$ was the only set in $\mathcal{NSR}$ the length of which was equal to $|AT|-1$. Hence, the ag-core was determined as $AT\setminus\{abce\} = \{d\}$. Since the new candidates were to be supersets of the ag-core, all sets from $\mathcal{NSR}$ that were not supersets of the ag-core were pruned and $\mathcal{NSR}$ became equal to $\{\{de\}\}$. The ag-core $\{d\}$ is not an ag-reduct, as it was not present in the set of the positively evaluated candidates $\mathcal{R}_1$ (here: $\mathcal{R}_1 = \varnothing$).

New candidates were created by merging the ag-core with the remaining attributes in AT resulting in the following 4 attribute candidates: $\{ad\}, \{bd\}, \{cd\}, \{de\}$. One of them ($\{de\}$) was known a priori not to be an ag-reduct as a subset of the known non-ag-super-reduct $\{de\}$ in $\mathcal{NSR}$. From now on, *CoreGRA* proceeded as *GRA*. The execution of the *CoreGRA* algorithm resulted in enumeration of 9 attribute sets instead of 21 (see Section 7.5).

Table 7. $\mathcal{R}_k,$ $\mathcal{A}_k,$ and $\mathcal{NSR}$ in subsequent iterations of *CoreGRA*.

k	$\mathcal{R}_k$ before validation	$\mathcal{A}_k$ before validation	$\mathcal{R}_k$ after validation	$\mathcal{A}_k$ after validation	$\mathcal{NSR}'$	$\mathcal{NSR}$
1	$\{a\}_{[id:1]}, \{b\}_{[id:1]}, \{c\}_{[id:1]},$ $\{d\}_{[id:1]}, \{e\}_{[id:1]}$			$\{a\}_{[id:2]}, \{b\}_{[id:1]}, \{c\}_{[id:3]},$ $\{d\}_{[id:2]}, \{e\}_{[id:2]}$	$\{abc\}, \{bc\}, \{c\},$ $\{de\}, \{abce\}$	$\{abce\},$ $\{de\}$
2	$\{ad\}_{[id:2]}, \{bd\}_{[id:2]},$ $\{cd\}_{[id:3]}$	$\{de\}_{[id:2]}$	$\{ad\}_{[id:8]},$ $\{cd\}_{[id:8]}$	$\{bd\}_{[id:2]}, \{de\}_{[id:2]}$	$\{bde\}$	$\{bde\}$

9 Discovering Approximate Certain Reducts

Approximate certain reducts of *DT* are defined by means of generalized decisions only of objects in *DT* with singleton *AT*-generalized decisions. This observation suggests that the *GRA* and *CoreGRA* algorithms shall calculate *ac*-reducts of *DT* correctly, provided the candidate attribute sets are evaluated only against the objects in *DT* with singleton *AT*-generalized decisions. This can be achieved in two ways:

a) either the initialization of candidates in the *GRA* procedure should be preceded by an additional operation that removes all objects from *DT* (or *DT'*) that have non-singleton *AT*-generalized decisions and renumbers the remaining objects;

b) or the evaluation of candidates should be modified so that to ignore objects with non-singleton *AT*-generalized decisions safely (please, see [13]).

10 Conclusion

In the article, we have offered two new algorithms: *RAD* and *CoreRAD* for discovering all exact generalized (and by this also possible) and certain reducts from decision tables. In addition, *CoreRAD* determines the core. Both algorithms require the calculation of all maximal attribute sets $\mathcal{MNSR}$ that are not super-reducts. An *Apriori*-like method of determining reducts based on $\mathcal{MNSR}$ was proposed. Our method of determining $\mathcal{MNSR}$ is orthogonal to the methods that determine a discernibility matrix ($\mathcal{DM}$), which stores information on sets of attributes each of which discerns at least one pair of objects that should be discerned, and return the family of all such minimal sets ($\mathcal{MDM}$). The reducts are then found from $\mathcal{MDM}$ by applying Boolean reasoning.

The calculation of $\mathcal{MNSR}$ (as well as $\mathcal{MDM}$) requires comparing each pair of objects in the decision table and finding maximal (minimal) attribute sets among those that are the result of the objects' comparison. This operation is very costly when the number of objects in a decision table is large. In order to overcome this problem one may use a reduced table $(AT, \{\partial_{AT}\})$, which stores one object instead of many original objects that are indiscernible on *AT* and ∂_{AT}. Nevertheless, when the number of objects in the reduced table is still large or the number of $\mathcal{MNSR}$ ($\mathcal{MDM}$) is large, the calculation of reducts may be infeasible. Our preliminary experiments indicate that the determination of $\mathcal{MNSR}$ is a bottleneck of the proposed *RAD*-like algorithms in such cases. To the contrary, the proposed *Apriori*-like method of determining reducts based on $\mathcal{MNSR}$ is very efficient.

In the case, when the determination of $\mathcal{MNSR}$ is infeasible, we advocate to search approximate reducts. In the article, we have defined such approximate reducts based on the properties of a generalized decision function. We have shown that for each *A*-generalized decision one may derive its upper bound (*A*-approximate generalized decision) from elementary *a*-generalized decisions, where $a \in A$. Whereas exact generalized (certain) reducts preserve the *AT*-generalized decision for all objects (for objects with singleton generalized decisions), each approximate generalized (certain) reduct *A* guarantees that *A*-approximate generalized decision is equal to the

AT-generalized decision for all objects (for objects with singleton generalized decisions). An exception to the rule is the case, when there is an object for which the approximate *AT*-generalized decision differs from the actual *AT*-generalized decision. Then the entire set of conditional attributes *AT* is defined as a reduct. We have proved that approximate generalized and certain reducts are supersets of exact reducts of respective types. In addition, approximate generalized reducts are supersets of exact possible reducts.

We have presented *GRA* and *CoreGRA* algorithms for discovering approximate generalized (and by this also possible) reducts and certain reducts from very large decision tables. The experiments we have carried out and reported in [13] prove that the *GRA*-like algorithms are scalable with respect to the number of objects in a decision table and that *CoreGRA* tends to outperform *GRA* with increasing number of conditional attributes. For a few conditional attributes, however, *GRA* may find reducts faster. Nevertheless, the experiments need to be continued to fully recognize the performance characteristics of particular *GRA*-like algorithms.

Finally, we note that all the proposed algorithms are capable to discover all discussed types of reducts from incomplete decision tables as well. The only difference consists in a slightly different determination of generalized decision value for atomic descriptors, namely $\partial_A(x) = \{d(y)|\ y \in S_A(x)\}$, where $S_A(x) = \{y \in O\ |\ \forall a \in A,\ (a(x) = a(y)) \vee (a(x)$ is NULL$) \vee (a(y)$ is NULL$)\}$ (see e.g. [12]). In the future, we intend to develop scalable algorithms for discovering all exact reducts.

References

1. Agrawal, R., Mannila, H., Srikant, R., Toivonen, H., Verkamo, A.I.: Fast Discovery of Association Rules. In: Advances in KDD. AAAI, Menlo Park, California (1996) 307-328
2. Bazan, J., Skowron, A., Synak, P.: Dynamic Reducts as a Tool for Extracting Laws from Decision Tables. In: Proc. of ISMIS '94, Charlotte, USA. LNAI, Vol. 869, Springer-Verlag, (1994) 346–355
3. Bazan, J., Nguyen, H.S., Nguyen, S.H., Synak, P., Wróblewski, J.: Rough Set Algorithms in Classification Problem. In: L. Polkowski, S. Tsumoto and T.Y. Lin (eds.): Rough Set Methods and Applications. Physica-Verlag, Heidelberg, New York (2000) 49 - 88
4. Jelonek, J., Krawiec, K., Stefanowski, J.: Comparative Study of Feature Subset Selection Techniques for Machine Learning Tasks. Proc. of IIS '98, Malbork, Poland (1998) 68–77
5. John, H.G., Kohavi, R., Pfleger, K.: Irrelevant Features and the Subset Selection Problem. In: Machine Learning: Proc. of the Eleventh International Conference, Morgan Kaufmann Publishers, San Francisco, CA, (1994) 121–129
6. Kohavi, R., Frasca, B.: Useful Feature Subsets and Rough Set Reducts. In: Proc. of the Third International Workshop on Rough Sets and Soft Computing, San Jose, CA (1994)
7. Kryszkiewicz, M.: The Algorithms of Knowledge Reduction in Information Systems, Ph.D. Thesis, Warsaw University of Technology, Institute of Computer Science (1994)
8. Kryszkiewicz, M., Rybinski, H.: Finding Reducts in Composed Information Systems. Fundamenta Informaticae Vol. 27, No. 2–3 (1996) 183–196
9. Kryszkiewicz, M.: Strong Rules in Large Databases. In: Proc. of IPMU' 98, Paris, France, Vol. 2 (1998) 1520–1527
10. Kryszkiewicz M., Rybinski H.: Knowledge Discovery from Large Databases using Rough Sets. In: Proc. of EUFIT '98, Aachen, Germany, Vol. 1 (1998) 85-89

11. Kryszkiewicz, M.: Comparative Study of Alternative Types of Knowledge Reduction in Inconsistent Systems. International Journal of Intelligent Systems, Wiley, Vol. 16, No. 1 (2001) 105–120

12. Kryszkiewicz, M.: Rough Set Approach to Rules Generation from Incomplete Information Systems. In: The Encyclopedia of Computer Science and Technology. Marcel Dekker, Inc., New York, Vol. 44 (2001) 319-346

13. Kryszkiewicz, M., Cichoń K.: Scalable Methods of Discovering Rough Sets Reducts. ICS Research Report 28/2003, Warsaw University of Technology (2003)

14. Lin, T.Y.: Rough Set Theory in Very Large Databases. In: Proc. of CESA IMACS '96, Lille, France Vol. 2 (1996) 936-941

15. Modrzejewski, M.: Feature Selection using Rough Sets Theory. In: Proc. of the European Conference on Machine Learning (1993) 213–226

16. Nguyen, S.H., Skowron, A., Synak, P., Wróblewski, J.: Knowledge Discovery in Databases: Rough Set Approach. In: Proc. of IFSA '97, Prague, Vol. II (1997) 204-209

17. Pawlak, Z.: Rough Sets: Theoretical Aspects of Reasoning about Data, Kluwer Academic Publishers, Vol. 9 (1991)

18. Pawlak, Z., Skowron, A.: A Rough Set Approach to Decision Rules Generation, ICS Research Report 23/93, Warsaw University of Technology (1993)

19. Romanski, S., Operations on Families of Sets for Exhaustive Search, Given a Monotonic Boolean Function. In: Proc. of Intl' Conf. on Data and Knowledge Bases, Israel (1988)

20. Skowron, A., Rauszer, C.: The Discernibility Matrices and Functions in Information Systems. In: Intelligent Decision Support: Handbook of Applications and Advances of Rough Sets Theory. Kluwer Academic Publishers (1992) 331-362

21. Skowron, A., Swiniarski, R.W.: Information Granulation and Pattern Recognition. In: S.K. Pal, L. Polkowski, A. Skowron (eds.): Rough-Neural Computing. Techniques for Computing with Words. Heidelberg: Springer-Verlag (2004)

22. Slezak, D.: Approximate Reducts in Decision Tables. In: Proc. of IPMU '96, Granada, Spain, Vol. 3 (1996) 1159-1164

23. Slezak, D.: Searching for Frequential Reducts in Decision Tables with Uncertain Objects. In: Proc. of RSCTC '98, Warsaw. Springer-Verlag, Berlin (1998) 52–59

24. Slowiński, R. (ed.): Intelligent Decision Support, Handbook of Applications and Advances of the Rough Sets Theory. Kluwer Academic Publishers, Vol 11 (1992)

25. Stepaniuk, J.: Approximation Spaces, Reducts and Representatives. In: Skowron, A., Polkowski, L. (eds.): Rough Sets in Data Mining and Knowledge Discovery, Springer-Verlag, Berlin (1998)

26. Susmaga, R.: Experiments in Incremental Computation of Reducts. In: Skowron, A., Polkowski, L., (eds.): Rough Sets in Data Mining and Knowledge Discovery, Springer-Verlag, Berlin (1998)

27. Susmaga, R.: Parallel Computation of Reducts. In: Proc. of RSCTC '98, Warsaw. Springer-Verlag, Berlin (1998) 450–457

28. Susmaga, R.: Computation of Shortest Reducts. In: Foundations of Computing and Decision Sciences, Poznan, Poland, Vol. 2, No. 23 (1998)

29. Susmaga, R.: Effective Tests for Inclusion Minimality in Reduct Generation. In: Foundations of Computing and Decision Sciences, Vol. 4, No. 23 (1998) 219–240

30. Tannhäuser, M.: Efficient Reduct Computation. M.Sc. Thesis, Institute of Mathematics, Warsaw University, Warsaw (1994)

31. Wroblewski, J.: Finding Minimal Reducts Using Genetic Algorithms. In: Proc. of the 2nd Annual Join Conference on Information Sc., Wrightsville Beach, NC, (1995) 186–189

Variable Precision Fuzzy Rough Sets

Alicja Mieszkowicz-Rolka and Leszek Rolka

Department of Avionics and Control
Rzeszów University of Technology
ul. W. Pola 2, 35-959 Rzeszów, Poland
{alicjamr,leszekr}@prz.edu.pl

Abstract. In this paper the variable precision fuzzy rough sets (VPFRS) concept will be considered. The notions of the fuzzy inclusion set and the α-inclusion error based on the residual implicators will be introduced. The level of misclassification will be expressed by means of α-cuts of the fuzzy inclusion set. Next, the use of the mean fuzzy rough approximations will be postulated and discussed. The concept of VPFRS will be defined using the extended version of the variable precision rough sets (VPRS) model, which utilises a general allowance for levels of misclassification expressed by two parameters: lower (l) and upper (u) limit. Remarks concerning the variable precision rough fuzzy sets (VPRFS) idea will be given. An example will illustrate the proposed VPFRS model.

1 Introduction

The rough sets theory [15] was originally based on the notions of classical sets theory. Dubois and Prade [3] and Nakamura [14] were among the first who showed that the basic idea of rough set given in the form of lower and upper approximation can be extended in order to approximate fuzzy sets defined in terms of membership functions. This makes it possible to analyse information systems with fuzzy attributes. The idea of fuzzy rough sets was pursued and investigated in many papers e.g. [1, 2, 4–6, 9, 16, 17].

An important extension of the rough sets theory, helpful in analysis of inconsistent decision tables, is the variable precision rough sets model (VPRS).

It seems natural and valuable to combine the concepts of VPRS and fuzzy rough sets. The motivation for doing this is supported by the fact that the extended fuzzy rough approximations defined by Dubois and Prade have the same disadvantages as their counterparts in the original (crisp) rough set theory [12]. Even a relative small inclusion error of a similarity class results in rejection (membership value equal to zero) of that class from the lower approximation. A small inclusion degree can also lead to an excessive increase of the upper approximation. These properties can be important especially in case of large universes, e.g. generated from dynamic processes.

In order to overcome the described drawbacks we generalised the idea of Ziarko for expressing the inclusion error of one fuzzy set in another.

If we want to determine the lower and upper approximation using real data sets, then we must take into account the quality of the data which is usually

J.F. Peters et al. (Eds.): Transactions on Rough Sets I, LNCS 3100, pp. 144–160, 2004.

influenced by noise and errors. The VPRS concept admits some level of mis-classification, but we go one step further and propose additionally an alternative way of evaluating the variable precision fuzzy rough approximations. We suggest determination of the mean membership degree. This is contrary to using only limit values of membership functions and disregarding the statistical properties of analysed large information system.

We start by recalling the basic notions of VPRS.

2 Variable Precision Rough Sets Model

The concept of VPRS has proven to be particularly useful in analysis of incon-sistent decision tables obtained from dynamic control processes [10]. We have utilised it in order to identify the decision model of a military pilot [11].

The idea of VPRS is based on a changed relation of set inclusion given in (1) and (2) [18], defined for any nonempty crisp subsets A and B of the universe X. We say that the set A is included in the set B with an inclusion error β:

$$A \overset{\beta}{\subseteq} B \iff e(A,B) \le \beta \tag{1}$$

$$e(A,B) = 1 - \frac{\text{card}(A \cap B)}{\text{card}(A)}. \tag{2}$$

The quantity $e(A,B)$ is called the inclusion error of A in B. The value of β should be limited: $0 \le \beta < 0.5$.

Katzberg and Ziarko proposed later [7] an extended version of VPRS with asymmetric bounds l and u for required inclusion degree instead of admissible inclusion error β which satisfy the following inequality:

$$0 \le l < u \le 1. \tag{3}$$

By using the lower limit l and the upper limit u we can express the u-lower and the l-upper approximation of a crisp set A by an indiscernibility (eqiuvalence) relation R. In the following, we prefer to retain the notion of inclusion error e.

For any crisp set $A \subseteq X$ and an indiscernibility relation R the u-lower ap-proximation of A by R, denoted as $\underline{R}_u A$ is a set

$$\underline{R}_u A = \{x \in X\colon\ [x]_R \overset{1-u}{\subseteq} A\} \tag{4}$$

which is equivalent to

$$\underline{R}_u A = \{x \in X\colon\ e([x]_R, A) \le 1 - u\} \tag{5}$$

where $[x]_R$ denotes an equivalence class of R containing the element x.

The l-upper approximation of A by R, denoted as $\overline{R}_l A$ is a set

$$\overline{R}_l A = \{x \in X\colon\ e([x]_R, A) < 1 - l\}. \tag{6}$$

3   Fuzzy Rough Sets

The notion of fuzzy rough sets was proposed by Dubois and Prade [3].

For a given fuzzy set F and a family $\Phi = \{F_1, F_2, \ldots, F_n\}$ of fuzzy sets on the universe X the membership functions of the lower and upper approximation of F by Φ are defined as follows:

$$\mu_{\underline{\Phi}F}(F_i) = \inf_{x \in X} \mu_{F_i}(x) \to \mu_F(x) \tag{7}$$

$$\mu_{\overline{\Phi}F}(F_i) = \sup_{x \in X} \mu_{F_i}(x) * \mu_F(x) \tag{8}$$

where: $*$ denotes a t-norm operator, $\to$ is an S-implication operator for that $x \to y = 1 - x * (1 - y)$.

The pair of sets $(\underline{\Phi}F, \overline{\Phi}F)$ is called a fuzzy rough set [3].

It is requested that the fuzzy partition Φ satisfies the property of covering X sufficiently

$$\inf_{x \in X} \max_{i=1,\ldots,n} \mu_{F_i}(x) > 0 \tag{9}$$

and the property of disjointness [3]

$$\forall i, j, \quad i \neq j, \quad \sup_{x \in X} \min(\mu_{F_i}(x), \mu_{F_j}(x)) < 1 . \tag{10}$$

We should use a mapping ω from the domain Φ into the domain of the universe X, if we want to express in X the membership functions of the lower and upper approximation given by (7) and (8). Assuming that Φ is equal to the quotient set of X by a fuzzy similarity relation R, we can determine the membership functions of the fuzzy extension of the lower and upper approximation of a fuzzy set F by R [3]:

$$\forall x \in X, \quad \mu_{\omega(\underline{R}F)}(x) = \inf_{y \in X} \mu_R(x, y) \to \mu_F(y) \tag{11}$$

$$\forall x \in X, \quad \mu_{\omega(\overline{R}F)}(x) = \sup_{y \in X} \mu_R(x, y) * \mu_F(y) . \tag{12}$$

In such a case the fuzzy extension $\omega(A)$ of a fuzzy set A on X/R can be expressed as follows:

$$\mu_{\omega(A)}(x) = \mu_A(F_i), \quad \text{if } \mu_{F_i}(x) = 1 . \tag{13}$$

We use later in this paper a fuzzy compatibility relation which is symmetric and reflexive. One can easy show in this case that the definitions (7) and (8) are equivalent to the definitions (11) and (12). Indeed, by using a symmetric and reflexive fuzzy relation we obtain a family of fuzzy compatibility classes. Any elements x and y of the universe X, for which $\mu_R(x, y) = 1$, belong with a membership degree equal to 1 to the same fuzzy compatibility class. In order to determine the membership degrees (11) and (12) for some x, we can take merely the membership degrees (7) and (8) obtained for that compatibility class, to which x belongs with a membership degree equal to 1.

Another general approach was given by Greco, Matarazzo and Słowiński, who proposed [5] approximation of fuzzy sets by means of fuzzy relations which are only reflexive.

An important issue is the choice of implicators used in the definitions (7), (8), (11) and (12). Apart from applying S-implications, Dubois and Prade considered also the R-implication variant of fuzzy rough sets [3].

A comprehensive study concerning a general concept of fuzzy rough sets was done more recently by Radzikowska and Kerre in [17]. They analysed the properties of fuzzy rough approximations based on three classes of fuzzy implicators: S-implicators, R-implicators and QL-implicators. As we state below, R-implicators constitute a good base for constructing the variable precision fuzzy rough sets model.

4   Variable Precision Fuzzy Rough Sets Model

An extension of the fuzzy rough sets concept in the sense of Ziarko requires a method of determination of the lower and upper approximation, in which only a significant part of the approximating set is taken into account. In other words, we should evaluate the membership degree of the approximating set in the lower or upper approximation by regarding only those of its elements, which are included to a sufficiently high degree in the approximated set. This way we allow some level of misclassification.

Before we try to express the inclusion error of one fuzzy set in another we will first recall the classical definition of fuzzy set inclusion [8].

For any fuzzy sets A and B defined on the universe X, we say that the set A is included in the set B:

$$A \subseteq B \iff \forall x \in X, \quad \mu_A(x) \leq \mu_B(x) . \tag{14}$$

If the condition (14) is satisfied, then we should say that the degree of inclusion of A in B is equal to 1 (the inclusion error is equal to 0).

In our approach we want to evaluate the inclusion degree of a fuzzy set A in a fuzzy set B regarding particular elements of A. We obtain in such a way a new fuzzy set, which we call the fuzzy inclusion set of A in B and denote by A^B. To this end we apply an implication operator $\rightarrow$ as follows:

$$\mu_{A^B}(x) = \begin{cases} \mu_A(x) \rightarrow \mu_B(x) & \text{if } \mu_A(x) > 0 \\ 0 & \text{otherwise} \end{cases} \tag{15}$$

Only the proper elements of A (support of A) are considered as relevant. The definition (15) is based on implication operator $\rightarrow$ in order to maintain the compatibility between the the approach of Dubois and Prade and the VPFRS model in limit cases. This will be stated later in this section by the propositions 2 and 3.

Examples of inclusion sets are given in the section 7. The Table 2 contains the membership functions of the approximating set X_1, the approximated set F_1 and the inclusion sets $X_1^{F_1}$ that are evaluated using implication operators of Gaines and Łukasiewicz (discussed below).

We should consider the choice of a suitable implication operator $\rightarrow$. Basing on (14) we put a requirement on the degree of inclusion of A in B with respect to any element x belonging to the support of the set A ($\mu_A(x) > 0$).

We assume that the degree of inclusion with respect to x should always be equal to 1, if the inequality $\mu_A(x) \leq \mu_B(x)$ for that x is satisfied:

$$\mu_A(x) \rightarrow \mu_B(x) = 1, \quad \text{if } \mu_A(x) \leq \mu_B(x) . \tag{16}$$

In general, not all implicators satisfy this requirement. For example, by applying the Kleene-Dienes S-implicator: $x \rightarrow y = \max(1 - x, y)$ we obtain the value 0.6, and for Early Zadeh QL-implicator: $x \rightarrow y = \max(1 - x, \min(x, y))$ the value 0.5, if we take $x = 0.5 < y = 0.6$.

Let us consider the definition of R-implicators (residual implicators) which are based on a t-norm $*$

$$x \rightarrow y = \sup\{\lambda \in [0, 1] : \ x * \lambda \leq y\} . \tag{17}$$

One can easy prove that any R-implicator satisfies the requirement (16). In the last section we demonstrate an example where two popular R-implicators were used:

- the implicator of Łukasiewicz: $x \rightarrow y = \min(1, 1 - x + y)$,
- the Gaines implicator: $x \rightarrow y = 1$ if $x \leq y$ and y/x otherwise.

Radzikowska and Kerre proved that fuzzy rough approximations based on the Łukasiewicz implicator satisfied all properties which were considered in [17]. This is because the Łukasiewicz implicator is both an S-implicator and a residual implicator.

In order to extend the idea of Ziarko on fuzzy sets we should express the error that would be made, when the weakest elements of approximating set, in the sense of their membership in the fuzzy inclusion set A^B, were discarded. We apply to this end the well known notion of α-cut [8], by which for any given fuzzy set A, a crisp set A_α is obtained as follows:

$$A_\alpha = \{x \in X : \ \mu_A(x) \geq \alpha\} \tag{18}$$

where $\alpha \in [0, 1]$.

We introduce the measure of α-inclusion error $e_\alpha(A, B)$ of any nonempty fuzzy set A in a fuzzy set B:

$$e_\alpha(A, B) = 1 - \frac{\text{power}(A \cap A_\alpha^B)}{\text{power}(A)} . \tag{19}$$

Power denotes here the cardinality of a fuzzy set. For any finite fuzzy set F defined on X

$$\text{power}(F) = \sum_{i=1}^{n} \mu_F(x_i) . \tag{20}$$

Now, we show that the measure of inclusion error (2) given by Ziarko is a special case of the proposed measure (19).

Proposition 1. *For any nonempty crisp sets A and B, and for $\alpha \in (0,1]$ the α-inclusion error $e_\alpha(A, B)$ is equivalent to the inclusion error $e(A, B)$.*

Proof. First, we show that for any crisp sets A and B the inclusion set A^B is equal to the intersection $A \cap B$.

For any crisp set C

$$\mu_C(x) = \begin{cases} 1 & \text{for } x \in C \\ 0 & \text{for } x \notin C \end{cases} \tag{21}$$

Every implicator $\rightarrow$ is a function satisfying: $1 \rightarrow 0 = 0$, and $1 \rightarrow 1 = 1$, $0 \rightarrow 1 = 1$, $0 \rightarrow 0 = 1$.

Thus, applying the definition (15), we get

$$\mu_{A^B}(x) = \mu_{A \cap B}(x) = \begin{cases} 1 & \text{if } x \in A \text{ and } x \in B \\ 0 & \text{otherwise} \end{cases} \tag{22}$$

Taking into account (20) and (21), we get for any finite crisp set C

$$\text{power}(C) = \text{card}(C) \ . \tag{23}$$

Furthermore, applying (18) for any $\alpha \in (0,1]$ we obtain

$$C_\alpha = C \ . \tag{24}$$

By the equations (22), (23) and (24), we finally have

$$\frac{\text{power}(A \cap A_\alpha^B)}{\text{power}(A)} = \frac{\text{power}(A \cap (A \cap B)_\alpha)}{\text{power}(A)} = \frac{\text{card}(A \cap B)}{\text{card}(A)} \ .$$

Hence, we obtain $e_\alpha(A, B) = e(A, B)$ for any $\alpha \in (0,1]$. $\qquad \square$

The use of α-cuts gives us the possibility to change gradually the level, at which some of the members of the approximating set are discarded. The evaluation of the membership degree of the whole approximating set in the lower and upper approximation will then be done by respecting only the remaining elements of the approximating set.

The level α can adopt any value from the infinite set $(0,1]$. In practice, only a finite subset of $(0,1]$ will be applied. In our illustrative examples we used values of α obtained with a resolution equal to 0.01.

Let us now consider a partition of the universe X which is generated by a fuzzy compatibility relation R. We denote by X_i some compatibility class on X, where $i = 1 \ldots n$. Any given fuzzy set F defined on the universe X can be approximated by the obtained compatibility classes.

The u-lower approximation of the set F by R is a fuzzy set on X/R with the membership function which we define as follows:

$$\mu_{\underline{R}_u F}(X_i) = \begin{cases} f_{i_u} & \text{if } \exists \alpha_u = \sup\{\alpha \in (0,1] : \ e_\alpha(X_i, F) \leq 1 - u\} \\ 0 & \text{otherwise} \end{cases} \tag{25}$$

where

$$f_{i_u} = \inf_{x \in S_{i_u}} \mu_{X_i}(x) \rightarrow \mu_F(x)$$

$$S_{i_u} = \text{supp}(X_i \cap X_{i_{\alpha_u}}^F) .$$

The set S_{i_u} contains those elements of the approximating class X_i that are included in F at least to the degree α_u, provided that such α_u exists. The membership f_{i_u} is then determined using the "better" elements from S_{i_u} instead of the whole class X_i. The given definition helps to prevent the situation when a few "bad" elements of a large class X_i significantly reduce the lower approximation of the set F.

Furthermore, we suggest the use of R-implicators both for evaluation of $e_\alpha(X_i, F)$ and in place of the operator $\rightarrow$ in (25).

The l-upper approximation of the set F by R can be defined similarly, as a fuzzy set on X/R with the membership function given by:

$$\mu_{\overline{R}_l F}(X_i) = \begin{cases} f_{i_l} & \text{if } \exists \alpha_l = \sup\{\alpha \in (0,1]: \ e'_\alpha(X_i, F) < 1 - l\} \\ 0 & \text{otherwise} \end{cases} \qquad (26)$$

where

$$f_{i_l} = \sup_{x \in S_{i_l}} \mu_{X_i}(x) * \mu_F(x)$$

$$S_{i_l} = \text{supp}(X_i \cap (X_i \cap F)_{\alpha_l})$$

$$e'_\alpha(X_i, F) = 1 - \frac{\text{power}(X_i \cap (X_i \cap F)_\alpha)}{\text{power}(X_i)} .$$

For the l-upper approximation a similar explanation as for the u-lower approximation can be given. Conversely, we want to prevent the situation when a few "good" elements of a large class X_i significantly increase the upper approximation of F. The inclusion error is now based on the intersection $X_i \cap F$ (t-norm operator $*$) and denoted by $e'_\alpha(X_i, F)$. It can be shown in the same way, as for the inclusion error e_α, that $e'_\alpha(A, B) = e(A, B)$ for any nonempty crisp sets A and B and $\alpha \in (0, 1]$.

Now, we demonstrate that the fuzzy rough sets of Dubois and Prade constitute a special case of the proposed variable precision fuzzy rough sets, if no inclusion error is allowed ($u = 1$ and $l = 0$).

Proposition 2. $\mu_{\underline{R}_1 F}(X_i) = \mu_{\underline{R}F}(X_i)$ *for any fuzzy set F and $X_i \in X/R$.*

Proof. For $u = 1$, it is required that $e_{\alpha_1}(X_i, F) = 0$. This means that no elements of an approximating compatibility class X_i can be discarded.

I. Assume that

$$\mu_{\underline{R}F}(X_i) = \inf_{x \in X} \mu_{X_i}(x) \rightarrow \mu_F(x) = c \in (0, 1] .$$

In that case there exists $\alpha_1 = c$ which is the largest possible value of α for that $e_\alpha(X_i, F) = 0$. This is because the same function $\mu_{X_i}(x) \rightarrow \mu_F(x)$ is used for determination of the inclusion set X_i^F. We evaluate f_{i_1} using the set S_{i_1}, which is equal to the class X_i since no elements of X_i are discarded. Hence, we have $\mu_{\underline{R}_1 F}(X_i) = \mu_{\underline{R}F}(X_i) = c$.

II. Assume now that

$$\mu_{\underline{R}F}(X_i) = \inf_{x \in X} \mu_{X_i}(x) \to \mu_F(x) = 0 .$$

There does not exist $\alpha \in (0, 1]$ for which $e_\alpha(X_i, F) = 0$. Any $\alpha \in (0, 1]$ would cause discarding some $x \in X_i$. In consequence, we get $\mu_{\underline{R}_1 F}(X_i) = \mu_{\underline{R}F}(X_i) = 0$ according to the definition (25).   □

Similarly, one can prove the next proposition which holds for the l-upper fuzzy rough approximation.

Proposition 3. $\mu_{\overline{R}_0 F}(X_i) = \mu_{\overline{R}F}(X_i)$ *for any fuzzy set* F *and* $X_i \in X/R$.

The fuzzy rough approximations based on limit values of membership functions are not always suitable for analysis of real data. This can be particulary justified in case of large universes. The obtained results should correspond to the statistical properties of analysed information systems. We need an approach that takes into account the overall set inclusion, and not merely uses a single value of membership function (often determined from noisy data). Therefore, we propose additionally an alternative definition of fuzzy rough approximations, in which the mean value of membership (in the fuzzy inclusion set) for all used elements of the approximating class is utilised.

The mean u-lower approximation of the set F by R is a fuzzy set on X/R with the membership function which we define as follows:

$$\mu_{\underline{R}_u F}(X_i) = \begin{cases} f_{i_u} & \text{if } \exists \alpha_u = \sup\{\alpha \in (0, 1] : \ e_\alpha(X_i, F) \le 1 - u\} \\ 0 & \text{otherwise} \end{cases} \qquad (27)$$

where

$$f_{i_u} = \frac{\text{power}(X_i^F \cap X_{i_{\alpha_u}}^F)}{\text{card}(X_{i_{\alpha_u}}^F)} .$$

The mean l-upper approximation of the set F by R is a fuzzy set on X/R with the membership function defined by:

$$\mu_{\overline{R}_l F}(X_i) = \begin{cases} f_{i_l} & \text{if } \exists \alpha_l = \sup\{\alpha \in (0, 1] : \ e_\alpha(X_i, F) < 1 - l\} \\ 0 & \text{otherwise} \end{cases} \qquad (28)$$

where

$$f_{i_l} = \frac{\text{power}(X_i^F \cap X_{i_{\alpha_l}}^F)}{\text{card}(X_{i_{\alpha_l}}^F)} .$$

The quantities f_{i_u} and f_{i_l} express the mean value of inclusion degree of X_i in F, determined by using only those elements of X_i, which are included in F at least to the degree α_u and α_l respectively.

Observe that we admit only $\alpha \in (0, 1]$. If the admissible inclusion error $(1-u)$ is equal to 0 and there exists any x with $\mu_{X_i}(x) > 0$ for that $\mu_{X_i}(x) \to \mu_F(x) = 0$, then the α-inclusion error $e_\alpha(X_i, F) = 0$ only for $\alpha = 0$. The use of $\alpha = 0$ would result in the same value of the membership function (27) for the admissible inclusion error equal to 0 and for some value of it greater than 0. Moreover, by avoiding $\alpha = 0$ we achieve full accordance with the original definitions of Ziarko in case of crisp sets and crisp equivalence relation R. In such a case the values of f_{i_u} and f_{i_l} are always equal to 1.

Proposition 4. *For any crisp set A and crisp equivalence relation R the mean variable precision fuzzy rough approximations of A by R are equal to the variable precision rough approximations of A by R.*

Proof. The equivalence relation R generates a partition of the universe X into crisp equivalence classes X_i, $i = 1 \ldots n$. By the proposition 1 and its proof: $e_\alpha(X_i, A) = e(X_i, A)$, $X_{i_{\alpha_u}}^A = X_i^A$, $\text{power}(X_i^A) = \text{card}(X_i^A)$ for $\alpha \in (0, 1]$. Thus, we get for the mean u-lower approximation of A by R

$$f_{i_u} = \frac{\text{power}(X_i^A \cap X_{i_{\alpha_u}}^A)}{\text{card}(X_{i_{\alpha_u}}^A)} = \frac{\text{card}(X_i^A)}{\text{card}(X_i^A)} = 1$$

$$\mu_{\underline{R}_u A}(X_i) = \begin{cases} 1 & \text{if } e(X_i, A) \leq 1 - u\} \\ 0 & \text{otherwise} \end{cases} \tag{29}$$

and for the mean l-upper approximation of A by R

$$f_{i_l} = \frac{\text{power}(X_i^A \cap X_{i_{\alpha_l}}^A)}{\text{card}(X_{i_{\alpha_l}}^A)} = \frac{\text{card}(X_i^A)}{\text{card}(X_i^A)} = 1$$

$$\mu_{\overline{R}_l A}(X_i) = \begin{cases} 1 & \text{if } e(X_i, A) < 1 - l\} \\ 0 & \text{otherwise} \end{cases} \tag{30}$$

Taking into account all approximating equivalence classes X_i and applying (13) we obtain from (29) and (30) the VPRS approximations (5) and (6) on the domain X. $\qquad\square$

5   Variable Precision Rough Fuzzy Sets Model

The idea of rough fuzzy sets was introduced by Dubois and Prade in order to approximate fuzzy concepts by means of equivalence classes X_i, $i = 1 \ldots n$, generated by a crisp equivalence relation R defined on X.

The lower and upper approximations of a fuzzy set F by R are fuzzy sets on X/R with membership functions defined as follows [3]:

$$\mu_{\underline{R}F}(X_i) = \inf\{\mu_F(x): \ x \in X_i\} \tag{31}$$

$$\mu_{\overline{R}F}(X_i) = \sup\{\mu_F(x): \ x \in X_i\} \, . \tag{32}$$

The pair of sets $(\underline{R}F, \overline{R}F)$ is called a rough fuzzy set [3].

Proposition 5. *For every implication operator* $\rightarrow$, *every t-norm* $*$, *and crisp equivalence relation R fuzzy rough sets are equivalent to rough fuzzy sets.*

Proof. Since we use crisp equivalence classes X_i we have $\mu_{X_i}(x) = 1$ for all elements $x \in X_i$. Every R-implicator, S-implicator, and QL-implicator is a border implicator [17] which satisfies the condition: $1 \rightarrow x = x$ for all $x \in [0,1]$. Every t-norm $*$ satisfies the boundary condition: $1 * x = x$. Thus, we get

$$\mu_{X_i}(x) \rightarrow \mu_F(x) = \mu_F(x)$$

$$\mu_{X_i}(x) * \mu_F(x) = \mu_F(x) .$$

Therefore, the definitions (31) and (32) are special case of (7) and (8).   □

Basing on the proposition 5 we can easy adopt the variable precision fuzzy rough approximations from the previous section in order to obtain a simpler form of the variable precision rough fuzzy approximations.

In [12] we proposed a concept of variable precision rough fuzzy sets in case of symmetrical bounds (admissible inclusion error β). We defined [12] the β-lower and β-upper approximation of a fuzzy set F respectively as follows:

$$\mu_{\underline{R}_\beta F}(X_i) = \begin{cases} \inf\{\mu_F(x): \ x \in S_i\} & \text{if } e_s(X_i, F) \le \beta \\ 0 & \text{otherwise} \end{cases} \tag{33}$$

$$\mu_{\overline{R}_\beta F}(X_i) = \begin{cases} \sup\{\mu_F(x): \ x \in S_i\} & \text{if } e_s(X_i, F) < 1 - \beta \\ 0 & \text{otherwise} \end{cases} \tag{34}$$

where $S_i = \text{supp}(X_i \cap F)$ is the support set of the intersection of X_i and F, and e_s is the support inclusion error, which can be defined for any nonempty fuzzy sets A and B

$$e_s(A, B) = 1 - \frac{\text{card}(\text{supp}(A \cap B))}{\text{card}(\text{supp}(A))} . \tag{35}$$

The mean rough fuzzy β-approximations were defined [12] as follows:

$$\mu_{\underline{R}_\beta F}(X_i) = \begin{cases} f_i & \text{if } e_s(X_i, F) \le \beta \\ 0 & \text{otherwise} \end{cases} \tag{36}$$

$$\mu_{\overline{R}_\beta F}(X_i) = \begin{cases} f_i & \text{if } e_s(X_i, F) < 1 - \beta \\ 0 & \text{otherwise} \end{cases} \tag{37}$$

$$f_i = \frac{\text{power}(X_i \cap F)}{\text{card}(\text{supp}(X_i \cap F))} . \tag{38}$$

By comparing (33),(34), (36) and (37) with the definitions (25), (26), (27) and (28) respectively and taking into account the proposition 5, one can easy show that the former definitions constitute a restricted version of the new ones. This can be done by setting $u = 1 - \beta$ and $l = \beta$, and by narrowing the interval $(0,1]$ of α so that only those elements of the approximating crisp class X_i are eliminated which do not belong to the fuzzy set F at all. This is the worst case $(1 \rightarrow 0)$, in which implication produces the value of 0 by definition.

We will use further only the refined variable precision fuzzy rough approximations given in the current paper.

6   Decision Tables with Fuzzy Attributes

In order to analyse decision tables with fuzzy attributes we defined in [12] a fuzzy compatibility relation R. We introduced furthermore the notion of fuzzy information system S with the following formal description

$$S = \langle X, Q, V, f \rangle \tag{39}$$

where:

X – a nonempty set, called the universe,

Q – a finite set of attributes,

V – a set of fuzzy values of attributes.

    $V = \bigcup_{q \in Q} V_q$, where: V_q is the fuzzy domain of the attribute q,

    V_q is the fuzzy (linguistic) value given by a membership function μ_{V_q}

    defined on the original domain U_q of the attribute q,

f – an information function, $f\colon X \times Q \to V$,

    $f(x, q) \in V_q$, $\forall q \in Q$, and $\forall x \in X$.

A compatibility relation R, for comparing any elements $x, y \in X$ with fuzzy values of attributes, is defined as follows [12]:

$$\mu_R(x, y) = \min_{q \in Q} \sup_{u \in U_q} \min(\mu_{V_q(x)}(u), \mu_{V_q(y)}(u)) \tag{40}$$

where $V_q(x), V_q(y)$ are fuzzy values of the attribute q for x and y respectively.

The relation given by (40) is reflexive and symmetric (tolerance relation). If the intersection of any two different fuzzy values of each attribute equals to an empty fuzzy set, then the relation (40) is additionally transitive (fuzzy similarity relation). In such a case the decision table can be analysed using the original measures of the rough sets theory. For crisp attributes the relation (40) is an equivalence relation.

Another form of fuzzy decision tables was considered by Bodjanova [1]. In that approach the attributes represented degree of membership in fuzzy condition and fuzzy decision concepts.

An important measure, often used for evaluating the consistency of decision tables, is the approximation quality, which was originally defined for a given family of crisp sets $Y = \{Y_1, Y_2, \ldots, Y_n\}$ and a crisp indiscernibility relation R:

$$\gamma_R(Y) = \frac{\text{card}(\text{Pos}_R(Y))}{\text{card}(X)} \tag{41}$$

$$\text{Pos}_R(Y) = \bigcup_{Y_i \in Y} \underline{R}Y_i \,. \tag{42}$$

We modified the measure of approximation quality in order to deal with fuzzy sets and fuzzy relations [12].

For a family $\Phi = \{F_1, F_2, \ldots, F_n\}$ of fuzzy sets and a fuzzy compatibility relation R the approximation quality of Φ by R is defined as follows:

$$\gamma_R(\Phi) = \frac{\text{power}(\text{Pos}_R(\Phi))}{\text{card}(X)} \tag{43}$$

$$\mathrm{Pos}_R(\Phi) = \bigcup_{F_i \in \Phi} \omega(\underline{R}F_i) \,. \tag{44}$$

The equation (43) is a generalised definition of approximation quality (mapping ω is explained in the section 3). If the family Φ and the relation R are crisp, then the generalised approximation quality (43) is equivalent to (41).

In the next section we will need the measure (43) for evaluating the quality of approximation of compatibility classes obtained with respect to fuzzy decision attributes by compatibility classes obtained with respect to fuzzy condition attributes. Because the positive area of classification (44) in the VPFRS model will be obtained by allowing some inclusion error $(1 - u)$ we use a measure, which is called u-lower approximation quality.

7   Examples

In the following example we apply the proposed concept of variable precision fuzzy rough approximations to analysis of a decision table with fuzzy attributes (Table 1). We use a compatibility relation (40) for comparing elements of the universe.

Table 1. Decision table with fuzzy attributes

x	c_1	c_2	c_3	d
x_1	A_1	B_1	C_1	D_1
x_2	A_2	B_2	C_2	D_2
x_3	A_1	B_2	C_2	D_2
x_4	A_1	B_1	C_1	D_1
x_5	A_1	B_1	C_1	D_3
x_6	A_2	B_2	C_2	D_2
x_7	A_1	B_2	C_1	D_3
x_8	A_1	B_1	C_1	D_1
x_9	A_1	B_2	C_1	D_3
x_{10}	A_1	B_1	C_1	D_1

For all attributes typical triangular fuzzy membership functions were chosen. The intersection levels of different linguistic values for attributes are assumed as follows:

for A_1 and A_2: 0.3,   for B_1 and B_2: 0.2,   for C_1 and C_2: 0.25,
for D_1 and D_2: 0.2,   for D_2 and D_3: 0.2,
otherwise: 0.

We obtain a family $\Phi = \{F_1, F_2, F_3\}$ of compatibility classes with respect to the fuzzy decision attribute d:

$$F_1 = \{\ 1.00/x_1,\quad 0.20/x_2,\quad 0.20/x_3,\quad 1.00/x_4,\quad 0.00/x_5,\quad 0.20/x_6,\quad 0.00/x_7,$$
$$1.00/x_8,\quad 0.00/x_9,\quad 1.00/x_{10}\ \},$$
$$F_2 = \{\ 0.20/x_1,\quad 1.00/x_2,\quad 1.00/x_3,\quad 0.20/x_4,\quad 0.20/x_5,\quad 1.00/x_6,\quad 0.20/x_7,$$
$$0.20/x_8,\quad 0.20/x_9,\quad 0.20/x_{10}\ \},$$
$$F_3 = \{\ 0.00/x_1,\quad 0.20/x_2,\quad 0.20/x_3,\quad 0.00/x_4,\quad 1.00/x_5,\quad 0.20/x_6,\quad 1.00/x_7,$$
$$0.00/x_8,\quad 1.00/x_9,\quad 0.00/x_{10}\ \},$$

and the following family $\Psi = \{X_1, X_2, X_3, X_4\}$ of compatibility classes with respect to the fuzzy condition attributes c_1, c_2, c_3:

$$X_1 = \{\ 1.00/x_1,\quad 0.20/x_2,\quad 0.20/x_3,\quad 1.00/x_4,\quad 1.00/x_5,\quad 0.20/x_6,\quad 0.20/x_7,$$
$$1.00/x_8,\quad 0.20/x_9,\quad 1.00/x_{10}\ \},$$
$$X_2 = \{\ 0.20/x_1,\quad 1.00/x_2,\quad 0.30/x_3,\quad 0.20/x_4,\quad 0.20/x_5,\quad 1.00/x_6,\quad 0.25/x_7,$$
$$0.20/x_8,\quad 0.25/x_9,\quad 0.20/x_{10}\ \},$$
$$X_3 = \{\ 0.20/x_1,\quad 0.30/x_2,\quad 1.00/x_3,\quad 0.20/x_4,\quad 0.20/x_5,\quad 0.30/x_6,\quad 0.25/x_7,$$
$$0.20/x_8,\quad 0.25/x_9,\quad 0.20/x_{10}\ \},$$
$$X_4 = \{\ 0.20/x_1,\quad 0.25/x_2,\quad 0.25/x_3,\quad 0.20/x_4,\quad 0.20/x_5,\quad 0.25/x_6,\quad 1.00/x_7,$$
$$0.20/x_8,\quad 1.00/x_9,\quad 0.20/x_{10}\ \}.$$

Table 2. Membership functions of X_1, F_1, $X_1^{F_1}$

x	$\mu_{X_1}(x)$	$\mu_{F_1}(x)$	$\mu_{X_1^{F_1}}(x)$ (G)	$\mu_{X_1^{F_1}}(x)$ (L)
x_1	1.00	1.00	1.00	1.00
x_2	0.20	0.20	1.00	1.00
x_3	0.20	0.20	1.00	1.00
x_4	1.00	1.00	1.00	1.00
x_5	1.00	0.00	0.00	0.00
x_6	0.20	0.20	1.00	1.00
x_7	0.20	0.00	0.00	0.80
x_8	1.00	1.00	1.00	1.00
x_9	0.20	0.00	0.00	0.80
x_{10}	1.00	1.00	1.00	1.00

The Table 2 contains the membership functions of the fuzzy inclusion sets $X_1^{F_1}$ obtained for the Gaines R-implicator and the Łukasiewicz RS-implicator respectively.

We can observe for x_7 and x_9 a big difference between the values of the membership functions $\mu_{X_1^{F_1}}(x)$ obtained for the Gaines and the Łukasiewicz implicator. If $\mu_{F_1}(x) = 0$ the Gaines implicator ($x \to y = 1$ if $x \le y$ and y/x otherwise) produces always 0. The implicator of Łukasiewicz ($x \to y = \min(1, 1 - x + y)$) is more suitable in that case because its value is proportional to the difference $x - y$, when $x > y$. It will be easier, on a later stage, to obtain the largest possible α-cut of the fuzzy inclusion set for a given value of the admissible inclusion error, if we apply the Łukasiewicz implicator.

Table 3. u-lower approximation of F_1

	u	G-inf	G-mean	L-inf	L-mean
			Method		
$\mu_{\underline{R}_u F_1}(X_1)$	1	0.00	0.00	0.00	0.00
	0.8	0.00	0.00	0.80	0.96
	0.75	1.00	1.00	1.00	1.00
$\mu_{\underline{R}_u F_1}(X_2)$	1	0.00	0.00	0.20	0.76
	0.8	0.20	0.72	0.20	0.76
	0.75	0.20	0.72	0.20	0.76
$\mu_{\underline{R}_u F_1}(X_3)$	1	0.00	0.00	0.20	0.83
	0.8	0.00	0.00	0.20	0.83
	0.75	0.20	0.79	0.20	0.83
$\mu_{\underline{R}_u F_1}(X_4)$	1	0.00	0.00	0.00	0.00
	0.8	0.00	0.00	0.00	0.00
	0.75	0.00	0.00	0.00	0.00

The results for the u-lower approximation of F_1 by the family Ψ are given in the Table 3.

Let us analyse for example the case, where the upper limit $u = 0.80$. The admissible inclusion error is equal to $1 - u = 0.20$.

We see that the membership degrees $\mu_{F_1}(X_1)$ for the Gaines implicator are equal to 0, whereas for the Lukasiewicz implicator we obtain $\mu_{F_1}(X_1) = 0.80$ for the infimum and $\mu_{F_1}(X_1) = 0.96$ for the mean u-lower approximation.

Only by using a larger value of the admissible inclusion error $1 - u = 0.25$ we obtain better results for the Gaines implicator: $\mu_{F_1}(X_1) = 1$ for both the infimum and the mean u-lower approximation.

The results for the u-lower approximation of F_2 and F_3 are given in the Tables 4 and 5. The u-lower approximations of the whole family Φ are given in the Tables 6, 7 and 8.

The obtained differences between the Gaines and the Lukasiewicz implicator have significant influence on the approximation quality for the considered fuzzy information system (see the Table 9) especially for the infimum u-lower approximation. We obtain smaller differences for the Gaines and Lukasiewicz implicator in case of the mean u-lower approximation.

The results given in the Table 9 validate the necessity and usefulness of the introduced VPFRS model. Allowing some level of misclassification leads to a significant increase of the u-approximation quality (important measure used in analysis of information systems). The mean based VPFRS model produce higher values of the u-approximation quality than the limit based VPFRS model.

It must be emphasised here that the strength of the variable precision rough set model can be observed especially for large universes. We had to choose larger values of the admissible inclusion error in the above example, in order to show

Table 4. u-lower approximation of F_2

	u	G-inf	G-mean	L-inf	L-mean
		\multicolumn{4}{c}{Method}		
$\mu_{\underline{R}_u F_2}(X_1)$	1	0.20	0.60	0.20	0.60
	0.75	0.20	0.60	0.20	0.60
$\mu_{\underline{R}_u F_2}(X_2)$	1	0.80	0.96	0.95	0.99
	0.85	1.00	1.00	1.00	1.00
$\mu_{\underline{R}_u F_2}(X_3)$	1	0.80	0.96	0.95	0.99
	0.8	1.00	1.00	1.00	1.00
$\mu_{\underline{R}_u F_2}(X_4)$	1	0.20	0.84	0.20	0.84
	0.75	0.20	0.84	0.20	0.84

Table 5. u-lower approximation of F_3

	u	G-inf	G-mean	L-inf	L-mean
$\mu_{\underline{R}_u F_3}(X_1)$	1	0.00	0.00	0.00	0.00
	0.75	0.00	0.00	0.00	0.00
$\mu_{\underline{R}_u F_3}(X_2)$	1	0.00	0.00	0.20	0.75
	0.75	0.20	0.68	0.20	0.75
$\mu_{\underline{R}_u F_3}(X_3)$	1	0.00	0.00	0.20	0.82
	0.75	0.00	0.00	0.20	0.82
$\mu_{\underline{R}_u F_3}(X_4)$	1	0.00	0.00	0.80	0.91
	0.75	0.80	0.90	0.95	0.98

Table 6. u-lower approximation of Φ for $u = 1$

G-inf	{ $0.20/X_1$,	$0.80/X_2$,	$0.80/X_3$,	$0.20/X_4$ }
G-mean	{ $0.60/X_1$,	$0.96/X_2$,	$0.96/X_3$,	$0.84/X_4$ }
L-inf	{ $0.20/X_1$,	$0.95/X_2$,	$0.95/X_3$,	$0.80/X_4$ }
L-mean	{ $0.60/X_1$,	$0.99/X_2$,	$0.99/X_3$,	$0.91/X_4$ }

Table 7. u-lower approximation of Φ for $u = 0.8$

G-inf	{ $0.20/X_1$,	$1.00/X_2$,	$1.00/X_3$,	$0.20/X_4$ }
G-mean	{ $0.60/X_1$,	$1.00/X_2$,	$1.00/X_3$,	$0.84/X_4$ }
L-inf	{ $0.80/X_1$,	$1.00/X_2$,	$1.00/X_3$,	$0.80/X_4$ }
L-mean	{ $0.96/X_1$,	$1.00/X_2$,	$1.00/X_3$,	$0.91/X_3$ }

Table 8. u-lower approximation of Φ for $u = 0.75$

G-inf	{ 1.00/X_1, 1.00/X_2, 1.00/X_3, 0.80/X_4 }
G-mean	{ 1.00/X_1, 1.00/X_2, 1.00/X_3, 0.90/X_4 }
L-inf	{ 1.00/X_1, 1.00/X_2, 1.00/X_3, 0.95/X_4 }
L-mean	{ 1.00/X_1, 1.00/X_2, 1.00/X_3, 0.98/X_4 }

Table 9. u-approximation quality of Φ

		Method			
	u	G-inf	G-mean	L-inf	L-mean
	1	0.380	0.756	0.553	0.779
$\gamma_R(\Phi)$	0.8	0.440	0.768	0.860	0.962
	0.75	0.960	0.980	0.990	0.996

the properties of the proposed approach. Nevertheless, the admissible inclusion error of about 0.2 turned out to be reasonable for analysing large universes obtained from dynamic processes [10, 11, 13].

8   Conclusions

In this paper a concept of variable precision fuzzy rough sets model (VPFRS) was proposed. The VPRS model with asymmetric bounds (l, u) was used. The starting point of the VPFRS idea was introduction of the notion of fuzzy inclusion set that should be based on R-implicators. A generalised notion of α-inclusion error was defined, expressed by means of α-cuts of the fuzzy inclusion set. The idea of mean fuzzy rough approximations was proposed, which helps to obtain results that better correspond to the statistical properties of analysed large information systems. We suggest to use it particularly for small values of the admissible inclusion error. Furthermore, it turns out that application of the Łukasiewicz R-implicator is a good choice for determination of fuzzy rough approximations. The presented generalised approach to VPFRS can be helpful especially in case of analysing fuzzy information systems obtained from real (dynamic) processes. In future work we will concentrate on axiomatisation and further development of the proposed VPFRS model.

References

1. Bodjanova, S.: Approximation of Fuzzy Concepts in Decision Making. Fuzzy Sets and Systems, Vol. 85 (1997)
2. Chakrabarty, K., Biswas, R., Nanda, S.: Fuzziness in Rough Sets. Fuzzy Sets and Systems, Vol. 110 (2000)
3. Dubois, D., Prade H.: Putting Rough Sets and Fuzzy Sets Together. In: Słowiński, R. (ed.): Intelligent Decision Support. Handbook of Applications and Advances of the Rough Sets. Kluwer Academic Publishers, Boston Dordrecht London (1992)

4. Greco, S., Matarazzo B., Słowiński R.: The use of rough sets and fuzzy sets in MCDM. In: Gal, T., Stewart, T., Hanne, T. (eds.): Advances in Multiple Criteria Decision Making. Kluwer Academic Publishers, Boston Dordrecht London (1999)
5. Greco, S., Matarazzo B., Słowiński R.: Rough set processing of vague information using fuzzy similarity relations. In: Calude, C.S., Paun, G. (eds.): Finite Versus Infinite – Contributions to an Eternal Dilemma. Springer-Verlag, Berlin Heidelberg New York (2000)
6. Inuiguchi, M., Tanino, T.: New Fuzzy Rough Sets Based on Certainty Qualification. In: Pal, S. K., Polkowski, L., Skowron, A. (eds.): Rough-Neuro-Computing: Techniques for Computing with Words. Springer-Verlag, Berlin Heidelberg New York (2002)
7. Katzberg, J.D., Ziarko, W.: Variable Precision Extension of Rough Sets. Fundamenta Informaticae, Vol. 27 (1996)
8. Klir, J., Folger, T. A.: Fuzzy Stets Unertainty and Information. Prentice Hall, Englewood, New Jersey (1988)
9. Lin, T.Y.: Topological and Fuzzy Rough Sets. In: Słowiński, R. (ed.): Intelligent Decision Support. Handbook of Applications and Advances of the Rough Sets. Kluwer Academic Publishers, Boston Dordrecht London (1992)
10. Mieszkowicz-Rolka, A., Rolka, L.: Variable Precision Rough Sets in Analysis of Inconsistent Decision Tables. In: Rutkowski, L., Kacprzyk, J. (eds.): Advances in Soft Computing. Physica-Verlag, Heidelberg (2003)
11. Mieszkowicz-Rolka, A., Rolka, L.: Variable Precision Rough Sets. Evaluation of Human Operator's Decision Model. In: Sołdek, J., Drobiazgiewicz, L. (eds.): Artificial Intelligence and Security in Computing Systems. Kluwer Academic Publishers, Boston Dordrecht London (2003)
12. Mieszkowicz-Rolka, A., Rolka, L.: Fuziness in Information Systems. Electronic Notes in Theoretical Computer Science, Vol. 82, Issue No. 4. http://www.elsevier.nl/locate/entcs/volume82.html
13. Mieszkowicz-Rolka, A., Rolka, L.: Studying System Properties with Rough Sets. Lectures Notes in Computer Science, Vol. 2657. Springer Verlag, Berlin Heidelberg New York (2003)
14. Nakamura, A.: Application of Fuzzy-Rough Classifications to Logics. In: Słowiński, R. (ed.): Intelligent Decision Support. Handbook of Applications and Advances of the Rough Sets. Kluwer Academic Publishers, Boston Dordrecht London (1992)
15. Pawlak, Z.: Rough Sets. Theoretical Aspects of Reasoning about Data. Kluwer Academic Publishers, Boston Dordrecht London (1991)
16. Polkowski, L.: Rough Sets: Mathematical Foundations. Physica-Verlag, Heidelberg (2002)
17. Radzikowska, A.M., Kerre, E.E.: A Comparative Study of Fuzzy Rough Sets. Fuzzy Sets and Systems, Vol. 126 (2002)
18. Ziarko, W.: Variable Precision Rough Sets Model. Journal of Computer and System Sciences, Vol. 40 (1993)

Greedy Algorithm of Decision Tree Construction for Real Data Tables

Mikhail Ju. Moshkov[1,2]

[1] Faculty of Computing Mathematics and
Cybernetics, Nizhny Novgorod State University
23, Gagarina Ave., Nizhny Novgorod, 603950, Russia
moshkov@unn.ac.ru
[2] Institute of Computer Science, University of Silesia
39, Będzińska St., Sosnowiec, 41-200, Poland

Abstract. In the paper a greedy algorithm for minimization of decision tree depth is described and bounds on the algorithm precision are considered. This algorithm is applicable to data tables with both discrete and continuous variables which can have missing values. Under some natural assumptions on the class NP and on the class of considered tables, the algorithm is, apparently, close to best approximate polynomial algorithms for minimization of decision tree depth.

Keywords: data table, decision table, decision tree, depth

1   Introduction

Decision trees are widely used in different applications as algorithms for task solving and as a way for knowledge representation. Problems of decision tree optimization are very complicated.

In this paper we consider approximate algorithm for decision tree depth minimization which can be applied to real data tables with both discrete and continuous variables having missing values. First, we transform given data table into a decision table, possibly, with many-valued decisions (i.e. pass to the model which is usual for rough set theory [7, 8]). Later, we apply to this table a greedy algorithm which is similar to algorithms for decision tables with one-valued decisions [3], but uses more complicated uncertainty measure.

We obtain bounds on precision for this algorithm and, based on results from [2], show that under some natural assumptions on the class NP and on the class of considered tables, the algorithm is, apparently, close to best approximate polynomial algorithms for minimization of decision tree depth.

Note that [6] contains some similar results without proofs.

The results of the paper were obtained partially in the frameworks of joint research project of Intel Nizhny Novgorod Laboratory and Nizhny Novgorod State University.

J.F. Peters et al. (Eds.): Transactions on Rough Sets I, LNCS 3100, pp. 161–168, 2004.

2   Data Tables and Attributes

A data table D is a rectangular table with t columns which correspond to variables $x_1, \ldots, x_t$. The rows of D are t-tuples of variable $x_1, \ldots, x_t$ values. Values of some variables in some rows can be missed. The table D can contain equal rows. The variables are separated into discrete and continuous. A discrete variable x_i takes values from an unordered finite set A_i. A continuous variable x_j takes values from the set $\mathbb{R}$ of real numbers. Each row r of the table D is labelled by an element $y(r)$ from a finite set C. One can interpret these elements as values of a new variable y. The problem connected with the table D is to predict the value of y using variables $x_1, \ldots, x_t$. To this end we will not use values of $x_1, \ldots, x_t$ directly. We will use values of some attributes depending on variables from the set $\{x_1, \ldots, x_t\}$.

An attribute is a function f depending on variables $x_{i_1}, \ldots, x_{i_m} \in \{x_1, \ldots, x_t\}$ and taking values from the set $E = \{0, 1, *\}$. Let r be a row of D. If values of all variables $x_{i_1}, \ldots, x_{i_m}$ are definite in r then for this row the value of $f(x_{i_1}, \ldots, x_{i_m})$ belongs to the set $\{0, 1\}$. If value of at least one of variables $x_{i_1}, \ldots, x_{i_m}$ is missed in r then for this row the value of $f(x_{i_1}, \ldots, x_{i_m})$ is equal to $*$.

Consider some examples of attributes. In the system CART [1] the attributes are considered (in the main) each of which depends on one variable x_i. Let x_i be a continuous variable, and a be a real number. Then the considered attribute takes value 0 if $x_i < a$, takes value 1 if $x_i \geq a$, and takes value $*$ if the value of x_i is missed. Let x_i be a discrete variable, which takes values from the set A_i, and B be a subset of A_i. Then the considered attribute takes value 0 if $x_i \notin B$, takes value 1 if $x_i \in B$, and takes value $*$ if the value of x_i is missed. It is possible to consider attributes depending on many variables. For example, let φ be a polynomial depending on continuous variables $x_{i_1}, \ldots, x_{i_m}$. Then the considered attribute takes value 0 if $\varphi(x_{i_1}, \ldots, x_{i_m}) < 0$, takes value 1 if $\varphi(x_{i_1}, \ldots, x_{i_m}) \geq 0$, and takes value $*$ if the value of at least one of variables $x_{i_1}, \ldots, x_{i_m}$ is missed.

Let $F = \{f_1, \ldots, f_k\}$ be a set of attributes which will be used for prediction of the variable y value. We will say that two rows r_1 and r_2 are equivalent relatively F if each attribute f_i from F takes the same value on r_1 and r_2. The considered equivalence relation divides the set of rows of the table D into equivalence classes $S_1, \ldots, S_q$. Let $j \in \{1, \ldots, q\}$. The rows from the equivalence class S_j are indiscernible from the point of view of values of attributes from F. So when we will predict the value y using only attributes from F we will give the same answer (element from C) for any row from S_j. Denote by $C(S_j)$ the set of elements d from C such that

$$|\{r : r \in S_j, y(r) = d\}| = \max_{c \in C} |\{r : r \in S_j, y(r) = c\}| \ .$$

It is clear that only answers from the set $C(S_j)$ minimize the number of mistakes for rows from the class S_j. For any $r \in S_j$ denote $C(r) = C(S_j)$.

Now we can formulate exactly the problem $\mathrm{Pred}(D, F)$ of prediction of the variable y value: for a given row r of the data table D we must find a number from the set $C(r)$ using values of attributes from F.

Note that in [5] another setting of the problem of prediction was considered: for a given row r of the data table D we must find the set $\{y(r') : r' \in S_j\}$, where S_j is the equivalence class containing r.

3   Decision Trees

As algorithms for the problem $\mathrm{Pred}(D, F)$ solving we will consider decision trees with attributes from the set F. Such decision tree is finite directed tree with the root in which each terminal node is labelled either by an element from the set C or by nothing, each non-terminal node is labelled by an attribute from the set F. Three edges start in each non-terminal node. These edges are labelled by 0, 1 and $*$ respectively.

The functioning of a decision tree Γ on a row of the data table D is defined in the natural way. We will say that the decision tree Γ solves the problem $\mathrm{Pred}(D, F)$ if for any row r of D the computation is finished in a terminal node of Γ which is labelled by an element of the set $C(r)$.

The depth of a decision tree is the maximal length of a path from the root to a terminal node of the tree. We denote by $h(\Gamma)$ the depth of a decision tree Γ. By $h(D, F)$ we denote the minimal depth of a decision tree with attributes from F which solves the problem $\mathrm{Pred}(D, F)$.

4   Decision Tables with Many-Valued Decisions

We will assume that the information about the problem $P(D, F)$ is represented in the form of a decision table $T = T(D, F)$. The table T has k columns corresponding to the attributes $f_1, \ldots, f_k$ and q rows corresponding to the equivalence classes $S_1, \ldots, S_q$. The value $f_j(r_i)$ is on the intersection of a row S_i and a column f_j, where r_i is an arbitrary row from the equivalence class S_i. For $i = 1, \ldots, q$ the row S_i of the table T is labelled by the subset $C(S_i)$ of the set C.

We will consider sub-tables of the table T which can be obtained from T by removal of some rows. Let T' be a sub-table of T.

Denote by $\mathrm{Row}(T')$ the set of rows of the table T'. The table T' will be called degenerate if $\mathrm{Row}(T') = \emptyset$ or $\bigcap_{S_i \in \mathrm{Row}(T')} C(S_i) \neq \emptyset$.

Let $i_1, \ldots, i_m \in \{1, \ldots, k\}$ and $\delta_1, \ldots, \delta_m \in E = \{0, 1, *\}$. We denote by $T'(i_1, \delta_1) \ldots (i_m, \delta_m)$ the sub-table of the table T' that consists of rows each of which on the intersections with columns $f_{i_1}, \ldots, f_{i_m}$ has elements $\delta_1, \ldots, \delta_m$ respectively.

We define the parameter $M(T')$ of the table T' as follows. If T' is a degenerate table then $M(T') = 0$. Let T' be a non-degenerate table. Then $M(T')$ is minimal natural m such that for any $(\delta_1, \ldots, \delta_k) \in E^k$ there exist numbers $i_1, \ldots, i_n \in \{1, \ldots, k\}$, for which $T'(i_1, \delta_{i_1}) \ldots (i_n, \delta_{i_n})$ is a degenerate table, and $n \leq m$.

A nonempty subset B of the set $\mathrm{Row}(T')$ will be called boundary set if $\bigcap_{S_i \in B} C(S_i) = \emptyset$ and $\bigcap_{S_i \in B'} C(S_i) \neq \emptyset$ for any nonempty subset B' of the set B such that $B' \neq B$. We denote by $R(T')$ the number of boundary subsets of the set $\mathrm{Row}(T')$. It is clear that $R(T') = 0$ if and only if T' is a degenerate table.

5   Algorithm U for Decision Tree Construction

For decision table $T = T(D, F)$ we construct a decision tree $U(T)$ which solves the problem $\text{Pred}(D, F)$.

We begin the construction from the tree that consists of one node v which is not labelled. If T has no rows then we finish the construction. Let T have rows and $\bigcap_{S_i \in \text{Row}(T)} C(S_i) \neq \emptyset$. Then we mark the node v by an element from the set $\bigcap_{S_i \in \text{Row}(T)} C(S_i)$ and finish the construction.

Let T have rows and $\bigcap_{S_i \in \text{Row}(T)} C(S_i) = \emptyset$. For $i = 1, \ldots, k$ we compute the value $Q_i = \max\{R(T(i, \delta)) : \delta \in E\}$. We mark the node v by the attribute f_{i_0} where i_0 is the minimal i for which Q_i has minimal value. For each $\delta \in E$ we add to the tree the node $v(\delta)$, draw the edge from v to $v(\delta)$, and mark this edge by element δ. For the node $v(\delta)$ we will make the same operations as for the node v, but instead of the table T we will consider the table $T(i_0, \delta)$, etc.

6   Bounds on Algorithm U Precision

If T is a degenerate table then the decision tree $U(T)$ consists of one node. The depth of this tree is equal to 0. Consider now the case when T is a non-degenerate table.

Theorem 1. *Let the decision table $T = T(D, F)$ be non-degenerate. Then*

$$h(U(T)) \leq M(T) \ln R(T) + 1 \ .$$

Later we will show (see Lemma 3) that $M(T) \leq h(D, F)$. So we have the following

Corollary 1. *Let the decision table $T = T(D, F)$ be non-degenerate. Then*

$$h(U(T)) \leq h(D, F) \ln R(T) + 1 \ .$$

Let t be a natural number. Denote by $\text{Tab}(t)$ the set of decision tables T such that $|C(S_i)| \leq t$ for any row $S_i \in \text{Row}(T)$. Let $T \in \text{Tab}(t)$. One can show that each boundary subset of the set $\text{Row}(T)$ has at most $t + 1$ rows. Using this fact it is not difficult to show that the algorithm U has polynomial time complexity on the set $\text{Tab}(t)$.

Using results from [2] on precision of approximate polynomial algorithms for set covering problem it is possible to prove that if $NP \not\subseteq DTIME(n^{O(\log \log n)})$ then for any ε, $0 < \varepsilon < 1$, there is no polynomial algorithm which for a given decision table $T = T(D, F)$ from $\text{Tab}(t)$ constructs a decision tree Γ such that Γ solves the problem $\text{Pred}(D, F)$ and $h(\Gamma) \leq (1 - \varepsilon)h(D, F) \ln R(T)$. We omit the proof of this statement. Proof of a similar result can be found in [4].

Using Corollary 1 we conclude that if $NP \not\subseteq DTIME(n^{O(\log \log n)})$ then the algorithm U is, apparently, close to best (from the point of view of precision) approximate polynomial algorithms for minimization of decision tree depth for decision tables from $\text{Tab}(t)$ (at least for small values of t).

7   Proof of Precision Bounds

Lemma 1. *Let Γ be a decision tree which solves the problem* $\mathrm{Pred}(D, F)$, $T = T(D, F)$ *and τ be a path of the length n from the root to a terminal node of Γ, in which non-terminal nodes are labelled by attributes $f_{i_1}, ..., f_{i_n}$, and edges are labelled by elements $\delta_1, ..., \delta_n$. Then $T(i_1, \delta_1) \ldots (i_n, \delta_n)$ is a degenerate table.*

Proof. Assume the contrary: let the table $T' = T(i_1, \delta_1) \ldots (i_n, \delta_n)$ be non-degenerate. Let the terminal node v of the path τ be labelled by an element $c \in C$. Since T' is a non-degenerate table, it has a row (equivalence class) S_i such that $c \notin C(S_i)$. Evidently, $c \notin C(r)$ for any row $r \in S_i$. It is clear that for any row $r \in S_i$ the computation in the tree Γ moves along the path τ and finishes in the node v which is impossible since Γ is a decision tree solving the problem $\mathrm{Pred}(D, F)$, and $c \notin C(r)$. Therefore $T(i_1, \delta_1) \ldots (i_n, \delta_n)$ is a degenerate table. $\qquad\square$

Lemma 2. *Let $T = T(D, F)$ and T_1 be a sub-table of T. Then $M(T_1) \leq M(T)$.*

Proof. Let $i_1, \ldots, i_n \in \{1, \ldots, k\}$ and $\delta_1, \ldots, \delta_n \in E$. If $T(i_1, \delta_1) \ldots (i_n, \delta_n)$ is a degenerate table then $T_1(i_1, \delta_1) \ldots (i_n, \delta_n)$ is a degenerate table too. From here and from the definition of the parameter M the statement of the lemma follows. $\qquad\square$

Lemma 3. *Let $T = T(D, F)$. Then $h(D, F) \geq M(T)$.*

Proof. Let T be a degenerate table. Then, evidently, $M(T) = 0$ and $h(D, F) = 0$. Let T be a non-degenerate table, and Γ be a decision tree, which solves the problem $\mathrm{Pred}(D, F)$ and for which $h(\Gamma) = h(D, F)$. Consider a tuple $(\delta_1, \ldots, \delta_k) \in E^k$, which satisfies the following condition: if $i_1, \ldots, i_n \in \{1, \ldots, k\}$ and $T(i_1, \delta_{i_1}) \ldots (i_n, \delta_{i_n})$ is a degenerate table then $n \geq M(T)$. The existence of such tuple follows from the definition of the parameter $M(T)$.

Consider a path τ from the root to a terminal node of Γ, which satisfies the following conditions. Let the length of τ be equal to m and non-terminal nodes of τ be labelled by attributes $f_{i_1}, \ldots, f_{i_m}$. Then the edges of τ are labelled by elements $\delta_{i_1}, \ldots, \delta_{i_m}$ respectively. From Lemma 1 follows that $T(i_1, \delta_{i_1}) \ldots (i_m, \delta_{i_m})$ is a degenerate table. Therefore $m \geq M(T)$, and $h(\Gamma) \geq M(T)$. Since $h(D, F) = h(\Gamma)$, we obtain $h(D, F) \geq M(T)$. $\qquad\square$

Lemma 4. *Let $T = T(D, F)$, T_1 be be a sub-table of T, $i, i_1, \ldots, i_m \in \{1, \ldots, k\}$ and $\delta, \delta_1, \ldots, \delta_m \in E$. Then*

$$R(T_1) - R(T_1(i, \delta)) \geq R(T_1(i_1, \delta_1) \ldots (i_m, \delta_m))$$
$$-R(T_1(i_1, \delta_1) \ldots (i_m, \delta_m)(i, \delta)) \ .$$

Proof. Let $T_2 = T_1(i_1, \delta_1) \ldots (i_m, \delta_m)$. We denote by P_1 (respectively by P_2) the set of boundary sets of rows from T_1 (respectively from T_2) in each of which at least one row has in the column f_i an element, which is not equal to δ. One can show that $P_2 \subseteq P_1$, $|P_1| = R(T_1) - R(T_1(i, \delta))$ and $|P_2| = R(T_2) - R(T_2(i, \delta))$. $\qquad\square$

Proof (of Theorem 1). Consider a most long path in the tree $U(T)$ from the root to a terminal node. Let its length be equal to n, its non-terminal nodes be labelled by attributes $f_{l_1}, \ldots, f_{l_n}$, and its edges be labelled by elements $\delta_1, \ldots, \delta_n$. Consider the tables $T_1, \ldots, T_{n+1}$, where $T_1 = T$ and $T_{p+1} = T_p(l_p, \delta_p)$ for $p = 1, \ldots, n$. Let us prove that for any $p \in \{1, \ldots, n\}$ the following inequality holds:

$$R(T_{p+1}) \le \frac{M(T_p) - 1}{M(T_p)} R(T_p) \ . \tag{1}$$

From the description of the algorithm U follows that T_p is a non-degenerate table. Therefore $M(T_p) > 0$. For $i = 1, \ldots, k$ we denote by σ_i an element from E such that $R(T_p(i, \sigma_i)) = \max\{R(T_p(i, \sigma)) : \sigma \in E\}$. From the description of the algorithm U follows that l_p is the minimal number from $\{1, \ldots, k\}$ such that

$$R(T_p(l_p, \sigma_{l_p})) = \min\{R(T_p(i, \sigma_i)) : i = 1, \ldots, k\} \ .$$

Consider the tuple $(\sigma_1, \ldots, \sigma_k)$. From the definition of $M(T_p)$ follows that there exists numbers $i_1, \ldots, i_m \in \{1, \ldots, k\}$ for which $m \le M(T_p)$ and $T_p(i_1, \sigma_{i_1}) \ldots (i_m, \sigma_{i_m})$ is a degenerate table. Therefore $R(T_p(i_1, \sigma_{i_1}) \ldots (i_m, \sigma_{i_m})) = 0$. Hence

$$R(T_p) - [R(T_p) - R(T_p(i_1, \sigma_{i_1}))] - [R(T_p(i_1, \sigma_{i_1}))$$
$$-R(T_p(i_1, \sigma_{i_1})(i_2, \sigma_{i_2}))] - \ldots - [R(T_p(i_1, \sigma_{i_1}) \ldots (i_{m-1}, \sigma_{i_{m-1}}))$$
$$-R(T_p(i_1, \sigma_{i_1}) \ldots (i_m, \sigma_{i_m}))] = R(T_p(i_1, \sigma_{i_1}) \ldots (i_m, \sigma_{i_m})) = 0 \ .$$

Using Lemma 4 we conclude that for $j = 1, \ldots, m$ the inequality

$$R(T_p(i_1, \sigma_{i_1}) \ldots (i_{j-1}, \sigma_{i_{j-1}})) - R(T_p(i_1, \sigma_{i_1}) \ldots (i_j, \sigma_{i_j})) \le R(T_p) - R(T_p(i_j, \sigma_{i_j}))$$

holds. Therefore $R(T_p) - \sum_{j=1}^m (R(T_p) - R(T_p(i_j, \sigma_{i_j}))) \le 0$ and

$$\sum_{j=1}^m R(T_p(i_j, \sigma_{i_j})) \le (m - 1) R(T_p) \ .$$

Let $s \in \{1, \ldots, m\}$ and $R(T_p(i_s, \sigma_{i_s})) = \min\{R(T_p(i_j, \sigma_{i_j})) : j = 1, \ldots, m\}$. Then $mR(T_p(i_s, \sigma_{i_s})) \le (m - 1) R(T_p)$ and $R(T_p(i_s, \sigma_{i_s})) \le \frac{m-1}{m} R(T_p)$.

Taking into account that $R(T_p(l_p, \sigma_{l_p})) \le R(T_p(i_s, \sigma_{i_s}))$ and $m \le M(T_p)$ we obtain

$$R(T_p(l_p, \sigma_{l_p})) \le \frac{M(T_p) - 1}{M(T_p)} R(T_p) \ . \tag{2}$$

From the inequality $R(T_p(l_p, \delta_p)) \le R(T_p(l_p, \sigma_{l_p}))$ and from (2) follows that the inequality (1) holds.

From the inequality (2) in the case $p = 1$ and from description of the algorithm U follows that if $M(T) = 1$ then $h(U(T)) = 1$, and the statement of the theorem holds. Let $M(T) \ge 2$. From (1) follows that

$$R(T_n) \le R(T_1) \frac{M(T_1) - 1}{M(T_1)} \cdot \frac{M(T_2) - 1}{M(T_2)} \cdot \ldots \cdot \frac{M(T_{n-1}) - 1}{M(T_{n-1})} \ . \tag{3}$$

From the description of the algorithm U follows that T_n is a non-degenerate table. Consequently,

$$R(T_n) \geq 1 . \tag{4}$$

From Lemma 2 follows that for $p = 1, \ldots, n-1$ the inequality

$$M(T_p) \leq M(T) \tag{5}$$

holds. From (3)-(5) follows that $1 \leq R(T) \left(\frac{M(T)-1}{M(T)}\right)^{n-1}$. Therefore

$$\left(1 + \frac{1}{M(T)-1}\right)^{n-1} \leq R(T) .$$

If we take the natural logarithm of both sides of this inequality we conclude that $(n-1)\ln\left(1 + \frac{1}{M(T)-1}\right) \leq \ln R(T)$. It is known that for any natural m the inequality $\ln(1 + \frac{1}{m}) > \frac{1}{m+1}$ holds. Taking into account that $M(T) \geq 2$, we obtain the inequality $\frac{(n-1)}{M(T)} < \ln R(T)$. Hence $n < M(T)\ln R(T) + 1$. Taking into account that $h(U(T)) = n$ we obtain $h(U(T)) < M(T)\ln R(T) + 1$.   □

8   Conclusion

A greedy algorithm for minimization of decision tree depth is described. This algorithm is applicable to real data tables, which are transformed into decision tables.

The structure of the algorithm is so simple that it is possible to obtain bounds on precision of this algorithm. These bounds show that under some natural assumptions on the class NP and on the class of considered decision tables the algorithm is close, apparently, to best approximate polynomial algorithms for minimization of decision tree depth.

The second peculiarity of the algorithm is the way of the work with missing values: if we compute the value of an attribute $f(x_{i_1}, \ldots, x_{i_m})$, and the value of at least one of variables $x_{i_1}, \ldots, x_{i_m}$ is missed then the computation will go along the special edge labelled by $*$. This peculiarity may be helpful if we will see on the constructed decision tree as on a way for representation of knowledge about data table D.

References

1. Breiman, L., Friedman, J.H., Olshen, R.A., Stone, C.J.: Classification and Regression Trees. Wadsworth & Brooks, 1984
2. Feige, U.: A threshold of $\ln n$ for approximating set cover (Preliminary version). Proceedings of 28th Annual ACM Symposium on the Theory of Computing (1996) 314–318
3. Moshkov, M.Ju.: Conditional tests. Problems of Cybernetics **40**, Edited by S.V. Yablonskii. Nauka Publishers, Moscow (1983) 131–170 (in Russian)

4. Moshkov, M.Ju.: About works of R.G. Nigmatullin on approximate algorithms for solving of discrete extremal problems. Discrete Analysis and Operations Research (Series 1) **7**(1) (2000) 6–17 (in Russian)
5. Moshkov, M.Ju.: Approximate algorithm for minimization of decision tree depth. Proceedings of the Ninth International Conference Rough Sets, Fuzzy Sets, Data Mining, and Granular Computing. Chongqing, China. Lecture Notes in Computer Science **2639**, Springer-Verlag (2003) 611–614
6. Moshkov, M.Ju.: On minimization of decision tree depth for real data tables. Proceedings of the Workshop Concurrency Specification and Programming. Czarna, Poland (2003)
7. Pawlak, Z.: Rough Sets - Theoretical Aspects of Reasoning about Data. Kluwer Academic Publishers, Dordrecht, Boston, London, 1991
8. Skowron, A., Rauszer, C.: The discernibility matrices and functions in information systems. Intelligent Decision Support. Handbook of Applications and Advances of the Rough Set Theory. Edited by R. Slowinski. Kluwer Academic Publishers, Dordrecht, Boston, London (1992) 331–362

Consistency Measures for Conflict Profiles

Ngoc Thanh Nguyen and Michal Malowiecki

Department of Information Systems, Wroclaw University of Technology
Wyb. Wyspianskiego 27, 50-370 Wroclaw, Poland
{thanh,malowiecki}@pwr.wroc.pl

Abstract. The formal definition of conflict was formulated and analyzed by Pawlak. In Pawlak's works the author presented the concept and structure of conflicts. In this concept a conflict may be represented by an information system (U,A), where U is a set of agents taking part in conflict and A is a set of attributes representing conflict issues. On the basis of the information system Pawlak defined also various measures describing conflicts, for example the measure of military potential of the conflict sites. Next the concept has been developed by other authors. In their works the authors defined a multi-valued structure of conflict and proposed using consensus methods to their solving. In this paper the authors present the definition of consistency functions which should enable to measure the degree of consistency of conflict profiles. A conflict profile is defined as a set of opinions of agents referring to the subject of the conflict. Using this degree one could make choice of the method for solving the conflict, for example, a negotiation method or a consensus method. A set of postulates for consistency functions are defined and analyzed. Besides, some concrete consistency functions are formulated and their properties referring to postulates are included.

1 Introduction

In Pawlak's concept [20] a conflict is defined by an information system (U,A) in which U is the set of agents being in conflict, A is a set of attributes representing conflict issues, and the information table contains the conflict content, i.e. the opinions of the agents on particular issues. Each agent for each issue has three possibilities for presenting his opinion: (+) - *yes*, (–) - *no*, and (0) - *neutral*. For example Table 1 below represents the content of a conflict [20].

Within a conflict one can determine several conflict profiles. A conflict profile is the set of opinions generated by the agents on an issue. In the conflict represented by Table 1 we have 5 profiles P_a, P_b, P_c, P_d and P_e, where for example $P_b = \{+,+,-,-,-,+\}$ and $P_c = \{+,-,-,-,-,-\}$.

Referring to opinions belonging to these profiles one can observe that the opinions of a certain profile are more similar to each other (that is more convergent or more consistent) than opinions of some other profile. For example, opinions in profile P_c seem to be more consistent than opinions in profile P_b. Below we present another, more practical example.

J.F. Peters et al. (Eds.): Transactions on Rough Sets I, LNCS 3100, pp. 169–186, 2004.
© Springer-Verlag Berlin Heidelberg 2004

Table 1. The content of a conflict.

U	a	b	c	d	e
1	–	+	+	+	+
2	+	+	–	–	–
3	+	–	–	–	0
4	0	–	–	0	–
5	+	–	–	–	–
6	0	+	–	0	+

Inconsistency often occurs in testimonies of crime witnesses. Witnesses often have different versions of the same event or on the subject of the same suspect. In this example we consider an investigator who has gathered the following testimonies from four witnesses describing a suspect:

- Witness A said: It was a very high and black man with the long black hairs;
- Witness B said: It was a high and brown eyes man with the medium long hairs;
- Witness C said: Skin: dark; Hairs: short and dark; Height: medium; Eyes: blue;
- Witness D said: The suspect was a bald and short man, his skin was brown.

As we can noticed, the opinions of the witnesses are not identical, thus they are in conflict. With reference to deposition of witnesses we can create 5 issues of the conflict: *Colour of the skin*; *Colour of eyes*; *Colour of hairs*; *Length of hairs* and *Height*.

The following conflict profiles are determined:

P_1. Colour of the skin: {A: black, C: dark, D: brown}

P_2. Colour of eyes: {B: brown, C: blue}

P_3. Colour of hairs: {A: black, C: dark}

P_4. Length of hairs: {A: long, B: medium long, C: short, D: bald}

P_5. Height: {A: very high, B: high, C: high, D: short}

Let us notice that in each profile the opinions are different, thus knowledge of the investigator about the suspect is inconsistent. He may not be sure of the proper value in each subject of the description, but the degrees of his uncertainty are not identical referring to all the conflict issues. It seems that the more consistent are witnesses' opinions the smaller is the uncertainty degree. In the profiles presented above one may conclude that the elements of profile P_3 are most consistent because black and dark colours are very similar, while the elements of profile P_4 are the least consistent. In this profile the four witnesses mention all four possible values for the length of hairs, thus the uncertainty degree of the investigator should be large in this case.

These examples show that it is needed to determine a value which would represent the degree (or level) of consistency (or inconsistency) of a conflict profile. This value may be very useful in evaluating if a conflict is "solvable" or not.

In this paper we propose to define this parameter of conflict profiles as consistency functions. We show a set of postulates which should be satisfied by these functions. We define also several consistency functions and show which postulates they fulfil. The paper is organized as follows. After Introduction, Section 2 presents some aspects of knowledge consistency and inconsistency. Section 3 outlines conflict theories which are the base of consistency measures. Section 4 includes an overview of consensus methods applied for conflict solving. Definition, postulates and their analysis for consistency functions are given in Section 5. Section 6 presents the definitions of four concrete consistency functions and their analysis referring to defined postulates. Section 7 describes several practical aspects of consistency measures, and some conclusions and the description of the future works are given in Section 8.

2 Consistency and Inconsistency of Knowledge

The "consistency" term seems to be well-known and often used notion. This is caused by the fact of intuitive using of this word. Many authors use this term to describe some divergences in various occurrences. The consistency of knowledge notion appears more seldom in knowledge engineering context, but in most of cases it is still used intuitively in order to name some divergences in scientific research. Authors usually use the term, but they do not define what it means. Thus the following questions may arise: *What does it mean that the knowledge is consistent or inconsistent? Is there any way to compare levels (or degrees) of consistency or inconsistency? Is there any way to measure up them?*

During creating publications authors often ignore answers on these questions, because all they need is divalent definition. Either all versions of knowledge are identical (that is it is consistent), or not. However, there exist situations, in which it is needed to know the knowledge consistency level. One of these situations is related to solving knowledge conflicts in multiagent environments. In this kind of conflicts the consistency level (or degree) is very useful because it could help to decide what to do with the agent knowledge states. If these states are different in small degree (the consistency level is high) then the agents could make a compromise (consensus) for reconciling their knowledge. If they are different in large degree (that is the consistency level is low), then it is necessary to gather other knowledge states for more precise reconciling.

The need for measures of knowledge consistency has been announced earlier in the aspect, if looking for consensus as the way to solve knowledge conflict of agents is rational [14]. Indeed in multiagent systems, where sources of knowledge acquisition are as various as methods of its acquisition, the inconsistency of knowledge leads to conflict arising. In 1990 Ng and Abramson [12] asked: *Is it possible to perform consistency checking on knowledge bases mechanically? If so, how?* They claim that this is very important that the knowledge base is consistent because inconsistent knowledge bases may provide inconsistent conclusions. Getting back to conflicts, we can notice that there have been worked out a lot of methods of solving conflicts invented [4][14][15][17], but level of divergence has been usually described in divalent way. Either something was consistent or not. Notified need has led to raising some measures [2]. For looking up for the consensus or the knowledge consistency it is necessary to agree on some formal representation of this knowledge based on distance functions between knowledge states [14].

A multiagent system is a case of distributed environments. Knowledge possessed by agents usually comes from different sources and the problem of their integration may arise. The knowledge of agents may be not only true or false, but also undefined or inconsistent [8]. Knowledge is undefined if there is a case in which the agents do not have any information (absence of information) referring to some subject and the knowledge is inconsistent if the agents have different versions of it. For integrating this kind of knowledge Loyer, Spyratos and Stamate [8] use Belnap's four-valued logics which uses *true, false, undefined* and *inconsistent* values. They use multi-valued logic for describing acquired knowledge and its reasoning. However, the knowledge consistency is divalent. If all versions of knowledge are identical then the knowledge is consistent, else it is inconsistent.

The notion of consistency is also very important in enforcing consistency in knowledge-based systems. The rule-based consistency enforcement in knowledge-based systems has been presented by Eick and Werstein [5]. These authors deal with enforcing the consistency constraints which are also called semantic integrity constraints. Actually consistency is one of the most important issues in database system, but it is considered in other sense and there is no need to measure it up.

In literature the term of knowledge consistency has been used very often, but we still need an answer to the question about a definition of knowledge consistency. We found some measures, by means of which we can estimate it and we can use it to solve many problems, but we still need a definition, which will translate intuitive approach into formal definition. This definition is provided by Neiger [10]: *Knowledge consistency is a property that a knowledge interpretation has with respect to a particular system.* Neiger refers to the definition of internal knowledge consistency defined by Helpern and Moses [6]. He formalizes this and other forms of knowledge consistency. After giving the definition, Neiger shows some cases in which knowledge consistency can be applied in distributed systems. In this way he shows how consistent knowledge interpretation can be used to simplify the design of knowledge-based protocols.

In other paper [11] Neiger presents how to use knowledge consistency for useful suspension of disbelief. He considers alternative interpretation of knowledge and explores the notion of consistent interpretation. Neiger shows how it can be used to circumvent the known impossibility results in a number of cases.

There are of course a lot of applications of knowledge consistency. The authors use this term in each case where it is some kind of divergence, for example between some pieces of knowledge. But there are some situations where we need to know exactly the level of consistency or inconsistency. So we need good measures and tools to estimate the quality of these parameters. These tools have been introduced in [9] and in this paper we are going to present some results of their analysis.

3   Outline of Conflict Theories

The simplest conflict takes place when two bodies generate different opinions on the same subject. In works [18][19][20] Pawlak specifies the following elements of a conflict: a set of agents, a set of issues, and a set of opinions of these agents on these issues. The agents and the issues are related with one another in some social or political context. Then we say that a conflict should take place if there are at least two agents whose opinions on some issue differ from each other.

Generally, one can distinguish the following 3 constrains of a conflict:

- *Conflict body*: specifies the direct participants of the conflict.
- *Conflict subject*: specifies to whom (or what) the conflict refers and its topic.
- *Conflict content*: specifies the opinions of the participants on the conflict topic.

In Pawlak's approach the body of conflict is a set of agents, the conflict subject consists of contentious issues and the conflict content is a collection of tuples representing the participants' opinions. Information system tools [4][21] seem to be very good for representing conflicts.

In works [14][15] the authors have defined conflicts in distributed systems in the similar way. However, we have built a system which can include more than one conflict, and within one conflict values of the attributes representing agents' opinions should more precisely describe their opinions. This aim has been realized by assuming that values of attributes representing conflict contents are not atomic as in Pawlak's approach, but sets of elementary values, where an elementary value is not necessarily an atomic one. Thus we accept the assumption that attributes are multi-valued, similarly like in Pawlak's concept of multi-valued information systems. Besides, the conflict content in our model is partitioned into three groups. The first group includes opinions of type "*Yes, the fact should take place*", the second includes opinions of type "*No, the following fact should not take place*", and to the last group contains the opinions of type "*I do not know if the fact takes place*". For example, making the forecast of sunshine for tomorrow a meteorological agent can present its opinion as "*(Certainly) it will sunny between 10a.m. and 12a.m. and will be cloudy between 3p.m. and 6p.m.*", that means during the rest of the day the agent does not know if it will be sunny or not. This type of knowledge should be taken into account in the system because the set of all possible states of the real world in which the system is placed, is large and an agent having limited possibilities is not assumed to "know everything". We call the above three kinds of knowledge as positive, negative and uncertain, respectively. In Pawlak's approach positive knowledge is represented by value "+", and negative knowledge by value "–". Certain difference occurs between the semantics of Pawlak's "neutrality" and the semantics of "uncertainty" of agents presented in mentioned works. Namely, most often neutrality appears in voting processes and does not mean uncertainty, while uncertainty means that an agent is not competent to present its opinions on some matter.

It is worth to note that rough set theory is a very useful tool for conflict analysis. In works [4][21] the authors present an enhancement of the model proposed by Pawlak. With using rough sets tools they explain the nature of conflict and define the conflict situation model in such way that encapsulates the conflict components. Such approach also enables to choose consensus as the conflicts solution, although it is still assumed that attribute values are atomic.

In the next section we present an approach to conflict solving, which is based on determining consensus for conflict profiles.

4   The Roles of Consensus Methods in Solving Conflicts

Consensus theory has a root in choice theory. A choice from some set A of alternatives is based on a relation α called a preference relation. Owing to it the choice function may be defined as follows:

$$C(A) = \{x \in A : (\forall y \in A)((x,y) \in \alpha)\}$$

Many works have dealt with the special case, where the preference relation is determined on the basis of a linear order on A. The most popular were the Condorcet choice functions. A choice function is called a Condorcet function if:

$$x \in C(A) \Leftrightarrow (\forall y \in A)(x \in C(\{x,y\}))$$

In the consensus-based approaches, however, it is assumed that the chosen alternatives do not have to be included in the set presented for choice, thus $C(A)$ need not be a subset of A. On the beginning of this research the authors have dealt only with simple structures of the set A (named *macrostructure*), such as linear or partial order. Later with the development of computing techniques the structure of each alternative (named *microstructure*) have been also investigated. Most often the authors assume that all the alternatives have the same microstructure [3]. On the basis of the microstructure one can determine a macrostructure of the set A. Among others, following microstructures have been investigated: linear orders, ordered set partitions, non–ordered set partitions, n–trees, time intervals. The following macrostructures have been considered: linear orders and distance (or similarity) functions. Consensus of the set A is most often determined on the basis of its macrostructure by some optimality rules. If the macrostructure is a distance (or similarity) function then the Kemeny's median [1] is very often used to choose the consensus. According to Kemeny's rule the consensus should be nearest to the elements of the set A.

Now, we are trying to analyze what are the roles of consensus in conflict resolution in distributed environments. Before the analysis we should consider what is represented by the conflict content (i.e. the opinions generated by the conflict participations). We may notice that the opinions included in the conflict content represent a unknown solution of some problem. The following two cases may take place [16]:

1. *The solution is independent on the opinions of the conflict participants.*

As an example of this kind of conflicts we can consider different forecasts generated by different meteorological stations referring to the same region for a period of time. The problem is then relied on determining the proper scenario of weather which is unambiguous and really known only when the time comes, and is independent on given forecasts.

A conflict in which the solution is independent on opinions of the conflict participants is called *independent conflict*.

In independent conflicts the independence means that the solution of the problem exists but it is not known for the conflict participants. The reasons of this phenomenon may follow from many aspects, among others, from the ignorance of the conflict participations or the random characteristics of the solution which may make the solution impossible to be calculated in a deterministic way. Thus the content of the solution is independent on the conflict content and the conflict participations for some interest have to "guess" it. In this case their solutions have to reflect the proper solution but it is not known if in a valid and complete way. In this case the natural solution of the conflict is relied on determining the proper version of data on the basis of given opinions of the participants. This final version should satisfy the following condition: *It should best reflect the given versions.*

The above defined condition should be suitable to this kind of conflicts because the versions given by the conflict participations reflect the "hidden" and independent solution but it is not known in what degree. Thus in advance each of them is treated as partially valid and partially invalid (which of its part is valid and which of its part is invalid – it is not known). The degree in which an opinion is treated as valid is the same for each opinion. This degree may not be equal to 100%. The reason for which all the opinions should be taken into account is that it is not known how large is the degree. It is known only to be greater than 0% and smaller than 100%. In this way the consensus should at best reflect these opinions. In other words, it should at best represent them.

For independent conflicts resolution the solution of the problem my be determined by consensus methods. Here for consensus calculation one should use the criterion for minimal sum of distances between the consensus and elements of the profile representing opinions of the conflict participants. This criterion guarantees satisfying the condition mentioned above.

2. *The solution is dependent on the opinions of the conflict participants.*

Conflicts of this kind are called *dependent conflicts*. In this case this is the opinions of conflict participants, which decide about the solution. As an example let us consider votes at an election. The result of the election is determined only on the basis of these votes. In general this case has a social or political character and the diversity between opinions of the participants most often follows from differences of choice criteria or their hierarchy.

For dependent conflicts the natural resolution is relied on determining a version of data on the basis of given opinions. This final version (consensus) should satisfy the following conditions: *It should be a good compromise which could be acceptable by the conflict participants.* Thus consensus should not only at best represent the opinions but also should reflect each of them in the same degree (with the assumption that each of them is treated in the same way). The condition "acceptable compromise" means that any of opinions should neither be "harmed" nor "favored". Consider the following example: From a set of candidates (denoted by symbols $X, Y, Z...$) 4 voters have to choose a committee (as a subset of the candidates' set). In this aim each of voter votes on such committee which in his opinion is the best one. Assume that the votes are the following: $\{X, Y, Z\}$, $\{X, Y, Z\}$, $\{X, Y, Z\}$ and $\{T\}$. Let the distance between 2 sets of candidates is equal to the cardinality of their symmetrical difference. If the consensus choice is made only by the first condition then committee $\{X, Y, Z\}$ should be determined because the sum of distances between it and the votes is minimal. However, one can note that it prefers the first 3 votes while totally ignoring the fourth (the distances from this committee to the votes are: 0, 0, 0 and 4, respectively). Now, if we take committee $\{X, Y, Z, T\}$ as the consensus then the distances would be 1, 1, 1 and 3, respectively. In this case the consensus neither is too far from the votes nor "harms" any of them. It has been proved that these conditions in general may not be satisfied simultaneously [13]. It is true that the choice based on the criterion of minimization of the sum of squared distances between consensus and the profile' elements gives a consensus more uniform than the consensus chosen by minimization of the distances' sum. Therefore, the criterion of the minimal sum of squared distances is also very important. However, the squared distances' minimal sum criterion often generates computationally complex problems (NP-hard problems), which demand working out heuristic algorithms.

Figure 1 below presents the scheme of using consensus methods in the above mentioned cases.

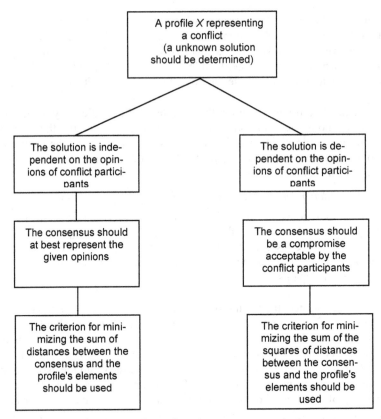

Fig. 1. The scheme for using consensus methods.

5   Postulates for Consistency Measures

Formally, let U denote a finite universe of objects (alternatives), and $\Pi(U)$ denote the set of subsets of U. By $\hat{\Pi}_k(U)$ we denote the set of k-element subsets (with repetitions) of the set U for $k \in N$, and let $\hat{\Pi}(U) = \bigcup_{k>0} \hat{\Pi}_k(U)$. Each element of set $\hat{\Pi}(U)$ is called a *profile*. In this work we do not use the formalism often used in the consensus theory [1], in which the domain of consensus is defined as $U^* = \bigcup_{k>0} U^k$, where U^k is the k-fold Cartesian product of U. In this way we specify how many times an object can occur in a profile and ensure that the order of profile elements is not important. We also accept in this paper an algebra of sets with repetitions (multisets) given by Lipski and Marek [7]. Some of its elements are as follows: An expression $A=(x,x,y,y,y,z)$ is called a set with repetitions with cardinality equal to 6. In the set A element x appears 2 times, y 3 times and z one time. Set A can also be written as $A=(2*x,3*y,1*z)$. The sum of sets with repetitions is denoted by the symbol $\cup$ and is defined in the following way: if element x appears in set A n times and in B n' times, then in their sum

$A \cup B$ the same element should appear $n+n'$ times. The difference of sets with partitions is denoted by symbol "–", its definition follows from the following example: $(6*x,5*y,1*z) - (2*x,3*y,1*z) = (4*x,2*y)$. For example, if $A=(2*x,3*y,1*z)$ and $B=(4*x,2*y)$, then $A \cup B=(6*x,5*y,1*z)$. A set A with repetitions is a subset of a set B with repetitions $(A \subseteq B)$ if each element from A does not have a greater number of occurrences than it has got in set B. For example $(2*x,3*y,1*z) \subseteq (2*x,4*y,1*z)$.

In this paper we only assume that the macrostructure of the set U is known as a distance function $d: U \times U \rightarrow \Re$, which is

a) *Nonnegative*: $(\forall x,y \in U)[d(x,y) \geq 0]$

b) *Reflexive*: $(\forall x,y \in U)[d(x,y) = 0 \text{ iff } x = y]$

c) *Symmetrical*: $(\forall x,y \in U)[d(x,y) = d(y,x)]$.

For normalization process we can assume that values of function d belong to interval $[0,1]$ and the maximal distance between elements of universe U is equal 1.

Let us notice, that the above conditions are only a part of metric conditions. Metric is a good measure of distance, but its conditions are too strong. A space (U,d) defined in this way does not need to be a metric space. Therefore we will call it a *distance space* [13].

A profile X is called *homogeneous* if all its elements are identical, that is $X=\{n*x\}$ for some $x \in U$ and n being a natural number. A profile jest *heterogeneous* if it is not homogeneous. A profile is called *distinguishable* if all its elements are different from each other. A profile X is *multiple* referring to a profile Y (or X is a *multiple* of Y), if $X=\{n*x_1,..., n*x_k\}$ and $Y = \{x_1,...,x_k\}$. A profile X is *regular* if it is a multiple of some distinguishable profile.

By symbol c we denote the consistency function of profiles. This function has the following signature:

$$c: \hat{\Pi}(U) \rightarrow [0,1].$$

where $[0,1]$ is the closed interval of real numbers between 0 and 1.

The idea of this function is relied on measuring the consistency degree of profile's elements. The consistency degree of a profile mentions the degrees of indiscernibility (discernibility) defined for an information system [22]. However, they are different conceptions. The difference is based on that the consistency degree represents the coherence level of the profile elements and for its measuring one should firstly define the distances between these elements.

The requirements for consistency are expressed in the following postulates:

P1a. *Postulate for maximal consistency*:
   If X is a homogeneous profile then $c(X)=1$.

P1b. *Extended postulate for maximal consistency*:
   For $X^{(n)} = \{n*x, k_1*x_1, ..., k_m*x_m\}$ being a profile such that element x occurs n times, and element x_i occurs k_i times, where k_i is a constant for $i=1,2,...,m$. The following equation should be true:
   $$\lim_{n \rightarrow +\infty} c(X^{(n)}) = 1.$$

P2a. *Postulate for minimal consistency:*
   If $X=\{a,b\}$ and $d(a,b)= \max\limits_{x,y \in U} d(x,y)$ then $c(X)=0$.

P2b. *Extended postulate for minimal consistency:*
   For $X^{(n)} = \{n*a, k_1*x_1, ..., k_m*x_m, n*b\}$ being a profile such that elements a and b occur n times, element x_i occurs k_i times, where k_i is a constant for $i=1,2,...,m$ and $d(a,b)= \max\limits_{x,y \in U} d(x,y)$. The following equation should be true:

$$\lim\limits_{n \to +\infty} c(X^{(n)}) = 0 .$$

P2c. *Alternative postulate for minimal consistency:*
   If $X=U$ then $c(X)=0$.

P3. *Postulate for non-zero consistency:*
   If there exist $a,b \in X$ such that $d(a,b) < \max\limits_{x,y \in U} d(x,y)$ then $c(X)>0$.

P4. *Postulate for heterogeneous profiles:*
   If X is a heterogeneous profile then $c(X)<1$.

P5. *Postulate for multiple profiles:*
   If profile X is a multiple of profile Y then $c(X)=c(Y)$.

P6. *Postulate for greater consistency:*
   Let $d(x,X)= \Sigma_{y \in X} d(x,y)$ denote the sum of distances between an element x of universe U and the elements of profile X. Let $D(X)=\{d(x,X): x \in U\}$ be the set of all such sums. For any profiles $X,Y \in \hat{\prod}(U)$ the following dependency should be true:

$$[\frac{\min (D(X))}{card(X)} \leq \frac{\min (D(Y))}{card(Y)}] \Rightarrow [c(X) \geq c(Y)].$$

P7a. *Postulate for consistency improvement:*
   Let a and b be such elements of profile X that $d(a,X) = \min (D(X))$ and $d(b,X) = \max (D(X))$, then
$$c(X-\{a\}) \leq c(X) \leq c(X-\{b\}).$$

P7b. *Second postulate for consistency improvement:*
   Let a and b be such elements of universe U, that $d(a,X) = \min (D(X))$ and $d(b,X) = \max (D(X))$, then
$$c(X \cup \{b\}) \leq c(X) \leq c(X \cup \{a\}).$$

Commentaries for postulates: The above mentioned postulates illustrate intuitive conditions for consistency functions. However, some commentaries should be given.

   • *Postulate for maximal consistency:* This postulate requires that the consistency of a homogeneous profile is equal 1. A homogeneous profile represents a conflict in which conflict bodies have the same opinion on some issue. Thus it is not a conflict, and for this situation the consistency should be maximal. In the example of testimonies of crime witnesses (Introduction), if each witness says that the assassin has white skin then the consistency of knowledge in this subject should be maximal.

- *Extended postulate for maximal consistency*: If in given profile some opinion is dominant (by large number of occurrences) in such degree that the number of other opinions is insignificant, then the consistency should be near to the maximal.
- *Postulate for minimal consistency* and *extended postulate for minimal consistency*: For the first of them if profile $X=\{a,b\}$ and $d(a,b)= \max\limits_{x,y\in U} d(x,y)$ then it represents the "worst conflict", that is the opinions of two conflict bodies are maximally different from each other. In this case the consistency should be 0. The second postulate characterizes also a "very bad" conflict, in which two maximally differencing from each other opinions dominate. For this conflict the consistency should be near 0.
- *Alternative postulate for minimal consistency*: This postulate is also a condition for consistency in the "worst conflicts", but its sense is different from those of the two above. Concretely, a "worst conflict" in this sense is the one, in which all possible opinions occur. Additionally, each of them appears exactly one time. If a profile is understood as a set of information about some reality, then occurrences of all possible versions mean that the information quantity of this profile is 0.
- *Postulate for non-zero consistency*: A profile is understood to contain convergent opinions if it contains at least two such elements that their distance is smaller than the maximal distance in the universe. For such profiles the consistency should be greater than 0.
- *Postulate for heterogeneous profiles*: If a profile is heterogeneous then its consistency may not be maximal, because the opinions represented by its elements are not identical.
- *Postulate for multiple profiles*: If profile X is a multiple of profile Y, then it means that the proportions of occurrences numbers of elements are identical in these profiles. Thus one can say that the superiority of an element x referring to an element y $(x,y\in U)$ is the same in profiles X and Y. In this case they should have the same consistency.
- *Postulate for greater consistency*: The fact that $min(D(X)) \leq min(D(Y))$ means that the elements of profile X are more concentrated than the elements of profile X. In other words, profile X is denser than profile Y. Thus there should be $c(X) \geq c(Y)$.
- *Postulates for consistency improvement*: These postulates show when it is possible to improve the consistency. According postulate 7a the consistency should be improved if one removes from the profile the element which has the maximal sum of distances to the profile' elements and the consistency should be worse if one removes from the profile the element which has the minimal sum of distances to the profile' elements. Dually, according postulate 7b the consistency should be improved if one adds to the profile the element which has the minimal sum of distances to the profile' element and it will be worse if one adds to the profile the element which has the minimal sum of distances to the profile' elements.

Postulates P7a and P7b are very important because they present the way for improvement of the consistency. According to them the consistency should be larger if we add to the profile the element of universe U, which generates the minimal sum of distances to the elements of the profile, or if we remove from the profile the element which generates maximal sum of distances. For example, let $U = \{a, b, c*, d, e, f\}$ be such universe that the distances between element f and other elements are much greater than the distances among a, b, c^*, d and e. (see Figure 2 below). Let profile X = $\{a, b, d, e, f\}$, $d(c^*,X)$ = min $(D(X))$ and $d(f,X)$ = max $(D(X))$. According to postulates P7a and P7b we can improve the consistency by moving element f or adding element c^*, that is $c(X) \leq c(X\setminus\{f\})$ and $c(X) \leq c(X\cup\{c^*\})$.

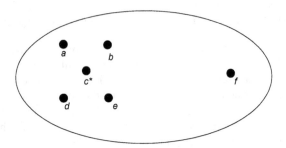

Fig. 2. An example of universe U.

Let P = $\{$P1a, P1b, P2a, P2b, P2c, P3, P4, P5, P6, P7a, P7b$\}$ be the set of all postulates. By C_p we denote the set of all consistency functions which fulfil postulate p where $p \in P$. These postulates are independent on each other, that is $C_p \not\subset C_{p'}$ for $p,p' \in P$ and $p \neq p'$. Postulates P2c and P3 are contradictory, that is consistency functions satisfying simultaneously these postulates do not exist. It follows also that there does not exist a function which satisfies all the postulates.

Below we present some other properties of the classes of consistency functions:

Proposition 1. *Let $c \in C_{P6}$, and let a be such element of universe U that $d(a,X)$ = min $(D(X))$ and b be such element of profile X that $d(a,b)$ = $\max\limits_{x \in X} d(a,x)$. The following dependence is true*

$$c(X) \leq c(X - \{b\}).$$

Proof.
Let $Y=X-\{b\}$ and $card(X)=n$, then $card(Y)=n-1$. We can assume that $n>1$ because if $n=1$ then function c could be indefinite for Y. Let a' be such element of universe U that $d(a',Y)$ = min $(D(Y))$. It implies that $d(a',Y) \leq d(a,Y)$. Besides, from $d(a,b)$ = $\max\limits_{x \in X} d(a,x)$ it implies that $(n-1) \cdot d(a,b) \geq d(a,Y)$. Then we have

$$\frac{\min(D(X))}{card(X)} = \frac{d(a,X)}{n} = \frac{d(a,Y)+d(a,b)}{n} \geq \frac{d(a,Y)}{n-1} \geq \frac{d(a',Y)}{n-1} = \frac{\min(D(Y))}{card(Y)}.$$

Because function c satisfies postulate P6 then there should be $c(X) \leq c(Y)$.

This property allows to improve the consistency by removing from the profile the element which generates the maximal distance to the element with minimal sum of distances to the profile's elements. It also shows that if a consistency function satisfies postulate P6 then it should also partially satisfy postulate P7a.

Proposition 2. *Let* $c \in C_{P6}$, *and let a be such an element of universe U that* $d(a,X) =$ min $(D(X))$. *The following dependence is true*

$$c(X) \leq c(X \dot\cup \{a\}).$$

Proof.

Let $Y = X \dot\cup \{a\}$. From $d(a,X) =$ min $(D(X))$ it implies that $d(a,Y) =$ min $(D(Y))$. Besides, $d(a,X) = d(a,Y)$, thus $\dfrac{\min (D(X))}{card(X)} \geq \dfrac{\min (D(Y))}{card(Y)}$ because $card$ $(Y) =$ $card(X)+1$. Using the assumption that $c \in C_{P6}$ we have $c(X) \leq c(X \dot\cup \{a\})$.

This property allows to improve the consistency by adding to the profile an element which generates the minimal sum of distances to the profile's elements. It also shows that if a consistency function satisfies postulate P6 then it should also partially satisfy postulate P7b.

Propositions 3-5 below show the independence of postulates P7a and P7b from some other postulates.

Proposition 3. *Postulates P1a and P2a are inconsistent with postulate P7a, that is*

$$C_{P1a} \cap C_{P2a} \cap C_{P7a} = \varnothing.$$

Proof.

We show that if a consistency function c satisfies postulates P1a and P2a then it can not satisfy postulate P7a. Let $c \in C_{P1a} \cap C_{P2a}$, let $X = U = \{a,b\}$ and $d(a,b)$ $= \max\limits_{x,y \in U} d(x,y) > 0$, then $c(X) = 0$ according postulate P2a. Because c satisfies postulate P1a we have $c(X - \{a\}) = c(\{b\}) = 1$. Besides, we have $d(a,X) =$ min $(D(X))$ and $c(X - \{a\}) = 1 > c(X) = 0$, so function c can not satisfy postulate P7a. That means postulate P7a is independent on postulates P1a and P2a.

Proposition 4. *Postulates P1a and P4 are inconsistent with postulate P7a, that is*

$$C_{P1a} \cap C_{P4} \cap C_{P7a} = \varnothing.$$

Proof.

We show that if a consistency function c satisfies postulates P1a and P4 then it can not satisfy postulate P7a. Let $c \in C_{P1a} \cap C_{P4}$, let $X = U = \{a,b\}$ and $d(a,b)$ $= \max\limits_{x,y \in U} d(x,y) > 0$, then $c(X) < 1$ according postulate P4. Because c satisfies postulate P1a we have $c(X - \{a\}) = c(\{b\}) = 1$. Besides, we have $d(a,X) =$ min $(D(X))$ and $c(X - \{a\}) > c(X)$, so function c can not satisfy postulate P7a. That means postulate P7a is independent on postulates P1a and P4.

Proposition 5. *Postulates P2a and P3 are inconsistent with postulate P7b, that is*

$$C_{P2a} \cap C_{P3} \cap C_{P7b} = \varnothing.$$

Proof.

We show that if a consistency function c satisfies postulates P2a and P3 then it can not satisfy postulate P7b. Let $c \in C_{P2a} \cap C_{P3}$, let $X = U = \{a,b\}$ and $d(a,b)$ $= \max\limits_{x,y \in U} d(x,y) > 0$, we have $d(a,X) =$ min $(D(X))$ and $d(b,X) =$ max $(D(X))$. Then $c(X)$

= 0 because $c \in C_{P2a}$. But $c(X \cup \{b\}) = c(\{a,b,b\}) > 0$ because $d(b,b)=0$ and function c satisfies postulate P3, so it may not satisfy postulate P7b. Here we have the independence of postulate P7b on postulates P2a and P3.

6 Consistency Functions Analysis

In this section we present the analysis of 4 consistency functions. These functions are defined as follows:

Let $X=\{x_1, \ldots, x_M\}$ be a profile. We assume that $M>1$, because if $M=1$ then the profile X is a homogeneous one. We introduce the following parameters:

- The matrix of distances between the elements of profile X:

$$D^X = \left[d_{ij}^X\right] = \begin{bmatrix} d(x_1,x_1) & \cdots & d(x_1,x_M) \\ \vdots & \ddots & \vdots \\ d(x_M,x_1) & \cdots & d(x_M,x_M) \end{bmatrix},$$

- The vector of average distances between an element to the rest:

$$W^X = \left[w_i^X\right] = \left(\frac{1}{M-1}\sum_{j=1}^{M} d_{j1}^X, \frac{1}{M-1}\sum_{j=1}^{M} d_{j2}^X, \ldots, \frac{1}{M-1}\sum_{j=1}^{M} d_{jM}^X \right),$$

- Diameters of sets X and U:

$$Diam(X) = \max_{x,y \in X} d(x,y),$$
$$Diam(U) = \max_{x,y \in U} d(x,y) = 1,$$

and the maximal element of vector W^X:

$$Diam(W^X) = \max_{1 \le i \le M} w_i^X,$$

representing the element of profile X, which generate the maximal sum of distances to other elements,

- The average distance in profile X:

$$\overline{d}(X) = \frac{1}{M(M-1)}\sum_{i=1}^{M}\sum_{j=1}^{M} d_{ij}^X = \frac{1}{M}\sum_{i=1}^{M} w_i^X,$$

- The sum of distances between an element x of universe U and the elements of set X:

$$d(x,X) = \Sigma_{y \in X} d(x,y),$$

- The set of all sums of distances:

$$D(X) = \{d(x,X): x \in U\},$$

- The minimal sum of distances from an object to the elements of profile X:

$$d_{min}(X) = \min(D(X)).$$

These parameters are now applied for the defining the following consistency functions:

$$c_1(X) = 1 - Diam(X),$$

$$c_2(X) = 1 - Diam(W^X),$$

$$c_3(X) = 1 - \overline{d}(X),$$

$$c_4(X) = 1 - \frac{1}{M} d_{min}(X).$$

Values of functions c_1, c_2, c_3 and c_4 reflect accordingly:

- $c_1(X)$ – the maximal distance between two elements of profile. The intuitive sense of this function is based on the fact that if this maximal distance is equal 0 then consistency is maximal (that is 1).
- $c_2(X)$ – the maximal average distance between an element of profile X and other elements of this profile. If the value of this maximal average distance is small, that is the elements of profile X are near from each other, then the consistency should be high.
- $c_3(X)$ – the average distance between elements of X. This parameter seems to be most representative for consistency. The larger is this value the smaller is the consistency and vice versa.
- $c_4(X)$ – the minimal average distance between an element of universe U and elements of X. The element of universe U, which generates the minimal average distance to elements of profile X, may be the consensus for this profile. The profile have a good consensus (that is a good solution for the conflict) if this consensus generates small average distance to the elements of the profile. In this case the consistency should be large.

Table 2 presented below shows result of analysing functions. The columns represent postulates and the rows represent the defined functions. Symbol '+' means that presented function satisfies the postulate, symbol '–' means that presented function does not satisfy the postulate and symbol ± means partial satisfying given postulate. From these results it implies that function c_4 satisfies partially postulates P7a and P7b. The reason is based on the fact that function c_4 satisfies postulate P6 and Propositions 1 and 2.

Table 2. Results of consistency functions analysis.

	P1a	P1b	P2a	P2b	P2c	P3	P4	P5	P6	P7a	P7b
c_1	+	−	+	+	+	−	+	+	−	−	−
c_2	+	−	+	−	−	−	+	+	−	+	+
c_3	+	+	+	−	−	+	+	−	−	+	+
c_4	+	+	−	−	−	+	+	+	+	±	±

Satisfying some postulates and non-satisfying other postulates of each consistency function show many its properties. Below we present some another property of functions c_2 and c_3 [2].

Proposition 6. *If* $X' \subseteq X$, *I is the set of indexes of elements from* X' *and* $w_i^X = Diam(W^X)$ *for* $i \in I$ *then profile* $X \backslash X'$ *should not have smaller consistency than* X, *that is* $c(X \backslash X') \geq c(X)$ *where* $c \in \{c_2, c_3\}$.

Proof.

a) For function c_2 the proof follows immediately from the notice that $Diam(W^X) \leq Diam(W^{X \backslash X'})$.

b) For function c_3 we have $\overline{d}(X) = \dfrac{1}{M} \sum\limits_{i=1}^{M} w_i^X$, with the assumption that $w_i^X = Diam(W^X)$ for $i \in I$ it follows $\overline{d}(X) \geq \overline{d}(X \backslash X')$, that is $c_3(X \backslash X') \geq c_3(X)$.

This property shows a way to improve the consistency by moving from the profile these elements which generate maximal average distance. This way for consistency improvement is simple and therefore is a useful property of functions c_2, c_3. The way for consistency improvement using function c_4 which satisfies postulate P6, has been presented by means of Propositions 1 and 2.

7 Practical Aspects of Consistency Measures

One of the practical aspects of consistency measures is their applications for choice of the best method for solving conflicts in distributed environments. Some methods for conflict solving have been developed, each of them is suitable for a given kind of conflicts. But before one decides to select a method for a conflict, he should takes into account the degree of consistency of the opinions which occur in the conflict. This measure should be useful in evaluating if the conflict is already "mature" for solving or not yet.

Let us consider an example which illustrates the statement that it is good to know the consistency level before using consensus algorithms. Assume that one is collecting information from meteorological institutes about the weather for the city of Zakopane during the weekend. He want to know if it will be snowing during the weekend. Five institutes say *yes* and five another institutes say *no*. Thus a conflict appears. The profile of the conflict looks as follows: $X = \{5*yes, 5*no\}$. The consensus, if determined, should be *yes* or *no*. However, neither *yes* nor *no* seems to be a good conflict solution. The consistency of the profile is low, according to postulates P2a and P5 it should be equal 0. It is the reason why the solution is not good.

Consensus algorithms are usually very complex. Therefore, it is worth to check out the consistency of conflict profile before determining the consensus. Evaluating consistency measure before using consensus algorithms may eliminate these situations in which the consensus is not a good conflict solution. This will surely increase the effectiveness of solving conflicts systems. In the above example there is no good conflict solution at all and one has to collect more information or choose other conflict solution method.

Consistency measures are also very helpful for investigators during investigations. Witnesses' opinions about a suspect can be very inconsistent. The consistency degree of suspect evidences may by used for determining the reliability of witness.

Another interesting application of consistency measures is some kind of explorative system, where the menstruation results are collected in some interval of time. The results may be inconsistent, but when the consistency of results equals 1 then the alert can be sent. In this way we can measure for example the concentration of sulfur oxide.

The scheme for application of a consistency measure in a conflict situation may look as follows:

- First we should define the universe of all possible opinions on some subject,
- Then we should determine a conflict profile on this subject,
- After this, we have to find proper distance function, and calculate the distances between elements in the created profile,
- Next, we choose the most proper consistency measure, which depends on the postulates that we want them to be satisfied,
- We use chosen measure to calculate the consistency degree,
- Now, we can use this level in decision process.

As a matter of fact there is a lot of practical aspects of consistency measures. We can use the consistency degrees in multiagent systems and in all kinds of information systems where knowledge is processed by autonomous programs; in distributed database systems where the data consistency is a one of the key factors, and also in reasoning systems and many others.

8 Conclusions

In this paper the concept of measuring consistency degrees of conflict profiles is presented. The authors formulate the conditions (postulates) which should be satisfied by consistency functions. These postulates are independent on the structure of conflict profiles. Some consistency functions have been defined and analyzed referring to the postulates. The future works should concern the solid analysis of presented postulates, which should allow to choose appropriate consistency functions for concrete practical conflict situations. Besides, some implementation should be performed for justifying the sense of introduced postulates and consistency functions.

References

1. Barthelemy, J.P., Janowitz M.F.: A Formal Theory of Consensus. SIAM J. Discrete Math. 4 (1991) 305-322.
2. Danilowicz, C., Nguyen, N.T., Jankowski, Ł.: Methods for selection of representation of agent knowledge states in multi-agent systems. Wroclaw University of Technology Press (2002) (in Polish).

3. Day, W.H.E.: Consensus Methods as Tools for Data Analysis. In: Bock, H.H. (ed.): Classification and Related Methods for Data Analysis. North-Holland (1988) 312-324.
4. Deja, R.: Using Rough Set Theory in Conflicts Analysis, Ph.D. Thesis (Advisor: A. Skowron), Institute of Computer Science, Polish Academy of Sciences, Warsaw 2000.
5. Eick, C.F., Werstein, P.: In: Rule-Based Consistency Enforcement for Knowledge-Based Systems, IEEE Transactions on Knowledge and Data Engineering **5** (1993) 52-64.
6. Helpern, J. Y., Moses, Y.: Knowledge and common knowledge in distributed environment. Journal of the Association for Computing Machinery **37** (2001) 549-587.
7. Lipski, W., Marek, W.: Combinatorial Analysis. WNT Warsaw (1986).
8. Loyer, Y., Spyratos, N., Stamate, D.: Integration of Information in Four-Valued Logics under Non-Uniform Assumption. In: Proceedings of 30th IEEE International Symposium on Multiple-Valued Logic (2000) 180-193.
9. Malowiecki, M., Nguyen, N.T.: Consistency Measures of Agent Knowledge in Multiagent Systems. In: Proceedings of 8th National Conference on Knowledge Engineering and Expert Systems, Wroclaw Univ. of Tech. Press vol. 2 (2003) 245-252.
10. Neiger, G.: Simplifying the Design of Knowledge-based Algorithms Using Knowledge Consistency. Information & Computation **119** (1995) 283-293.
11. Neiger, G.: Knowledge Consistency: A Useful Suspension of Disbelief. In: Proceedings of the Second Conference on Theoretical Aspects of Reasoning about Knowledge. Morgan Kaufmann. Los Altos, CA, USA (1988) 295-308.
12. Ng, K. Ch., Abramson, B.: Uncertainty Management in Expert Systems. In: IEEE Expert: Intelligent Systems and Their Applications (1990) 29-48.
13. Nguyen, N.T.: Using Distance Functions to Solve Representation Choice Problems. Fundamenta Informaticae **48**(4) (2001) 295-314.
14. Nguyen, N.T.: Consensus Choice Methods and their Application to Solving Conflicts in Distributed Systems. Wroclaw University of Technology Press (2002) (in Polish).
15. Nguyen, N.T.: Consensus System for Solving Conflicts in Distributed Systems. Journal of Information Sciences **147** (2002) 91-122.
16. Nguyen, N.T., Sobecki, J.: Consensus versus Conflicts – Methodology and Applications. In: Proceedings of RSFDGrC 2003, Lecture Notes in Artificial Intelligence **2639**, 565-572.
17. Nguyen, N.T.: Susceptibility to Consensus of Conflict Profiles in Consensus Systems. Bulletin of International Rough Sets Society **5**(1/2) (2001) 217-224.
18. Pawlak, Z.: On Conflicts, Int. J. Man-Machine Studies **21** (1984) 127-134.
19. Pawlak, Z.: Anatomy of Conflicts, Bull. EATCS **50** (1993) 234-246.
20. Pawlak, Z.: An Inquiry into Anatomy of Conflicts, Journal of Information Sciences **109** (1998) 65-78.
21. Skowron, A., Deja, R.: On Some Conflict Models and Conflict Resolution. Romanian Journal of Information Science and Technology **5**(1-2) (2002) 69-82.
22. Skowron, A., Rauszer, C.: The Discernibility Matrices and Functions in Information Systems. In: E. Słowi ski (ed.): Intelligent Decision Support, Handbook of Applications and Advances of the Rough Sets Theory. Kluwer Academic Publishers (1992) 331-362.

Layered Learning for Concept Synthesis

Sinh Hoa Nguyen[1], Jan Bazan[2], Andrzej Skowron[3], and Hung Son Nguyen[3]

[1] Polish-Japanese Institute of Information Technology
Koszykowa 86, 02-008, Warsaw, Poland
[2] Institute of Mathematics,University of Rzeszów
Rejtana 16A, 35-959 Rzeszów, Poland
[3] Institute of Mathematics, Warsaw University
Banacha 2, 02-097 Warsaw, Poland
{hoa,bazan,skowron,son}@mimuw.edu.pl

Abstract. We present a hierarchical scheme for synthesis of concept approximations based on given data and domain knowledge. We also propose a solution, founded on rough set theory, to the problem of constructing the approximation of higher level concepts by composing the approximation of lower level concepts. We examine the effectiveness of the layered learning approach by comparing it with the standard learning approach. Experiments are carried out on artificial data sets generated by a road traffic simulator.

Keywords: Concept synthesis, hierarchical schema, layered learning, rough sets.

1 Introduction

Concept approximation is an important problem in data mining [10]. In a typical process of concept approximation we assume that there is given information consisting of values of conditional and decision attributes on objects from a finite subset (training set, sample) of the object universe and on the basis of this information one should induce approximations of the concept over the whole universe. In many practical applications, this standard approach may show some limitations. Learning algorithms may go wrong if the following issues are not taken into account:

Hardness of Approximation: A target concept, being a compositions of some simpler one, is too complex, and cannot be approximated directly from feature value vectors. The simpler concepts may be either approximated directly from data (by attribute values) or given as domain knowledge acquired from experts. For example, in the hand-written digit recognition problem, the raw input data are $n \times n$ images, where $n \in [32, 1024]$ for typical applications. It is very hard to find an approximation of the target concept (digits) directly from values of n^2 pixels (attributes). The most popular approach to this problem is based on defining some additional, e.g., basic shapes, skeleton graph. These features must be easily extracted from images, and they are used to describe the target concept.

J.F. Peters et al. (Eds.): Transactions on Rough Sets I, LNCS 3100, pp. 187–208, 2004.

Efficiency: The fact that the complex concept can be decomposed into simpler one allows to decrease the complexity of the learning process. Each component can be learned separately on a piece of a data set and independent components can be learned in parallel. Moreover, dependencies between component concepts and their consequences can be approximated using domain knowledge and experimental data.

Expressiveness: Sometime, one can increase the readability of concept description by introducing some additional concepts. The description is more understandable, if it is expressed in natural language. For example, one can compare the readability of the following decision rules:

if *car speed is high* **and** *a distance to a preceding car is small* **then** *a traffic situation is dangerous*	**if** $car\_speed(X) > 176.7km/h$ **and** $distance\_to\_front\_car(X) < 11.4m$ **then** *a traffic situation is dangerous*

Layered learning [25] is an alternative approach to concept approximation. Given a hierarchical concept decomposition, the main idea is to synthesize a target concept gradually from simpler ones. One can imagine the decomposition hierarchy as a treelike structure (or acyclic graph structure) containing the target concept in the root. A learning process is performed through the hierarchy, from leaves to the root, layer by layer. At the lowest layer, basic concepts are approximated using feature values available from a data set. At the next layer more complex concepts are synthesized from basic concepts. This process is repeated for successive layers until the target concept is achieved.

The importance of hierarchical concept synthesis is now well recognized by researchers (see, e.g., [15, 14, 12]). An idea of hierarchical concept synthesis, in the rough mereological and granular computing frameworks has been developed (see, e.g., [15, 17, 18, 21]) and problems related to approximation compound concept are discussed, e.g., in [18, 22, 5, 24].

In this paper we concentrate on concepts that are specified by decision classes in decision systems [13]. The crucial factor in inducing concept approximations is to create the concepts in a way that makes it possible to maintain the acceptable level of precision all along the way from basic attributes to final decision. In this paper we discuss some strategies for concept composing founded on the rough set approach. We also examine effectiveness of the layered learning approach by comparison with the standard rule-based learning approach. The quality of the new approach will be verified relative to the following criteria: generality of concept approximation, preciseness of concept approximation, computation time required for concept induction, and concept description lengths. Experiments are carried out on an artificial data set generated by a road traffic simulator.

2   Concept Approximation Problem

In many real life situations, we are not able to give an exact definition of the concept. For example, frequently we are using adjectives such as "good", "nice", "young", to describe some classes of peoples, but no one can give their exact definition. The concept "young person" appears be easy to define by age, e.g., with the rule:

$$\textbf{if}\, age(X) \leq 30 \;\textbf{then}\; X \text{ is young,}$$

but it is very unnatural to explain that "Andy is not young because yesterday was his 30th birthday". Such uncertain situations are caused either by the lack of information about the concept or by the richness of natural language.

Let us assume that there exists a concept X defined over the universe $\mathcal{U}$ of objects ($X \subseteq \mathcal{U}$). The problem is to find a description of the concept X, that can be expressed in a predefined descriptive language $\mathcal{L}$ consisting of formulas that are interpretable as subsets of $\mathcal{U}$. In general, the problem is to find a description of a concept X in a language $\mathcal{L}$ (e.g., consisting of boolean formulae defined over subset of attributes) assuming the concept is definable in another language $\mathcal{L}'$ (e.g., natural language, or defined by other attributes, called decision attributes).

Inductive learning is one of the most important approaches to concept approximation. This approach assumes that the concept X is specified partially, i.e., values of characteristic function of X are given only for objects from a *training sample* $U \subseteq \mathcal{U}$. Such information makes it possible to search for patterns in a given language $\mathcal{L}$ defined on the training sample sets included (or sufficiently included) into a given concept (or its complement). Observe that the approximations of a concept can not be defined uniquely from a given sample of objects. The approximations of the whole concept X are induced from given information on a sample U of objects (containing some positive examples from $X \cap U$ and negative examples from $U - X$). Hence, the quality of such approximations should be verified on new testing objects.

One should also consider uncertainty that may be caused by methods of object representation. Objects are perceived by some features (attributes). Hence, some objects become indiscernible with respect to these features. In practice, objects from $\mathcal{U}$ are perceived by means of vectors of attribute values (called information vectors or information signature). In this case, the language $\mathcal{L}$ consists of boolean formulas defined over accessible attributes such that their values are effectively measurable on objects. We assume that $\mathcal{L}$ is a set of formulas defining subsets of $\mathcal{U}$ and boolean combinations of formulas from $\mathcal{L}$ are expressible in $\mathcal{L}$.

Due to bounds on expressiveness of language $\mathcal{L}$ in the universe $\mathcal{U}$, we are forced to find some approximate rather than exact description of a given concept. There are different approaches to deal with uncertain and vague concepts like multi-valued logics, fuzzy set theory, or rough set theory. Using those approaches, concepts are defined by "multi-valued membership function" instead of classical "binary (crisp) membership relations" (set characteristic functions). In particular, rough set approach offers a way to establish membership functions that are data-grounded and significantly different from others.

In this paper, the input data set is represented in a form of information system or decision system. An *information system* [13] is a pair $\mathbb{S} = (U, A)$, where U is a non-empty, finite set of *objects* and A is a non-empty, finite set, of *attributes*. Each $a \in A$ corresponds to the function $a : U \rightarrow V_a$ called an *evaluation function*, where V_a is called the *value set* of a. Elements of U can be interpreted as cases, states, patients, or observations.

For a given information system $\mathbb{S} = (U, A)$, we associate with any non-empty set of attributes $B \subseteq A$ *the B-information signature* for any object $x \in U$ by $inf_B(x) = \{(a, a(x)) : a \in B\}$. The set $\{inf_A(x) : x \in U\}$ is called the *A-information set* and it is denoted by $INF(\mathbb{S})$.

The above formal definition of information systems is very general and it covers many different systems such as database systems, or *information table* which is a two–dimensional array (matrix). In an information table, we usually associate its rows with objects (more precisely information vectors of objects), its columns with attributes and its cells with values of attributes.

In supervised learning, objects from a training set are pre-classified into several *categories* or *classes*. To deal with this type of data we use a special information systems called *decision systems* that are information systems of the form $\mathbb{S} = (U, A, dec)$, where $dec \notin A$ is a distinguished attribute called *decision*. The elements of attribute set A are called *conditions*.

In practice, decision systems contain a description of a finite sample U of objects from a larger (may be infinite) universe $\mathcal{U}$. Usually decision attribute is a characteristic function of an unknown concept or concepts (in the case of several classes). The main problem of learning theory is to generalize the decision function (concept description) partially defined on the sample U, to the universe $\mathcal{U}$. Without loss of generality for our considerations we assume that the domain V_{dec} of the decision dec is equal to $\{1, \ldots, d\}$. The decision dec determines a partition

$$U = CLASS_1 \cup \ldots \cup CLASS_d$$

of the universe U, where $CLASS_k = \{x \in U : dec(x) = k\}$ is called the kth *decision class* of $\mathbb{S}$ for $1 \leq k \leq d$.

3   Concept Approximation Based on Rough Set Theory

Rough set methodology for concept approximation can be described as follows (see [5]).

Definition 1. *Let $X \subseteq \mathcal{U}$ be a concept and let $U \subseteq \mathcal{U}$ be a finite sample of $\mathcal{U}$. Assume that for any $x \in U$ there is given information whether $x \in X \cap U$ or $x \in U - X$. A rough approximation of the concept X in a given language $\mathcal{L}$ (induced by the sample U) is any pair $(\mathbf{L}_{\mathcal{L}}, \mathbf{U}_{\mathcal{L}})$ satisfying the following conditions:*

1. *$\mathbf{L}_{\mathcal{L}} \subseteq \mathbf{U}_{\mathcal{L}} \subseteq \mathcal{U}$,*
2. *$\mathbf{L}_{\mathcal{L}}, \mathbf{U}_{\mathcal{L}}$ are expressible in the language $\mathcal{L}$, i.e., there exist two formulas $\phi_L, \phi_U \in \mathcal{L}$ such that $\mathbf{L}_{\mathcal{L}} = \{x \in \mathcal{U} : x \text{ satisfies } \phi_L\}$ and $\mathbf{U}_{\mathcal{L}} = \{x \in \mathcal{U} : x \text{ satisfies } \phi_U\}$,*
3. *$\mathbf{L}_{\mathcal{L}} \cap U \subseteq X \cap U \subseteq \mathbf{U}_{\mathcal{L}} \cap U$;*
4. *the set $\mathbf{L}_{\mathcal{L}}$ ($\mathbf{U}_{\mathcal{L}}$) is maximal (minimal) in the family of sets definable in $\mathcal{L}$ satisfying (3).*

The sets $\mathbf{L}_{\mathcal{L}}$ and $\mathbf{U}_{\mathcal{L}}$ are called the *lower approximation* and the *upper approximation* of the concept $X \subseteq \mathcal{U}$, respectively. The set $\mathbf{BN} = \mathbf{U}_{\mathcal{L}} \setminus \mathbf{L}_{\mathcal{L}}$ is called the *boundary region of approximation* of X. The set X is called *rough* with respect to its approximations $(\mathbf{L}_{\mathcal{L}}, \mathbf{U}_{\mathcal{L}})$ if $\mathbf{L}_{\mathcal{L}} \neq \mathbf{U}_{\mathcal{L}}$, otherwise X is called *crisp* in $\mathcal{U}$. The pair $(\mathbf{L}_{\mathcal{L}}, \mathbf{U}_{\mathcal{L}})$ is also called the *rough set* (for the concept X). Condition (3) in the above list can be substituted by inclusion to a degree to make it possible to induce approximations of higher quality of the concept on the whole universe $\mathcal{U}$. In practical applications the last condition in the above definition can be hard to satisfy. Hence, by using some heuristics we construct sub-optimal instead of maximal or minimal sets. Also, since during the process of approximation construction we only know U it may be necessary to change the approximation after we gain more information about new objects from $\mathcal{U}$. The rough approximation of concept can be also defined by means of a rough membership function.

Definition 2. *Let $X \subseteq \mathcal{U}$ be a concept and let $U \subseteq \mathcal{U}$ be a finite sample. A function $f : \mathcal{U} \to [0,1]$ is called a rough membership function of the concept $X \subseteq \mathcal{U}$ if and only if $(\mathbf{L}_f, \mathbf{U}_f)$ is an approximation of X (induced from sample U) where $\mathbf{L}_f = \{x \in \mathcal{U} : f(x) = 1\}$ and $\mathbf{U}_f = \{x \in \mathcal{U} : f(x) > 0\}$.*

Note that the proposed approximations are not defined uniquely from information about X on the sample U. They are obtained by inducing the concept $X \subseteq \mathcal{U}$ approximations from such information. Hence, the quality of approximations should be verified on new objects and information about classifier performance on new objects can be used to improve gradually concept approximations. Parameterizations of rough membership functions corresponding to classifiers make it possible to discover new relevant patterns on the object universe extended by adding new (testing) objects. In the following sections we present illustrative examples of such parameterized patterns. By tuning parameters of such patterns one can obtain patterns relevant for concept approximation on the extended training sample by some testing objects.

3.1   Case-Based Rough Approximations

For case-base reasoning methods, like kNN (k nearest neighbors) classifier [1, 6], we define a distance (similarity) function between objects $\delta : \mathcal{U} \times \mathcal{U} \to [0, \infty)$. The problem of determining the distance function from given data set is not trivial, but in this paper, we assume that such a distance function has been already defined for all object pairs.

In kNN classification methods (kNN classifiers), the decision for a new object $x \in \mathcal{U} - U$ is made by decisions of k objects from U that are nearest to x with respect to the distance function δ. Usually, k is a parameter defined by an expert or automatically constructed by experiments from data. Let us denote by $NN(x; k)$ the set of k nearest neighbors of x, and by $n_i(x) = |NN(x; k) \cap CLASS_i|$ the number of objects from $NN(x; k)$ that belong to i^{th} decision class. The kNN classifiers often use a voting algorithm for decision making, i.e.,

$$dec(x) = Voting(\langle n_1(x), \ldots, n_d(x) \rangle) = \arg\max_i n_i(x),$$

In case of imbalanced data, the vector $\langle n_1(x), \dots, n_d(x) \rangle$ can be scaled with respect to the global class distribution before applying the voting algorithm.

Rough approximation based on the set $NN(x;k)$, that is, an extension of a kNN classifier can be defined as follows. Assume that $0 \le t_1 < t_2 < k$ and let us consider for i^{th} decision class $CLASS_i \subseteq \mathcal{U}$ a function with parameters t_1, t_2 defined on any object $x \in \mathcal{U}$ by:

$$\mu_{CLASS_i}^{t_1,t_2}(x) = \begin{cases} 1 & \text{if } n_i(x) \ge t_2 \\ \frac{n_i(x) - t_1}{t_2 - t_1} & \text{if } n_i(x) \in (t_1, t_2) \\ 0 & \text{if } n_i(x) \le t_1, \end{cases} \tag{1}$$

where $n_i(x)$ is the ith coordinate in the class distribution $ClassDist(NN(x;k)) = \langle n_1(x), \dots, n_d(x) \rangle$ of $NN(x;k)$.

Let us assume that parameters t_1^o, t_2^o have been chosen in such a way that the above function satisfies for every $x \in U$ the following conditions:

$$\text{if } \mu_{CLASS_i}^{t_1^o,t_2^o}(x) = 1 \text{ then } [x]_A \subseteq CLASS_i \cap U, \tag{2}$$

$$\text{if } \mu_{CLASS_i}^{t_1^o,t_2^o}(x) = 0 \text{ then } [x]_A \cap (CLASS_i \cap U) = \emptyset, \tag{3}$$

where $[x]_A = \{y \in U : inf_A(x) = inf_A(y)\}$ denotes the indiscernibility class defined by x relatively to a fixed set of attributes A.

Then the function $\mu_{CLASS_i}^{t_1^o,t_2^o}$ considered on $\mathcal{U}$ can be treated as the rough membership function of the ith decision class. It is the result of induction on $\mathcal{U}$ of the rough membership function of ith decision class restricted to the sample U. The function $\mu_{CLASS_i}^{t_1^o,t_2^o}$ defines a rough approximations $\mathbf{L}_{kNN}(CLASS_i)$ and $\mathbf{U}_{kNN}(CLASS_i)$ of ith decision class $CLASS_i$. For any object $x \in \mathcal{U}$ we have

$$x \in \mathbf{L}_{kNN}(CLASS_i) \Leftrightarrow n_i(x) \ge t_2^o \text{ and } x \in \mathbf{U}_{kNN}(CLASS_i) \Leftrightarrow n_i(x) \ge t_1^o.$$

Certainly, one can consider in conditions (2)-(3) an inclusion to a degree and equality to a degree instead of the crisp inclusion and the crisp equality. Such degrees parameterize additionally extracted patterns and by tuning them one can search for relevant patterns.

As we mentioned above kNN methods have some drawbacks. One of them is caused by the assumption that the distance function is defined *a priori* for all pairs of objects, that is not the case for many complex data sets. In the next section we present an alternative way to define rough approximations from data.

3.2   Rule-Based Rough Approximations

In this section we describe the rule-based rough set approach to approximations.

Let $\mathbb{S} = (U, A, dec)$ be a decision table. A *decision rule* for the kth decision class is any expression of the form

$$(a_{i_1} = v_1) \wedge \dots \wedge (a_{i_m} = v_m) \Rightarrow (dec = k), \tag{4}$$

where $a_{i_j} \in A$ and $v_j \in V_{a_{i_j}}$. Any decision rule $\mathbf{r}$ of the form (4) can be characterized by following parameters:

- *length*($\mathbf{r}$): the number of descriptors in the the left hand side of implication;
- [$\mathbf{r}$] = *carrier of* $\mathbf{r}$, i.e., the set of objects satisfying the premise of $\mathbf{r}$,
- *support*($\mathbf{r}$) = *card*([$\mathbf{r}$] ∩ $CLASS_k$),
- *confidence*($\mathbf{r}$) introduced to measure the truth degree of the decision rule:

$$confidence(\mathbf{r}) = \frac{support(\mathbf{r})}{card([\mathbf{r}])}, \qquad (5)$$

The decision rule $\mathbf{r}$ is called *consistent* with $\mathbb{S}$ if *confidence*($\mathbf{r}$) = 1.

Among decision rule generation methods developed using the rough set approach one of the most interesting is related to *minimal consistent decision rules*. Given a decision table $\mathbb{S} = (U, A, dec)$, the rule $\mathbf{r}$ is called a *minimal consistent decision rule* (with $\mathbb{S}$) if is consistent with $\mathbb{S}$ and any decision rule $\mathbf{r}'$ created from $\mathbf{r}$ by removing any of descriptors from the left hand side of $\mathbf{r}$ is not consistent with $\mathbb{S}$. The set of all minimal consistent decision rules for a given decision table $\mathbb{S}$, denoted by $Min\_Cons\_Rules(\mathbb{S})$, can be computed by extracting from the decision table *object oriented reducts* (called also local reducts relative to objects) [3, 9, 26].

The elements of $Min\_Cons\_Rules(\mathbb{S})$ can be treated as interesting, valuable and useful patterns in data and used as a knowledge base in classification systems. Unfortunately, the number of such patterns can be exponential with respect to the size of a given decision table [3, 9, 26, 23]. In practice, we must apply some heuristics, like rule filtering or object covering, for selection of subsets of decision rules

Given a decision table $\mathbb{S} = (U, A, dec)$. Let us assume that **RULES**($\mathbb{S}$) is a set of decision rules induced by some rule extraction method. For any object $x \in U$, let $MatchRules(\mathbb{S}, x)$ be the set of rules from **RULES**($\mathbb{S}$) supported by x. One can define the rough membership function $\mu_{CLASS_k} : U \rightarrow [0, 1]$ for the concept determined by $CLASS_k$ as follows:

1. Let $\mathbf{R}_{yes}(x)$ be the set of all decision rules from $MatchRules(\mathbb{S}, x)$ for k^{th} class and let $\mathbf{R}_{no}(x) \subset MatchRules(\mathbb{S}, x)$ be the set of decision rules for other classes.
2. We define two real values $w_{yes}(x), w_{no}(x)$, called "for" and "against" weights for the object x by:

$$w_{yes}(x) = \sum_{\mathbf{r} \in \mathbf{R}_{yes}(x)} strength(\mathbf{r}) \qquad w_{no}(x) = \sum_{\mathbf{r} \in \mathbf{R}_{no}(x)} strength(\mathbf{r}) \qquad (6)$$

where $strength(\mathbf{r})$ is a normalized function depending on *length*, *support*, *confidence* of $\mathbf{r}$ and some global information about the decision table $\mathbb{S}$ like table size, class distribution (see [3]).
3. One can define the value of $\mu_{CLASS_k}(x)$ by

$$\mu_{CLASS_k}(x) = \begin{cases} undetermined & \text{if } \max(w_{yes}(x), w_{no}(x)) < \omega \\ 0 & \text{if } w_{no}(x) - w_{yes}(x) \geq \theta \text{ and } w_{no}(x) > \omega \\ 1 & \text{if } w_{yes}(x) - w_{no}(x) \geq \theta \text{ and } w_{yes}(x) > \omega \\ \frac{\theta + (w_{yes}(x) - w_{no}(x))}{2\theta} & \text{in other cases} \end{cases}$$

where w, θ are parameters set by user. These parameters make it possible in a flexible way to control the size of boundary region for the approximations established according to Definition 2.

Let us assume that for $\theta = \theta_o > 0$ the above function satisfies for every $x \in U$ the following conditions:

$$\text{if } \mu_{CLASS_k}^{\theta_o}(x) = 1 \text{ then } [x]_A \subseteq CLASS_k \cap U, \tag{7}$$

$$\text{if } \mu_{CLASS_k}^{\theta_o}(x) = 0 \text{ then } [x]_A \cap (CLASS_k \cap U) = \emptyset, \tag{8}$$

where $[x]_A = \{y \in U : inf_A(x) = inf_A(y)\}$ denotes the indiscernibility class defined by x with respect to the set of attributes A.

Then the function $\mu_{CLASS_k}^{\theta_o}$ considered on $\mathcal{U}$ can be treated as the rough membership function of the kth decision class. It is the result of induction on $\mathcal{U}$ of the rough membership function of kth decision class restricted to the sample U. The function $\mu_{CLASS_k}^{\theta_o}$ defines a rough approximations $\mathbf{L}_{rule}(CLASS_k)$ and $\mathbf{U}_{rule}(CLASS_k)$ of kth decision class $CLASS_k$, where $\mathbf{L}_{rule}(CLASS_k) = \{x : w_{yes}(x) - w_{no}(x) \geq \theta_o\}$ and $\mathbf{U}_{rule}(CLASS_k) = \{x : w_{yes}(x) - w_{no}(x) \geq -\theta_o\}$.

4 Hierarchical Scheme for Concept Synthesis

In this section we present general layered learning scheme for concept synthesizing. We recall the main principles of the layered learning paradigm [25].

1. Layered learning is designed for domains that are too complex for learning a mapping directly from the input to the output representation. The layered learning approach consists of breaking a problem down into several task layers. At each layer, a concept needs to be acquired. A learning algorithm solves the local concept-learning task.
2. Layered learning uses a bottom-up incremental approach to hierarchical concept decomposition. Starting with low-level concepts, the process of creating new sub-concepts continues until the high-level concepts, that deal with the full domain complexity, are reached. The appropriate learning granularity and sub-concepts to be learned are determined as a function of the specific domain. Concept decomposition in layered learning is not automated. The layers and concept dependencies are given as a background knowledge of the domain.
3. Sub-concepts may be learned independently and in parallel. Learning algorithms may be different for different sub-concepts in the decomposition hierarchy. Layered learning is effective for huge data sets and it is useful for adaptation when a training set changes dynamically.
4. The key characteristic of layered learning is that each learned layer directly affects learning at the next layer.

When using the layered learning paradigm, we assume that the target concept can be decomposed into simpler ones called sub-concepts. A hierarchy of concepts has a treelike structure. A higher level concept is constructed from those concepts in lower levels. We assume that a concept decomposition hierarchy is given by domain knowledge [18, 21]. However, one should observe that concepts and dependencies among them represented in domain knowledge are expressed often in natural language. Hence, there is a need to approximate such concepts and such dependencies as well as the whole reasoning. This issue is directly related to the computing with words paradigm [27, 28] and to rough-neural approach [12], in particular to rough mereological calculi on information granules (see, e.g., [15–19]).

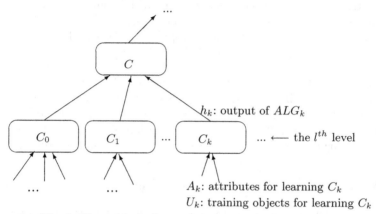

h_k: output of ALG_k

$\longleftarrow$ the lth level

A_k: attributes for learning C_k
U_k: training objects for learning C_k

Fig. 1. Hierarchical scheme for concept approximation

The goal of a layered learning algorithm is to construct a scheme for concept composition. This scheme is a structure consisting of levels. Each level consists of concepts $(C_0, C_1, ..., C_n)$. Each concept C_k is defined as a tuple

$$C_k = (U_k, A_k, O_k, ALG_k, h_k), \tag{9}$$

where (Figure 1):

- U_k is a set of objects used for learning the concept C_k,
- A_k is the set of attributes relevant for learning the concept C_k,
- O_k is the set of outputs used to define the concept C_k,
- ALG_k is the algorithm used for learning the function mapping vector values over A_k into O_k,
- h_k is the hypothesis returned by the algorithm ALG_k as a result of its run on the training set U_k.

The hypothesis h_k of the concept C_k in a current level directly affects the next level in the following ways:

1. h_k is used to construct a set of training examples U of a concept C in the next level, if C is a direct ancestor of C_k in the decomposition hierarchy.
2. h_k is used to construct a set of features A of a concept C in the next level, if C is a direct ancestor of C_k in the decomposition hierarchy.

To construct a layered learning algorithm, for any concept C_k in the concept decomposition hierarchy, one must solve the following problems:

1. Define a set of training examples U_k used for learning C_k. A training set in the lowest level are subsets of an input data set. The training set U_k at the higher level is composed from training sets of sub-concepts of C_k.
2. Define an attribute set A_k relevant to approximate the concept C_k. In the lowest level the attribute set A_k is a subset of an available attribute set. In higher levels the set A_k is created from attribute sets of sub-concepts of C_k, from an attribute set of input data and/or they are newly created attributes. The attribute set A_k is chosen depending on the domain of the concept C_k.
3. Define an output set to describe the concept C_k.
4. Choose an algorithm to learn the concept C_k that is based on a diven object set and on the defined attribute set.

In the next section we discuss in detail methods for concept synthesis. The foundation of our methods is rough set theory. We have already presented some preliminaries of rough set theory as well as parameterized methods for basic concepts approximation. They are a generalization of existing rough set based methods. Let us describe strategies for concept composing from sub-concepts.

4.1   Approximation of Compound Concept

We assume that a concept hierarchy H is given. A training set is represented by decision table $\mathbb{S}_S = (U, A, D)$. D is a set of decision attributes. Among them are decision attributes corresponding to all basic concepts and a decision attribute for the target concept. Decision values indicate if an object belong to basic concepts or to the target concept, respectively.

Using information available from a concept hierarchy for each basic concept C_b, one can create a training decision system $\mathbb{S}_{C_b} = (U, A_{C_b}, dec_{C_b})$, where $A_{C_b} \subseteq A$, and $dec_{C_b} \in D$. To approximate the concept C_b one can apply any classical method (e.g., k-NN, supervised clustering, or rule-based approach [7, 11]) to the table $\mathbb{S}_{C_b}$. For example, one can use a case-based reasoning approach presented in Section 3.1 or a rule-based reasoning approach proposed in Section 3.2 for basic concept approximation. In further discussion we assume that basic concepts are approximated by rule based classifiers derived from relevant decision tables.

To avoid overly complicated notation let us limit ourselves to the case of constructing compound concept approximation on the basis of two simpler concept approximations. Assume we have two concepts C_1 and C_2 that are given to us in the form of rule-based approximations derived from decision systems $\mathbb{S}_{C_1} = (U, A_{C_1}, dec_{C_1})$ and $\mathbb{S}_{C_1} = (U, A_{C_1}, dec_{C_1})$. Henceforth, we are given two rough membership functions $\mu_{C_1}(x)$, $\mu_{C_2}(x)$. These functions are determined with use of parameter sets $\{w_{yes}^{C_1}, w_{no}^{C_1}, \omega^{C_1}, \theta^{C_1}\}$ and $\{w_{yes}^{C_2}, w_{no}^{C_2}, \omega^{C_2}, \theta^{C_2}\}$, respectively. We want to establish a similar set of parameters $\{w_{yes}^{C}, w_{no}^{C}, \omega^{C}, \theta^{C}\}$ for the target concept C, which we want to describe with use of rough membership function μ_C. As previously noted, the parameters ω, θ controlling of the boundary region are user-configurable. But, we need to derive $\{w_{yes}^C, w_{no}^C\}$ from data. The issue is to define a decision system from which rules used to define approximations can be derived.

We assume that both simpler concepts C_1, C_2 and the target concept C are defined over the same universe of objects $\mathcal{U}$. Moreover, all of them are given on the same sample $U \subset \mathcal{U}$. To complete the construction of the decision system $\mathbb{S}_C = (U, A_C, dec_C)$, we need to specify the conditional attributes from A_C and the decision attribute dec_C. The decision attribute value $dec_C(x)$ is given for any object $x \in U$. For conditional attributes, we assume that they are either rough membership functions for simpler concepts (i.e., $A_C = \{\mu_{C_1}(x), \mu_{C_2}(x)\}$) or weights for simpler concepts (i.e., $A_C = \{w_{yes}^{C_1}, w_{no}^{C_1}, w_{yes}^{C_2}, w_{no}^{C_2}\}$). The output set O_i for each concept C_i, where $i = 1, 2$, consists of one attribute that is a rough membership function μ_{C_i} in the first case or two attributes $w_{yes}^{C_i}, w_{no}^{C_i}$ that describe fitting degrees of objects to the concept C_i and its complement, respectively. The rule-based approximations of the concept C are created by extracting rules from $\mathbb{S}_C$.

It is important to observe that such rules describing C use attributes that are in fact classifiers themselves. Therefore, in order to have a more readable and intuitively understandable description as well as more control over the quality of approximation (especially for new cases), it pays to stratify and interpret attribute domains for attributes in A_C. Instead of using just a value of a membership function or weight we would prefer to use linguistic statements such as *"the likelihood of the occurrence of C_1 is low"*. In order to do that we have to map the attribute value sets onto some limited family of subsets. Such subsets are then identified with notions such us *"certain"*, *"low"*, *"high"* etc. It is quite natural, especially in case of attributes being membership functions, to introduce linearly ordered subsets of attribute ranges, e.g., $\{negative, low, medium, high, positive\}$. That yields a fuzzy-like layout of attribute values. One may (and in some cases should) consider also the case when these subsets overlap. Then, there may be more linguistic value attached to attribute values since variables like *low* or *medium* appear.

Stratification of attribute values and introduction of a linguistic variable attached to the strata serves multiple purposes. First, it provides a way for representing knowledge in a more human-readable format since if we have a new situation (new object $x^* \in \mathcal{U} \setminus U$) to be classified (checked against compliance with concept C) we may use rules like:

If *compliance of x^* with C_1 is* high or medium **and** *compliance of x^* with C_2 is* high **then** $x^* \in C$.

Another advantage in imposing the division of attribute value sets lays in extended control over flexibility and validity of system constructed in this way. As we may define the linguistic variables and corresponding intervals, we gain the ability of making a system more stable and inductively correct. In this way we control the general layout of boundary regions for simpler concepts that contribute to construction of the target concept. The process of setting the intervals for attribute values may be performed by hand, especially when additional background information about the nature of the described problem is available. One may also rely on some automated methods for such interval construction, such as, e.g., clustering, template analysis and discretization. Some extended discussion on the foundations of this approach, which is related to rough-neural computing [12, 18] and computing with words can be found in [24, 20].

Algorithm 1 Layered learning algorithm

Input: Decision system $\mathbb{S} = (U, A, d)$, concept hierarchy H;
Output: Scheme for concept composition
1: **begin**
2:    **for** $l := 0$ to max_level **do**
3:       **for** (any concept C_k at the level l in H) **do**
4:          **if** $l = 0$ **then**
5:             $U_k := U$;
6:             $A_k := B$;
                 // where $B \subseteq A$ is a set relevant to define C_k
7:          **else**
8:             $U_k := U$;
9:             $A_k = \bigcup O_i$;
                 // for all sub-concepts C_i of C_k, where O_i is the output vector of C_i
10:            Generate a rule set $RULE(C_k)$ to determine the approximation of C_k;
11:            **for** any object $x \in U_k$ **do**
12:               generate the output vector $(w_{yes}^{C_k}(x), w_{no}^{C_k}(x))$;
                    // where $w_{yes}^{C_k}(x)$ is a fitting degree of x to the concept C_k
                    // and $w_{no}^{C_i}(x)$ is a fitting degree of x to the complement of C_k.
13:            **end for**
14:         **end if**
15:      **end for**
16:   **end for**
17: **end**

Algorithm 1 is the layered learning algorithm used in our experiments.

5   Experimental Results

To verify effectiveness of layered learning approach, we have implemented Algorithm 1 for concept composition presented in Section 4.1. The experiments were performed on data generated by road traffic simulator. In the following section we present a description of the simulator.

5.1   Road Traffic Simulator

The road simulator is a computer tool that generates data sets consisting of recording vehicle movements on the roads and at the crossroads. Such data sets are used to learn and test complex concept classifiers working on information coming from different devices and sensors monitoring the situation on the road.

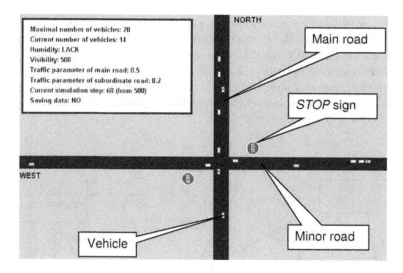

Fig. 2. Left: the board of simulation.

A driving simulation takes place on a board (see Figure 2) that presents a crossroads together with access roads. During the simulation the vehicles may enter the board from all four directions that is east, west, north and south. The vehicles coming to the crossroads form the south and north have the right of way in relation to the vehicles coming from the west and east. Each of the vehicles entering the board has only one goal - to drive through the crossroads safely and leave the board.

Both the entering and exiting roads of a given vehicle are determined at the beginning, that is, at the moment the vehicle enters the board. Each vehicle may perform the following maneuvers during the simulation like passing, overtaking, changing direction (at the crossroads), changing lane, entering the traffic from the minor road into the main road, stopping, pulling out.

Planning each vehicle's further steps takes place independently in each step of the simulation. Each vehicle, "observing" the surrounding situation on the road, keeping in mind its destination and its own parameters, makes an independent decision about its further steps; whether it should accelerate, decelerate and what (if any) maneuver should be commenced, continued, or stopped.

We associate the simulation parameters with the readouts of different measuring devices or technical equipment placed inside the vehicle or in the outside environment (e.g., by the road, in a helicopter observing the situation on the road, in a police car). These are devices and equipment playing the role of detecting devices or converters meaning sensors (e.g., a thermometer, range finder, video camera, radar, image and sound converter). The attributes taking the simulation parameter values, by analogy to devices providing their values will be called sensors. Here are exemplary sensors: distance from the crossroads (in screen units), vehicle speed, acceleration and deceleration, etc.

Apart from sensors the simulator registers a few more attributes, whose values are determined based on the sensor's values in a way determined by an expert. These parameters in the present simulator version take over the binary values and are therefore called concepts. Concepts definitions are very often in a form of a question which one can answer YES, NO or DOES NOT CONCERN (NULL value). In Figure 3 there is an exemplary relationship diagram for some concepts that are used in our experiments.

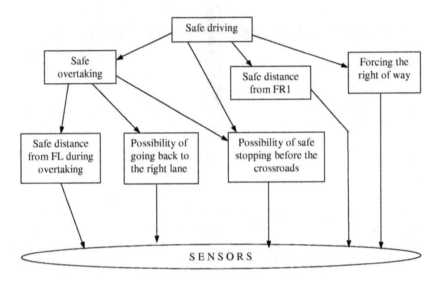

Fig. 3. The relationship diagram for exemplary concepts

During the simulation, when a new vehicle appears on the board, its so called driver's profile is determined and may not be changed until it disappears from the board. It may take one of the following values: a very careful driver, a careful driver and a careless driver. Driver's profile is the identity of the driver and according to this identity further decisions as to the way of driving are made. Depending on the driver's profile and weather conditions speed limits are determined, which cannot be exceeded. The humidity of the road influences the length of braking distance, for depending on humidity different speed changes take place within one simulation step, with the same braking mode. The driver's profile influences the speed limits dictated by visibility. If another vehicle is invisible for a given vehicle, this vehicle is not taken into consideration in the independent planning of further driving by a given car. Because this may cause dangerous situations, depending on the driver's profile, there are speed limits for the vehicle.

During the simulation data may be generated and stored in a text file. The generated data are in a form of information table. Each line of the board depicts the situation of a single vehicle and the sensors' and concepts' values are registered for a given vehicle and its neighboring vehicles. Within each simulation step descriptions of situations of all the vehicles are saved to file.

5.2   Experiment Description

A number of different data sets have been created with the road traffic simulator. They are named by cxx_syyy, where xx is the number of cars and yyy is the number of time units of the simulation process. The following data sets: $c10\_s100$, $c10\_s200$, $c10\_s300$, $c10\_s400$, $c10\_s500$, $c20\_s500$ have been generated for our experiments. Let us emphasize that the first data set consists of about 800 situations, whereas the last data set is the largest one, which can be generated by the simulator. This data set consists of about 10000 situations. Every data set has 100 attributes and has imbalanced class distribution, i.e., about $6\% \pm 2\%$ of situations are unsave.

Every data set cxx_syyy was divided randomly into two subsets cxx_syyy.trn and cxx_syyy.test with proportion 80% and 20%, respectively. The data sets of form cxx_syyy.trn are used in learning the concept approximations.

We consider two testing models called *testing for similar situations* and *testing for new situations*. They are described as follows:

Model I: *Testing for similar situations.* This model uses the data sets of the form cxx_syyy.test for testing the quality of approximation algorithms. The situations, which are used in this testing model, are generated from the same simulation process as the training situations.

Model II: *Testing for new situations.* This model uses data from a new simulation process. In this model, we create new data sets using the simulator. They are named by $c10\_s100N$, $c10\_s200N$, $c10\_s300N$, $c10\_s400N$, $c10\_s500N$, $c20\_s500N$, respectively.

We compare the quality of two learning approaches called *RS rule-based learning* (**RS**) and *RS-layered learning* (**RS-L**). In the first approach, we employed the RSES system [4] to generate the set of minimal decision rules and classified the situations from testing data. The conflicts are resolved by simple voting strategy. The comparison analysis is performed with respect to the following criteria:

1. accuracy of classification,
2. covering rate of new cases (generality),
3. computing time necessary for classifier synthesis, and
4. size of rule set used for target concept approximation.

In the layered learning approach, from training table we create five sub-tables to learn five basic concepts (see Figure 3): C_1: "*safe distance from FL during overtaking*," C_2: "*possibility of safe stopping before crossroads*," C_3: "*possibility of going back to the right lane*," C_4: "*safe distance from preceding car*," C_5: "*forcing the right of way*."

These tables are created using information available from the concept decomposition hierarchy. A concept in the next level is C_6: "*safe overtaking*". C_6 is located over the concepts C_1, C_2 and C_3 in the concept decomposition hierarchy. To approximate concept C_6, we create a table with three conditional attributes. These attributes describe fitting degrees of object to concepts C_1, C_2,

C_3, respectively. The decision attribute has three values YES, NO, or $NULL$ corresponding to the cases of overtaking made by car: safe, not safe, not applicable.

The target concept C_7: "*safe driving*" is located at the third level of the concept decomposition hierarchy. The concept C_7 is obtained by composition from concepts C_4, C_5 and C_6. To approximate C_7 we also create a decision table with three attributes, representing fitting degrees of objects to the concepts C_4, C_5, C_6, respectively. The decision attribute has two possible values YES or NO if a car is satisfying global safety condition, or not, respectively.

Classification Accuracy. As we mentioned before, the decision class "safe driving = YES" is dominating in all training data sets. It takes above 90% of training sets. Sets of training examples belonging to the "NO" class are small relative to the training set size. Searching for approximation of the "NO" class with the high precision and generality is a challenge for learning algorithms. In experiments we concentrate on approximation of the "NO" class.

In Table 1 we present the classification accuracy of RS and RS-L classifiers for the first of testing models. It means that training sets and test sets are disjoint and samples are chosen from the same simulation data set.

Table 1. Classification accuracy for the first testing model

Testing model I	Total accuracy		Accuracy of YES		Accuracy of NO	
	RS	RS-L	RS	RS-L	RS	RS-L
c10_s100	0.98	0.93	0.99	0.98	0.67	0
c10_s200	0.99	0.99	1	0.99	0.90	1
c10_s300	0.99	0.96	0.99	0.96	0.82	0.81
c10_s400	0.99	0.97	0.99	0.98	0.88	0.85
c10_s500	0.99	0.94	0.99	0.93	0.94	0.96
c20_s500	0.99	0.93	0.99	0.94	0.91	0.91
Average	**0.99**	**0.95**	**0.99**	**0.96**	**0.85**	**0.75**

One can observe that the classification accuracy of the testing model I is higher, because the testing the training sets are chosen from the same data set. Although accuracy of the "YES" class is better than the "NO" class but accuracy of the "NO" class is quite satisfactory. In those experiments, the standard classifier shows a little bit better performance than hierarchical classifier. One can observe that when training sets reach a sufficient size (over 2500 objects) the accuracy on the class "NO" of both classifiers are comparable. To verify if classifier approximations are of high precision and generality, we use the second testing model, where training tables and testing tables are chosen from the new generated simulation data sets. One can observe that accuracy of the "NO" class strongly decreased. In this case the hierarchical classifier shows much better performance. In Table 2 we present the accuracy of the standard classifier and the hierarchical classifier using the second testing model.

Table 2. Classification accuracy for the second testing model

Testing model II	Total accuracy		Accuracy of YES		Accuracy of NO	
	RS	RS-L	RS	RS-L	RS	RS-L
$c10\_s100N$	0.94	0.97	1	1	0	0
$c10\_s200N$	0.99	0.96	1	0.98	0.75	0.60
$c10\_s300N$	0.99	0.98	1	0.98	0	0.78
$c10\_s400N$	0.96	0.77	0.96	0.77	0.57	0.64
$c10\_s500N$	0.96	0.89	0.99	0.90	0.30	0.80
$c20\_s500N$	0.99	0.89	0.99	0.88	0.44	0.93
Average	**0.97**	**0.91**	**0.99**	**0.92**	**0.34**	**0.63**

Covering Rate. Generality of classifiers usually is evaluated by the recognition ability of unseen objects. In this section we analyze covering rate of classifiers for new objects. In Table 3 we present coverage degrees using the first testing model. One can observe that the coverage degrees of standard and hierarchical classifiers are comparable in this case.

Table 3. Covering rate for the first testing model

Testing model I	Total accuracy		Accuracy of YES		Accuracy of NO	
	RS	RS-L	RS	RS-L	RS	RS-L
$c10\_s100$	0.97	0.96	0.98	0.96	0.85	1
$c10\_s200$	0.95	0.95	0.96	0.96	0.67	0.80
$c10\_s300$	0.94	0.93	0.97	0.95	0.59	0.55
$c10\_s400$	0.96	0.94	0.96	0.94	0.91	0.87
$c10\_s500$	0.96	0.95	0.97	0.96	0.84	0.86
$c20\_s500$	0.93	0.97	0.94	0.98	0.79	0.92
Average	**0.95**	**0.95**	**0.96**	**0.96**	**0.77**	**0.83**

We also examined the coverage degrees using the second testing model. We obtained the similar scenarios to the accuracy degree. The coverage rate for the both decision classes strongly decreases. Again the hierarchical classifier shows to be more stable than the standard classifier. The results are presented in Table 4.

Computing Speed. A time computation necessary for concept approximation synthesis is one of the important features of learning algorithms. Quality of learning approach should be assessed not only by quality of the classifier. In many real-life situations it is necessary not only to make precise decisions but also to learn classifiers in a short time.

The layered learning approach shows a tremendous advantage in comparison with the standard learning approach with respect to computation time. In the case of standard classifier, computational time is measured as a time required for computing a rule set used for decision class approximation. In the case of

Table 4. Covering rate for the second testing model

Testing model II	Total accuracy		Accuracy of YES		Accuracy of NO	
	RS	RS-L	RS	RS-L	RS	RS-L
$c10\_s100N$	0.44	0.72	0.44	0.74	0.50	0.38
$c10\_s200N$	0.72	0.73	0.73	0.74	0.50	0.63
$c10\_s300N$	0.47	0.68	0.49	0.69	0.10	0.44
$c10\_s400N$	0.74	0.90	0.76	0.93	0.23	0.35
$c10\_s500N$	0.72	0.86	0.74	0.88	0.40	0.69
$c20\_s500N$	0.62	0.89	0.65	0.89	0.17	0.86
Average	**0.62**	**0.79**	**0.64**	**0.81**	**0.32**	**0.55**

Table 5. Time for standard and hierarchical classifier generation

Table names	RS	RS-L	Speed up ratio
$c10\_s100$	94 s	2.3 s	40
$c10\_s200$	714 s	6.7 s	106
$c10\_s300$	1450 s	10.6 s	136
$c10\_s400$	2103 s	34.4 s	60
$c10\_s500$	3586 s	38.9 s	92
$c20\_s500$	10209 s	98s	104
Average			**90**

hierarchical classifier computational time is equal to the total time required for all sub-concepts and target concept approximation. The experiments were performed on computer with processor AMD Athlon 1.4GHz. One can see in Table 5 that the speed up ratio of the layered learning approach over the standard one reaches from 40 to 130 times.

Description Size. Now, we consider the complexity of concept description. We approximate concepts using decision rule sets. The size of a rule set is characterized by rule lengths and its cardinality. In Table 6 we present rule lengths and the number of decision rules generated by the standard learning approach. One can observe that rules generated by the standard approach are quite long. They contain above 40 descriptors (on average).

Table 6. Rule set size for the standard learning approach

Tables	Rule length	# Rule set
$c10\_s100$	34.1	12
$c10\_s200$	39.1	45
$c10\_s300$	44.7	94
$c10\_s400$	42.9	85
$c10\_s500$	47.6	132
$c20\_s500$	60.9	426
Average	**44.9**	

Table 7. Description length: C_1, C_2, C_3 for the hierarchical learning approach

Tables	Concept C_1		Concept C_2		Concept C_3	
	Ave. rule l.	# Rules	Ave. rule l.	# Rules	Ave. rule l.	# Rules
c10_s100	5.0	10	5.3	22	4.5	22
c10_s200	5.1	16	4.5	27	4.6	41
c10_s300	5.2	18	6.6	61	4.1	78
c10_s400	7.3	47	7.2	131	4.9	71
c10_s500	5.6	21	7.5	101	4.7	87
c20_s500	6.5	255	7.7	1107	5.8	249
Average	**5.8**		**6.5**		**4.8**	

Table 8. Description length: C_4, C_5 for the hierarchical learning approach

Tables	Concept C_4		Concept C_5	
	Rule length	# Rule set	Rule length	# Rule set
c10_s100	4.5	22	1.0	2
c10_s200	4.6	42	4.7	14
c10_s300	5.2	90	3.4	9
c10_s400	6.0	98	4.7	16
c10_s500	5.8	146	4.9	15
c20_s500	5.4	554	5.3	25
Average	**5.2**		**4.0**	

Table 9. Description length: C_6, C_7, hierarchical learning approach

Tables	Concept C_6		Concept C_7	
	Rule length	# Rule set	Rule length	# Rule set
c10_s100	2.2	6	3.5	8
c10_s200	1.3	3	3.7	13
c10_s300	2.4	7	3.6	18
c10_s400	2.5	11	3.7	27
c10_s500	2.6	8	3.7	30
c20_s500	2.9	16	3.8	35
Average	**2.3**		**3.7**	

The size of rule sets generated by layered learning approach are presented in Tables 7, 8 and 9. One can notice that rules approximating sub-concepts are short. The average rule length is from 4 to 6.5 for the basic concepts and from 2 to 3.7 for the super-concepts. Therefore rules generated by layered learning approach are more understandable and easier to interpret than rules induced by the standard learning approach.

Two concepts C_2 and C_4 are more complex than the others. The rule set induced for C_2 takes 28% and the rule set induced for C_4 takes above 27% of the number of rules generated for all seven concepts in the traffic road problem.

6   Conclusion

We presented a method for concept synthesis based on the layered learning approach. Unlike the traditional learning approach, in the layered learning approach the concept approximations are induced not only from accessed data sets but also from expert's domain knowledge. In the paper, we assume that knowledge is represented by a concept dependency hierarchy. The layered learning approach proved to be promising for complex concept synthesis. Experimental results with road traffic simulation are showing advantages of this new approach in comparison to the standard learning approach. The main advantages of the layered learning approach can be summarized as follows:

1. High precision of concept approximation.
2. High generality of concept approximation.
3. Simplicity of concept description.
4. High computational speed.
5. Possibility of localization of sub-concepts that are difficult to approximate. It is important information, because is specifying a task on which we should concentrate to improve the quality of the target concept approximation.

In future we plan to investigate more advanced approaches for concept composition. One interesting possibility is to use patterns defined by rough approximations of concepts defined by different kinds of classifiers in synthesis of compound concepts. We also would like to develop methods for rough-fuzzy classifier's synthesis (see Section 4.1). In particular, the method mentioned in Section 4.1 based on rough-fuzzy classifiers introduces more flexibility for such composing because a richer class of patterns introduced by different layers of rough-fuzzy classifiers can lead to improving of the classifier quality [18]. On the other hand, such a process is more complex and efficient heuristics for synthesis of rough-fuzzy classifiers should be developed.

We also plan to apply the layered learning approach to real-life problems especially when domain knowledge is specified in natural language. This can make further links with the computing with words paradigm [27, 28, 12]. This is in particular linked with the rough mereological approach (see, e.g., [15, 17]) and with the rough set approach for approximate reasoning in distributed environments [20, 21], in particular with methods of information system composition [20, 2].

Acknowledgements

The research has been partially supported by the grant 3T11C00226 from Ministry of Scientific Research and Information Technology of the Republic of Poland.

References

1. Aha, D.W.. The omnipresence of case-based reasoning in science and application. Knowledge-Based Systems, 11 (5-6) (1998) 261-273.

2. Barwise, J., Seligman, J., eds.: Information Flow: The Logic of Distributed Systems. Volume 44 of Tracts in Theoretical Computer Scienc. Cambridge University Press, Cambridge, UK (1997)
3. Bazan, J.G.: A comparison of dynamic and non-dynamic rough set methods for extracting laws from decision tables. In Polkowski, L., Skowron, A., eds.: Rough Sets in Knowledge Discovery 1: Methodology and Applications. Physica-Verlag, Heidelberg, Germany (1998) 321–365
4. Bazan, J.G., Szczuka, M.: RSES and RSESlib - a collection of tools for rough set computations. In Ziarko, W., Yao, Y., eds.: Second International Conference on Rough Sets and Current Trends in Computing RSCTC. LNAI **2005**. Banff, Canada, Springer-Verlag (2000) 106–113
5. Bazan, J., Nguyen, H.S., Skowron, A., Szczuka, M.: A view on rough set concept approximation. In Wang, G., Liu, Q., Yao, Y., Skowron, A., eds.: Proceedings of the Ninth International Conference on Rough Sets, Fuzzy Sets, Data Mining and Granular Computing (RSFDGrC'2003),Chongqing, China. LNAI **2639**. Heidelberg, Germany, Springer-Verlag (2003) 181–188
6. Cover, T.M. and Hart, P.E.: Nearest neighbor pattern classification. IEEE Transactions on Information Theory, 13 (1967) 21-27.
7. Friedman, J., Hastie, T., Tibshirani, R.: The Elements of Statistical Learning: Data Mining, Inference, and Prediction. Springer-Verlag, Heidelberg, Germany (2001)
8. Grzymała-Busse, J.: A new version of the rule induction system lers. Fundamenta Informaticae **31(1)** (1997) 27–39
9. Komorowski, J., Pawlak, Z., Polkowski, L., Skowron, A.: Rough sets: a tutorial. In Pal, S.K., Skowron, A., eds.: Rough Fuzzy Hybridization: A New Trend in Decision-Making. Springer-Verlag, Singapore (1999) 3–98
10. Kloesgen, W., Żytkow, J., eds.: Handbook of Knowledge Discovery and Data Mining. Oxford University Press, Oxford (2002)
11. Mitchell, T.: Machine Learning. Mc Graw Hill (1998)
12. Pal, S.K., Polkowski, L., Skowron, A., eds.: Rough-Neural Computing: Techniques for Computing with Words. Cognitive Technologies. Springer-Verlag, Heidelberg, Germany (2003)
13. Pawlak, Z.: Rough Sets: Theoretical Aspects of Reasoning about Data. Volume 9 of System Theory, Knowledge Engineering and Problem Solving. Kluwer Academic Publishers, Dordrecht, The Netherlands (1991)
14. Poggio, T., Smale, S.: The mathematics of learning: Dealing with data. Notices of the AMS **50** (2003) 537–544
15. Polkowski, L., Skowron, A.: Rough mereology: A new paradigm for approximate reasoning. International Journal of Approximate Reasoning **15** (1996) 333–365
16. Polkowski, L., Skowron, A.: Rough mereological calculi of granules: A rough set approach to computation. Computational Intelligence **17** (2001) 472–492
17. Polkowski, L., Skowron, A.: Towards adaptive calculus of granules. In Zadeh, L.A., Kacprzyk, J., eds.: Computing with Words in Information/Intelligent Systems, Heidelberg, Germany, Physica-Verlag (1999) 201–227
18. Skowron, A., Stepaniuk, J.: Information granules and rough-neural computing. [12] 43–84
19. Skowron, A., Stepaniuk, J.: Information granules: Towards foundations of granular computing. International Journal of Intelligent Systems **16** (2001) 57–86
20. Skowron, A., Stepaniuk, J.: Information granule decomposition. Fundamenta Informaticae **47(3-4)** (2001) 337–350

21. Skowron, A.: Approximate reasoning by agents in distributed environments. In Zhong, N., Liu, J., Ohsuga, S., Bradshaw, J., eds.: Intelligent Agent Technology Research and Development: Proceedings of the 2nd Asia-Pacific Conference on Intelligent Agent Technology IAT01, Maebashi, Japan, October 23-26. World Scientific, Singapore (2001) 28–39

22. Skowron, A.: Approximation spaces in rough neurocomputing. In Inuiguchi, M., Tsumoto, S., Hirano, S., eds.: Rough Set Theory and Granular Computing. Volume 125 of Studies in Fuzziness and Soft Computing. Springer-Verlag, Heidelberg, Germany (2003) 13–22

23. Skowron, A., Rauszer, C.: The discernibility matrices and functions in information systems. In Słowiński, R., ed.: Intelligent Decision Support - Handbook of Applications and Advances of the Rough Sets Theory. Volume 11 of D: System Theory, Knowledge Engineering and Problem Solving. Kluwer Academic Publishers, Dordrecht, Netherlands (1992) 331–362

24. Skowron, A., Szczuka, M.: Approximate reasoning schemes: Classifiers for computing with words. In: Proceedings of SMPS 2002. Advances in Soft Computing, Heidelberg, Canada, Springer-Verlag (2002) 338–345

25. Stone, P.: Layered Learning in Multi-Agent Systems: A Winning Approach to Robotic Soccer. The MIT Press, Cambridge, MA (2000)

26. Wróblewski, J.: Covering with reducts - a fast algorithm for rule generation. In Polkowski, L., Skowron, A., eds.: Proceedings of the First International Conference on Rough Sets and Current Trends in Computing (RSCTC'98), Warsaw, Poland. LNAI **1424**, Heidelberg, Germany, Springer-Verlag (1998) 402–407

27. Zadeh, L.A.: Fuzzy logic = computing with words. IEEE Transactions on Fuzzy Systems **4** (1996) 103–111

28. Zadeh, L.A.: A new direction in AI: Toward a computational theory of perceptions. AI Magazine **22** (2001) 73–84

Basic Algorithms and Tools for Rough Non-deterministic Information Analysis

Hiroshi Sakai and Akimichi Okuma

Department of Computer Engineering
Kyushu Institute of Technology
Tobata, Kitakyushu 804, Japan
sakai@comp.kyutech.ac.jp

Abstract. *Rough non-deterministic information analysis* is a framework for handling the rough sets based concepts, which are defined in not only *DISs* (*Deterministic Information Systems*) but also *NISs* (*Non-deterministic Information Systems*), on computers. *NISs* were proposed for dealing with information incompleteness in *DISs*. In this paper, two modalities, i.e., the certainty and the possibility, are defined for each concept like the definability of a set, the consistency of an object, data dependency, rule generation, reduction of attributes, criterion of rules *support*, *accuracy* and *coverage*. Then, each algorithm for computing two modalities is investigated. An important problem is how to compute two modalities depending upon all derived *DISs*. A simple method, such that two modalities are sequentially computed in all derived *DISs*, is not suitable. Because the number of all derived *DISs* increases in exponential order. This problem is uniformly solved by means of applying either *inf* and *sup* information or *possible equivalence relations*. An information analysis tool for *NISs* is also presented.

1 Introduction

Rough set theory offers a new mathematical approach to vagueness and uncertainty, and the rough sets based concepts have been recognized to be very useful [1,2,3,4]. This theory usually handles tables with deterministic information, which we call *Deterministic Information Systems* (*DISs*). Many applications of this theory to data mining, rule generation, machine learning and knowledge discovery have been investigated [5–11].

Non-deterministic Information Systems (*NISs*) and *Incomplete Information Systems* have been proposed for handling information incompleteness in *DISs*, like null values, unknown values, missing values and etc. [12–16]. For any *NIS*, we usually suppose that there exists a *DIS* with unknown real information in a set of all derived *DISs*. Let DIS^{real} denote this deterministic information system from *NIS*. Of course, it is impossible to know DIS^{real} itself without additional information. However, if a formula α holds in every derived *DIS* from a *NIS*, α also holds in DIS^{real}. This formula α is not influenced by the information incompleteness in *NIS*. If a formula α holds in some derived *DISs*

J.F. Peters et al. (Eds.): Transactions on Rough Sets I, LNCS 3100, pp. 209–231, 2004.

from a NIS, there exists such a possibility that α holds in DIS^{real}. We call the former the *certainty* (of the formula α for DIS^{real}) and the latter the *possibility*, respectively. In $NISs$, such two modalities for DIS^{real} have been employed, and several work on logic in $NISs$ has been studied [12,14,15,17].

Very few work deals with algorithms for handling $NISs$ on computers. In [15,16], Lipski showed a question-answering system besides an axiomatization of logic. In [18,19], Grzymala-Busse surveyed the unknown attribute values, and studied the learning from examples with unknown attribute values. In [20,21,22], Kryszkiewicz investigated rules in incomplete information systems. These are the most important work for handling information incompleteness in $DISs$ on computers. This paper follows these two modalities for DIS^{real}, and focuses on the issues in the following.
(1) The definability of a set in $NISs$ and an algorithm for handling it on computers.
(2) The consistency of an object in $NISs$ and an algorithm for handling it on computers.
(3) Data dependency in $NISs$ and an algorithm for handling it on computers.
(4) Rules in $NISs$ and an algorithm for handling them on computers.
(5) Reduction of attributes in $NISs$ and an algorithm for handling it on computers.

An important problem is how to compute two modalities depending upon all derived $DISs$ from a NIS. A simple method, such that every definition is sequentially computed in all derived $DISs$ from a NIS, is not suitable. Because the number of derived $DISs$ from a NIS increases in exponential order. This problem is uniformly solved by means of applying either inf and sup information or possible equivalence relations in the subsequent sections.

In Preliminary, definitions in $DISs$ and rough sets based concepts are surveyed. Then, each algorithm for five issues is sequentially examined. Tool programs for these issues are also implemented, which are presented in appendixes.

2   Preliminary

This section surveys some definitions in $DISs$, and connects these definitions with equivalence relations.

2.1   Some Definitions in DISs

A *Deterministic Information System* (DIS) is a quadruplet $(OB, AT, \{VAL_A | A \in AT\}, f)$, where OB is a finite set whose elements are called *objects*, AT is a finite set whose elements are called *attributes*, VAL_A is a finite set whose elements are called *attribute values* and f is such a mapping that $f : OB \times AT \rightarrow \cup_{A \in AT} VAL_A$ which is called a *classification function*.

For $ATR = \{A_1, \cdots, A_n\} \subseteq AT$, we call $(f(x, A_1), \cdots, f(x, A_n))$ a *tuple* (for ATR) of $x \in OB$. If $f(x, A) = f(y, A)$ holds for every $A \in ATR \subseteq AT$, we see there is a relation between x and y for ATR. This relation is an equivalence relation over OB. Let $eq(ATR)$ denote this equivalence relation, and let $[x]_{ATR} \in eq(ATR)$ denote an equivalence class $\{y \in OB | f(y, A) = f(x, A)$ for every $A \in ATR\}$.

Now, let us show some rough sets based concepts defined in $DISs$ [1,3].

(D-i) The Definability of a Set: If a set $X \subseteq OB$ is the union of some equivalence classes in $eq(ATR)$, we say X is *definable* (for ATR) in DIS. Otherwise, we say X is *rough* (for ATR) in DIS.

(D-ii) The Consistency of an Object: Let us consider two disjoint sets $CON \subseteq AT$ which we call *condition attributes* and $DEC \subseteq AT$ which we call *decision attributes*. An object $x \in OB$ is *consistent* (with any other object $y \in OB$ in the relation from CON to DEC), if $f(x, A) = f(y, A)$ holds for every $A \in CON$ implies $f(x, A) = f(y, A)$ holds for every $A \in DEC$.

(D-iii) Dependencies among Attributes: We call a ratio $deg(CON, DEC) = |\{x \in OB | x$ is consistent in the relation from CON to $DEC \}| / |OB|$ the *degree of dependency* from CON to DEC. Clearly, $deg(CON, DEC) = 1$ holds if and only if every object $x \in OB$ is consistent.

(D-iv) Rules and Criteria (Support, Accuracy and Coverage): For any object $x \in OB$, let $imp(x, CON, DEC)$ denote a formula called an *implication*: $\wedge_{A \in CON} [A, f(x, A)] \Rightarrow \wedge_{A \in DEC}[A, f(x, A)]$, where a formula $[A, f(x, A)]$ implies that $f(x, A)$ is the value of the attribute A. This is called a *descriptor* in [15,22]. In most of work on rule generation, a rule is defined by an implication $\tau : imp(x, CON, DEC)$ satisfying some constraints. A constraint, such that $deg(CON, DEC) = 1$ holds from CON to DEC, has been proposed in [1]. Another familiar constraint is defined by three values in the following: $support(\tau) = |[x]_{CON} \cap [x]_{DEC}| / |OB|$, $accuracy(\tau) = |[x]_{CON} \cap [x]_{DEC}| / |[x]_{CON}|$ and *coverage* $(\tau) = |[x]_{CON} \cap [x]_{DEC}| / |[x]_{DEC}|$ [9].

(D-v) Reduction of Condition Attributes in Rules: Let us consider such an implication $imp(x, CON, DEC)$ that x is consistent in the relation from CON to DEC. An attribute $A \in CON$ is *dispensable* in CON, if x is consistent in the relation from $CON - \{A\}$ to DEC.

These are the definitions of rough sets based concepts in $DISs$. Several tools for $DISs$ have been realized according to these definitions [5,6,7,8,9,10,11].

2.2   Definitions from D-i to D-v and Equivalence Relations over OB

Rough set theory makes use of equivalence relations for solving problems. Each definition from D-i to D-v is solved by means of applying equivalence relations.

As for the definability of a set $X \subseteq OB$, X is definable (for ATR) in a DIS, if $\cup_{x \in K}[x]_{ATR} = X$ holds for a set $K \subseteq X \subseteq OB$. According to this definition, it is possible to derive such a necessary and sufficient condition that a set X is definable if and only if $\cup_{x \in X}[x]_{ATR} = X$ holds.

Now, let us show the most important proposition, which connects two equivalence classes $[x]_{CON}$ and $[x]_{DEC}$ with the consistency of x.

Proposition 1 [1]. For any DIS, (1) and (2) in the following are equivalent.
(1) An object $x \in OB$ is consistent in the relation from CON to DEC.
(2) $[x]_{CON} \subseteq [x]_{DEC}$.

According to Proposition 1, the degree of dependency from CON to DEC is equal to $|\{x \in OB|[x]_{CON} \subseteq [x]_{DEC}\}|/|OB|$. As for criteria *support*, *accuracy* and *coverage*, they are defined by equivalence classes $[x]_{CON}$ and $[x]_{DEC}$. As for the reduction of attributes values in rules, let us consider such an implication $imp(x, CON, DEC)$ that x is consistent in the relation from CON to DEC. Here, an attribute $A \in CON$ is dispensable, if $[x]_{CON-\{A\}} \subseteq [x]_{DEC}$ holds.

In this way, definitions from D-i to D-v are uniformly computed by means of applying equivalence relations in $DISs$.

3   A Framework
of Rough Non-deterministic Information Analysis

This section gives definitions in $NISs$ and two modalities due to the information incompleteness in $NISs$. Then, a framework of rough non-deterministic information analysis is proposed.

3.1   A Proposal of Rough Non-deterministic Information Analysis

A *Non-deterministic Information System* (NIS) is also a quadruplet $(OB, AT, \{VAL_A|A \in AT\}, g)$, where $g : OB \times AT \to P(\cup_{A \in AT} VAL_A)$ (a power set of $\cup_{A \in AT} VAL_A$). Every set $g(x, A)$ is interpreted as that there is a real value in this set but this value is not known [13,15,21]. Especially if the real value is not known at all, $g(x, A)$ is equal to VAL_A. This is called the *null value* interpretation [12].

Definition 1. Let us consider a $NIS=(OB, AT, \{VAL_A|A \in AT\}, g)$, a set $ATR \subseteq AT$ and a mapping $h : OB \times ATR \to \cup_{A \in ATR} VAL_A$ such that $h(x, A) \in g(x, A)$. We call a $DIS=(OB, ATR, \{VAL_A|A \in ATR\}, h)$ a *derived DIS (for ATR) from NIS*.

Example 1. Let us consider NIS_1 in Table 1, which is automatically produced by means of applying a random number program. There are $2176782336(=2^{12} \times 3^{12})$ derived $DISs$ for $ATR=\{A, B, C, D, E, F\}$. As for $ATR=\{A, B, C\}$, there are $1118744(=2^7 \times 3^4)$ derived $DISs$.

Definition 2. Let us consider a NIS. There exists a derived DIS with unknown real attribute values due to the interpretation of $g(x, A)$. So, let DIS^{real} denote a derived DIS with unknown real attribute values.

Of course, it is impossible to know DIS^{real} without additional information. However, some information based on DIS^{real} may be derived. Let us consider a relation from $CON=\{A, B\}$ to $DEC=\{C\}$ and object 2 in NIS_1. The tuple of object 2 is either (2,2,2) or (4,2,2). In both cases, object 2 is consistent. Thus, it is possible to conclude that object 2 is consistent in DIS^{real}, too. In order to handle

Table 1. A Table of NIS_1

OB	A	B	C	D	E	F
1	{3}	{1,3,4}	{3}	{2}	{5}	{5}
2	{2,4}	{2}	{2}	{3,4}	{1,3,4}	{4}
3	{1,2}	{2,4,5}	{2}	{3}	{4,5}	{5}
4	{1,5}	{5}	{2,4}	{2}	{1,4,5}	{5}
5	{3,4}	{4}	{3}	{1,2,3}	{1}	{2,5}
6	{3,5}	{4}	{1}	{2,3,5}	{5}	{2,3,4}
7	{1,5}	{4}	{5}	{1,4}	{3,5}	{1}
8	{4}	{2,4,5}	{2}	{1,2,3}	{2}	{1,2,5}
9	{2}	{5}	{3}	{5}	{4}	{2}
10	{2,3,5}	{1}	{2}	{3}	{1}	{1,2,3}

such information based on DIS^{real}, two modalities *certainty* and *possibility* are usually defined in most of work handling information incompleteness.

(Certainty). If a formula α holds in every derived DIS from a NIS, α also holds in DIS^{real}. In this case, we say α *certainly holds* in DIS^{real}.

(Possibility). If a formula α holds in some derived $DISs$ from a NIS, there exists such a possibility that α holds in DIS^{real}. In this case, we say α *possibly holds* in DIS^{real}.

According to two modalities for DIS^{real}, it is possible to extend definitions from D-i to D-v in $DISs$ to the definitions in $NISs$. In the subsequent sections, we sequentially give definitions from N-i to N-v in $NISs$. We name information analysis, which depends upon definitions from N-i to N-v and other extended definitions in $NISs$, *Rough Non-deterministic Information Analysis (RNIA)* from now on.

3.2   Incomplete Information Systems and NISs

Incomplete information systems in [21,22] and $NISs$ seem to be the same, but there exist some distinct differences. Example 2 clarifies the difference between incomplete information systems and $NISs$.

Example 2. Let us consider an incomplete information system in Table 2.

Table 2. A Table of an Incomplete Information System

OB	A	B
1	*	2
2	3	3

Table 3. A Table of a NIS

OB	A	B
1	$\{1, 2\}$	2
2	3	3

Here, let us suppose VAL_A be $\{1, 2, 3\}$, $CON=\{A\}$ and $DEC=\{B\}$. The attribute value of object 1 is not definite, and the $*$ symbol is employed for describing it. In this case, the null value interpretation is applied to this $*$, and $3 \in VAL_A$ may occur instead of $*$. Therefore, object 2 is not consistent in this case. According to the definition in [22], a formula $[A, 3] \Rightarrow [B, 3]$ is a possible rule. Now, let us consider a NIS in Table 3. The attribute value of object 1 is not definite, either. However in this NIS, object 2 is consistent in every derived DIS. So, a formula $[A, 3] \Rightarrow [B, 3]$ is a certain rule according to the definition in [22]. Thus, the meaning of the formula $[A, 3] \Rightarrow [B, 3]$ in Table 2 is different from that in Table 3. In incomplete information systems, each indefinite value is uniformly identified with unknown value $*$. However in $NISs$, each indefinite value is identified with a subset of VAL_A ($A \in AT$). Clearly, $NISs$ are more informative than incomplete information systems.

3.3   The Core Problem for RNIA and the Purpose of This Work

Definitions from N-i to N-v, which are sequentially given in the following sections, depend upon every derived $DISs$ from a NIS. Therefore, it is necessary to compute definitions from D-i to D-v in every derived DIS. The number of all derived $DISs$, which is the product $\prod_{x \in OB, A \in ATR} |g(x, A)|$ for $ATR \subseteq AT$, increases in exponential order. Even though each definition from D-i to D-iv can be solved in the polynomial time order for input data size [23], each definition from N-i to N-iv depends upon all derived $DISs$. The complexity for finding a minimal reduct in a DIS is also proved to be NP-hard [3]. Namely for handling $NISs$ with large number of derived $DISs$, it may take much execution time without effective algorithms. This is the core problem for $RNIA$.

This paper proposes the application of inf and sup information and possible equivalence relations, which are defined in the next subsection, to solving the above core problem. In Section 2.2, the connection between definitions from D-i to D-v and equivalence relations is proved. Analogically, we think about the connection between definitions from N-i to N-v and possible equivalence relations [24,25].

3.4   Basic Definitions for RNIA

Now, we give some basic definitions, which appear through this paper.

Definition 3. Let us consider a derived DIS (for ATR) from a NIS. We call an equivalence relation $eq(ATR)$ in DIS a *possible equivalence relation (pe-relation)* in NIS. We also call every element in $eq(ATR)$ a *possible equivalence class (pe-class)* in NIS.

For $ATR=\{C\}$ in NIS_1, there exist two derived $DISs$ and two pe-relations i.e., $\{\{1,5,9\},\{2,3,4,8,10\},\{6\},\{7\}\}$ and $\{\{1,5,9\},\{2,3,8,10\},\{4\},\{6\},\{7\}\}$. Every element in two pe-relations is a pe-class for $ATR=\{C\}$ in NIS_1.

Definition 4. Let us consider a NIS and a set $ATR=\{A_1,\cdots,A_n\} \subseteq AT$. For any $x \in OB$, let $PT(x,ATR)$ denote the Cartesian product $g(x,A_1) \times \cdots \times g(x,A_n)$. We call every element a *possible tuple (for ATR)* from x. For a possible tuple $\zeta=(\zeta_1,\cdots,\zeta_n) \in PT(x,ATR)$, let $[ATR,\zeta]$ denote a formula $\bigwedge_{1\leq i\leq n}[A_i,\zeta_i]$. Furthermore for disjoint sets $CON,DEC \subseteq AT$, and two possible tuples $\zeta=(\zeta_1,\cdots,\zeta_n) \in PT(x,CON)$ and $\eta=(\eta_1,\cdots,\eta_m) \in PT(x,DEC)$, let (ζ,η) denote a possible tuple $(\zeta_1,\cdots,\zeta_n,\eta_1,\cdots,\eta_m) \in PT(x,CON \cup DEC)$.

Definition 5. Let us consider a NIS and a set $ATR \subseteq AT$. For any $\zeta \in PT(x,ATR)$, let $DD(x,\zeta,ATR)$ denote a set $\{\varphi|\ \varphi$ is such a derived DIS for ATR that the tuple of x in φ is $\zeta\}$. Furthermore in this $DD(x,\zeta,ATR)$, we define (1) and (2) below.
(1) $inf(x,\zeta,ATR)=\{y \in OB|PT(y,ATR)=\{\zeta\}\}$,
(2) $sup(x,\zeta,ATR)=\{y \in OB|\zeta \in PT(y,ATR)\}$.

For object 1 and $ATR=\{A,B\}$ in NIS_1, $PT(1,\{A,B\})=\{(3,1),(3,3),(3,4)\}$ holds. The possible tuple $(3,1) \in PT(1,\{A,B\})$ appears $1/3$ derived $DISs$ for $ATR=\{A,B\}$. The number of elements in $DD(1,(3,1),\{A,B\})$ is $1728(=2^6 \times 3^3)$. In this set $DD(1,(3,1),\{A,B\})$, $inf(1,(3,1),\{A,B\})=\{1\}$ and $sup(1,(3,1),\{A,B\})=\{1,10\}$ hold.

These inf and sup in Definition 5 are key information for $RNIA$, and each algorithm in the following depends upon these two sets. The set sup is semantically equal to a set defined by the similarity relation SIM in [20,21]. In [20,21], some theorems are presented based on the relation SIM, and our theoretical results are closely related to those theorems. However, the set inf causes new properties, which hold just in $NISs$.

Now, let us consider a relation between a pe-class $[x]_{ATR}$ and two sets inf and sup. In every DIS, $PT(y,ATR)$ is a singleton set, so $[x]_{ATR}=inf(x,\zeta,ATR)=sup(x,\zeta,ATR)$ holds. However in every NIS, $[x]_{ATR}$ depends upon derived $DISs$, and $\{x\} \subseteq inf(x,\zeta,ATR) \subseteq [x]_{ATR} \subseteq sup(x,\zeta,ATR)$ holds. Proposition 2 in the following connects a pe-class $[x]_{ATR}$ with $inf(x,\zeta,ATR)$ and $sup(x,\zeta,ATR)$.

Proposition 2 [25]. For a NIS, an object x, $ATR \subseteq AT$ and $\zeta \in PT(x,ATR)$, conditions (1) and (2) in the following are equivalent.
(1) X is an equivalence class $[x]_{ATR}$ in some $\varphi \in DD(x,\zeta,ATR)$.
(2) $inf(x,\zeta,ATR) \subseteq X \subseteq sup(x,\zeta,ATR)$.

4   Algorithms and Tool Programs for the Definability of a Set in NISs

This section proposes algorithms and tool programs for the definability of a set. It is possible to obtain distinct pe-relations as a side effect of an algorithm. An algorithm for merging pe-relations is also proposed.

4.1 An Algorithm for Checking the Definability of a Set in NISs

The definability of a set in $NISs$ is given, and an algorithm is proposed.

Definition 6. (N-i. The Definability of a Set) We say $X \subseteq OB$ is *certainly definable* for $ATR \subseteq AT$ in DIS^{real}, if X is definable (for ATR) in every derived DIS. We say $X \subseteq OB$ is *possibly definable* for $ATR \subseteq AT$ in DIS^{real}, if X is definable (for ATR) in some derived $DISs$.

In a DIS, it is enough to check a formula $\cup_{x \in X}[x]_{ATR}=X$ for the definability of $X \subseteq OB$. However in every NIS, $[x]_{ATR}$ depends upon a derived DIS, and $inf(x, \zeta, ATR) \subseteq [x]_{ATR} \subseteq sup(x, \zeta, ATR)$ holds. Algorithm 1 in the following checks the formula $\cup_{x \in X}[x]_{ATR}=X$ according to these inclusion relations, and finds a subset of a *pe*-relation which makes the set X definable.

Algorithm 1.
Input: A NIS, a set $ATR \subseteq AT$ and a set $X \subseteq OB$.
Output: The definability of a set X for ATR.
   (1) $X^*=X$, $eq=\emptyset$, $count=0$ and $total=\prod_{x \in X, A \in ATR} |g(x, A)|$.
   (2) For any $x \in X^*$, find $[x]_{ATR}$ satisfying constraints (CL-1) and (CL-2).
      (CL-1) $[x]_{ATR} \subseteq X^*$,
      (CL-2) $eq \cup \{[x]_{ATR}\}$ is a subset of a *pe*-relation.
      (2-1) If there is a set $[x]_{ATR}$, $eq=eq \cup \{[x]_{ATR}\}$ and $X^*=X^* - [x]_{ATR}$.
         If $X^* \neq \emptyset$, go to (2). If $X^*=\emptyset$, X is definable in a derived DIS.
         Set $count=count+1$, and backtrack.
      (2-2) If there is no $[x]_{ATR}$, backtrack.
   (3) After finishing the search, X is certainly definable for ATR in DIS^{real}, if $count=total$. X is possibly definable for ATR in DIS^{real}, if $count \geq 1$.

Algorithm 1 tries to find a set of *pe*-classes, which satisfy constraints (CL-1) and (CL-2). Whenever $X^*=\emptyset$ holds in Algorithm 1, a subset of a *pe*-relation is stored in the variable eq. At the same time, a derived DIS (restricted to the set X) from NIS is also detected [24,25]. Because $X=\cup_{K \in eq}K$ holds for eq, X is definable in this detected DIS. In order to count such a case that $X^*=\emptyset$, the variable $count$ is employed. At the end of execution, if $count$ is equal to the number of derived $DISs$ (restricted to the set X), it is possible to conclude X is certainly definable.

The constraints (CL-1) and (CL-2) keep the correctness of this search. For example in Table 1, $inf(1, (3), \{A\})=\{1\}$ and $sup(1, (3), \{A\})=\{1, 5, 6, 10\}$ hold. So, $\{1\} \subseteq [1]_{\{A\}} \subseteq \{1, 5, 6, 10\}$ holds. Let us suppose $[1]_{\{A\}}=\{1, 5, 10\}$. Since $6 \notin [1]_{\{A\}}$ holds in this case, the tuple from object 6 is not (3). In a branch with $[1]_{\{A\}}=\{1, 5, 10\}$, the tuple from object 6 is implicitly fixed to $(5) \in PT(6, \{A\})= \{(3), (5)\}$. The details of (CL-1), (CL-2) and an illustrative example based on a previous version of Algorithm 1 are presented in [25].

Algorithm 1 is a solution to handle definition N-i. Algorithm 1 is extended to Algorithm 2 in the subsequent sections. A real execution of a tool, which simulates Algorithm 1, is shown in Appendix 1.

4.2   The Definability of a Set and Pe-relations in NISs

In Algorithm 1, let X be OB. Since every pe-relation is an equivalence relation over OB, OB is definable in every derived DIS. Thus, OB is certainly definable in DIS^{real}. In Algorithm 1, every pe-relation is asserted in the variable eq whenever $X^*=\emptyset$ is derived. In this way, it is possible to obtain all pe-relations.

However in this case, the number of the search branches with $X^*=\emptyset$ is equal to the number of all derived $DISs$. Therefore, it is hard to apply Algorithm 1 directly to $NISs$ with large number of derived $DISs$. We solve this problem by means of applying Proposition 3 in the following, which shows us a way to merge equivalence relations.

Proposition 3 [1]. Let us suppose $eq(A)$ and $eq(B)$ be equivalence relations for $A, B \subseteq AT$ in a DIS. The equivalence relation $eq(A \cup B)$ is $\{M \subseteq OB | M = [x]_A \cap [x]_B$ for $[x]_A \in eq(A)$ and $[x]_B \in eq(B)$, $x \in OB\}$.

4.3   A Property of Pe-relations in NISs

Before proposing another algorithm for producing pe-relations, we clarify a property of pe-relations. Because some pe-relations in distinct derived $DISs$ may be the same, generally the number of distinct pe-relations is smaller than the number of derived $DISs$. Let us consider Table 4, which shows the relation between the numbers of derived $DISs$ and distinct pe-relations. This result is computed by tool programs in the subsequent sections. For $ATR=\{A, B, C\}$, there are $10368(=2^7 \times 3^4)$ derived $DISs$. However in reality, there are 10 distinct pe-relations. For larger attributes set ATR, every object is much more discerned from other objects, i.e., every $[x]_{ATR}$ will become $\{x\}$. Therefore, every pe-relation will become a unique equivalence relation $\{\{1\}, \{2\}, \cdots, \{10\}\}$. For $ATR=\{A, B, C, D, E, F\}$, there exists in reality only 1 distinct pe-relation $\{\{1\}, \{2\}, \cdots, \{10\}\}$.

Table 4. The numbers of derived $DISs$ and distinct pe-relations in NIS_1

ATR	$\{A, B\}$	$\{A, B, C\}$	$\{A, B, C, D\}$	$\{A, B, C, D, E\}$	$\{A, B, C, D, E, F\}$
$derived\_DISs$	5184	10368	1118744	40310784	2176782336
$pe\_relations$	107	10	6	2	1

We examined several $NISs$, and we experimentally conclude such a property that *the number of distinct possible equivalence relations is generally much smaller than the number of all derived $DISs$*. We make use of this property for computing definitions from N-i to N-v.

4.4   A Revised Algorithm for Producing Pe-relations

Algorithm 1 produces pe-relations as a side effect of the search, but this algorithm is not suitable for $NISs$ with large number of derived $DISs$. This section revises Algorithm 1 by means of applying Proposition 3.

Algorithm 2.

Input: A NIS and a set $ATR \subseteq AT$.

Output: A set of distinct pe-relations for ATR: $pe\_rel(ATR)$.

(1) Produce a set of pe-relations $pe\_rel(\{A\})$ for every $A \in ATR$.

(2) Set $temp=\{\}$ and $pe\_rel(ATR)=\{\{\{1, 2, 3, \cdots, |OB|\}\}\}$.

(3) Repeat (4) to $pe\_rel(ATR)$ and $pe\_rel(\{K\})$ $(K \in ATR - temp)$ until $temp=ATR$.

(4) For each pair of $pe_i \in pe\_rel(ATR)$ and $pe_j \in pe\_rel(\{K\})$, apply Proposition 3 and produce $pe_{i,j}=\{M \subseteq OB | M = [x]_i \cap [x]_j$ for $[x]_i \in pe_i$ and $[x]_j \in pe_j$, $x \in OB\}$. Let $pe\_rel(ATR)$ be $\{pe_{i,j} | pe_i \in pe\_rel(ATR)$ $pe_j \in pe\_rel(\{K\})\}$, and set $temp=temp \cup \{K\}$.

In step (1), Algorithm 1 is applied to producing $pe\_rel(\{A\})$ for every $A \in ATR$. In steps (3) and (4), Proposition 3 is repeatedly applied to merging two sets of pe-relations.

For NIS_1, let us consider a case of $ATR=\{A, B, C, D\}$ in Algorithm 2. After finishing step (1), Table 5 is obtained. Table 5 shows the numbers of derived $DISs$ and distinct pe-relations in every attribute.

Table 5. The numbers of derived $DISs$ and distinct pe-relations in every attribute

Attribute	A	B	C	D	E	F
derived_DISs	192	27	2	108	36	54
pe_relations	176	27	2	96	36	54

For $ATR=\{A, B, C, D\}$, Algorithm 2 sequentially produces $pe\_rel(\{A, B\})$, $pe\_rel(\{A, B, C\})$ and $pe\_rel(\{A, B, C, D\})$. For producing $pe\_rel(\{A, B\})$, it is necessary to handle $4752(=176 \times 27)$ combinations of pe-relations, and it is possible to know $|pe\_rel(\{A, B\})|=107$ in Table 4. However after this execution, the number of these combinations is reduced due to the property of pe-relations. For producing $pe\_rel(\{A, B, C\})$, it is enough to handle $214(=107 \times 2)$ combinations for 10368 derived $DISs$, and $|pe\_rel(\{A, B, C\})|=10$ in Table 4 is obtained. For producing $pe\_rel(\{A, B, C, D\})$, it is enough to handle $960(=10 \times 96)$ combinations for 1118744 derived $DISs$.

Generally, Algorithm 1 depends upon the number $\prod_{x \in OB, A \in ATR} |g(x, A)|$, which is the number of derived $DISs$. However, Algorithm 2 at the most depends upon the number $(|ATR| - 1)| \times \prod_{x \in OB} |g(x, A)|^2$ for such an attribute A that $\prod_{x \in OB} |g(x, B)| \leq \prod_{x \in OB} |g(x, A)|$ for any $B \in ATR$. The product of $|g(x, A)|$ in the number for Algorithm 1 is almost corresponding to the sum of $|g(x, A)|^2$ in the number for Algorithm 2. Thus, in order to handle $NISs$ with large ATR and large number of derived $DISs$, Algorithm 2 will be more efficient than Algorithm 1. In reality, the result in Table 4 is calculated by Algorithm 2. It is hard to apply Algorithm 1 to calculating pe-relations for $ATR=\{A, B, C, D, E\}$ or $ATR=\{A, B, C, D, E, F\}$.

As for the implementation of Algorithm 2, the data structure of *pe*-relations and the program depending upon this structure are following the definitions in [23,25]. A real execution of a tool, which simulates Algorithm 2, is shown in Appendix 2.

4.5   Another Solution of the Definability of a Set

Algorithm 1 solves the definability of a set in $NISs$, and it is also possible to apply *pe*-relations for solving the definability of a set in $NISs$. After obtaining distinct *pe*-relations, we have only to check $\cup_{x \in X}[x]=X$ for every *pe*-relation.

Example 3. Let us consider NIS_1. For $ATR=\{A, B, C, D, E\}$, there are two distinct *pe*-relations $pe_1=\{\{1\}, \{2\}, \cdots, \{10\}\}$ and $pe_2=\{\{1\}, \{2, 3\}, \{4\}, \cdots, \{10\}\}$. A set $\{1, 2\}$ is definable in pe_1, but it is not definable in pe_2. Therefore, a set $\{1, 2\}$ is possibly definable in DIS^{real}. Since a set $\{1, 2, 3\}$ is definable in every *pe*-relation, this set is certainly definable in DIS^{real}.

In this way the load of the calculation, which depends upon all derived $DISs$, can be reduced.

5   The Necessary and Sufficient Condition for Checking the Consistency of an Object

This section examines the necessary and sufficient condition for checking the consistency of an object.

Definition 7. (N-ii. The Consistency of an Object) Let us consider two disjoint sets $CON, DEC \subseteq AT$ in a NIS. We say $x \in OB$ is *certainly consistent* (in the relation from CON to DEC) in DIS^{real}, if x is consistent (in the relation from CON to DEC) in every derived DIS from NIS. We say x is *possibly consistent* in DIS^{real}, if x is consistent in some derived $DISs$ from NIS.

According to *pe*-relations and Proposition 1, it is easy to check the consistency of x. Let us consider two sets of *pe*-relations $pe\_rel(CON)$ and $pe\_rel(DEC)$. An object x is certainly consistent in DIS^{real}, if and only if $[x]_{CON} \subseteq [x]_{DEC}$ ($[x]_{CON} \in pe_i$ and $[x]_{DEC} \in pe_j$) for any $pe_i \in pe\_rel(CON)$ and any $pe_j \in pe\_rel(DEC)$. An object x is possibly consistent in DIS^{real}, if and only if $[x]_{CON} \subseteq [x]_{DEC}$ ($[x]_{CON} \in pe_i$ and $[x]_{DEC} \in pe_j$) for some $pe_i \in pe\_rel(CON)$ and some $pe_j \in pe\_rel(DEC)$.

However, it is also possible to check the consistency of an object by means of applying inf and sup information in Definition 5.

Theorem 4. For a NIS and an object x, let CON be condition attributes and let DEC be decision attributes.
(1) x is certainly consistent in DIS^{real} if and only if $sup(x, \zeta, CON) \subseteq inf(x, \eta, DEC)$ holds for any $\zeta \in PT(x, CON)$ and any $\eta \in PT(x, DEC)$.
(2) x is possibly consistent in DIS^{real} if and only if $inf(x, \zeta, CON) \subseteq sup(x, \eta, DEC)$ holds for a pair of $\zeta \in PT(x, CON)$ and $\eta \in PT(x, DEC)$.

Proof. Let us consider pe-classes $[x]_{CON}$ and $[x]_{DEC}$ in $\varphi \in DD(x, (\zeta, \eta), CON \cup DEC)$. Then, $inf(x, \zeta, CON) \subseteq [x]_{CON} \subseteq sup(x, \zeta, CON)$ and $inf(x, \eta, DEC) \subseteq [x]_{DEC} \subseteq sup(x, \eta, DEC)$ hold according to Proposition 2.
(1) Let us suppose $sup(x, \zeta, CON) \subseteq inf(x, \eta, DEC)$ holds. Then, $[x]_{CON} \subseteq sup(x, \zeta, CON) \subseteq inf(x, \eta, DEC) \subseteq [x]_{DEC}$, and $[x]_{CON} \subseteq [x]_{DEC}$ is derived. According to Proposition 1, object x is consistent in any $\varphi \in DD(x, (\zeta, \eta), CON \cup DEC)$. This holds for any $\zeta \in PT(x, CON)$ and any $\eta \in PT(x, DEC)$. Thus, x is certainly consistent in DIS^{real}. Conversely, let us suppose $sup(x, \zeta, CON) \nsubseteq inf(x, \eta, DEC)$ holds for a pair of ζ and η. According to Proposition 2, $[x]_{CON} = sup(x, \zeta, CON)$ and $[x]_{DEC} = inf(x, \eta, DEC)$ hold in some $\varphi \in DD(x, (\zeta, \eta), CON \cup DEC)$. Since $[x]_{CON} \nsubseteq [x]_{DEC}$ holds in φ, x is not certainly consistent. By contraposition, the converse is also proved.
(2) Let us suppose $inf(x, \zeta, CON) \subseteq sup(x, \eta, DEC)$ holds for a pair of $\zeta \in PT(x, CON)$ and $\eta \in PT(x, DEC)$. According to Proposition 2, $[x]_{CON} = inf(x, \zeta, CON)$ and $[x]_{DEC} = sup(x, \eta, DEC)$ hold in some $\varphi \in DD(x, (\zeta, \eta), CON \cup DEC)$. Namely, x is consistent in this φ. Conversely, let us suppose $inf(x, \zeta, CON) \nsubseteq sup(x, \eta, DEC)$ holds. Since $inf(x, \zeta, CON) \subseteq [x]_{CON}$ and $[x]_{DEC} \subseteq sup(x, \eta, DEC)$ hold for any $[x]_{CON}$ and $[x]_{DEC}$, $[x]_{CON} \nsubseteq [x]_{DEC}$ is derived. Namely, x is not possibly consistent. By contraposition, the converse is also proved.

Theorem 4, which is one of the most important results in this paper, is an extension of Proposition 1 and results in [20,21]. In Proposition 1, $[x]_{CON}$ and $[x]_{DEC}$ are mutually unique. However in $NISs$, these pe-classes may not be unique. In order to check the consistency of objects in $NISs$, it is necessary to consider possible tuples and derived $DISs$.

Algorithm 1 and 2 produce pe-relations according to inf and sup information in Definition 5. Theorem 4 also characterizes the consistency of an object by means of applying inf and sup information, therefore inf and sup information in Definition 5 is the most essential information.

6   An Algorithm and Tool Programs for Data Dependency in NISs

The formal definition of data dependency in $NISs$ has not been established yet. This section extends the definition D-iii to N-iii in the following, and examines an algorithm and tool programs for data dependency in $NISs$.

Definition 8 [26] (N-iii. Data Dependencies among Attributes). Let us consider any NIS, condition attributes CON, decision attributes DEC and all derived $DIS_1, \cdots, DIS_m$ from NIS. For two threshold values val_1 and $val_2 (0 \leq val_1, val_2 \leq 1)$, if conditions (1) and (2) hold then we see DEC depends on CON in NIS.
(1) $|\{DIS_i | deg(CON, DEC) = 1 \text{ in } DIS_i (1 \leq i \leq m)\}|/m \geq val_1$.
(2) $min_i\{deg(CON, DEC) \text{ in } DIS_i\} \geq val_2$.

In Definition 8, condition (1) requires most of derived $DISs$ are consistent, i.e., every object is consistent in most of derived $DISs$. Condition (2) specifies the minimal value of the degree of dependency. If both two conditions are satisfied, it is expected that $deg(CON, DEC)$ in DIS^{real} will also be high.

The definition N-iii is easily computed according to $pe\_rel(CON)$ and $pe\_rel(DEC)$. For each pair of $pe_i \in pe\_rel(CON)$ and $pe_j \in pe\_rel(DEC)$, the degree of dependency is $|\{x \in OB|[x]_{CON} \subseteq [x]_{DEC}$ for $[x]_{CON} \in pe_i, [x]_{DEC} \in pe_j\}|/|OB|$. Namely, all kinds of the degrees of dependency are obtained by means of calculating all combinations of pairs. For example, let us consider $CON=\{A, B, C, D, E\}$ and $DEC=\{F\}$ in NIS_1. Since $|pe\_rel(\{A, B, C, D, E\})|$ $=2$ and $|pe\_rel(\{F\})|=54$, it is possible to obtain all kinds of degrees by means of examining $108(=2 \times 54)$ combinations. This calculation is the same as the calculation depending upon 2176782336 derived $DISs$. A real execution handling data dependency and the consistency of objects is shown in Appendix 3.

7   An Algorithm and Tool Programs for Rules in NISs

This section investigates an algorithm and tool programs [27] for rules in $NISs$.

7.1   Certain Rules and Possible Rules in NISs

Possible implications in $NISs$ are proposed, and certain rules and possible rules are defined by possible implications satisfying some constraints.

Definition 9. For any NIS, let CON be condition attributes and let DEC be decision attributes. For any $x \in OB$, let $PI(x, CON, DEC)$ denote a set $\{[CON, \zeta] \Rightarrow [DEC, \eta]|\zeta \in PT(x, CON), \eta \in PT(x, DEC)\}$. We call an element of $PI(x, CON, DEC)$ a *possible implication* (in the relation from CON to DEC) from x. We call a possible implication, which satisfies some constraints, a *rule* in NIS.

It is necessary to remark that a possible implication $\tau : [CON, \zeta] \Rightarrow [DEC, \eta]$ from x appears in every $\varphi \in DD(x, (\zeta, \eta), CON \cup DEC)$. This set $DD(x, (\zeta, \eta), CON \cup DEC)$ is a subset of all derived $DISs$ for $ATR=CON \cup DEC$. In NIS_1, $PT(1, \{A, B\})=\{(3, 1), (3, 3), (3, 4)\}$, $PT(1, \{C\})=\{(3)\}$ and $PI(1, \{A, B\}, \{C\})$ consists of three possible implications $[A, 3] \wedge [B, 1] \Rightarrow [C, 3]$, $[A, 3] \wedge [B, 3] \Rightarrow [C, 3]$ and $[A, 3] \wedge [B, 4] \Rightarrow [C, 3]$. The first possible implication appears in every $\varphi \in DD(1, (3, 1, 3), \{A, B, C\})$. This set $DD(1, (3, 1, 3), \{A, B, C\})$ consists of $1/3$ of derived $DISs$ for $\{A, B, C\}$.

Definition 10. Let us consider a NIS, condition attributes CON and decision attributes DEC. If $PI(x, CON, DEC)$ is a singleton set $\{\tau\}$ ($\tau : [CON, \zeta] \Rightarrow [DEC, \eta]$), we say τ (from x) is *definite*. Otherwise we say τ (from x) is *indefinite*. If a set $\{\varphi \in DD(x, (\zeta, \eta), CON \cup DEC)|$ x is consistent in $\varphi\}$ is equal to $DD(x, (\zeta, \eta), CON \cup DEC)$, we say τ is *globally consistent* (GC). If this set is equal to $\emptyset$, we say τ is *globally inconsistent* (GI). Otherwise we say τ is *marginal* (MA). According to two cases, i.e., '$D(efinite)$ or $I(ndefinite)$' and 'GC or MA or GI', we define six classes, D-GC, D-MA, D-GI, I-GC, I-MA, I-GI, for possible implications.

If a possible implication from x belongs to either D-GC, I-GC, D-MA or I-MA, x is consistent in some derived $DISs$. If a possible implication from x belongs to D-GC, x is consistent in every derived $DISs$. Thus, we give Definition 11 in the following.

Definition 11 (N-iv. Certain and Possible Rules).s For a NIS, let CON be condition attributes and let DEC be decision attributes. We say $\tau \in PI(x, CON, DEC)$ is a *possible rule* in DIS^{real}, if τ belongs to either D-GC, I-GC, D-MA or I-MA class. Especially, we say τ is a *certain rule* in DIS^{real}, if τ belongs to D-GC class.

Theorem 5 in the following characterizes certain and possible rules according to inf and sup information. Theorem 5 is also related to results in [20,21], but there exist some differences, which we have shown in Example 2.

Theorem 5 [27]. For a NIS, let CON be condition attributes and let DEC be decision attributes. For $\tau : [CON, \zeta] \Rightarrow [DEC, \eta] \in PI(x, CON, DEC)$, the following holds.
(1) τ is a possible rule if and only if $inf(x, \zeta, CON) \subseteq sup(x, \eta, DEC)$ holds.
(2) τ is a certain rule if and only if $PI(x, CON, DEC) = \{\tau\}$ and $sup(x, \zeta, CON) \subseteq inf(x, \eta, DEC)$ hold.

Proposition 6. For any NIS, let $ATR \subseteq AT$ be $\{A_1, \cdots, A_n\}$, and let a possible tuple $\zeta \in PT(x, ATR)$ be $(\zeta_1, \cdots, \zeta_n)$. Then, the following holds.
(1) $inf(x, \zeta, ATR) = \cap_i inf(x, (\zeta_i), \{A_i\})$.
(2) $sup(x, \zeta, ATR) = \cap_i sup(x, (\zeta_i), \{A_i\})$.

Proof of (1): For any $y \in inf(x, \zeta, ATR)$, $PT(y, ATR) = \{(\zeta_1, \cdots, \zeta_n)\}$ holds due to the definition of inf. Namely, $PT(y, \{A_i\}) = \{(\zeta_i)\}$ holds for every i, and $y \in inf(x, (\zeta_i), \{A_i\})$ for every i. Namely, $y \in \cap_i inf(x, (\zeta_i), \{A_i\})$. The converse of this proof clearly holds.

Proposition 6 shows us a way to manage inf and sup information in Definition 5. Namely, we first prepare inf and sup information for every $x \in OB$, $A_i \in AT$ and $(\zeta_{i,j}) \in PT(x, \{A_i\})$. Then, we produce inf and sup information by means of repeating the set intersection operations. For obtained $inf(x, \zeta, CON)$, $sup(x, \zeta, CON)$, $inf(x, \eta, DEC)$ and $sup(x, \eta, DEC)$, Theorem 5 is applied to checking the certainty or the possibility of $\tau : [CON, \zeta] \Rightarrow [DEC, \eta]$.

7.2   The Minimum and Maximum of Three Criterion Values

This section proposes the minimum and maximum of three criterion values for possible implications, and investigates an algorithm to calculate them.

Definition 12. For a NIS, let us consider a possible implication $\tau : [CON, \zeta] \Rightarrow [DEC, \eta] \in PI(x, CON, DEC)$ and $DD(x, (\zeta, \eta), CON \cup DEC)$. Let $minsup(\tau)$ denote $min_{\varphi \in DD(x, (\zeta, \eta), CON \cup DEC)}\{support(\tau)$ in $\varphi\}$, and let $maxsup(\tau)$ denote $max_{\varphi \in DD(x, (\zeta, \eta), CON \cup DEC)}\{support(\tau)$ in $\varphi\}$. As for *accuracy* and *coverage*, $minacc(\tau)$, $maxacc(\tau)$, $mincov(\tau)$ and $maxcov(\tau)$ are similarly defined.

Let us suppose $DIS^{real} \in DD(x, (\zeta, \eta), CON \cup DEC)$. According to Definition 12, clearly $minsup(\tau) \leq support(\tau)$ in $DIS^{real} \leq maxsup(\tau)$, $minacc(\tau) \leq$

$accuracy(\tau)$ in $DIS^{real} \leq maxacc(\tau)$ and $mincov(\tau) \leq coverage(\tau)$ in DIS^{real} $\leq maxcov(\tau)$ hold. For calculating each definition, it is necessary to examine every $\varphi \in DD(x, (\zeta, \eta), CON \cup DEC)$, and this calculation depends upon $|DD(x, (\zeta, \eta), CON \cup DEC)|$. However, these minimum and maximum values are also calculated by means of applying inf and sup information, again.

Theorem 7. For a NIS, let us consider a possible implication $\tau{:}[CON, \zeta] \Rightarrow [DEC, \eta] \in PI(x, CON, DEC)$. The following holds.
(1) $minsup(\tau)=|inf(x, \zeta, CON) \cap inf(x, \eta, DEC)|/|OB|$.
(2) $maxsup(\tau)=|sup(x, \zeta, CON) \cap sup(x, \eta, DEC)|/|OB|$.

Theorem 8. For a NIS, let us consider a possible implication $\tau{:}[CON, \zeta] \Rightarrow [DEC, \eta] \in PI(x, CON, DEC)$. Let $INACC$ denote a set $[sup(x, \zeta, CON) - inf(x, \zeta, CON)] \cap sup(x, \eta, DEC)$, and let $OUTACC$ denote a set $[sup(x, \zeta, CON) - inf(x, \zeta, CON)] - inf(x, \eta, DEC)$. Then, the following holds.
(1) $minacc(\tau)=\dfrac{|inf(x,\zeta,CON)\cap inf(x,\eta,DEC)|}{|inf(x,\zeta,CON)|+|OUTACC|}$.
(2) $maxacc(\tau)=\dfrac{|inf(x,\zeta,CON)\cap sup(x,\eta,DEC)|+|INACC|}{|inf(x,\zeta,CON)|+|INACC|}$.

Proof of (1) According to Proposition 2, $inf(x, \zeta, CON) \subseteq [x]_{CON} \subseteq sup(x, \zeta, CON)$ holds. Therefore, the denominator is in the form of $|inf(x, \zeta, CON)| + |K_1|$ ($K_1 \subseteq [sup(x, \zeta, CON) - inf(x, \zeta, CON)]$). Since $PI(y, CON, DEC)=\{\tau\}$ for any $y \in inf(x, \zeta, CON) \cap inf(x, \eta, DEC)$, the numerator is in the form of $|inf(x, \zeta, CON) \cap inf(x, \eta, DEC)| + |K_2| + |K_3|$ ($K_2 \subseteq inf(x, \zeta, CON) \cap [sup(x, \eta, DEC) - inf(x, \eta, DEC)]$ and $K_3 \subseteq K_1$). Thus, $accuracy(\tau)$ is in the form of $(|inf(x, \zeta, CON) \cap inf(x, \eta, DEC)| + |K_2| + |K_3|)/(|inf(x, \zeta, CON)| + |K_1|)$. In order to produce the $minacc(\tau)$, we show such $\varphi_1 \in DD(x, (\zeta, \eta), CON \cup DEC)$ that $K_2=K_3=\emptyset$ and $|K_1|$ is maximum. This is approved by such a formula that $b/(a + (k_1 - k_3)) \leq (b + k_2 + k_3)/(a + k_1)$ for any $0 \leq b \leq a$ ($a \neq 0$), any $0 \leq k_3 \leq k_1$ and any $0 \leq k_2$. Since $sup(x, \zeta, CON)-inf(x, \zeta, CON)$ is equal to the union of disjoint sets ($[sup(x, \zeta, CON) - inf(x, \zeta, CON)] - inf(x, \eta, DEC)\bigcup([sup(x, \zeta, CON) - inf(x, \zeta, CON)]\cap inf(x, \eta, DEC))$), let us consider two disjoint sets. The first set is $OUTACC$. For any $y \in OUTACC$, there exists a possible implication $\tau^* : [CON, \zeta] \Rightarrow [DEC, \eta^*] \in PI(y, CON, DEC)$ ($\eta^* \neq \eta$) by the definition of inf and sup. For any $y \in [sup(x, \zeta, CON) - inf(x, \zeta, CON)] \cap inf(x, \eta, DEC)$, $PT(y, DEC)=\{\eta\}$ holds, and there exists a possible implication $\tau^{**} : [CON, \zeta^*] \Rightarrow [DEC, \eta] \in PI(y, CON, DEC)$ ($\zeta^* \neq \zeta$). In $\varphi_1 \in DD(x, (\zeta, \eta), CON \cup DEC)$ with these τ^* and τ^{**}, the denominator is $|inf(x, \zeta, CON)| + |OUTACC|$ and the numerator is $|inf(x, \zeta, CON) \cap inf(x, \eta, DEC)|$.

Theorem 9. For a NIS, let us consider a possible implication $\tau{:}[CON, \zeta] \Rightarrow [DEC, \eta] \in PI(x, CON, DEC)$. Let $INCOV$ denote a set $[sup(x, \eta, DEC) - inf(x, \eta, DEC)] \cap sup(x, \zeta, CON)$, and let $OUTCOV$ denote a set $[sup(x, \eta, DEC) - inf(x, \eta, DEC)] - inf(x, \zeta, CON)$. Then, the following holds.
(1) $mincov(\tau)=\dfrac{|inf(x,\zeta,CON)\cap inf(x,\eta,DEC)|}{|inf(x,\eta,DEC)|+|OUTCOV|}$.
(2) $maxcov(\tau)=\dfrac{|sup(x,\zeta,CON)\cap inf(x,\eta,DEC)|+|INCOV|}{|inf(x,\eta,DEC)|+|INCOV|}$.

8   An Algorithm for Reduction of Attributes in NISs

This section gives an algorithm for reducing the condition attributes in rules.

Definition 13 (N-v. Reduction of Condition Attributes in Rules). Let us consider a certain rule $\tau : [CON, \zeta] \Rightarrow [DEC, \eta] \in PI(x, CON, DEC)$. We say $K \in CON$ is *certainly dispensable* from τ in DIS^{real}, if $\tau' : [CON - \{K\}, \zeta'] \Rightarrow [DEC, \eta]$ is a certain rule. We say $K \in CON$ is *possibly dispensable* from τ in DIS^{real}, if $\tau' : [CON - \{K\}, \zeta'] \Rightarrow [DEC, \eta]$ is a possible rule.

Let us consider a possible implication $\tau : [A, 3] \wedge [C, 3] \wedge [D, 2] \wedge [E, 5] \Rightarrow [F, 5]$ in NIS_1. This τ is definite, and τ belongs to $D\text{-}GC$ class, i.e., τ is a certain rule. For $ATR=\{A, C, E\}$, $inf(1, (3, 3, 5), \{A, C, E\})=inf(1, (3), \{A\}) \cap inf(1, (3), \{C\}) \cap inf(1, (5), \{E\})=\{1\} \cap \{1, 5, 9\} \cap \{1, 6\}=\{1\}$ and $sup(1, (3, 3, 5), \{A, C, E\})=sup(1, (3), \{A\}) \cap sup(1, (3), \{C\}) \cap sup(1, (5), \{E\})=\{1, 5, 6, 10\} \cap \{1, 5, 9\} \cap \{1, 3, 4, 6, 7\}=\{1\}$ hold according to Proposition 6. For $ATR=\{F\}$, $inf(1, (5), \{F\})=\{1, 3, 4\}$ and $sup(1, (5), \{F\})=\{1, 3, 4, 5, 8\}$. Because $sup(1, (3, 3, 5), \{A, C, E\})=\{1\} \subseteq \{1, 3, 4\}=inf(1, (5), \{F\})$ holds, $\tau' : [A, 3] \wedge [C, 3] \wedge [E, 5] \Rightarrow [F, 5]$ is also a certain rule by Theorem 5. Thus, an attribute D is certainly dispensable from τ. In this way, it is possible to examine the reduction of attributes. In this case also, inf and sup information in Definition 5 is essential.

An important problem on the reduction in $DISs$ is to find some minimal sets of condition attributes. Several work deals with reduction for finding minimal reducts. In [3], this problem is proved to be NP-hard, which means that to compute reducts is a non-trivial task. For solving this problem, a *discernibility function* is proposed also in [3], and this function is extended to a discernibility function in incomplete information systems [21,22]. In [19], an algorithm for finding a minimal complex is presented. In $NISs$, it is also important to deal with this problem on the minimal reducts.

Definition 14. For any NIS and any disjoint $CON, DEC \subseteq AT$, let us consider a possible implication $\tau:[CON, \zeta] \Rightarrow [DEC, \eta]$, which belongs to either $D\text{-}GC$, $I\text{-}GC$, $D\text{-}MA$ or $I\text{-}MA$ class. Furthermore, let Φ be a set $\{\varphi \in DD(x, (\zeta, \eta), CON \cup DEC)|$ x is consistent in $\varphi \}$. If there is no proper subset $CON^* \subseteq CON$ such that $\{\varphi \in DD(x, (\zeta, \eta), CON \cup DEC)|x$ is consistent (in the relation from CON^* to DEC) in $\varphi \}$ is equal to the set Φ, we say τ is *minimal* (in this class).

Problem 1. For any NIS, let DEC be decision attributes and let η be a tuple of decision attributes values for DEC. Then, find all minimal certain or minimal possible rules in the form of $[CON, \zeta] \Rightarrow [DEC, \eta]$. For additional information, calculate the minimum and maximum values of *support*, *accuracy* and *coverage* for every rule, too.

For solving Problem 1, we introduced a total order, which is defined by the significance of attributes, over $(AT\text{--}DEC)$, and we think about rules based on this order. Under this assumption, we have realized a tool for solving Problem 1, which is shown in Appendix 4. For example, let us suppose $\{A, B, C, D, E\}$

be an ordered set, and let $[A, \zeta_A] \wedge [B, \zeta_B] \wedge [C, \zeta_C] \wedge [D, \zeta_D] \Rightarrow [F, \eta_F]$ and $[B, \zeta_B'] \wedge [E, \zeta_E'] \Rightarrow [F, \eta_F]$ be certain rules. The latter seems simple, but we choose the former rule according to the order of significance. In this case, each attribute $A_i \in (AT\text{–}DEC)$ is sequentially picked up based on this order, and the necessity of the descriptor $[A_i, \zeta_{i,j}]$ is checked. Then, Proposition 6 and Theorem 5 are applied. Of course, the introduction of total order over attributes is too strong simplification of the problem. Therefore in the next step, it is necessary to solve the problem of reduction in $NISs$ without using any total order.

9   Concluding Remarks

A framework of $RNIA$ (rough non-deterministic information analysis) is proposed, and an overview of algorithms is presented. Throughout this paper, rough sets based concepts in $NISs$ and the application of either inf and sup information or equivalence relations are studied. Especially, inf and sup in Definition 5 are key information for $RNIA$. This paper also presented some tool programs for $RNIA$.

The authors would be grateful to Professor J.W. Grzymala-Busse and anonymous referees.

References

1. Pawlak, Z.: Rough Sets: Theoretical Aspects of Reasoning about Data. Kluwer Academic Publishers, Dordrecht, (1991)
2. Pawlak, Z.: New Look on Bayes' Theorem - The Rough Set Outlook. Bulletin of Int'l. Rough Set Society **5** (2001) 1–8
3. Komorowski, J., Pawlak, Z., Polkowski, L., Skowron, A.: Rough Sets: A Tutorial. Rough Fuzzy Hybridization. Springer (1999) 3–98
4. Nakamura, A., Tsumoto, S., Tanaka, H., Kobayashi, S.: Rough Set Theory and Its Applications. Journal of Japanese Society for AI **11** (1996) 209–215
5. Polkowski, L., Skowron, A.(eds.): Rough Sets in Knowledge Discovery 1. Studies in Fuzziness and Soft Computing, Vol.18. Physica-Verlag (1998)
6. Polkowski, L., Skowron, A.(eds.): Rough Sets in Knowledge Discovery 2. Studies in Fuzziness and Soft Computing, Vol.19. Physica-Verlag (1998)
7. Grzymala-Busse, J.: A New Version of the Rule Induction System LERS. Fundamenta Informaticae **31** (1997) 27–39
8. Ziarko, W.: Variable Precision Rough Set Model. Journal of Computer and System Sciences **46** (1993) 39–59
9. Tsumoto, S.: Knowledge Discovery in Clinical Databases and Evaluation of Discovered Knowledge in Outpatient Clinic. Information Sciences **124** (2000) 125–137
10. Zhong, N., Dong, J., Fujitsu, S., Ohsuga, S.: Soft Techniques to Rule Discovery in Data. Transactions of Information Processing Society of Japan **39** (1998) 2581–2592
11. Rough Set Software. Bulletin of Int'l. Rough Set Society **2** (1998) 15–46
12. Codd, E.: A Relational Model of Data for Large Shared Data Banks. Communication of the ACM **13** (1970) 377–387

13. Orłowska, E., Pawlak, Z.: Representation of Nondeterministic Information. Theoretical Computer Science **29** (1984) 27–39
14. Orłowska, E.: What You Always Wanted to Know about Rough Sets. Incomplete Information: Rough Set Analysis. Studies in Fuzziness and Soft Computing, Vol.13. Physica-Verlag (1998) 1–20
15. Lipski, W.: On Semantic Issues Connected with Incomplete Information Databases. ACM Transaction on Database Systems **4** (1979) 262–296
16. Lipski, W.: On Databases with Incomplete Information. Journal of the ACM **28** (1981) 41–70
17. Nakamura, A.: A Rough Logic based on Incomplete Information and Its Application. Int'l. Journal of Approximate Reasoning **15** (1996) 367-378
18. Grzymala-Busse, J.: On the Unknown Attribute Values in Learning from Examples. Lecture Notes in AI, Vol.542. Springer-Verlag (1991) 368–377
19. Grzymala-Busse, J., Werbrouck, P.: On the Best Search Method in the LEM1 and LEM2 Algorithms. Incomplete Information: Rough Set Analysis. Studies in Fuzziness and Soft Computing, Vol.13. Physica-Verlag (1998) 75–91
20. Kryszkiewicz, M.: Properties of Incomplete Information Systems in the Framework of Rough Sets. Rough Sets in Knowledge Discovery 1. Studies in Fuzziness and Soft Computing, Vol.18. Physica-Verlag (1998) 442-450
21. Kryszkiewicz, M.: Rough Set Approach to Incomplete Information Systems. Information Sciences **112** (1998) 39–49
22. Kryszkiewicz, M.: Rules in Incomplete Information Systems. Information Sciences **113** (1999) 271–292
23. Sakai, H.: Effective Procedures for Data Dependencies in Information Systems. Rough Set Theory and Granular Computing. Studies in Fuzziness and Soft Computing, Vol.125. Springer (2003) 167–176
24. Sakai, H., Okuma, A.: An Algorithm for Finding Equivalence Relations from Tables with Non-deterministic Information. Lecture Notes in AI, Vol.1711. Springer-Verlag (1999) 64–72
25. Sakai, H.: Effective Procedures for Handling Possible Equivalence Relations in Non-deterministic Information Systems. Fundamenta Informaticae **48** (2001) 343–362
26. Sakai, H., Okuma, A.: An Algorithm for Checking Dependencies of Attributes in a Table with Non-deterministic Information: A Rough Sets based Approach. Lecture Notes in AI, Vol.1886. Springer-Verlag (2000) 219–229
27. Sakai, H.: A Framework of Rough Sets based Rule Generation in Non-deterministic Information Systems. Lecture Notes in AI, Vol.2871. Springer-Verlag (2003) 143–151

Appendixes

Throughout the appendixes, every input to Unix system and programs is underlined. Furthermore, every attribute is identified with the ordinal number. For example, attributes A and C are identified with 1 and 3, respectively. These tool programs are implemented on a workstation with 450MHz Ultrasparc CPU.

Appendix 1.
```
% more nis1.pl ··· (A1-1)
object(10,6).
data(1,[3,[1,3,4],3,2,5,5]).
data(2,[[2,4],2,2,[3,4],[1,3,4],4]).
data(3,[[1,2],[2,4,5],2,3,[4,5],5]).
data(4,[[1,5],5,[2,4],2,[1,4,5],5]).
data(5,[[3,4],4,3,[1,2,3],1,[2,5]]).
data(6,[[3,5],4,1,[2,3,5],5,[2,3,4]]).
data(7,[[1,5],4,5,[1,4],[3,5],1]).
data(8,[4,[2,4,5],2,[1,2,3],2,[1,2,5]]).
data(9,[2,5,3,5,4,2]).
data(10,[[2,3,5],1,2,3,1,[1,2,3]]).
% more attrib1.pl ··· (A1-2)
condition([1,2,3]).
decision([6]).
% prolog ··· (A1-3)
K-Prolog Compiler version 4.11 (C).
?-consult(define.pl).
yes
?-translate1. ··· (A1-4)
Data File Name: 'nis1.pl'.
Attribute File Name: 'attrib1.pl'.
EXEC_TIME=0.073(sec)
yes
?-class(con,[4,5,6]). ··· (A1-5)
[1] Pe-classes: [4],[5],[6]
Positive Selection
Tuple from 4: [1,5,2] *
Tuple from 5: [3,4,3] *
Tuple from 6: [3,4,1] *
Negative Selection
Tuple from 1: [3,4,3] *
Tuple from 3: [1,5,2] *
[2] Pe-classes: [4],[5],[6]
    :      :     :
[16] Pe-classes: [4],[5],[6]
Positive Selection
Tuple from 4: [5,5,4] *
Tuple from 5: [4,4,3] *
Tuple from 6: [5,4,1] *
Negative Selection
Certainly Definable
EXEC_TIME=0.058(sec)
yes
```

In (A1-1), data NIS_1 is displayed. In (A1-2), condition attributes $\{A, B, C\}$ and decision attributes $\{F\}$ are displayed. In (A1-3), prolog interpreter is invoked. In (A1-4), inf and sup information is produced according to attribute file. In (A1-5), the definability of a set $\{4, 5, 6\}$ for $ATR=\{A, B, C\}$ is examined. In the first response, tuples from 4, 5 and 6 are fixed to (1,5,2), (3,4,3) and (3,4,1). At the same time, tuples (3,4,3) from object 1 and (1,5,2) from object 3 are implicitly rejected. There are 16 responses, and a set $\{4, 5, 6\}$ is proved to be certainly definable.

Appendix 2.

```
?-translate2. ··· (A2-1)
Data File Name: 'nis1.pl'.
EXEC_TIME=0.189(sec)
yes
?-pe. ··· (A2-2)
[1] Derived DISs: 192
    Distinct Pe-relations: 176
[2] Derived DISs: 27
    Distinct Pe-relations: 27
    :      :      :
[6] Derived DISs: 54
    Distinct Pe-relations: 54
EXEC_TIME=1.413(sec)
yes
% more 3.rs ··· (A2-3)
object(10).
attrib(3).
cond(1,3,1,3).
pos(1,3,1).
cond(2,3,1,2).
pos(2,3,1).
    :      :      :
inf([7,3,1],[7,3,1],[[7],[1]]).
sup([7,3,1],[7,3,1],[[7],[1]]).
% more 3.pe ··· (A2-4)
10 1 3 2 2 1 2 2 2 1 6 7 2 1 2 5 3 4 8 9 0 0 10 0 0 1 1 2 2 4 1 6
7 2 1 2 5 3 8 0 9 0 0 10 0 0 1
% merge ··· (A2-5)
EXEC_TIME=0.580(sec)
% more 12345.pe ··· (A2-6)
10 5 1 2 3 4 5 40310784 2 1 2 3 4 5 6 7 8 9 10 0 0 0 0 0 0 0 0 0 0
40030848 1 2 2 4 5 6 7 8 9 10 0 3 0 0 0 0 0 0 0 0 279936
```

In (A2-1), inf and sup information is produced for each attribute. In (A2-2), the definability of a set OB is examined for each attribute. As a side effect, every pe-relation is obtained. In (A2-3), inf and sup information for the attribute C

Table 6. Definitions of $NISs$

NIS	$|OB|$	$|AT|$	$Derived\_DISs$
NIS_2	30	5	$7558272(= 2^7 \times 3^{10})$
NIS_3	50	5	$120932352(= 2^{11} \times 3^{10})$
NIS_4	100	5	$1451188224(= 2^{13} \times 3^{11})$

is displayed, and the contents of $pe\_rel(\{C\})$ are displayed in (A2-4). In (A2-5), program $merge$ is invoked for merging five pe-relations $pe\_rel(\{A\})$, $pe\_rel(\{B\})$, $\cdots$, $pe\_rel(\{E\})$. The produced $pe\_rel(\{A, B, C, D, E\})$ are displayed in (A2-6).

Here, we show each execution time for the following $NISs$, which are automatically produced by means of applying a random number program.

Appendix 3.

```
% depend ··· (A3-1)
File Name for Condition: 1.pe
File Name for Decision: 6.pe
CRITERION 1
Derived DISs: 10368
Derived Consistent DISs: 0
Degree of Consistent DISs: 0.000
CRITERION 2
Minimum Degree of Dependency: 0.000
Maximum Degree of Dependency: 0.600
EXEC_TIME=0.030(sec)
% depratio ··· (A3-2)
File Name for Condition: 12345.pe
File Name for Decision: 6.pe
CRITERION 1
Derived DISs: 2176782336
Derived Consistent DISs: 2161665792
Degree of Consistent DISs: 0.993
CRITERION 2
Minimum Degree of Dependency: 0.800
Maximum Degree of Dependency: 1.000
Consistency Ratio
Object 1: 1.000(=2176782336/2176782336)
Object 2: 0.993(=2161665792/2176782336)
Object 3: 0.993(=2161665792/2176782336)
        :     :     :
Object 10: 1.000(=2176782336/2176782336)
EXEC_TIME=0.020(sec)
```

In (A3-1), the dependency from $\{A\}$ to $\{F\}$ is examined. Here, two pe-relations $pe\_rel(\{A\})$ and $pe\_rel(\{F\})$ are applied. There is no consistent derived DIS. Furthermore, the maximum value of the dependency is 0.6. Therefore, it will be difficult to recognize the dependency from $\{A\}$ to $\{F\}$. In (A3-2), the depen-

Table 7. Each execution time(sec) of *translate2*, *pe* and *merge* for $\{A, B, C\}$. $N1$ denotes the number of derived *DISs*, and $N2$ denotes the number of distinct *pe*-relations

NIS	translate2	pe	merge	N1	N2
NIS_2	0.308	1.415	0.690	5832	120
NIS_3	0.548	8.157	0.110	5184	2
NIS_4	1.032	16.950	2.270	20736	8

Table 8. Each execution time(sec) of *depend* and *depratio* from $\{A, B, C\}$ to $\{E\}$. $N3$ denotes the number of derived *DISs* for $\{A, B, C, E\}$, and $N4$ denotes the number of combined pairs $pe_i \in pe\_rel(\{A, B, C\})$ and $pe_j \in pe\_rel(\{E\})$

NIS	depend	depratio	N3	N4
NIS_2	0.020	0.080	104976	2160
NIS_3	0.010	0.060	279936	108
NIS_4	0.070	0.130	4478976	1728

dency from $\{A, B, C, D, E\}$ to $\{F\}$ is examined. This time, it will be possible to recognize the dependency from $\{A, B, C, D, E\}$ to $\{F\}$.

Appendix 4.
```
% more attrib2.pl ··· (A4-1)
decision([6]).
decval([5]).
order([1,2,3,4,5]).
?-translate3. ··· (A4-2)
Data File Name: 'nis1.pl'.
Attribute File Name: 'attrib2.pl'.
EXEC_TIME=0.066(sec)
yes
?-certain. ··· (A4-3)
DECLIST:<inf=[1,3,4],sup=[1,3,4,5,8]>
[A Certain Rule from Object 1]
   [1,3]&[3,3]&[5,5]=>[6,5] [746496/746496,DGC]
   [(0.1,0.1),(1.0,1.0),(0.2,0.333)]
[A Certain Rule from Object 3]
[A Certain Rule from Object 4]
EXEC_TIME=0.026(sec)
yes
?-possible. ··· (A4-4)
DECLIST: <inf=[1,3,4],sup=[1,3,4,5,8]>
[Possible Rules from Object 1]
   === One Attribute ===
   [1,3]=>[6,5] [10368/10368,DMA]
   [(0.1,0.2),(0.25,1.0),(0.2,0.5)]
```

```
[2,3]=>[6,5] [486/1458,IGC]
[(0.1,0.1),(1.0,1.0),(0.2,0.333)]
[4,2]=>[6,5] [5832/5832,DMA]
[(0.2,0.4),(0.4,1.0),(0.4,0.8)]
[Possible Rules from Object 3]
   === One Attribute ===
   [1,1]=>[6,5] [5184/10368,IMA]
   :    :    :
[Possible Rules from Object 8]
   === One Attribute ===
   [1,4]=>[6,5] [3456/10368,IMA]
   :    :    :
   [(0.3,0.4),(0.6,1.0),(0.6,0.8)]
   [5,2]=>[6,5] [648/1944,IGC]
   [(0.1,0.1),(1.0,1.0),(0.2,0.25)]
EXEC_TIME=0.118(sec)
yes
```

In order to handle rules, it is necessary to prepare a file like in (A4-1). In (A4-2), *inf* and *sup* information is produced according to *attrib2.pl*. Program *certain* extracts possible implications belonging to *D-GC* class in (A4-3). As an additional information, *minsup*, *maxsup*, *minacc*, *maxacc*, *mincov* and *maxcov* are sequentially displayed. Program *possible* extracts possible implications belonging to *I-GC* or *MA* classes in (A4-4). Table 9 shows each execution time for three *NISs*. Here, the order is sequentially *A*, *B*, *C* and *D* for the decision attribute $\{E\}$, and the decision attribute value is 1. This execution time depends upon the number of such object x that $1 \in PT(x, \{E\})$.

Table 9. Each execution time(sec) of *translate3*, *possible* and *certain*. N5 denotes the number of such object x that $1 \in PT(x, \{E\})$

NIS	translate3	possible	certain	N5
NIS_2	0.178	0.115	0.054	7
NIS_3	0.173	0.086	0.039	4
NIS_4	0.612	0.599	0.391	9

A Partition Model of Granular Computing

Yiyu Yao

Department of Computer Science, University of Regina
Regina, Saskatchewan, Canada S4S 0A2
yyao@cs.uregina.ca
http://www.cs.uregina.ca/~yyao

Abstract. There are two objectives of this chapter. One objective is
to examine the basic principles and issues of granular computing. We
focus on the tasks of granulation and computing with granules. From
semantic and algorithmic perspectives, we study the construction, interpretation, and representation of granules, as well as principles and
operations of computing and reasoning with granules. The other objective is to study a partition model of granular computing in a set-theoretic
setting. The model is based on the assumption that a finite set of universe is granulated through a family of pairwise disjoint subsets. A hierarchy of granulations is modeled by the notion of the partition lattice.
The model is developed by combining, reformulating, and reinterpret-
ing notions and results from several related fields, including theories of
granularity, abstraction and generalization (artificial intelligence), partition models of databases, coarsening and refining operations (evidential
theory), set approximations (rough set theory), and the quotient space
theory for problem solving.

1 Introduction

The basic ideas of granular computing, i.e., problem solving with different granularities, have been explored in many fields, such as artificial intelligence, interval
analysis, quantization, rough set theory, Dempster-Shafer theory of belief functions, divide and conquer, cluster analysis, machine learning, databases, and
many others [73]. There is a renewed and fast growing interest in granular computing [21, 30, 32, 33, 41, 43, 48, 50, 51, 58, 60, 70, 77].

The term "granular computing (GrC)" was first suggested by T.Y. Lin [74].
Although it may be difficult to have a precise and uncontroversial definition, we
can describe granular computing from several perspectives.

We may define granular computing by examining its major components and
topics. Granular computing is a label of theories, methodologies, techniques, and
tools that make use of granules, i.e., groups, classes, or clusters of a universe, in
the process of problem solving [60]. That is, granular computing is used as an
umbrella term to cover these topics that have been studied in various fields in
isolation. By examining existing studies in a unified framework of granular computing and extracting their commonalities, one may be able to develop a general
theory for problem solving. Alternatively, we may define granular computing by

J.F. Peters et al. (Eds.): Transactions on Rough Sets I, LNCS 3100, pp. 232–253, 2004.
© Springer-Verlag Berlin Heidelberg 2004

identifying its unique way of problem solving. Granular computing is a way of thinking that relies on our ability to perceive the real world under various grain sizes, to abstract and consider only those things that serve our present interest, and to switch among different granularities. By focusing on different levels of granularities, one can obtain various levels of knowledge, as well as inherent knowledge structure. Granular computing is essential to human problem solving, and hence has a very significant impact on the design and implementation of intelligent systems.

The ideas of granular computing have been investigated in artificial intelligence through the notions of granularity and abstraction. Hobbs proposed a theory of granularity based on the observation that "[w]e look at the world under various grain seizes and abstract from it only those things that serve our present interests" [18]. Furthermore, "[o]ur ability to conceptualize the world at different granularities and to switch among these granularities is fundamental to our intelligence and flexibility. It enables us to map the complexities of the world around us into simpler theories that are computationally tractable to reason in" [18]. Giunchigalia and Walsh proposed a theory of abstraction [14]. Abstraction can be thought of as "the process which allows people to consider what is *relevant* and to forget a lot of *irrelevant* details which would get in the way of what they are trying to do". They showed that the theory of abstraction captures and generalizes most previous work in the area.

The notions of granularity and abstraction are used in many subfields of artificial intelligence. The granulation of time and space leads naturally to temporal and spatial granularities. They play an important role in temporal and spatial reasoning [3, 4, 12, 19, 54]. Based on granularity and abstraction, many authors studied some fundamental topics of artificial intelligence, such as, for example, knowledge representation [14, 75], theorem proving [14], search [75, 76], planning [24], natural language understanding [35], intelligent tutoring systems [36], machine learning [44], and data mining [16].

Granular computing recently received much attention from computational intelligence community. The topic of fuzzy information granulation was first proposed and discussed by Zadeh in 1979 and further developed in the paper published in 1997 [71, 73]. In particular, Zadeh proposed a general framework of granular computing based on fuzzy set theory [73]. Granules are constructed and defined based on the concept of generalized constraints. Relationships between granules are represented in terms of fuzzy graphs or fuzzy if-then rules. The associated computation method is known as computing with words (CW) [72]. Although the formulation is different from the studies in artificial intelligence, the motivations and basic ideas are the same. Zadeh identified three basic concepts that underlie human cognition, namely, granulation, organization, and causation [73]. "Granulation involves decomposition of whole into parts, organization involves integration of parts into whole, and causation involves association of causes and effects." [73] Yager and Filev argued that "human beings have been developed a granular view of the world" and "... objects with which mankind perceives, measures, conceptualizes and reasons are granular" [58]. Therefore, as pointed out by Zadeh, "[t]he theory of fuzzy information granulation (TFIG) is inspired by the ways in which humans granulate information and reason with it."[73]

The necessity of information granulation and simplicity derived from information granulation in problem solving are perhaps some of the practical reasons for the popularity of granular computing. In many situations, when a problem involves incomplete, uncertain, or vague information, it may be difficult to differentiate distinct elements and one is forced to consider granules [38–40]. In some situations, although detailed information may be available, it may be sufficient to use granules in order to have an efficient and practical solution. In fact, very precise solutions may not be required at all for many practical problems. It may also happen that the acquisition of precise information is too costly, and coarse-grained information reduces cost [73]. They suggest a basic guiding principle of fuzzy logic: "*Exploit the tolerance for imprecision, uncertainty and partial truth to achieve tractability, robustness, low solution cost and better rapport with reality*" [73]. This principle offers a more practical philosophy for real world problem solving. Instead of searching for the optimal solution, one may search for good approximate solutions. One only needs to examine the problem at a finer granulation level with more detailed information when there is a need or benefit for doing so [60].

The popularity of granular computing is also due to the theory of rough sets [38, 39]. As a concrete theory of granular computing, rough set model enables us to precisely define and analyze many notions of granular computing. The results provide an in-depth understanding of granular computing.

The objectives of this chapter are two-fold based on investigations at two levels. Sections 2 and 3 focus on a high and abstract level development of granular computing, and Section 3 deals with a low and concrete level development by concentrating on a partition model of granular computing. The main results are summarized as follows.

In Section 2, we discuss in general terms the basic principles and issues of granular computing based on related studies, such as the theory of granularity, the theory of abstraction, and their applications. The tasks of granulation and computing with granules are examined and related to existing studies. We study the construction, interpretation, and representation of granules, as well as principles and operations of computing and reasoning with granules. In Section 3, we argue that granular computing is a way of thinking. This way of thinking is demonstrated based on three problem solving domains, i.e., concept formation, top-down programming, and top-down theorem proving.

In Section 4, we study a partition model of granular computing in a set-theoretic setting. The model is based on the assumption that a finite set of universe is granulated through a family of pairwise disjoint subsets. A hierarchy of granulations is modeled by the notion of the partition lattice. Results from rough sets [38], quotient space theory [75, 76], belief functions [46], databases[27], data mining [31, 34], and power algebra [6] are reformulated, re-interpreted, refined, extended and combined for granular computing. We introduce two basic operations called zooming-in and zooming-out operators. Zooming-in allows us to expand an element of the quotient universe into a subset of the universe, and hence reveals more detailed information. Zooming-out allows us to move to the quotient universe by ignoring some details. Computations in both universes can be connected through zooming operations.

2   Basic Issues of Granular Computing

Granular computing may be studied based on two related issues, namely granulation and computation [60]. The former deals with the construction, interpretation, and representation of granules, and the latter deals with the computing and reasoning with granules. They can be further divided with respect to algorithmic and semantic aspects [60]. The algorithmic study concerns the procedures for constructing granules and related computation, and the semantic study concerns the interpretation and physical meaningfulness of various algorithms. Studies from both aspects are necessary and important. The results from semantic study may provide not only interpretations and justifications for a particular granular computing model, but also guidelines that prevent possible misuses of the model. The results from algorithmic study may lead to efficient and effective granular computing methods and tools.

2.1   Granulations

Granulation of a universe involves dividing the universe into subsets or grouping individual objects into clusters. A granule may be viewed as a subset of the universe, which may be either fuzzy or crisp. A family of granules containing every object in the universe is called a granulation of the universe, which provides a coarse-grained view of the universe. A granulation may consist of a family of either disjoint or overlapping granules. There are many granulations of the same universe. Different views of the universe can be linked together, and a hierarchy of granulations can be established.

The notion of granulation can be studied in many different contexts. The granulation of the universe, particularly the semantics of granulation, is domain and application dependent. Nevertheless, one can still identify some domain independent issues [75]. Some of such issues are described in more detail below.

Granulation Criteria. A granulation criterion deals with the semantic interpretation of granules and addresses the question of *why* two objects are put into the same granule. It is domain specific and relies on the available knowledge. In many situations, objects are usually grouped together based on their relationships, such as indistinguishability, similarity, proximity, or functionality [73]. One needs to build models to provide both semantical and operational interpretations of these notions. A model enables us to define formally and precisely various notions involved, and to study systematically the meanings and rationalities of granulation criteria.

Granulation Structures. It is necessary to study granulation structures derivable from various granulations of the universe. Two structures can be observed, the structure of individual granules and structure of a granulation. Consider the case of crisp granulation. One can immediately define an order relation between granules based on the set inclusion. In general, a large granule may contain small granules, and small granules may be combined to form a large granule. The order relation can be extended to different granulations. This leads to multi-level granulations in a natural hierarchical structure. Various hierarchical granulation structures have been studied by many authors [22, 36, 54, 75, 76].

Granulation Methods. From the algorithmic aspect, a granulation method addresses the problem of *how* to put two objects into the same granule. It is necessary to develop algorithms for constructing granules and granulations efficiently based on a granulation criterion. The construction process can be modeled as either top-down or bottom-up. In a top-down process, the universe is decomposed into a family of subsets, each subset can be further decomposed into smaller subsets. In a bottom-up process, a subset of objects can be grouped into a granule, and smaller granules can be further grouped into larger granules. Both processes lead naturally to a hierarchical organization of granules and granulations [22, 61].

Representation/Description of Granules. Another semantics related issue is the interpretation of the results of a granulation method. Once constructed, it is necessary to describe, to name and to label granules using certain languages. This can be done in several ways. One may assign a name to a granule such that an element in the granule is an instance of the named category, as being done in classification [22]. One may also construct a certain type of center as the representative of a granule, as being done in information retrieval [45, 56]. Alternatively, one may select some typical objects from a granule as its representative. For example, in many search engines, the search results are clustered into granules and a few titles and some terms can be used as the description of a granule [8, 17].

Quantitative Characteristics of Granules and Granulations. One can associate quantitative measures to granules and granulations to capture their features. Consider again the case of crisp granulation. The cardinality of a granule, or Hartley information measure, can be used as a measure of the size or uncertainty of a granule [64]. The Shannon entropy measure can be used as a measure of the granularity of a partition [64].

These issues can be understood by examining a concrete example of granulation known as the cluster analysis [2]. This can be done by simply change granulation into clustering and granules into clusters. *Clustering structures* may be hierarchical or non-hierarchical, exclusive or overlapping. Typically, a similarity or distance function is used to define the relationships between objects. *Clustering criteria* may be defined based on the similarity or distance function, and the required cluster structures. For example, one would expect strong similarities between objects in the same cluster, and weak similarities between objects in different clusters. Many *clustering methods* have been proposed and studied, including the families of hierarchical agglomerative, hierarchical divisive, iterative partitioning, density search, factor analytic, clumping, and graph theoretic methods [1]. Cluster analysis can be used as an exploratory tool to interpret data and find regularities from data [2]. This requires the active participation of experts to interpret the results of clustering methods and judge their significance. A good *representation* of clusters and their *quantitative characterizations* may make the task of exploration much easier.

2.2 Computing and Reasoning with Granules

Computing and reasoning with granules depend on the previously discussed notion of granulations. They can be similarly studied from both the semantic and algorithmic perspectives. One needs to design and interpret various methods based on the interpretation of granules and relationships between granules, as well as to define and interpret operations of granular computing.

The two level structures, the granule level and the granulation level, provide the inherent relationships that can be explored in problem solving. The granulated view summarizes available information and knowledge about the universe. As a basic task of granular computing, one can examine and explore further relationships between granules at a lower level, and relationships between granulations at a higher level. The relationships include closeness, dependency, and association of granules and granulations [43]. Such relationships may not hold fully and certain measures can be employed to quantify the degree to which the relationships hold [64]. This allows the possibility to extract, analyze and organize information and knowledge through relationships between granules and between granulations [62, 63].

The problem of computing and reasoning with granules is domain and application dependent. Some general domain independent principles and issues are listed below.

Mappings between Different Level of Granulations. In the granulation hierarchy, the connections between different levels of granulations can be described by mappings. Giunchglia and Walsh view an abstraction as a mapping between a pair of formal systems in the development of a theory of abstraction [14]. One system is referred to as the ground space, and the other system is referred to as the abstract space. At each level of granulation, a problem is represented with respect to the granularity of the level. The mapping links different representations of the same problem at different levels of details. In general, one can classify and study different types of granulations by focusing on the properties of the mappings [14].

Granularity Conversion. A basic task of granular computing is to change views with respect to different levels of granularity. As we move from one level of details to another, we need to convert the representation of a problem accordingly [12, 14]. A move to a more detailed view may reveal information that otherwise cannot be seen, and a move to a simpler view can improve the high level understanding by omitting irrelevant details of the problem [12, 14, 18, 19, 73, 75, 76]. The change between grain-sized views may be metaphorically stated as the change between the forest and trees.

Property Preservation. Granulation allows the different representations of the same problem in different levels of details. It is naturally expected that the same problem must be consistently represented [12]. A granulation and its related computing methods are meaningful only they preserve certain desired properties [14, 30, 75]. For example, Zhang and Zhang studied the "false-preserving" property, which states that if a coarse-grained space has no solution for a problem then the original fine-grained space has no solution [75, 76]. Such a property can be explored to improve the efficiency of problem solving by eliminating a more detailed study in a coarse-grained space. One may require that the structure of a solution in a coarse-grained space is similar to the solution in a fine-grained space. Such a property is used in top-down problem solving techniques. More specifically, one starts with a sketched solution and successively refines it into a full solution. In the context of hierarchical planning, one may impose similar properties, such as upward solution property, downward solution property, monotonicity property, etc. [24].

Operators. The relationship between granules at different levels and conversion of granularity can be precisely defined by operators [12, 36]. They serve as the basic build blocks of granular computing. There are at least two types of operators that can be defined. One type deals with the shift from a fine granularity to a coarse granularity. A characteristics of such an operator is that it will discard certain details, which makes distinct objects no longer differentiable. Depending on the context, many interpretations and definitions are available, such as abstraction, simplification, generalization, coarsening, zooming-out, etc. [14, 18, 19, 36, 46, 66, 75]. The other type deals with the change from a coarse granularity to a fine granularity. A characteristics of such an operator is that it will provide more details, so that a group of objects can be further classified. They can be defined and interpreted differently, such as articulation, specification, expanding, refining, zooming-in, etc. [14, 18, 19, 36, 46, 66, 75]. Other types of operators may also be defined. For example, with the granulation, one may not be able to exactly characterize an arbitrary subset of a fine-grained universe in a coarse-grained universe. This leads to the introduction of approximation operators in rough set theory [39, 59].

The notion of granulation describes our ability to perceive the real world under various grain sizes, and to abstract and consider only those things that serve our present interest. Granular computing methods describe our ability to switch among different granularities in problem solving. Detailed and domain specific methods can be developed by elaborating these issues with explicit reference to an application. For example, concrete domain specific conversion methods and operators can be defined. In spite of the differences between various methods, they are all governed by the same underlying principles of granular computing.

3   Granular Computing as a Way of Thinking

The underlying ideas of granular computing have been used either explicitly or implicitly for solving a wide diversity of problems. Their effectiveness and merits may be difficult to study and analyze based on some kind of formal proofs. They may be judged based on the powerful and yet imprecise and subjective tools of our experience, intuition, reflections and observations [28]. As pointed out by Leron [28], a good way of activating these tools is to carry out some case studies. For such a purpose, the general ideas, principles, and methodologies of granular computing are further examined with respect to several different fields in the rest of this section. It should be noted that analytical and experimental results on the effectiveness of granular computing in specific domains, though will not be discussed in this chapter, are available [20, 24, 75].

3.1   Concept Formation

From philosophical point of view, granular computing can be understood as a way of thinking in terms of the notion of concepts that underlie the human knowledge.

Every concept is understood as a unit of thoughts consisting of two parts, the intension and the extension of the concept [9, 52, 53, 55, 57]. The intension (comprehension) of a concept consists of all properties or attributes that are valid for all those objects to which the concept applies. The extension of a concept is the set of objects or entities which are instances of the concept. All objects in the extension have the same properties that characterize the concept. In other words, the intension of a concept is an abstract description of common features or properties shared by elements in the extension, and the extension consists of concrete examples of the concept. A concept is thus described jointly by its intension and extension. This formulation enables us to study concepts in a logic setting in terms of intensions and also in a set-theoretic setting in terms of extensions. The description of granules characterize concepts from the intension point of view, while granules themselves characterize concepts from the extension point of view. Through the connections between extensions of concepts, one may establish relationships between concepts [62, 63].

In characterizing human knowledge, one needs to consider two topics, namely, context and hierarchy [42, 47]. Knowledge is contextual and hierarchical. A context in which concepts are formed provides meaningful interpretation of the concepts. Knowledge is organized in a tower or a partial ordering. The base-level, or first-level, concepts are the most fundamental concepts, and higher-level concepts depend on lower-level concepts. To some extent, granulation and inherent hierarchical granulation structures reflect naturally the way in which human knowledge is organized. The construction, interpretation, and description of granules and granulations are of fundamental importance in the understanding, representation, organization and synthesis of data, information, and knowledge.

3.2   Top-Down Programming

The top-down programming is an effective technique to deal with the complex problem of programming, which is based on the notions of structured programming and stepwise refinement [26]. The principles and characteristics of the top-down design and stepwise refinement, as discussed by Ledgard, Gueras and Nagin [26], provide a convincing demonstration that granular computing is a way of thinking.

According to Ledgard, Gueras and Nagin [26], the top-down programming approach has the following characteristics:

Design in Levels. A level consists of a set of modules. At higher levels, only a brief description of a module is provided. The details of the module are to be refined, divided into smaller modules, and developed in lower levels.

Initial Language Independence. The high-level representations at initial levels focus on expressions that are relevant to the problem solution, without explicit reference to machine and language dependent features.

Postponement of Details to Lower Levels. The initial levels concern critical broad issues and the structure of the problem solution. The details such as the choice of specific algorithms and data structures are postponed to lower, implementation levels.

Formalization of Each Level. Before proceeding to a lower level, one needs to obtain a formal and precise description of the current level. This will ensure a full understanding of the structure of the current sketched solution.

Verification of Each Level. The sketched solution at each level must be verified, so that errors pertinent to the current level will be detected.

Successive Refinements. Top-down programming is a successive refinement process. Starting from the top level, each level is redefined, formalized, and verified until one obtains a full program.

In terms of granular computing, program modules correspond to granules, and levels of the top-down programming correspond to different granularities. One can immediately see that those characteristics also hold for granular computing in general.

3.3   Top-Down Theorem Proving

Another demonstration of granular computing as a way of thinking is the approach of top-down theorem proving, which is used by computer systems and human experts. The PROLOG interpreter basically employs a top-down, depth-first search strategy to solve problem through theorem proving [5]. It has also been suggested that the top-down approach is effective for developing, communicating and writing mathematical proofs [13, 14, 25, 28].

PROLOG is a logic programming language widely used in artificial intelligence. It is based on the first-order predicate logic and models problem solving as theorem proving [5]. A PROLOG program consists of a set of facts and rules in the form of Horn clauses that describe the objects and relations in a problem domain. The PROLOG interpreter answers a query, referred to as goals, by finding out whether the query is a logical consequence of the facts and rules of a PROLOG program. The inference is performed in a top-down, left to right, depth-first manner. A query is a sequence of one or more goals. At the top level, the leftmost goal is reduced to a sequence of subgoals to be tried by using a clause whose head unifies with the leftmost goal. The PROLOG interpreter then continues by trying to reduce the leftmost goal of the new sequence of goals. Eventually the lestmost goal is satisfied by a fact, and the second leftmost goal is tried in the same manner. Backtracking is used when the interpreter fails to find a unification that solves a goal, so that other clauses can be tried.

A proof found by the PROLOG interpreter can be expressed naturally in a hierarchical structure, with the proofs of subgoals as the children of a goal. In the process of reducing a goal to a sequence of subgoals, one obtains more details of the proof. The strategy can be applied to general theorem proving. This may be carried out by abstracting the goal, proving its abstracted version and then using the structure of the resulting proof to help construct the proof of the original goal [14].

By observing the systematic way of top-down programming, some authors suggest that the similar approach can be used in developing, teaching and communicating mathematical proofs [13, 28]. Leron proposed a structured method for presenting mathematical proofs [28]. The main objective is to increase the comprehensibility of mathematical presentations, and at the same time, retain their rigor. The traditional linear fashion presents a proof step-by-step from hypotheses to conclusion. In contrast, the structured method arranges the proof in levels and proceeds in a top-down manner. Like the top-down, step-wise refinement programming approach, a level consists of short autonomous modules, each embodying one major idea of the proof to be further developed in the subsequent levels. The top level is a very general description of the main line of the proof. The second level elaborates on the generalities of the top level by supplying proofs of unsubstantiated statements, details of general descriptions, and so on. For some more complicated tasks, the second level only gives brief descriptions and the details are postponed to the lower levels. The process continues by supplying more details of the higher levels until a complete proof is reached. Such a development of proofs procedure is similar to the strategy used by the PROLOG interpreter. A complicated proof task is successively divided into smaller and easier subtasks. The inherent structures of those tasks not only improve the comprehensibility of the proof, but also increase our understanding of the problem.

Lamport proposed a proof style, a refinement of natural deduction, for developing and writing structured proofs [25]. It is also based on hierarchical structuring, and divides proofs into levels. By using a numbering scheme to label various parts of a proof, one can explicitly show the structures of the proof. Furthermore, such a structure can be conveniently expressed using a computer-based hypertext system. One can concentrate on a particular level in the structure and suppress lower level details. In principle, the top-down design and step-wise refinement strategy of programming can be applied in developing proofs to eliminate possible errors.

3.4   Granular Computing Approach of Problem Solving

In their book on research methods, Granziano and Raulin make a clear separation of research process and content [11]. They state, "... the basic processes and the systematic way of studying problems are common elements of science, regardless of each discipline's particular subject matter. It is the process and not the content that distinguishes science from other ways of knowing, and it is the content – the particular phenomena and fact of interest – that distinguishes one scientific discipline from another." [11] From the discussion of the previous examples, we can make a similar separation of the granular computing process and content (i.e., domains of applications). The systematic way of granular computing is generally applicable to different domains, and can be studied based on the basic issues and principles discussed in the last section.

In general, granular computing approach can be divided into top-down and bottom-up modes. They present two directions of switch between levels of granularities. The concept formation can be viewed as a combination of top-down and bottom-up. One can combine specific concepts to produce a general concept in a bottom-up manner, and divide a concept into more specific subconcepts in top-down manner. Top-down programming and top-down theorem proving are typical examples of top-down approaches. Independent of the modes, step-wise (successive) refinement plays an important role. One needs to fully understand all notions of a particular level before moving up or down to another level.

From the case studies, we can abstract some common features by ignoring irrelevant formulation details. It is easy to arrive at a conclusion that granular computing is a way of thinking and a philosophy for problem solving. At an abstract level, it captures and reflects our ability to solve a problem by focusing on different levels of details, and move easily from different levels at various stages. The principles of granular computing are the same and applicable to many domains.

4   A Partition Model

A partition model is developed by focusing on the basic issues of granular computing. The partition model has been studied extensively in rough set theory [39].

4.1   Granulation by Partition and Partition Lattice

A simple granulation of the universe can be defined based on an equivalence relation or a partition. Let U denote a finite and non-empty set called the universe. Suppose $E \subseteq U \times U$ denote an equivalence relation on U, where $\times$ denotes the Cartesian product of sets. That is, E is reflective, symmetric, and transitive.

The equivalence relation E divides the set U into a family of disjoint subsets called the partition of the universe induced by E and denoted by $\pi_E = U/E$. The subsets in a partition are also called blocks. Conversely, given a partition π of the universe, one can uniquely define an equivalence relation E_π:

$$xE_\pi y \iff x \text{ and } y \text{ are in the same block of the partition } \pi. \qquad (1)$$

Due to the one to one relationship between equivalence relations and partitions, one may use them interchangeably.

One can define an order relation on the set of all partitions of U, or equivalently the set of all equivalence relations on U. A partition π_1 is a refinement of another partition π_2, or equivalently, π_2 is a coarsening of π_1, denoted by $\pi_1 \preceq \pi_2$, if every block of π_1 is contained in some block of π_2. In terms of equivalence relations, we have $E_{\pi_1} \subseteq E_{\pi_2}$. The refinement relation $\preceq$ is a partial order, namely, it is reflexive, antisymmetric, and transitive. It defines a partition lattice $\Pi(U)$. Given two partitions π_1 and π_2, their meet, $\pi_1 \wedge \pi_2$, is the largest partition that is a refinement of both π_1 and π_2, their join, $\pi_1 \vee \pi_2$, is the smallest partition that is a coarsening of both π_1 and π_2. The blocks of the meet are all nonempty intersections of a block from π_1 and a block from π_2. The blocks of the join are the smallest subsets which are exactly a union of blocks from π_1 and π_2. In terms of equivalence relations, for two equivalence relations R_1 and R_2, their meet is defined by $R_1 \cap R_2$, and their join is defined by $(R_1 \cup R_2)^*$, the transitive closure of the relation $R_1 \cup R_2$.

The lattice $\Pi(U)$ contains all possible partition based granulations of the universe. The refinement partial order on partitions provides a natural hierarchy of granulations. The partition model of granular computing is based on the partition lattice or subsystems of the partition lattice.

4.2   Partition Lattice in an Information Table

Information tables provide a simple and convenient way to represent a set of objects by a finite set of attributes [39, 70]. Formally, an information table is defined as the following tuple:

$$(U, At, \{V_a \mid a \in At\}, \{I_a \mid a \in At\}), \qquad (2)$$

where U is a finite set of objects called the universe, At is a finite set of attributes or features, V_a is a set of values for each attribute $a \in At$, and $I_a : U \longrightarrow V_a$ is an information function for each attribute $a \in At$. A database is an example of information tables. Information tables give a specific and concrete interpretation of equivalence relations used in granulation.

With respect to an attribute $a \in At$, an object $x \in U$ takes only one value from the domain V_a of a. Let $a(x) = I_a(x)$ denote the value of x on a. By extending to a subset of attributes $A \subseteq At$, $A(x)$ denotes the value of x on attributes A, which can be viewed as a vector with each $a(x)$, $a \in A$, as one of its components. For an attribute $a \in At$, an equivalence relation E_a is given by: for $x, y \in U$,

$$xE_a y \iff a(x) = a(y). \qquad (3)$$

Two objects are considered to be indiscernible, in the view of a single attribute a, if and only if they have exactly the same value. For a subset of attributes $A \subseteq At$, an equivalence relation E_A is defined by:

$$
\begin{aligned}
xE_Ay &\iff A(x) = A(y) \\
&\iff (\forall a \in A)a(x) = a(y) \\
&\iff \bigcap_{a \in A} E_a.
\end{aligned} \tag{4}
$$

With respect to all attributes in A, x and y are indiscernible, if and only if they have the same value for every attribute in A.

The empty set $\emptyset$ produces the coarsest relation, i.e., $E_\emptyset = U \times U$. If the entire attribute set is used, one obtains the finest relation E_{At}. Moreover, if no two objects have the same description, E_{At} becomes the identity relation. The algebra $(\{E_A\}_{A \subseteq At}, \cap)$ is a lower semilattice with the zero element E_{At} [37]. The family of partitions $\Pi(At(U)) = \{\pi_{E_A} \mid A \subseteq At\}$ has been studied in databases [27]. In fact, $\Pi(At(U))$ is a lattice on its own right. While the meet of $\Pi(At(U))$ is the same as the meet of $\Pi(U)$, their joins are different [27]. The lattice $\Pi(At(U))$ can be used to develop a partition model of databases.

A useful result from the constructive definition of the equivalence relation is that one can associate a precise description with each equivalence class. This is done through the introduction of a decision logic language DL in an information table [39, 43, 65]. In the language DL, an atomic formula is given by $a = v$, where $a \in At$ and $v \in V_a$. If ϕ and ψ are formulas, then so are $\neg\phi$, $\phi \wedge \psi$, $\phi \vee \psi$, $\phi \to \psi$, and $\phi \equiv \psi$. The semantics of the language DL can be defined in the Tarski's style through the notions of a model and satisfiability. The model is an information table, which provides interpretation for symbols and formulas of DL. The satisfiability of a formula ϕ by an object x, written $x \models \phi$, is given by the following conditions:

(1) $x \models a = v$ iff $a(x) = v$,

(2) $x \models \neg\phi$ iff not $x \models \phi$,

(3) $x \models \phi \wedge \psi$ iff $x \models \phi$ and $x \models \psi$,

(4) $x \models \phi \vee \psi$ iff $x \models \phi$ or $x \models \psi$,

(5) $x \models \phi \to \psi$ iff $x \models \neg\phi \vee \psi$,

(6) $x \models \phi \equiv \psi$ iff $x \models \phi \to \psi$ and $x \models \psi \to \phi$.

If ϕ is a formula, the set $m(\phi)$ defined by:

$$
m(\phi) = \{x \in U \mid x \models \phi\}, \tag{5}
$$

is called the meaning of the formula ϕ. For an equivalence class of E_A, it can be described by a formula of the form, $\bigwedge_{a \in A} a = v_a$, where $v_a \in V_a$. Furthermore, $[x]_{E_A} = m(\bigwedge_{a \in A} a = a(x))$, where $a(x)$ is the value of x on attribute a.

4.3   Mappings between Two Universes

Given an equivalence relation E on U, we obtain a coarse-grained universe U/E called the quotient set of U. The relation E can be conveniently represented by a mapping from U to 2^U, where 2^U is the power set of U. The mapping $[\cdot]_E : U \longrightarrow 2^U$ is given by:

$$[x]_E = \{y \in U \mid xEy\}. \tag{6}$$

The equivalence class $[x]_E$ containing x plays dual roles. It is a subset of U and an element of U/E. That is, in U, $[x]_E$ is subset of objects, and in U/E, $[x]_E$ is considered to be a whole instead of many individuals [61]. In cluster analysis, one typically associates a name with a cluster such that elements of the cluster are instances of the named category or concept [22]. Lin [29], following Dubois and Prade [10], explicitly used $[x]_E$ for representing a subset of U and $Name([x]_E)$ for representing an element of U/E. In subsequent discussion, we use this convention.

With a partition or an equivalence relation, we have two views of the same universe, a coarse-grained view U/E and a detailed view U. Their relationship can be defined by a pair of mappings, $r : U/E \longrightarrow U$ and $c : U \longrightarrow U/E$. More specifically, we have:

$$r(Name([x]_E)) = [x]_E,$$
$$c(x) = Name([x]_E). \tag{7}$$

A concept, represented as a subset of a universe, is described differently under different views. As we move from one view to the other, we change our perceptions and representations of the same concept. In order to achieve this, we define zooming-in and zooming-out operators based on the pair of mappings [66].

4.4   Zooming-in Operator for Refinement

Formally, zooming-in can be defined by an operator $\omega : 2^{U/E} \longrightarrow 2^U$. Shafer referred to the zooming-in operation as refining [46]. For a singleton subset $\{X_i\} \in 2^{U/E}$, we define [10]:

$$\omega(\{X_i\}) = [x]_E, \quad X_i = Name([x]_E). \tag{8}$$

For an arbitrary subset $X \subseteq U/E$, we have:

$$\omega(X) = \bigcup_{X_i \in X} \omega(\{X_i\}). \tag{9}$$

By zooming-in on a subset $X \subseteq U/E$, we obtain a unique subset $\omega(X) \subseteq U$. The set $\omega(X) \subseteq U$ is called the refinement of X.

The zooming-in operation has the following properties [46]:

$$
\begin{array}{ll}
\text{(zi1)} & \omega(\emptyset) = \emptyset, \\
\text{(zi2)} & \omega(U/E) = U, \\
\text{(zi3)} & \omega(X^c) = (\omega(X))^c, \\
\text{(zi4)} & \omega(X \cap Y) = \omega(X) \cap \omega(Y), \\
\text{(zi5)} & \omega(X \cup Y) = \omega(X) \cup \omega(Y), \\
\text{(zi6)} & X \subseteq Y \Longleftrightarrow \omega(X) \subseteq \omega(Y),
\end{array}
$$

where ^{c} denotes the set complement operator, the set-theoretic operators on the left hand side apply to the elements of $2^{U/E}$, and the same operators on the right hand side apply to the elements of 2^U. From these properties, it can be seen that any relationships of subsets observed under coarse-grained view would hold under the detailed view, and vice versa. For example, in addition to (zi6), we have $X \cap Y = \emptyset$ if and only if $\omega(X) \cap \omega(Y) = \emptyset$, and $X \cup Y = U/E$ if and only if $\omega(X) \cup \omega(Y) = U$. Therefore, conclusions drawn based on the coarse-grained elements in U/E can be carried over to the universe U.

4.5   Zooming-out Operators for Approximation

The change of views from U to U/E is called a zooming-out operation. By zooming-out, a subset of the universe is considered as a whole rather than many individuals. This leads to a loss of information. Zooming-out on a subset $A \subseteq U$ may induce an inexact representation in the coarse-grained universe U/E.

The theory of rough sets focuses on the zooming-out operation. For a subset $A \subseteq U$, we have a pair of lower and upper approximations in the coarse-grained universe [7, 10, 59]:

$$
\begin{aligned}
\underline{apr}(A) &= \{Name([x]_E) \mid x \in U, [x]_E \subseteq A\}, \\
\overline{apr}(A) &= \{Name([x]_E) \mid x \in U, [x]_E \cap A \neq \emptyset\}.
\end{aligned} \tag{10}
$$

The expression of lower and upper approximations as subsets of U/E, rather than subsets of U, has only been considered by a few researchers in rough set community [7, 10, 30, 59, 69]. On the other hand, such notions have been considered in other contexts. Shafer [46] introduced those notions in the study of belief functions and called them the inner and outer reductions of $A \subseteq U$ in U/E. The connections between notions introduced by Pawlak in rough set theory and these introduced by Shafer in belief function theory have been pointed out by Dubois and Prade [10].

The expression of approximations in terms of elements of U/E clearly shows that representation of A in the coarse-grained universe U/E. By zooming-out, we only obtain an approximate representation. The lower and upper approximations satisfy the following properties [46, 69]:

$$
\text{(zo1)} \quad \underline{apr}(\emptyset) = \overline{apr}(\emptyset) = \emptyset,
$$

(zo2) $\underline{apr}(U) = \overline{apr}(U) = U/E,$

(zo3) $\underline{apr}(A) = (\overline{apr}(A^c))^c,$

$\overline{apr}(A) = (\underline{apr}(A^c))^c;$

(zo4) $\underline{apr}(A \cap B) = \underline{apr}(A) \cap \underline{apr}(B),$

$\overline{apr}(A \cap B) \subseteq \overline{apr}(A) \cap \overline{apr}(B),$

(zo5) $\underline{apr}(A) \cup \underline{apr}(B) \subseteq \underline{apr}(A \cup B),$

$\overline{apr}(A \cup B) = \overline{apr}(A) \cup \overline{apr}(B),$

(zo6) $A \subseteq B \Longrightarrow [\underline{apr}(A) \subseteq \underline{apr}(B), \overline{apr}(A) \subseteq \overline{apr}(B)],$

(zo7) $\underline{apr}(A) \subseteq \overline{apr}(A).$

According to properties (zo4)-(zo6), relationships between subsets of U may not be carried over to U/E through the zooming-out operation. It may happen that $A \cap B \neq \emptyset$, but $\underline{apr}(A \cap B) = \emptyset$, or $A \cup B \neq U$, but $\overline{apr}(A \cup B) = U/E$. Similarly, we may have $A \neq B$, but $\underline{apr}(A) = \underline{apr}(B)$ and $\overline{apr}(A) = \overline{apr}(B)$. Nevertheless, we can draw the following inferences:

(i1) $\underline{apr}(A) \cap \underline{apr}(B) \neq \emptyset \Longrightarrow A \cap B \neq \emptyset,$

(i2) $\overline{apr}(A) \cap \overline{apr}(B) = \emptyset \Longrightarrow A \cap B = \emptyset,$

(i3) $\underline{apr}(A) \cup \underline{apr}(B) = U/E \Longrightarrow A \cup B = U,$

(i4) $\overline{apr}(A) \cup \overline{apr}(B) \neq U/E \Longrightarrow A \cup B \neq U.$

If $\underline{apr}(A) \cap \underline{apr}(B) \neq \emptyset$, by property (zo4) we know that $\underline{apr}(A \cap B) \neq \emptyset$. We say that A and B have a non-empty overlap, and hence are related, in U/E. By (i1), A and B must have a non-empty overlap, and hence are related, in U. Similar explanations can be associated with other inference rules.

The approximation of a set can be easily extended to the approximation of a partition, also called a classification [39]. Let $\pi = \{X_1, \ldots, X_n\}$ be a partition of the universe U. Its approximations are a pair of families of sets, the family of lower approximations $\underline{apr}(\pi) = \{\underline{apr}(X_1), \ldots, \underline{apr}(X_n)\}$ and the family of upper approximations $\overline{apr}(\pi) = \{\overline{apr}(X_1), \ldots, \overline{apr}(X_n)\}$.

4.6   Classical Rough Set Approximations by a Combination of Zooming-out and Zooming-in

Traditionally, lower and upper approximations of a set are also subsets of the same universe. One can easily obtain the classical definition by performing a combination of zooming-out and zooming-in operators as follows [66]:

$$\omega(\underline{apr}(A)) = \bigcup_{X_i \in \underline{apr}(A)} \omega(\{X_i\})$$

$$= \bigcup \{[x]_E \mid x \in U, [x]_E \subseteq A\},$$

$$\omega(\overline{apr}(A)) = \bigcup_{X_i \in \overline{apr}(A)} \omega(\{X_i\})$$

$$= \bigcup \{[x]_E \mid x \in U, [x]_E \cap A \neq \emptyset\}. \tag{11}$$

For a subset $X \subseteq U/E$ we can zoom-in and obtain a subset $\omega(X) \subseteq U$, and then zoom-out to obtain a pair of subsets $\underline{apr}(\omega(X))$ and $\overline{apr}(\omega(X))$. The compositions of zooming-in and zooming-out operations have the properties [46]: for $X \subseteq U/E$ and $A \subseteq U$,

$$\text{(zio1)} \quad \omega(\underline{apr}(A)) \subseteq A \subseteq \omega(\overline{apr}(A)),$$
$$\text{(zio2)} \quad \underline{apr}(\omega(X)) = \overline{apr}(\omega(X)) = X.$$

The composition of zooming-out and zooming-in cannot recover the original set $A \subseteq U$. The composition zooming-in and zooming-out produces the original set $X \subseteq U/E$. A connection between the zooming-in and zooming-out operations can be established. For a pair of subsets $X \subseteq U/E$ and $A \subseteq U$, we have [46]:

$$\text{(1)} \quad w(X) \subseteq A \Longleftrightarrow X \subseteq \underline{apr}(A),$$
$$\text{(2)} \quad A \subseteq \omega(X) \Longleftrightarrow \overline{apr}(A) \subseteq X.$$

Property (1) can be understood as follows. Any subset $X \subseteq U/E$, whose refinement is a subset of A, is a subset of the lower approximation of A. Only a subset of the lower approximation of A has a refinement that is a subset of A. It follows that $\underline{apr}(A)$ is the largest subset of U/E whose refinement is contained in A, and $\overline{apr}(A)$ is the smallest subset of U/E whose refinement containing A.

4.7 Consistent Computations in the Two Universes

Computation in the original universe is normally based on elements of U. When zooming-out to the coarse-grained universe U/E, we need to find the consistent computational methods. The zooming-in operator can be used for achieving this purpose.

Suppose $f : U \longrightarrow \Re$ is a real-valued function on U. One can lift the function f to U/E by performing set-based computations [67]. The lifted function f^+ is a set-valued function that maps an element of U/E to a subset of real numbers. More specifically, for an element $X_i \in U/E$, the value of function is given by:

$$f^+(X_i) = \{f(x) \mid x \in \omega(\{X_i\})\}. \tag{12}$$

The function f^+ can be changed into a single-valued function f_0^+ in a number of ways. For example, Zhang and Zhang [75] suggested the following methods:

$$f_0^+(X_i) = \min f^+(X_i) = \min\{f(x) \mid x \in \omega(\{X_i\})\},$$
$$f_0^+(X_i) = \max f^+(X_i) = \max\{f(x) \mid x \in \omega(\{X_i\})\},$$
$$f_0^+(X_i) = \operatorname{average} f^+(X_i) = \operatorname{average}\{f(x) \mid x \in \omega(\{X_i\})\}. \tag{13}$$

The minimum, maximum, and average definitions may be regarded as the most permissive, the most optimistic, and the balanced view in moving functions from U to U/E. More methods can be found in the book by Zhang and Zhang [75].

For a binary operation $\circ$ on U, a binary operation $\circ^+$ on U/E is defined by [6, 67]:

$$X_i \circ^+ X_j = \{x_i \circ x_j \mid x_i \in \omega(\{X_i\}), x_j \in \omega(\{X_j\})\}, \tag{14}$$

In general, one may lift any operation p on U to an operation p^+ on U/E, called the power operation of p. Suppose $p : U^n \longrightarrow U$ $(n \geq 1)$ is an n-ary operation on U. Its power operation $p^+ : (U/E)^n \longrightarrow 2^U$ is defined by [6]:

$$p^+(X_0, \ldots, X_{n-1}) = \{p(x_0, \ldots, x_{n-1}) \mid x_i \in \omega(\{X_i\}) \text{ for } i = 0, \ldots, n-1\}, \quad (15)$$

for any $X_0, \ldots, X_{n-1} \in U/E$. This provides a universal-algebraic construction approach. For any algebra $(U, p_1, \ldots, p_k)$ with base set U and operations $p_1, \ldots, p_k$, its quotient algebra is given by $(U/E, p_1^+, \ldots, p_k^+)$.

The power operation p^+ may carry some properties of p. For example, for a binary operation $p : U^2 \longrightarrow U$, if p is commutative and associative, p^+ is commutative and associative, respectively. If e is an identity for some operation p, the set $\{e\}$ is an identity for p^+. Many properties of p are not carried over by p^+. For instance, if a binary operation p is idempotent, i.e., $p(x, x) = x$, p^+ may not be idempotent. If a binary operation g is distributive over p, g^+ may not be distributive over p^+.

In some situations, we need to carry information from the quotient set U/E to U. This can be done through the zooming-out operators. A simple example is used to illustrate the basic idea.

Suppose $\mu : 2^{U/E} \longrightarrow [0,1]$ is a set function on U/E. If μ satisfies the conditions:

(i) $\mu(\emptyset) = 0$,

(ii) $\mu(U/E) = 1$,

(iii) $X \subseteq Y \Longrightarrow \mu(X) \leq \mu(Y)$,

μ is called a fuzzy measure [23]. Examples of fuzzy measures are probability functions, possibility and necessity functions, and belief and plausibility functions. Information about subsets in U can be obtained from μ on U/E and the zooming-out operation. For a subset $A \subseteq U$, we can define a pair of inner and outer fuzzy measures [68]:

$$\underline{\mu}(A) = \mu(\underline{apr}(A)),$$
$$\overline{\mu}(A) = \mu(\overline{apr}(A)). \quad (16)$$

They are fuzzy measures. If μ is a probability function, $\underline{\mu}$ and $\overline{\mu}$ are a pair of belief and plausibility functions [15, 49, 46, 68]. If μ is a belief function, $\underline{\mu}$ is a belief function, and if μ is a plausibility function, $\overline{\mu}$ is a plausibility [68].

5   Conclusion

Granular computing, as a way of thinking, has been explored in many fields. It captures and reflects our ability to perceive the world at different granularity and to change granularities in problem solving. In this chapter, the same approach is used to study the granular computing itself in two levels. In the first part of the chapter, we consider the fundamental issues of granular computing in general terms. The objective is to present a domain-independent way of thinking without details of any specific formulation. The second part of the chapter concretizes the high level investigations by considering a partition model of granular computing. To a large extent, the model is based on the theory of rough sets. However, results from other theories, such as the quotient space theory, belief functions, databases, and power algebras, are incorporated.

In the development of different research fields, each field may develop its theories and methodologies in isolation. However, one may find that these theories and methodologies share the same or similar underlying principles and only differ in their formulation. It is evident that granular computing may be a basic principle that guides many problem solving methods.

The results of rough set theory have drawn our attention to granular computing. On the other hand, the study of rough set theory in the wide context of granular computing may result in an in-depth understanding of rough set theory.

References

1. Aldenderfer, M.S., Blashfield, R.K.: Cluster Analysis. Sage Publications, The International Professional Publishers, London (1984)
2. Anderberg, M.R.: Cluster Analysis for Applications. Academic Press, New York (1973)
3. Bettini, C., Montanari, A. (Eds.): Spatial and Temporal Granularity: Papers from the AAAI Workshop. Technical Report WS-00-08. The AAAI Press, Menlo Park, CA. (2000)
4. Bettini, C., Montanari, A.: Research issues and trends in spatial and temporal granularities. Annals of Mathematics and Artificial Intelligence 36 (2002) 1-4
5. Bratko, I.: PROLOG: Programming for Artificial Intelligence, Second edition. Addison-Wesley, New York (1990)
6. Brink, C.: Power structures. Algebra Universalis 30 (1993) 177-216
7. Bryniarski, E.: A calculus of rough sets of the first order. Bulletin of the Polish Academy of Sciences, Mathematics 37 (1989) 71-77
8. de Loupy, C., Bellot, P., El-Bèze, M., Marteau, P.F.: Query expansion and classification of retrieved documents. Proceedings of the Seventh Text REtrieval Conference (TREC-7) (1998) 382-389
9. Demri, S, Orlowska, E.: Logical analysis of indiscernibility. In: Incomplete Information: Rough Set Analysis, Orlowska, E. (Ed.). Physica-Verlag, Heidelberg (1998) 347-380
10. Dubois, D., Prade, P.: Fuzzy rough sets and rough fuzzy sets. International Journal of General Systems 17 (1990) 191-209
11. Graziano, A.M., Raulin, M.L.: Research Methods: A Process of Inquiry, 4th edition. Allyn and Bacon, Boston (2000)
12. Euzenat, J.: Granularity in relational formalisms - with application to time and space representation. Computational Intelligence 17 (2001) 703-737
13. Friske, M.: Teaching proofs: a lesson from software engineering. American Mathematical Monthly 92 (1995) 142-144
14. Giunchglia, F., Walsh, T.: A theory of abstraction. Artificial Intelligence 56 (1992) 323-390

15. Grzymala-Busse, J.W.: Rough-set and Dempster-Shafer approaches to knowledge acquisition under uncertainty – a comparison. Manuscript. Department of Computer Science, University of Kansas (1987)
16. Han, J., Cai, Y., Cercone, N.: Data-driven discovery of quantitative rules in data bases. IEEE Transactions on Knowledge and Data Engineering **5** (1993) 29-40
17. Hearst, M.A., Pedersen, J.O.: Reexamining the cluster hypothesis: Scatter/Gather on retrieval results. Proceedings of SIGIR'96 (1996) 76-84
18. Hobbs, J.R.: Granularity. Proceedings of the Ninth Internation Joint Conference on Artificial Intelligence (1985) 432-435
19. Hornsby, K.: Temporal zooming. Transactions in GIS **5** (2001) 255-272
20. Imielinski, T.: Domain abstraction and limited reasoning. Proceedings of the 10th International Joint Conference on Artificial Intelligence (1987) 997-1003
21. Inuiguchi, M., Hirano, S., Tsumoto, S. (Eds.): Rough Set Theory and Granular Computing. Springer, Berlin (2003)
22. Jardine, N., Sibson, R.: Mathematical Taxonomy. Wiley, New York (1971)
23. Klir, G.J., Folger, T.A.: Fuzzy Sets, Uncertainty, and Information. Prentice Hall, Englewood Cliffs (1988)
24. Knoblock, C.A.: Generating Abstraction Hierarchies: an Automated Approach to Reducing Search in Planning. Kluwer Academic Publishers, Boston (1993)
25. Lamport, L.: How to write a proof. American Mathematical Monthly **102** (1995) 600-608
26. Ledgard, H.F., Gueras, J.F., Nagin, P.A.: PASCAL with Style: Programming Proverbs. Hayden Book Company, Inc., Rechelle Park, New Jersey (1979)
27. Lee, T.T.: An information-theoretic analysis of relational databases – part I: data dependencies and information metric. IEEE Transactions on Software Engineering **SE-13** (1987) 1049-1061
28. Leron, U.: Structuring mathematical proofs. American Mathematical Monthly **90** (1983) 174-185
29. Lin, T.Y.: Topological and fuzzy rough sets. In: Intelligent Decision Support: Handbook of Applications and Advances of the Rough Sets Theory, Slowinski, R. (Ed.). Kluwer Academic Publishers, Boston (1992) 287-304
30. Lin, T.Y.: Granular computing on binary relations I: data mining and neighborhood systems, II: rough set representations and belief functions. In: Rough Sets in Knowledge Discovery 1. Polkowski, L., Skowron, A. (Eds.). Physica-Verlag, Heidelberg (1998) 107-140
31. Lin, T.Y.: Generating concept hierarchies/networks: mining additional semantics in relational data. Advances in Knowledge Discovery and Data Mining, Proceedings of the 5th Pacific-Asia Conference, Lecture Notes on Artificial Intelligence 2035 (2001) 174-185
32. Lin, T.Y.: Granular computing. Rough Sets, Fuzzy Sets, Data Mining, and Granular Computing, Proceedings of the 9th International Conference, Lecture Notes in Artificial Intelligence 2639 (2003) 16-24.
33. Lin. T.Y., Yao, Y.Y., Zadeh, L.A. (Eds.): Rough Sets, Granular Computing and Data Mining. Physica-Verlag, Heidelberg (2002)
34. Lin, T.Y., Zhong, N., Dong, J., Ohsuga, S.: Frameworks for mining binary relations in data. Rough sets and Current Trends in Computing, Proceedings of the 1st International Conference, Lecture Notes in Artificial Intelligence 1424 (1998) 387-393
35. Mani, I.: A theory of granularity and its application to problems of polysemy and underspecification of meaning. Principles of Knowledge Representation and Reasoning, Proceedings of the Sixth International Conference (1998) 245-255

36. McCalla, G., Greer, J., Barrie, J., Pospisil, P.: Granularity hierarchies. Computers and Mathematics with Applications **23** (1992) 363-375
37. Orlowska, E.: Logic of indiscernibility relations. Bulletin of the Polish Academy of Sciences, Mathematics **33** (1985) 475-485
38. Pawlak, Z.: Rough sets. International Journal of Computer and Information Sciences **11** (1982) 341-356.
39. Pawlak, Z.: Rough Sets: Theoretical Aspects of Reasoning about Data. Kluwer Academic Publishers, Boston (1991)
40. Pawlak, Z.: Granularity of knowledge, indiscernibility and rough sets. Proceedings of 1998 IEEE International Conference on Fuzzy Systems (1998) 106-110
41. Pedrycz, W.: Granular Computing: An Emerging Paradigm. Springer-Verlag, Berlin (2001)
42. Peikoff, L.: Objectivism: the Philosophy of Ayn Rand. Dutton, New York (1991)
43. Polkowski, L., Skowron, A.: Towards adaptive calculus of granules. Proceedings of 1998 IEEE International Conference on Fuzzy Systems (1998) 111-116
44. Saitta, L., Zucker, J.-D.: Semantic abstraction for concept representation and learning. Proceedings of the Symposium on Abstraction, Reformulation and Approximation (1998) 103-120
    http://www.cs.vassar.edu/~ellman/sara98/papers/. retrieved on December 14, 2003.
45. Salton, G., McGill, M.: Introduction to Modern Information Retrieval. McGraw Hill, New York (1983)
46. Shafer, G.: A Mathematical Theory of Evidence. Princeton University Press, Princeton (1976)
47. Simpson, S.G.: What is foundations of mathematics? (1996).
    http://www.math.psu.edu/simpson/hierarchy.html. retrieved November 21, 2003.
48. Skowron, A.: Toward intelligent systems: calculi of information granules. Bulletin of International Rough Set Society **5** (2001) 9-30
49. Skowron, A., Grzymala-Busse, J.: From rough set theory to evidence theory. In: Advances in the Dempster-Shafer Theory of Evidence, Yager, R.R., Fedrizzi, M., Kacprzyk, J. (Eds.). Wiley, New York (1994) 193-236
50. Skowron, A., Stepaniuk, J.: Information granules and approximation spaces. Proceedings of 7th International Conference on Information Processing and Management of Uncertainty in Knowledge-Based Systems (1998) 354-361
51. Skowron, A., Stepaniuk, J.: Information granules: towards foundations of granular computing. International Journal of Intelligent Systems **16** (2001) 57-85
52. Smith, E.E.: Concepts and induction. In: Foundations of Cognitive Science, Posner, M.I. (Ed.). The MIT Press, Cambridge (1989) 501-526
53. Sowa, J.F.: Conceptual Structures, Information Processing in Mind and Machine. Addison-Wesley, Reading (1984)
54. Stell, J.G., Worboys, M.F.: Stratified map spaces: a formal basis for multi-resolution spatial databases. Proceedings of the 8th International Symposium on Spatial Data Handling (1998) 180-189
55. van Mechelen, I., Hampton, J., Michalski, R.S., Theuns, P. (Eds.): Categories and Concepts: Theoretical Views and Inductive Data Analysis. Academic Press, New York (1993)
56. van Rijsbergen, C.J.: Information Retrieval. Butterworths, London (1979)
57. Wille, R.: Concept lattices and conceptual knowledge systems. Computers Mathematics with Applications **23** (1992) 493-515

58. Yager, R.R., Filev,D.: Operations for granular computing: mixing words with numbers. Proceedings of 1998 IEEE International Conference on Fuzzy Systems (1998) 123-128
59. Yao, Y.Y.: Two views of the theory of rough sets in finite universes. International Journal of Approximation Reasoning **15** (1996) 291-317
60. Yao, Y.Y.: Granular computing: basic issues and possible solutions. Proceedings of the 5th Joint Conference on Information Sciences (2000) 186-189
61. Yao, Y.Y.: Information granulation and rough set approximation. International Journal of Intelligent Systems **16** (2001) 87-104
62. Yao, Y.Y.: Modeling data mining with granular computing. Proceedings of the 25th Annual International Computer Software and Applications Conference (COMPSAC 2001) (2001) 638-643
63. Yao, Y.Y.: A step towards the foundations of data mining. In: Data Mining and Knowledge Discovery: Theory, Tools, and Technology V, Dasarathy, B.V. (Ed.). The International Society for Optical Engineering (2003) 254-263
64. Yao, Y.Y.: Probabilistic approaches to rough sets. Expert Systems **20** (2003) 287-297
65. Yao, Y.Y., Liau, C.-J.: A generalized decision logic language for granular computing. Proceedings of FUZZ-IEEE'02 in the 2002 IEEE World Congress on Computational Intelligence, (2002) 1092-1097
66. Yao, Y.Y., Liau, C.-J., Zhong, N.: Granular computing based on rough sets, quotient space theory, and belief functions. Proceedings of ISMIS'03 (2003) 152-159
67. Yao, Y.Y., Noroozi, N.: A unified framework for set-based computations. Proceedings of the 3rd International Workshop on Rough Sets and Soft Computing. The Society for Computer Simulation (1995) 252-255
68. Yao, Y.Y., Wong, S.K.M.: Representation, propagation and combination of uncertain information. International Journal of General Systems **23** (1994) 59-83
69. Yao, Y.Y., Wong, S.K.M., Lin, T.Y.: A review of rough set models. In: Rough Sets and Data Mining: Analysis for Imprecise Data, Lin, T.Y., Cercone, N. (Eds.). Kluwer Academic Publishers, Boston (1997) 47-75
70. Yao, Y.Y., Zhong, N.: Granular computing using information tables. In: Data Mining, Rough Sets and Granular Computing, Lin, T.Y., Yao, Y.Y., Zadeh, L.A. (Eds.). Physica-Verlag, Heidelberg (2002) 102-124
71. Zadeh, L.A.: Fuzzy sets and information granularity. In: Advances in Fuzzy Set Theory and Applications, Gupta, N., Ragade, R., Yager, R. (Eds.). North-Holland, Amsterdam (1979) 3-18
72. Zadeh, L.A.: Fuzzy logic = computing with words. IEEE Transactions on Fuzzy Systems **4** (1996) 103-111
73. Zadeh, L.A.: Towards a theory of fuzzy information granulation and its centrality in human reasoning and fuzzy logic. Fuzzy Sets and Systems **19** (1997) 111-127
74. Zadeh, L.A.: Some reflections on soft computing, granular computing and their roles in the conception, design and utilization of information/intelligent systems. Soft Computing **2** (1998) 23-25
75. Zhang, B., Zhang, L.: Theory and Applications of Problem Solving, North-Holland, Amsterdam (1992)
76. Zhang, L., Zhang, B.: The quotient space theory of problem solving. Proceedings of International Conference on Rough Sets, Fuzzy Set, Data Mining and Granular Computing, Lecture Notes in Artrifical Intelligence 2639 (2003) 11-15
77. Zhong, N., Skowron, A., Ohsuga S. (Eds.): New Directions in Rough Sets, Data Mining, and Granular-Soft Computing. Springer-Verlag, Berlin (1999)

Musical Phrase Representation and Recognition by Means of Neural Networks and Rough Sets

Andrzej Czyzewski, Marek Szczerba, and Bozena Kostek

Multimedia Systems Department, Gdansk University of Technology
Narutowicza 11/12, 80-952 Gdansk, Poland
{andcz,marek,bozenka}@sound.eti.pg.gda.pl
http://sound.eti.pg.gda.pl

Abstract. This paper discusses various musical phrase representations that can be used to classify musical phrases with a considerable accuracy. Musical phrase analysis plays an important role in music information retrieval domain. In the paper various representations of a musical phrase are described and analyzed. Also the experiments were designed to facilitate pitch prediction within a musical phrase by means of entropy-coding of music. We used the concept of predictive data coding introduced by Shannon. Encoded music representations, stored in the database, are then used for automatic recognition of musical phrases by means of Neural Networks (NN) and rough sets (RS). A discussion on obtained results is carried out and conclusions are included.

1 Introduction

The ability to analyze musical phrases in the context of automatic retrieval is still not a fully achieved objective [11]. It should be however stated that such an objective depends both on a quality of a musical phrase representation and the inference engine utilized. Thorough analysis of musical phrase features would make possible searching for a particular melody in the musical databases, but also might reveal features that for example characterize music of the same epoch. Recognizing similarities between music of a particular epoch or particular genre would enable searching the Internet according to music taxonomy.

For the purpose of this study a collection of MIDI encoded musical phrases was gathered containing Bach's fugues. Musical phrase could be stored in various formats, such as mono- or polyphonic signal, MIDI code, and musical score. Any of such formats may be accompanied by textual information. In the study presented Bach's fugues from the *"Well Tempered Clavier"* were played on a MIDI keyboard and then transferred to the computer hard disk through the MIDI card and Cubase VST 3.5 program. The automatic recognition of musical phrase patterns required some preliminary stages, such as MIDI data conversion, parametrization of musical phrases, and discretization of parameter values in the case of rule-based decision systems [4], [23]. These tasks resulted in the creation of musical phrase database containing feature vectors.

J.F. Peters et al. (Eds.): Transactions on Rough Sets I, LNCS 3100, pp. 254–278, 2004.
© Springer-Verlag Berlin Heidelberg 2004

Experiments performed consisted in preparing various representations on the basis of gathered musical phrases and then analyzing them in the context of automatic music information retrieval. Both Neural Networks (NNs) and Rough Set (RS) method were used to this end. NNs were also used for feature quality evaluation, this issue will be explained later on. The decision systems were used both as a classifier and a comparator.

2 Musical Phrase Description

In the experiments it was assumed that musical phrases considered are single-voice, only. This means that at the moment t one musical event is occurring in the phrase. A musical event is defined as a single sound of defined pitch, amplitude and duration [26]. A musical pause – absence of sound – is a musical event as well. For practical reasons musical pause was assumed to be a musical event of pitch equal to the pitch of the preceding sound, but of amplitude equals zero. A single-voice musical phrase fr can be expressed as a sequence of musical events:

$$fr = \{e_1, e_2, ..., e_n\} \tag{1}$$

Musical event e_i can be described as a pair of values denoting sound pitch h_i (in the case of a pause, pitch of the previous sound), and sound duration t_i:

$$e_i = \{h_i, t_i\} \tag{2}$$

One can therefore express a musical phrase by a sequence of pitches being a function of time $fr(t)$. Sample illustration of the function $fr(t)$ is presented in Fig. 1. Sound pitch is defined according to the MIDI standard, i.e. as a difference from the C0 sound measured in semitones [2].

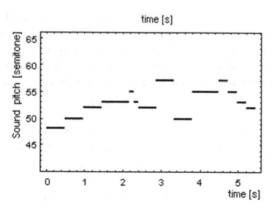

Fig. 1. Sequence of pitches as a function of time. Sound pitch is expressed according to the MIDI standard.

One of the basic composer's and performer's tools is transforming musical phrases according to rules specific to music perception, and aesthetic and cultural conventions and constraints [1]. Generally, listeners perceive a modified musical phrase as identical to the unmodified original phrase. Modifications of musical phrases involve sound pitch shifting (transposition), time changes (e.g. augmentation), changes of ornament and/or transposition, shifting pitches of individual sounds etc. [27]. A formal definition of such modifications may be presented on the example of a transposed musical phrase, expressed as follows:

$$fr_{mod}(t) = fr_{ref}(t) + c \tag{3}$$

where $fr_{ref}(t)$ denotes unmodified, original musical phrase, $fr_{mod}(t)$ modified musical phrase, and c is component expressing in semitones shift of individual sounds of the phrase (for $|c|=12n$ there is an octave shift).

A musical phrase with changed tempo can be expressed as follows:

$$fr_{mod}(t) = fr_{ref}(kt) \tag{4}$$

where k is tempo change factor.

Phrase tempo is slowed down for values of factor $k < 1$. Tempo increase is obtained for values of factor $k > 1$. A transposed musical phrase with changed tempo can be expressed as follows:

$$fr_{mod}(t) = fr_{ref}(kt) + c \tag{5}$$

Tempo variations in respects to the score can result mostly from inexactness in performance, which is often related to performer's expressiveness [7], [8], [22]. Tempo changes can be expressed as function $\Delta k(t)$. A musical phrase with varying tempo can be expressed as follows:

$$fr_{mod}(t) = fr_{ref}[t \cdot \Delta k(t)] \tag{6}$$

Modifications of musical phrase melodic content are often used. One can discern such modifications as: ornament, transposition, inversion, retrograde, scale change (major – minor), change of pitch of individual sounds (e.g. harmonic adjustment), etc. In general, they can be described by melodic modification function $\psi(t)$. Therefore, a musical phrase with melodic content modifications can be expressed as follows:

$$fr_{mod}(t) = fr_{ref}(t) + \psi(t) \tag{7}$$

In consequence, a musical phrase modified by transposition, tempo change, tempo fluctuation and melodic content modification can be expressed as follows:

$$fr_{mod}(t) = fr_{ref}[kt + t \cdot \Delta k(t)] + \psi[kt + t \cdot \Delta k(t)] + c \tag{8}$$

Above given formalism allows for defining the research problem of automatic classification of musical phrases. Let fr_{mod} be a modified musical phrase being classified and let FR be a set of unmodified reference phrases:

$$FR = \{fr_{1ref}, fr_{2ref}, ..., fr_{Nref}\} \tag{9}$$

The task of recognizing musical phrase fr_{mod} can therefore be described as finding in the set FR such a phrase fr_{nref}, for which the musical phrase modification formula is fulfilled.

If the applied modifications are limited to transposition and uniform tempo change, modification can be described using two constants: transposition constant c and tempo change constant k. In the discussed case the task of classifying a musical phrase is limited to determining such vales of constants c and k that the formula is fulfilled. If function $\Delta k(t) \neq 0$, then classification algorithm should minimize the influence of the function $\Delta k(t)$ on the expression. Small values of the function $\Delta k(t)$ indicate slight changes resulting from articulation inexactness and moderate performer's expressiveness [6]. Such changes can be easily corrected by using time quantization. Larger values of the function $\Delta k(t)$ indicate major temporal fluctuations resulting chiefly from performer's expressiveness. Such changes can be corrected using advanced methods of time quantization [8].

Function $\psi(t)$ describes a wide range of musical phrase modifications characteristic for composer, such as performer's style and technique. Values of function $\psi(t)$, which describe qualitatively the character of the above constants, are difficult or impossible to determine in a hard-defined manner. The last mentioned problem is the main issue associated with the task of automatic classification of musical phrases.

3 Parametrization of Musical Phrases

A fundamental quality of decision systems is the ability to classify data that is not precisely defined or cannot be modeled mathematically. This quality allows for using intelligent decision algorithms for automatically classifying musical phrases in conditions when the character of the $\psi(t)$ and $\Delta k(t)$ functions is rather qualitative. Parametrization can be considered as a part of feature selection, the latter process meaning finding a subset of features, from the original set of pattern features, optimally according to the defined criterion [25].

The data to be classified can be represented by a vector $\mathbf{P}$ of the form:

$$\mathbf{P} = [p_1, p_2, ..., p_N] \tag{10}$$

The constant number N of elements of vector $\mathbf{P}$ requires the musical phrase fr to be represented by N parameters, independent of number of notes in phrase fr. Converting a musical phrase fr of the form of $\{e_1, e_2, ..., e_n\}$ into N-element vector of parameters allows for representing only the distinctive features of musical phrase fr. As shown above, transposition of a musical phrase and uniform proportional tempo change can be represented as alteration of values: c and k. It would therefore be advantageous to design such method of musical phrase parameterization, for which:

$$\mathbf{P}(fr_{mod}) = \mathbf{P}(fr_{ref}) \tag{11}$$

where:

$$fr_{mod}(t) = fr_{ref}(kt) + c \tag{12}$$

Creating numerical representation of musical structures to be used in automatic classification and prediction systems requires among others defining the following characteristics of musical phrases: sequence length, method of representing sound pitch, methods of representing time-scale and frequency properties, methods of representing other musical properties by feature vectors. In addition, having defined various subsets of features, feature selection should be performed. Typically, this process consists in finding an optimal feature subset from a whole original feature set, which guarantees the accomplishment of a processing goal while minimizing a defined feature selection criterion [25]. Feature relevance may be evaluated on the basis of open-loop or closed-loop methods. In the first approach separability criteria are used. To this end Fisher criterion is often employed. The closed-loop methods are based on feature selection using a predictor performance. This means that the feedback from the predictor quality is used for the feature selection process [25]. On the other hand, here we deal with the situation, in which feature set contains several disjoint feature subsets. The feature selection defined for the purpose of this study consists in eliminating the less effective method of parametrization according to the processing goal, first, and then reducing number of parameters to the optimal one. Both, open- and closed-loop methods were used in the study performed.

Individual musical structures may show significant differences in number of elements, i.e. sounds or other musical units. In an extreme case one can imagine that the classifier can be fed with the melody or the whole musical piece. It is therefore necessary to limit the number of elements in the numerical representation vector.

Sound pitch can be expressed as absolute or relative value. An <u>absolute representation</u> is characterized by the exact definition of the reference sound (e.g. C1 sound). In the case of absolute representation the number of possible values defining a given note in a sequence is equal to the number of possible sound pitch values restricted to musical scale. A disadvantage of this representation is the fact of shifting the values for sequence elements by a constant in the case of transposition. In the case of <u>relative representation</u> the reference point is being updated all the time. The reference point may be e.g. the previous sound, a sound at the previously accented time part or a sound at the beginning of the onset. The number of possible values defining a given note in a sequence is equal to the number of possible intervals. An advantage of relative representation is absence of change of musical structures caused by transposition as well as the ability to limit the scope of available intervals without limiting the available musical scales. Its disadvantage is sensitivity to small structure modifications resulting in shifting the reference sound.

3.1 Parametric Representation

Research performed so far resulted in designing a number of parametric representations of musical phrases. Some of these methods were described in detail in earlier authors' publications [13], [14], [15], [16], [17], [24], therefore only their brief characteristics are given below. At the earlier stage of this study, both Fisher criterion and correlation coefficient for evaluation of parameter quality were used [17].

Statistical Parametrization. The designed statistical parameterization approach is aimed at describing structural features of a musical phrase based on music theory [27]. Statistical parametrization introduced by authors involves representing a musical phrase with five parameters [13], [15]:

- P_1 – difference between weighted average sound pitch and pitch of the lowest sound of phrase, where T is phrase duration, h_n denotes pitch of n-th sound, t_n is duration of n-th sound, and N is number of sounds in phrase.

$$P_1 = \left[\frac{1}{T} \sum_{n=1}^{N} h_n t_n \right] - \min_n (h_n) \qquad (13)$$

- P_2 – ambitus – difference between pitches of the highest and the lowest sounds of phrase. Typically, the term ambitus denotes a range of pitches for a given voice in a part of music. It also may denote the pitch range that a musical instrument is capable of playing, however, in our experiments, the first meaning is closer to the definition given below:

$$P_2 = \max_n (h_n) - \min_n (h_n) \qquad (14)$$

- P_3 – average absolute difference of pitch of subsequent sounds:

$$P_3 = \frac{1}{N-1} \sum_{n=1}^{N-1} |h_n - h_{n+1}| \qquad (15)$$

- P_4 – duration of the longest sound of phrase:

$$P_4 = \max_n (t_n) \qquad (16)$$

- P_5 – average sound duration:

$$P_5 = \frac{1}{N} \sum_{n=1}^{N} t_n \qquad (17)$$

Statistical parameters representing a musical phrase can be divided into two groups: parameters describing melodic features of musical phrase (P_1, P_2, P_3) and ones describing rhythmical features of musical phrase (P_4, P_5).

Trigonometric Parametrization. Trigonometric parametrization involves representing the shape of a musical phrase with a vector of parameters $\mathbf{P} = [p_1, p_2, ..., p_M]$ in the form of a series of cosines [15]:

$$fr^*(t) = p_1 \cos\left[\left(t - \frac{1}{2}\right)\frac{\pi}{T}\right] + p_2 \cos\left[2\left(t - \frac{1}{2}\right)\frac{\pi}{T}\right] + ... + p_M \cos\left[M\left(t - \frac{1}{2}\right)\frac{\pi}{T}\right] \qquad (18)$$

where M is the number of trigonometric parameters representing the musical phrase.

For discrete time domain it is assumed that the sampling period is a common denominator of durations of all rhythmic units of a musical phrase. Elements p_m of the trigonometric parameter vector **P** are calculated according to the following formula:

$$p_m = \sum_{k=1}^{l} h_k \cos[m(k - \frac{1}{2})\frac{\pi}{l}] \tag{19}$$

where p_m is m-th element of the feature vector, $l = \dfrac{T}{snt}$ denotes phrase length expressed as a multiple of sampling period, snt is the shortest note duration, and h_k denotes pitch of sound in k-th sample.

According to the above assumption each rhythmic value being a multiple of the sampling period is transformed into a series of rhythmic values equal to the sampling period. This leads to loss of information on rhythmic structure of the phrase. Absolute changes of values concerning sound pitch and proportional time changes do not affect the values of trigonometric parameters.

Trigonometric parameters allow for reconstructing the shape of the musical phrase they represent. Phrase shape is reconstructed using vector $\mathbf{K}=[k_1, k_2, ..., k_N]$. Elements of vector **K** are calculated according to the following formula:

$$k_n = \frac{1}{N} \sum_{m=1}^{M} 2 p_m \cos\left(\frac{mn\pi}{N}\right) \tag{20}$$

where M is the number of trigonometric parameters representing the musical phrase, and p_m denotes m-th element of parameters vector.

Values of elements k_n express in semitones the difference between the current and the average sound pitch in the musical phrase being reconstructed.

Polynomial Parametrization. Single-voice musical phrase fr can be represented by function $fr(t)$, whose time domain is either discrete or continuous. In discrete time domain musical phrase fr can be represented as a set of points in two-dimensional space of time and sound pitch. A musical phrase can be represented in discrete time domain by means of points denoting sound pitch at time t, and by points denoting note onset. If tempo varies in time (function $\Delta k(t) \neq 0$) or musical phrase includes additional sounds of duration inconsistent with the general rhythmic pattern (e.g. ornament or augmentation), sampling period can be determined by minimizing the quantization error defined by the formula:

$$\varepsilon(b) = \frac{1}{N-1} \sum_{i=1}^{N-1} \left| \frac{t_i - t_{i-1}}{b} - Round\left(\frac{t_i - t_{i-1}}{b}\right) \right| \tag{21}$$

where b denotes sampling period, and $Round$ is rounding function.

On the basis of representation of a musical phrase in discrete time domain one can approximate a musical phrase by a polynomial of order M:

$$fr^*(t) = a_0 + a_1 t + a_2 t^2 + ... a_M t^M \qquad (22)$$

Coefficients a_0, a_1,...a_M are found numerically by means of mean-square approximation, i.e. by minimizing the error ε of form:

$$\varepsilon^2 = \int_0^T \left| fr^*(t) - fr(t) \right|^2 dt \quad \text{- for continuous case}$$

$$(23)$$

$$\varepsilon^2 = \sum_{i=0}^N \left| fr_i^* - fr_i \right|^2 \quad \text{- for discrete case}$$

One can also express the error in semitones per sample, which facilitates approximation evaluation, according to the formula:

$$\chi = \frac{1}{N} \sum_{i=1}^N \left| fr_i^* - fr_i \right| \qquad (24)$$

3.2 Binary Representation

Binary representation is based on dividing the time window W into n equal time sections T, where n is consistent with metric division and T corresponds to the smallest, basic rhythmic unit in the music material being represented. Each time section T is assigned a bit of information b_T in the vector of rhythmic units. Bit b_T takes the value of 1, if a sound begins in the given time section T. If time section T covers a sound started in a previous section or a pause, the rhythmic information bit b_T assumes the value of 0.

An advantage of binary representation of rhythmic structures is fixed length of the sequence representation vector. On the other hand, the disadvantages are: large size of vector length in comparison to other representation methods and the possibility of errors resulting from time quantization.

On the basis of methods of representing values of individual musical parameters one can distinguish three types of representations: local, distributed and global ones. In the case of a local representation every musical unit e_n is represented by a vector of n bits, where n is the number of all possible values of a musical unit e_n. Current value of musical unit e_n is represented by ascribing the value of 1 to the bit of representation vector corresponding to this value. Other bits of the representation vector take the value of 0 (unipolar activation) or -1 (bipolar activation). This type of representation was used e.g. by Hörnel [10] and Todd [28].

The system of representing musical sounds proposed by Hörnel and co-workers is an example of parametric representation [9]. In this system each subsequent note p is represented by the following parameters: consonance of note p with respect to its harmony, relation of note p towards its successor and predecessor in the case of dissonance against the harmonic content, direction of p (up, down to next pitch), dis-

Table 1. Distributed representation of sound pitches according to Mozer.

Sound pitch	Mozer's distributed representation					
C	−1	−1	−1	−1	−1	−1
$C^{\#}$	−1	−1	−1	−1	−1	+1
D	−1	−1	−1	−1	+1	+1
$D^{\#}$	−1	−1	−1	+1	+1	+1
E	−1	−1	+1	+1	+1	+1
F	−1	+1	+1	+1	+1	+1
$F^{\#}$	+1	+1	+1	+1	+1	+1
G	+1	+1	+1	+1	+1	−1
$G^{\#}$	+1	+1	+1	+1	−1	−1
A	+1	+1	+1	−1	−1	−1
$A^{\#}$	+1	+1	−1	−1	−1	−1
B	+1	−1	−1	−1	−1	−1

tance of note p to base note (if p is consonant), octave, *tenuto* – if p is an extension of the previous note of the same pitch.

The presented method of coding does not employ direct representation of sound pitch, it is distributed with respect to pitch. Sound pitch is coded as a function of harmony. Such distributed representation was used among others by Mozer [19].

In the case of a <u>distributed representation</u> the value of musical unit E is encoded with m bits according to the formula:

$$m = \lfloor \log_2 N \rfloor \qquad (24)$$

where N – number of possible values of musical unit e_n.

An example of representing the sounds of the chromatic scale using a distributed representation is presented in Table 1. In the case of a <u>global representation</u> the value of a musical unit is represented by a real value.

The above methods of representing values of individual musical units imply their suitability for processing certain types of music material, for certain tasks and analysis tools, classifiers and predictors.

3.3 Prediction of Musical Events

Our experiments were aimed at designing a method of predicting and entropy-coding of music. We used the concept of predictive data coding presented by Shannon and later employed for investigating entropy coding of English text by Moradi, Grzymala–Busse and Roberts [18]. The engineered method was used as a musical event predictor in order to enhance a system of pitch detection of a musical sound. The block scheme of a prediction coding system for music is presented in Fig. 2.

The idea of entropy coding involves using two identical predictors in the modules of data coding and decoding. The process of coding consists in determining the number of prediction attempts k required for correct prediction of event e_{n+1}. Prediction is based on parameters of musical events collected in data buffer. The number of prediction attempts k is sent to the decoder. The decoder module determines the value of

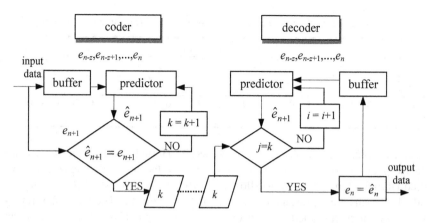

Fig. 2. Block diagram of prediction coder and decoder.

event e_{n+1} by repeating k prediction attempts. Subsequent values for samples – musical events – are then collected in a buffer.

Two types of data buffers were implemented: fixed-size buffer, and fading memory model. In the first case the buffer stores data on z musical events; each event is represented by a separate vector. That means that z vectors representing z individual musical events are supplied to the predictor input. In the carried out experiments the value z was limited to 5, 10 and 20 samples (music events). On the other hand, the fading memory model involves storing preceding values of the vector elements and summing them with current ones according to the formula:

$$b_n = \sum_{k=1}^{n} e_k r^{n-k} \tag{25}$$

where r is the fading factor from the range $(0,1)$.

In the case of using the fading memory model a single vector of parameters of musical events is supplied to the predictor input. This means a z-fold reduction of the number of input parameters compared with the buffer of size z.

For the needs of investigating the music predictor a set of musical data, consisted of fugues from the set *Well-Tempered Clavier* by J. S. Bach, were used as musical material. In the experiments performed a neural network-based predictor was employed. A series of experiments aimed at optimizing the predictor structure, data buffer parameters and prediction algorithm parameters were performed. In the training process we utilized all voices of the individual fugues except the uppermost ones. The highest voices were used for testing the predictor. Three methods of parametric representation of sound pitch: binary method, a so-called modified Hörnel's representation and modified Mozer's representation were utilized. In all cases relative representation was used, i.e. differences between pitch of subsequent sounds were coded.

In the case of binary representation individual musical intervals (differences between pitch of subsequent sounds) are represented as 27-bit vectors. The utilized representation of sound pitch is presented in Table. 2.

Table 2. Illustration of binary representation of musical interval (example – 2 semitones up).

-octave												Interval [in semitones]													+octave
	-12	-11	-10	-9	-8	-7	-6	-5	-4	-3	-2	-1	0	+1	+2	+3	+4	+5	+6	+7	+8	+9	+10	+11	+12	
0	0	0	0	0	0	0	0	0	0	0	0	0	0	0	1	0	0	0	0	0	0	0	0	0	0	

Presented representation of intervals designed by Hörnel is a diatonic representation (corresponding to seven-step musical scale). For the needs of our research we modified Hörnel's representation to allow for chromatic (twelve-step) representation. Individual intervals are represented by means of 11 parameters. The method of representing sound pitch designed by Mozer characterizes pitch as an absolute value. Within the scope of our research we modified Mozer's representation to allow relative representation of interval size. The representation was complemented by adding direction and octave bits. An individual musical event is therefore coded by means of 8 parameters.

A relative binary representation was designed for coding rhythmic values. Rhythmic values are coded by a parameters vector:

$$\mathbf{p}^r = \left\{ p_1^r, p_2^r, p_3^r, p_4^r, p_5^r \right\} \tag{26}$$

where individual parameters assume the values:

$$p_1^r = \frac{\left| \dfrac{e_{n-1}^r}{e_n^r} - 2 \right| + \dfrac{e_{n-1}^r}{e_n^r} - 2}{12} \tag{27}$$

$$p_2^r = \begin{cases} \dfrac{\left| 8 - \dfrac{e_{n-1}^r}{e_n^r} \right| + 8 - \dfrac{e_{n-1}^r}{e_n^r}}{12} & \text{for } \dfrac{e_{n-1}^r}{e_n^r} \geq 2 \\[4mm] \dfrac{\left| \dfrac{e_{n-1}^r}{e_n^r} - 1 \right| + \dfrac{e_{n-1}^r}{e_n^r} - 1}{2} & \text{for } \dfrac{e_{n-1}^r}{e_n^r} < 2 \end{cases} \tag{28}$$

$$p_3^r = \begin{cases} \dfrac{\left| 2 - \dfrac{e_n^r}{e_{n-1}^r} \right| + \left(2 - \dfrac{e_n^r}{e_{n-1}^r} \right)}{2} & \text{for } \dfrac{e_n^r}{e_{n-1}^r} \geq 1 \\[4mm] \dfrac{\left| 2 - \dfrac{e_{n-1}^r}{e_n^r} \right| + \left(2 - \dfrac{e_{n-1}^r}{e_n^r} \right)}{2} & \text{for } \dfrac{e_n^r}{e_{n-1}^r} < 1 \end{cases} \tag{29}$$

$$p_4^r = \begin{cases} \dfrac{\left|8 - \dfrac{e_n^r}{e_{n-1}^r}\right| + 8 - \dfrac{e_n^r}{e_{n-1}^r}}{12} & \text{for} \quad \dfrac{e_n^r}{e_{n-1}^r} \geq 2 \\[4mm] \dfrac{\left|\dfrac{e_n^r}{e_{n-1}^r} - 1\right| + \dfrac{e_n^r}{e_{n-1}^r} - 1}{2} & \text{for} \quad \dfrac{e_n^r}{e_{n-1}^r} < 2 \end{cases} \tag{30}$$

$$p_5^r = \dfrac{\left|\dfrac{e_n^r}{e_{n-1}^r} - 2\right| + \dfrac{e_n^r}{e_{n-1}^r} - 2}{12} \tag{31}$$

where e_n^r denotes rhythmic value (duration) of musical event e_n.

Values of parameters of rhythmic representation $\mathbf{p}^r$ using the $\dfrac{e_n^r}{e_{n-1}^r}$ ratio are presented in Table 3.

Table 3. Relative representation of rhythmic values.

$\dfrac{e_n^r}{e_{n-1}^r}$	p_1^r	p_2^r	p_3^r	p_4^r	p_5^r
$\frac{1}{3}$	1	0	0	0	0
$\frac{1}{4}$	0.5	0.5	0	0	0
$\frac{1}{2}$	0	1	0	0	0
1	0	0	1	0	0
2	0	0	0	1	0
4	0	0	0	0.5	0.5
8	0	0	0	0	1

Above methods of representing sound pitch and rhythmic values were used to prepare a set of data for investigating the neural musical predictor. Specifications of data sets are presented in Tab. 4. A buffer of fixed-size of 5, 10 and 20 samples, respectively, and the values of parameter r for the fading memory model from the set $r=\{0.2; 0.5; 0.8\}$ were used in experiments.

4 Neural Musical Predictor

The neural musical predictor was implemented using the *Stuttgart Neural Network Simulator* (SNNS) integrated system for emulating NN [29]. Experiments were performed for data sets presented in Tab. 4 and for different data buffer size.

Table 4. Specification of musical data used for investigating the neural musical predictor.

Database indicator	pitch representation		time representation		total number of parameters /sample
	representation	parameters/sample	representation	parameters/sample	
bin	binary	27	NO	0	27
mhor	modified Hörnel	11	NO	0	11
mmoz	modified Mozer	8	NO	0	8
bin_rel	binary - relative	27	relative	5	32
mhor_rel	modified Hörnel - relative	11	relative	5	16
mhor_rel	modified Mozer - relative	8	relative	5	13

Experiments were divided into two stages:

- investigating the predictor for individual methods of representing data and buffer parameters for data limited to 6 fugues chosen randomly (No. 1, 5, 6, 8, 15 and 17); the number of training elements, depending on buffer size, ranged from 5038 to 5318 samples, while the number of test elements ranged from 2105 to 2225 samples, respectively.
- investigating chosen representation methods and buffer parameters for all 48 fugues from the *"Well-Tempered Clavier"* collection.

Description of the developed predictor structure and its training process as well as the obtained results of musical data prediction are presented further on.

4.1 Structures and Training

A *feed-forward* neural network model with a single hidden layer was used in the experiments. In the cases of binary representation and modified Hörnel's representation we used a unipolar, sigmoidal shaping function, while in the case of modified Mozer's representation we used hiperbolic tangent bipolar function. The choice of the NN activation function was determined on the basis of pilot tests performed before the main experiments started. In the first stage of our research the number of neurons in the hidden layer was arbitrarily limited to the set {20, 50, 100}. We conducted a series of training the neural predictor for individual methods of representing musical events and data buffer parameters. Due to practical considerations the number of iterations was arbitrarily limited to 1000. We applied the error back-propagation algorithm augmented by the *momentum* method. On the basis of pilot tests we assumed constant values of training process coefficient $\eta=0.5$ and the *momentum* coefficient $\alpha=0.2$.

In general, in the cases of binary representation and modified Hörnel's representation we reached the mean-square error (MSE) value as early as after ca 100 iterations. Conversely, in the case of modified Mozer's representation (without rhythm representation) the training process did not lead to obtaining the mean-square error value lower than 1.

In order to evaluate the functioning of the neural musical predictor the following parameters were considered:

- measure of first-attempt prediction correctness:

$$pc = \frac{1}{N} \sum_{n=1}^{N} id(e_{n+1}, \hat{e}_{n+1}) \tag{32}$$

where e_{n+1} denotes actual value of event $n+1$, and $\hat{e}_{n+1}$ is the estimated value of event $n+1$,

$$id(e_{n+1}, \hat{e}_{n+1}) = \begin{cases} 1 & dla \quad \hat{e}_{n+1} = e_{n+1} \\ 0 & dla \quad \hat{e}_{n+1} \neq e_{n+1} \end{cases} \tag{33}$$

- average number of prediction attempts required for correct estimation of the value of event $n+1$,
- lower and upper bounds of entropy of musical data F_N according to:

$$\sum_{i=1}^{M} i(q_i^N - q_{i+1}^N) \log_2 i \leq F_N \leq \sum_{i=1}^{M} -q_i^N \log_2 iq_i^N \tag{34}$$

where q_i^N denotes frequency of correct estimations of the value of event e_{n+1} for i-th prediction attempt, and M is number of possible values of event e_n.

Subsequent prediction attempts and identification of event identity are emulated by ordering the list of possible values of event e_{n+1} by distance:

$$o_m = \sum_{i=1}^{N} |e_{n+1,i} - e_{m,i}| \tag{35}$$

where $m = \{1, 2, ..., M\}$, i is representation parameter index, N denotes number of representation parameters, and e_m is predicted value of event e_{n+1},

On the basis of results obtained in the first stage of our investigation the following conclusions concerning operation of the developed predictor can be presented:

- we obtained very high effectiveness of musical data prediction (above 97%) in the cases of using binary representation and modified Hörnel's representation together with fixed-size buffer, irrespectively of representation of rhythmic data,
- application of the fading memory model leads to degrading prediction effects (max. ca 75%) with simultaneous reduction of data and computational complexity of the training process,
- application of modified Mozer's method of representation results in low prediction effectiveness,
- the developed method of rhythm coding shows high effectiveness of representation and allows for obtaining high correctness of rhythm prediction,
- utilization of rhythmic data allows for improving effectiveness of prediction of musical data.

4.2 Tests Employing Whole Collection

In the next stage, we performed tests of the music predictor for the whole range of music sets, i.e. for 48 fugues from the *Well-Tempered Clavier* collection. Neural networks trained using six randomly-chosen fugues were employed. Neural network parameters and obtained results are presented in Tab. 5 (number of neurons in the hidden layer was equal to 50).

Obtained results lead to the conclusion that the developed system allows for effective prediction of musical data. The highest prediction effectiveness was obtained for binary representation of sound pitch and modified Hörnel's representation. Employing the fixed-size buffer allows for more effective prediction compared to the fading memory model. The last conclusion may be surprising: musical data in J. S. Bach's fugues possess low entropy.

Table 5. Predictor parameters and prediction results for the whole collection (denotations as previously introduced, r denotes fading coefficient).

Database indicator	Buffer type	Buffer size	No. of input parameters	No. of output parameters	Prediction effectiveness			Average number of predictions			Entropy upper bound		
					melody	rhythm	total	melody	rhythm	total	melody	rhythm	total
bin	fixed-size	10	270	27	0.54	-	0.54	4.12	-	4.12	1.62	-	1.62
bin_rel	fixed-size	10	320	32	0.6	0.91	0.58	5.21	1.15	8.31	1.44	0.34	1.4
bin	fixed-size	1	27	27	0.32	-	0.32	8.24	-	8.24	1.92	-	1.92
bin_rel	(r=0.8)	1	32	32	0.35	0.84	0.31	6.94	1.42	12.4	1.83	0.65	1.78
mhor	fixed-size	10	110	11	0.53	-	0.53	5.27	-	5.27	1.62	-	1.62
mhor_rel	fixed-size	10	160	16	0.57	0.85	0.52	4.76	1.3	9.14	1.49	0.63	1.28
mhor	(r=0.8)	1	11	11	0.28	-	0.28	5.62	-	5.62	1.62	-	1.62
mhor_rel	(r=0.8)	1	16	16	0.33	0.82	0.3	7.12	1.42	16.4	1.82	0.67	1.51
mmoz	fixed-size	10	80	8	0.12	-	0.12	8.56	-	8.56	1.73	-	1.73
mmoz_rel	fixed-size	10	130	13	0.24	0.72	0.19	6.12	1.85	18.4	1.62	0.72	1.51
mmoz	(r=0.8)	1	8	8	0.07	-	0.07	11.2	-	11.2	1.82	-	1.82
mmoz_rel	(r=0.8)	1	13	13	0.14	0.61	0.11	13.8	3.18	24.6	1.91	0.91	1.46

5 Classification of Musical Phrases

Within the scope of our experiments we developed two methods of classifying musical phrases using NN and rough sets:

- method of classifying phrases on the basis of sequences of musical data,
- method of classifying phrases on the basis of parameters of musical data.

Musical phrases can be classified on the basis of sequences of musical data. Similarly to prediction of musical events, classification is based on parameters collected in data buffer. In order to unify the size of input data we used the fading memory approach. Experiments were performed on the basis of binary representation of melody and of modified Hörnel's representation. We also performed tests using binary rhythm representation. Repetitions of musical phrases in data stream were identified in time windows. Window size can be adapted dynamically along a musical piece using the histogram of rhythmic values.

Musical phrases can also be classified on the basis of parameters of musical phrases. Our research utilized three methods of parametric representation: statistical, trigonometric and polynomial. Similar to the case of analysis of musical data sequences, repetitions of musical phrases were identified in time windows.

Two types of NNs were developed for classifying musical phrases:

- neural classifier (NCL), assigning phrases to objects from the reference phase set,
- neural comparator (NCP), analyzing similarities between musical phrases.

Both classifier types are illustrated in Fig. 3.

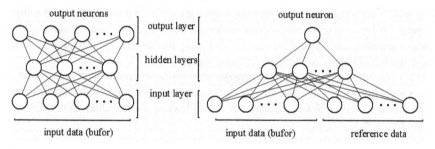

Fig. 3. Neural classifier and comparator of musical phrases.

In the case of NC of musical phrases classification involves determining phrase identity on the basis of knowledge modeled using the neural network. Individual neurons of the output layer correspond to individual classes of musical phrases. Identity of input phrase is determined on the basis of output signals of neurons of the output layer. In the case of NCP of musical phrases classification involves determining the identity relation between the test phrase and a reference phrase. In this case the neural network models the knowledge of relations between musical phrase modification and preservation of phrase identity. Its application is therefore limited by the scope of phrase modifications used for network training.

Our studies involved also classifying musical phrases using the rough set-based approach [21]. The rough set approach was also used for classifying musical phrases on the basis of musical data sequences and of vector features. In addition, the rough set-based system was used both as a classifier, and a comparator. In our investigation we utilized the ROSETTA system developed by teams of Warsaw University and the University of Trondheim [12], [20].

5.1 Musical Databases

In order to investigate classification methods sets of musical data were prepared. Fugues from the set *Well-Tempered Clavier* by J. S. Bach were used as musical material for our research. Musical data were grouped in two databases:

- TOFFEE (*Themes of Fugues from Well-Tempered Clavier*) musical database containing fugue themes [14],
- DWK (*Well-Tempered Clavier*) musical database containing full musical material.

The DWK database was based on MIDI files prepared by M. Reyto and available from *Classical MIDI Archives* [5]. In order to allow for analyzing classification algorithms in the DWK database, the fugue theme repetitions were marked.

The TOFFEE musical database was created in our Department on the basis of fugue themes [16], [17], [24]. The used musical phrase modifications are consistent with music stylistics and performance technique [3]. Each of the 48 themes was included in the TOFFEE database in the following forms: reference form, form with a performance error, transposed form with a performance error and an ornament, transposed form with a performance error, form with an additional note at the beginning of the phrase, form with the omission of the first note of the phrase + an ornament, form with an ornament, transposed forms (5 types), augmented form (individual rhythmic values elongated twice), and transposed augmented form.

Table 6 summarizes methods of representing musical data as well as methods of classifying phrases. Used classification methods were denoted with corresponding symbols: NN – Neural Networks, and RS – rough set approach. Other denotations are indicated in Table.

Table 6. Methods of representing musical phrases for individual classification methods.

Data representation		No. of parameters	Classification algorithm
melodic	rhythmic		
binary relative (bin_rel)	-	27	NN
Hörnel relative, (mhorn_rel)	-	11	NN
binary relative, (bin_rel)	relative, binary	32	NN
Hörnel relative, (mhorn_rel)	relative, binary	16	NN
(rel) relative	-	1	RS
(rel) relative	relative	2	RS
statistical parameters (stat)		5	NN/RS
trigonometric parameters (trig)		5, 10, 15, 20	NN/RS
polynomial parameters (poly)		5, 10, 15	NN/RS

In the case of non-parametric representations the classification of musical phrases using NNs and the rough set approach is based on parameters collected in data buffer. The fading memory model was utilized. The classifier structure and setting along with obtained results will be presented later on in subsequent paragraphs.

5.2   Classification on the Basis of NN and Sequences of Musical Events

Neural Classifier. Research utilizing the neural classifier was divided into two stages:

- investigating classification in the case of training with reference forms only,
- *leave-one-out* tests for individual musical phrase modifications.

Melodic data were represented using binary representation, modified Hörnel's method and modified Mozer's method. Tests were run for collections both with and without representation of rhythmical data. Rhythm was represented using the relative representation approach. During the training process the value of buffer fading coefficient was observed to influence the course of training. For the greatest value of fading coefficient (0.8) the mean-square error reached the value of 0.1 faster in the case of sets with rhythm representation (ca 300 iterations) than in the case of sets without rhythm representation (ca 700 iterations). In the case of fading coefficient of 0.2 the mean-square error of 0.1 was reached in the case of sets with rhythm representation after ca 1700 iterations, while this value could not be achieved in the case of sets without rhythm representation. In general, higher effectiveness was achieved after taking into account rhythm representation; in the cases of binary representation and modified Hörnel's representation values of over 90% were achieved. Maximum classification effectiveness (92.63%) was achieved for binary representation of melodic data with relative rhythm representation, for neural network containing 20 neurons in the hidden layer. On the basis of this stage of experiments it may be said that NN-based classifier was also used for feature parameterization method selection, and eliminated for example Mozer's approach to musical phrase description.

Leave-one-out tests were performed for modified Hörnel's method of representing melodic data. Classification effectiveness as function of rhythmic data representation was analyzed. For practical reasons the process of network training was limited to 1000 iterations. Tab. 7 presents data on structures of employed neural networks as well as on the obtained classification results. Maximum classification effectiveness (91.67%) was achieved for data including rhythmic information, buffer fading coefficient $r=0.5$ and for neural network containing 50 neurons in the hidden layer (see Tab. 7). The obtained results confirm the usefulness of representing rhythmic data for correct classification of musical phrases.

Neural Comparator. Using the neural comparator we investigated the classification of musical phrases on the basis of representation of sequences of musical data, Set of 12 randomly-chosen fugue themes from the TOFFEE database (No. 2, 4, 11, 14, 16, 19, 21, 25, 26, 32, 34, 46) was used for training. The other fugue themes were used for testing the system. The number of iteration during the training process was limited to 2000. Musical data were represented using modified Hörnel's method and relative rhythm representation. For each neural network contained in the comparator we analyzed classification effectiveness for the arbitrary classification threshold $th=0.5$. For each neural network we also optimized the threshold value. Effectiveness both of positive classification (correct identification of a form) and negative classification (identification that not a correct form was presented to the classifier) was analyzed. Tab. 8 presents data on structures of employed NNs as well as on the obtained results.

Table 7. Parameters of neural networks and classification results for leave-one-out tests. Best results for individual musical phrase modifications are marked in bold (denotations as previously introduced).

Database	Fading coeff. r	Number of neurons in hidden layer	Musical phrase modification/ Classification effectiveness [%]									
			No modif.	Error	Error+ transp.+ ornament	Error + transp.	Addit. note	Omission of the first note + ornam.	Ornam.	Transp.	Augm.	Augm. transp.
mhor	0.8	20	100	85.4	68.8	79.2	93.8	85.4	68.8	97.9	97.9	100
mhor	0.8	50	**100**	81.3	79.2	83.3	93.8	**85.4**	70.8	97.9	**100**	**100**
mhor	0.5	20	97.9	87.5	79.2	83.3	95.8	77.1	62.5	95.8	97.9	95.8
mhor	0.5	50	95.8	85.4	81.3	83.3	93.8	83.3	68.8	95.8	97.9	97.9
mhor	0.2	20	91.7	85.4	68.8	62.5	77.1	72.9	64.6	83.3	87.5	95.8
mhor	0.2	50	87.5	79.2	72.9	64.6	77.1	70.8	66.7	91.7	95.8	93.8
mhor_rel	0.8	20	97.9	89.6	72.9	81.3	95.8	**85.4**	58.3	**100**	**100**	**100**
mhor_rel	0.8	50	**100**	89.6	81.3	89.6	93.8	**85.4**	68.8	**100**	**100**	97.9
mhor_rel	0.5	20	**100**	**93.8**	**87.5**	89.6	**97.9**	83.3	64.6	**100**	95.8	**100**
mhor_rel	0.5	50	**100**	91.7	85.4	**91.7**	95.8	**85.4**	66.7	**100**	**100**	**100**
mhor_rel	0.2	20	95.8	87.5	79.2	81.3	87.5	81.3	**72.9**	95.8	93.8	**100**
mhor_rel	0.2	50	95.8	91.7	79.2	81.3	89.6	**85.4**	**72.9**	97.9	95.8	97.9

Table 8. Parameters of NNs and classification results for neural comparator of musical phrases (denotations as previously introduced).

Database indicator	Fading coefficient	Number of neurons in the hidden layer	$th=0.5$			Optimization of the th threshold value			
			classification effectiveness (positive) [%]	classification effectiveness (negative)[%]	classification effectiveness - average [%]	threshold value of th	classification effectiveness (positive) [%]	classification effectiveness (negative) [%]	classification effectiveness - average [%]
mhor_rel	0.8	20	65.67	98.09	81.88	0.12	75.99	96.52	86.26
mhor_rel	0.8	50	73.41	97.16	85.29	0.02	80.95	91.44	86.20
mhor_rel	0.5	20	70.24	97.57	83.90	0.2	82.54	96.13	89.33
mhor_rel	0.5	50	80.16	98.46	89.31	0.08	90.67	97.09	93.88
mhor_rel	0.2	20	65.08	96.99	81.03	0.06	77.98	95.53	86.75
mhor_rel	0.2	50	78.57	95.26	86.92	0.02	94.05	91.46	92.75

5.3 Classification on the Basis of NN and Parameters of Musical Phrases

Both a classifier and a neural comparator were used in these experiments. In the first stage we investigated the classification in the case of training with reference musical forms only. Tab. 9 presents data on employed musical phrase parametrization and structures of employed NNs as well as the obtained classification results. Presented results were chosen within the larger set of experiments performed. On the other hand, *leave-one-out* tests were performed for trigonometric parametrization, only. This is because this kind of parametrization occurred to be most effective (apart from one case when the joint representation was used consisted of statistical and trigonometric parameters). The process of network training was limited to 1000 iterations. Tab. 10 presents data on structures of employed NNs as well as on the obtained classification results for *leave-one-out* tests.

Table 9. Classification in the case of training with reference forms only (denotations as previously introduced).

Parametrization method	Number of parameters	Number of neurons in the hidden layer	1000 MSE	1000 Classification effectiveness - [%]	2000 MSE	2000 Classification effectiveness - average [%]
stat	5	20	0.94633	6.89103	0.92082	12.3397
stat	5	50	0,90056	14.2628	0.90214	13.6218
trig	5	20	0.10128	78.5256	0.09826	78.5256
trig	5	50	0.08625	85.2564	0.06021	87.0192
trig	10	20	0.10887	79.9679	0.06554	83.8141
trig	10	50	0.12693	80.609	0.10465	82.0513
trig	15	20	0.04078	87.0192	0.03516	87.0192
trig	15	50	0.14704	78.8462	0.10481	81.891
poly	5	20	0.20356	60.8974	0.12456	68.75
poly	5	50	0.13287	68.5897	0.08023	69.7115
poly	10	20	0.02090	67.6282	0.00542	66.1859
poly	10	50	0.00987	64.9038	0.00297	64.7436
stat+trig	5+10	20	0.11657	80.7692	0.09324	82.8526
stat+trig	5+10	50	0.06371	87.8205	0.04290	**89.4231**
stat+poly	5+10	20	0.74441	23.2372	0.73575	24.5192
stat+poly	5+10	50	0.64372	29.1667	0.68451	26.4423
trig+poly	10+10	20	0.09703	81.25	0.07381	83.1731
trig+poly	10+10	50	0.08508	84.6154	0.04229	87.8205
stat+trig+poly	5+10+10	20	0.07649	82.6923	0.05310	84.2949
stat+trig+poly	5+10+10	50	0.18616	76.4423	0.10475	82.2115

Table 10. Parameters of NNs and classification results for *leave-one-out* tests on the basis of parametric representation of musical phrases. Best results for individual musical phrase modifications are marked in bold (denotations as previously introduced).

Musical phrase modification/ Classification effectiveness [%]

Param. method	No. of par.	Neurons in HL	No modif.	Error	Error+ transp.+ ornament	Error + transp.	Addit. note (beginning of the phrase)	Omission of the first note + ornament	Ornament	Transp.	Augm.	+ Augm. transp.
trig	5	20	89.6	58.3	79.2	75	43.8	45.8	83.3	89.6	68.8	87.5
trig	5	50	**100**	87.5	85.4	93.8	50	62.5	97.9	**100**	91.7	95.8
trig	10	20	97.9	83.3	91.7	85.4	62.5	62.5	**100**	**100**	93.8	95.8
trig	10	50	97.9	89.6	95.8	91.7	70.8	68.8	93.8	**100**	95.8	93.8
trig	15	20	**100**	79.2	**97.9**	91.7	62.5	**70.8**	**100**	**100**	95.8	**100**
trig	15	50	97.9	**91.7**	93.8	**100**	58.3	58.3	95.8	**100**	91.7	97.9
trig	20	20	**100**	89.6	95.8	91.7	54.2	62.5	**100**	**100**	95.8	**100**
trig	20	50	**100**	**91.7**	93.8	**100**	**79.2**	68.8	**100**	**100**	95.8	**100**

5.4 Rough Set-Based Approach

Investigations employing the rough set-based approach were performed in the analogous way to that of investigations concerning NNs. Experiments were performed for the TOFFEE database only. As said before, investigations employed the ROSETTA system. On the basis of pilot tests system parameters were limited to the following values:

- EFIM (*Equal Frequency Interval Method*) quantization, 10 intervals,
- generation of reducts and rules by means of a genetic algorithm.

Likewise in the case of investigations concerning NNs, intervals were represented using the binary method as well as modified Hörnel's representation. Tests were run for collections both without representation of rhythmical data and with rhythm encoded using the binary representation. Values were buffered using the fading memory model. Table 11 summarizes classification results obtained in the first stage of our investigation.

Also, *leave-one-out* tests were performed for modified Hörnel's representation by means of rough sets. Classification effectiveness as function of rhythmic data representation was analyzed. Obtained classification results are presented in Tab. 12.

Classification on the Basis of Parameters of Musical Phrases. Similar to investigations concerning NNs, a number of *leave-one-out* tests were performed for musical phrases represented by means of trigonometric parameters and RS method. Obtained results are presented in Tab. 13.

Using decision system as a comparator, we also investigated the classification effectiveness in the case of trigonometric parameters. *Leave-one-out* tests were performed for phrase representations employing 5 and 10 trigonometric parameters. Data

Table 11. Classification using the rough set approach in the case of training with reference forms only (denotations as previously introduced).

	No rhythm representation				Rhythm representation		
No.	Database indicator	Damping coefficient	Classification effectiveness [%]	No.	Database indicator	Damping coefficient	Classification effectiveness [%]
1.	Bin	0.8	85.58	7.	mbin_rel	0.8	89.42
2.	Bin	0.5	86.22	8.	mbin_rel	0.5	88.3
3.	Bin	0.2	82.21	9.	mbin_rel	0.2	83.81
4.	Mhor	0.8	90.06	10.	mhor_rel	0.8	**92.62**
5.	Mhor	0.5	87.82	11.	mhor_rel	0.5	89.42
6.	Mhor	0.2	80.61	12.	mhor_rel	0.2	81.57

Table 12. Classification using the rough set approach in the case of training with reference forms only (denotations as previously introduced).

Database indicator	Fading coeff. r	Musical phrase modification/Classification effectiveness [%]										
		no modification	error	error transposition ornament	error transposition	additional note (beginning of the phrase)	omission of the first note ornament	ornament	transposition	augmentation	augmentation transposition
mhor	0.8	**100**	81.3	79.2	83.3	95.8	83.3	70.8	**100**	**100**	**100**
mhor	0.5	95.8	87.5	79.2	83.3	95.8	83.3	68.8	95.8	95.8	95.8
mhor	0.2	91.7	81.3	92.9	68.8	91.7	70.8	64.6	91.7	91.7	91.7
mhor_rel	0.8	**100**	89.6	81.3	**89.6**	97.9	**85.4**	66.7	**100**	**100**	**100**
mhor_rel	0.5	**100**	**95.7**	**95.4**	**89.6**	97.9	**85.4**	66.7	**100**	**100**	**100**
mhor_rel	0.2	91.7	91.7	79.2	81.3	87.5	**85.4**	72.9	91.7	91.7	91.7

Table 13. Classification using the rough set approach in the case of training with reference forms only (denotations as previously introduced).

Parametrization method	Number of parameters	Musical phrase modification/Classification effectiveness [%]										
		no modification	error	error transposition ornament	error transposition	additional note (beginning of the phrase)	omission of the first note ornament	ornament	transposition	augmentation	augmentation transposition
trig	5	97.92	70.83	77.08	81.25	39.58	43.75	97.92	97.92	95.83	97.92
trig	10	**100**	87.5	85.42	81.25	43.75	52.08	**100**	**100**	97.91	97.92
trig	15	**100**	**89.58**	87.5	**93.75**	54.17	62.5	**100**	**100**	95.83	97.92
trig	20	**100**	**89.58**	91.67	91.67	**56.25**	64.58	**100**	**100**	**100**	**100**

set was limited to six fugue themes randomly-chosen from the collection. Again, both positive and negative identification was evaluated. Obtained results are presented in Tab. 14.

Table 14. Results of *leave-one-out* tests for musical phrase comparator - rough set approach (denotations as previously introduced).

Modification	Classification effectiveness [%]					
	5 parameters			10 parameters		
	negative	positive	average	negative	positive	average
error	100	100	100	100	83.33	98.48
error + transposition + ornament	100	83.33	98.48	100	100	100
error + transposition	100	100	100	100	100	100
additional note (phrase beginning)	100	66.67	96.97	100	83.33	98.48
omission of the first note + ornament	100	100	100	100	50	95.45
ornament	100	66.67	96.97	100	100	100
transposition	100	83.33	97.62	100	83.33	97.62
augmentation	100	100	100	100	100	100
augmentation + transposition	100	100	100	100	100	100

5.5 Discussion of Results

Performed investigations and obtained results allow us to present a number of conclusions concerning the developed methods of representing and automatically classifying musical phrases:

- developed methods allow for effective classification of musical phrases in presence of phrase modifications,
- NNs and the RS-based approach show comparable suitability for classifying musical phrases on the basis of sequence data and of musical phrase parameters,
- process of training the classifying system which utilizes the RS approach shows much lower computational complexity in comparison with training of NNs,
- in the case of sequence data, highest classification effectiveness was obtained for modified Hörnel's representation and relative rhythm representation,
- in the case of phrase parameters, highest classification effectiveness was obtained for trigonometric parametrization.

6 Conclusions

Prediction of musical events using NNs was investigated. The obtained results indicate changes of data entropy depending on musical form and composer's style. The obtained results indicate also low entropy of musical data and therefore the possibility of predicting musical data with high accuracy in the case of exact polyphonic forms.

Three methods of representing musical phrases parametrically were developed. Methods for automatic classification of musical phrases were designed and implemented: they utilized intelligent decision algorithms – NNs and the RS-based approach. In conjunction with utilizing intelligent decision algorithms two methods of classification of musical phrases were implemented: assigning test objects to one of the reference objects (classifier) and analyzing the identity of test phrase and reference phrases (comparator). Resulted from the feature selection process, based on closed-loop approach, only a few parametrization methods were classified as an effective description of musical phrases, others were discarded after receiving a feedback from the NN predictor.

Obtained results prove high effectiveness of classifying musical phrases using intelligent decision algorithms and trigonometric parameters. High classification correctness obtained with intelligent decision algorithms and musical sequence representations using the fading memory model was negatively verified in conditions of musical material homogeneity. The decision process employing rough sets revealed best ratio of classification accuracy to computational complexity.

Acknowledgements

The research is sponsored by the Committee for Scientific Research, Warsaw, Grant No. 4T11D 014 22, and by the Foundation for Polish Science, Poland.

References

1. Bharucha, J. J., Todd, P. M.: Modeling the Perception of Tonal Structure with Neural Nets, Music and Connectionism. In: Todd, P. M. & Loy, D. G. (eds.), The MIT Press, Cambridge, Massachusetts, London, England (1991)128-137
2. Braut, Ch.: The Musician's Guide to MIDI. SYBEX, San Francisco (1994)
3. Bullivant, R.: Fugue, The New Grove Dictionary of Music and Musicians. Macmillan Publishers Limited, London, Vol. 7 (1980) 9-21
4. Chmielewski, M. R, Grzymała-Busse, J. W.: Global Discretization of Continuous Attributes as Preprocessing for Machine Learning. 3rd International Workshop on Rough Sets and Soft Computing, San Jose, California, USA, Nov. 10-12 (1994)
5. Classical MIDI Archives, http://www.prs.net
6. Desain, P.: A (de)composable theory of rhythm perception. Music Perception, Vol. 9 (4) (1992) 439-454
7. Desain P., Honing, H.: Music, Mind, Machine: Computational Modeling of Temporal Structure in Musical Knowledge and Music Cognition. http://www.nici.kun.nl/PAPERS/DH-95-C.HTML
8. Desain, P., Honing, H.: The Quantization of Musical Time: A Connectionist Approach, Music and Connectionism. Todd, P. M., Loy, D. G. (eds.). The MIT Press, Cambridge, Massachusetts, London, England (1991) 150-169
9. Feulner, J., Hörnel, D.: MELONET: Neural Networks that Learn Harmony-Based Melodic Variations. Proc. International Computer Music Conference, San Francisco: International Computer Music Association (1994) 121-124

10. Hörnel, D.: MELONET I: Neural Nets for Inventing Baroque-Style Chorale Variations. Advances in Neural Information Processing 10 (NIPS 10), M. I. Jordan, M. J. Kearns, S. A. Solla (eds.), MIT Press (1997)
11. http://www.ismir.net
12. Komorowski, J., Pawlak, Z., Polkowski, L., Skowron, A.: Rough Sets: A Tutorial, Rough Fuzzy Hybridization: A New Trend in Decision-Making. Pal, S. K., Skowron, A. (eds.), Springer-Verlag (1998) 3-98
13. Kostek, B.: Computer Based Recognition of Musical Phrases Using the Rough Set Approach, 2nd Annual Joint Conference on Inform. Sciences, NC, USA, Sept. 28-Oct. 1 (1995) 401-404
14. Kostek, B., Szczerba, M.: MIDI Database for the Automatic Recognition of Musical Phrases. 100th AES Convention, preprint 4169, Copenhagen, May 11-14 (1996) J. Audio Eng. Soc., (Abstr.), Vol. 44, 10 (1996)
15. Kostek, B., Szczerba, M.: Parametric Representation of Musical Phrases, 101st AES Convention, preprint 4337, Los Angeles (1996) J. Audio Eng. Soc., (Abstr.), Vol. 44, 12 (1996) 1158
16. Kostek, B.: Computer-Based Recognition of Musical Phrases Using the Rough-Set Approach. J. Information Sciences, Vol. 104 (1998) 15-30
17. Kostek B.: Soft Computing in Acoustics, Applications of Neural Networks, Fuzzy Logic and Rough Sets to Musical Acoustics, Studies in Fuzziness and Soft Computing, Physica Verlag, Heidelberg, New York (1999)
18. Moradi, H., Grzymała-Busse, J. W., Roberts, J. A.: Entropy of English Text: Experiments with Humans and a Machine Learning System Based on Rough Sets. J. Information Sciences, Vol. 104 (1-2) (1998) 31-47
19. Mozer, M. C.: Connectionist Music Composition Based on Melodic, Stylistic, and Psychophysical Constraints. Music and Connectionism, Todd, P. M., Loy, D. G. (eds.), The MIT Press, Cambridge, Massachusetts, London, England (1991) 195-211
20. Øhrm, A.: Discernibility and Rough Sets in Medicine: Tools and Applications, Ph.D. Thesis, Department of Computer and Information Science, Norwegian University of Science and Technology, Trondheim, NTNU Report 1999:133, IDI Report (1999)
21. Pawlak, Z.: Rough Sets, International J. Computer and Information Sciences, No. 11 (5) (1982)
22. Repp, B. H.: Patterns of note asynchronies in expressive piano performance, J. Acoust. Soc. Am., Vol. 100 (6) (1996) 3917-3932
23. Skowron, A., Nguyen, S. H.: Quantization of Real Value Attributes, Rough Set and Boolean Approach, ICS Research Report 11/95, Warsaw University of Technology (1995)
24. Szczerba, M.: Recognition and Prediction of Music: A Machine Learning Approach, Proc. of 106th AES Convention, Munich (1999)
25. Swiniarski, R.: Rough sets methods in feature reduction and classification, Int. J. Applied Math. Comp. Sci., Vol. 11 (3) (2001) 565-582
26. Tanguiane, A. S.: Artificial Perception and Music Recognition, Lecture Notes in Artificial Intelligence, No. 746, Springer Verlag (1991)
27. The New Grove Dictionary of Music and Musicians, Macmillan Publishers Limited, Vol. 1, London, (1980) 774-877
28. Todd, P. M.: A Connectionist Approach to Algorithmic Composition, Music and Connectionism, Todd, P. M. & Loy, D. G. (eds.), The MIT Press, Cambridge, Massachusetts, London, England (1991) 173-194
29. Zell, A.: SNNS – Stuttgart Neural Network Simulator User Manual, Ver. 4.1.

Processing of Musical Metadata
Employing Pawlak's Flow Graphs

Bozena Kostek and Andrzej Czyzewski

Gdansk University of Technology, Multimedia Systems Department
Narutowicza 11/12, 80-952 Gdansk, Poland
{bozenka,andcz}@sound.eti.pg.gda.pl
http://www.multimed.org

Abstract. The objective of the presented research is enabling music retrieval based on intelligent analysis of metadata contained in musical databases. A database was constructed for the purpose of this study including textual data related to approximately 500 compact discs representing various categories of music. The description format of musical recordings stored in the database is compatible to the format of the widely-used CDDB database available in the Internet. An advanced query algorithm was prepared employing the concept of inference rule derivation from flow graphs introduced recently by Pawlak. The created database searching engine utilizes knowledge acquired in advance and stored in flow graphs in order to enable searching CD records.

1 Introduction

A broadened interest in music information retrieval from music databases, which are most often heterogeneous and distributed information sources, is based on the fact that they provide, apart from music, a machine-processable semantic description. The semantic description is becoming a basis of the next web generation, i.e., the Semantic Web. Several important concepts were introduced recently by the researchers associated with the rough set community with regard to semantic data processing including techniques for computing with words [16], [22]. Moreover, Zdzislaw Pawlak in his recent papers [23], [24] promotes his new mathematical model of flow networks which can be used to mining knowledge in databases. Given the increasing amount of music information available online, the aim is to enable effective and efficient access to such information sources. The authors introduce a concept of how to organize such a database, what kind of searching engine should be applied and demonstrate how to apply this conceptual framework based on flow graphs to improve music information retrieval efficiency.

The experiments that were performed consisted in constructing a music database collecting music recordings along with semantic description. A searching engine is designed, which enables searching for a particular musical piece utilizing the knowledge on the entire database content and relations among its elements contained in the flow graphs constructed following Pawlak's ideas.

J.F. Peters et al. (Eds.): Transactions on Rough Sets I, LNCS 3100, pp. 279–298, 2004.

Generally, the paper addresses the capabilities that should be expected from intelligent Web search tools in order to respond properly to user's multimedia information retrieval needs. Two features, seem to be of great importance for searching engines: the capability of properly ordering the retrieved documents and the capability to draw user's attention to next interesting documents (intelligent navigation concept). As results from hitherto performed experiments, these goals could be achieved efficiently provided the searching engine uses the knowledge of database content acquired a priori and represented by distribution ratios between branches of the flow graph which in turn can be treated as a prototype of a rule-based decision algorithm.

2 Database Organization

2.1 Data Description in CDs

One of most important music archiving format is data format of CDs (Compact Disks). According to the so-called Red Book specifications (ICE 908), a CD is divided into a lead-in area, which contains the table of contents (TOC), a program area, which contains the audio data, and a lead-out area, which contains no data. An audio CD can hold up to 74 minutes of recorded sound, and up to 99 separate tracks. Data on a CD is organized into sectors (the smallest possible separately addressable block) of information. The audio information is stored in frames of 1/75 second length. 44,100 16-bit samples per second are stored, and there are two channels (left and right). This gives a sector size of 2,352 bytes per frame, which is the total size of a physical block on a CD. Moreover, CD data is not arranged in distinct physical units; data is organized into frames (consisting of 24 bytes of user data, plus synchronization, error correction, and control and display bits), which are interleaved [25].

2.2 CDDB Service

CDDB service is the industry standard for music recognition services. It contains the largest online database of music information in the world (currently more than 22 million tracks), and is used by over 30 million people in more than 130 countries every month. Seamless handling of soundtracks data provide music listeners both professional and amateurs with access to a huge store of information on recorded music [8], [9]. The large database queried so frequently by users from all over the world provides also a very interesting material for research experiments in the domain of searching engine optimizing. The organization of metadata related to compact discs in the CDDB database is presented in Tab. 1.

   The content of the world-wide CDDB was targeted in our experiments as the principal material for experiments. However, because of the large volume of this database and expected high computational cost we decided that at initially we will construct and use much smaller local database utilizing the CDDB data format. Consequently, a database was constructed especially for the purpose of this study containing approximately 500 compact discs textual data stored together with fragments of music corresponding to various categories. This database provided a material for initial experiments with searching music employing the proposed method. Subsequently, the huge CDDB database containing metadata related to majority of compact disks hitherto produced in the world was utilized.

Table 1. Metadata fields in the CDDB database.

Album Data Fields:

Album Title	Can be a multi-word expression (string)
Album Artist	As above
Record Label	The label or publisher of the CD
Year	The year the CD was recorded or published
Genre	Every album can have both a primary and a secondary genre
Compilation	Indicates whether tracks have different artists
Number/Total Set	Can identify a CD as a member of a box sets
Language	Used to help display in appropriate character set
Region	To identify where the CD was released
Certifier	Authorized party (artist or label) who has certified the data accuracy
Notes	General notes such as dedications, etc.

Track Data Fields:

Track Title	Can be a multi-word expression (string)
Track Artist	Vital for compilations, such as soundtracks or samplers
Record Label	May be different from track to track for compilations
Year	May be different from track to track
Beats/Minute	Used for special purposes (synchronizing with special devices)
Credits	E.g. guest musicians, songwriter, etc.
Genre	Every track can have both a primary and a secondary genre
ISRC	The International Standard Recording Code number for the CD track
Notes	General track notes such as "recorded in ...", etc.
Credits	Can be entered for entire album, for individual tracks or segments
Credit Name	Can be person, company, or place such as recording location
Credit Role	Instrument, composer, songwriter, producer, recording place, etc.
Credit Notes	E.g. to specify unusual instruments, etc.
Genres	Can be entered for entire album or applied to individual tracks
Metagenres	General classification. e.g. Rock; Classical; New Age; Jazz
Subgenres	More specific style. e.g. Ska; Baroque, Choral; Ambient; Bebop, Ragtime
Segments	Each segment can have its own name, notes, and credits

2.3 Extended Tagging System

The information presented in this paragraph is not directly related to our current experimental phase. Nevertheless, we decided to discuss it, because it provides an extension to current standards, illustrating the future trends and expected developments of methods for advanced searching music in databases using metadata associated with musical recordings.

The ID3v2 is a currently used tagging system that allows to enrich and to include extended information about audio files within them (Fig. 1). It represents data prepended to the binary audio data. Each ID3v2 tag holds one or more frames. These frames can contain any kind of information and data, such as title, album, performer, website, lyrics, equalizer presets, pictures etc. Since each frame can be 16MB and the entire tag can be 256MB thus there is a lot of place to write a useful comment. The ID3v2 supports Unicode, so that the comments can be written in the user's native language. Also information in which language the comment is written can be included [10]. In addition the main characteristics of this system are as follows:

- It is a container format allowing new frames to be included.
- It has several new text fields such as composer, conductor, media type, copyright message, etc. and the possibility to design user's own fields.
- It can contain lyrics as well as music-synced lyrics (karaoke) in many languages.
- It could be linked to CD-databases such as CDDB [8].
- It supports enciphered information, linked information and weblinks.

An example of the internal layout of an ID3v2 tagged file is presented in Fig. 1.

Fig. 1. Example of the internal layout of an ID3v2 tagged file [10].

The much larger and more diversified metadata set covered with ID3v2 standard in comparison to CDDB format opens a way towards future experiments in the domain of advanced music searching including application of the method proposed in this paper.

3   CDDB Database Organization and Searching Tools

A sample record from the CDDB database is presented in Fig. 2. The field denoted as "frames" needs some explanation. It contains the frame numbers, because the CDDB protocol defines beginning of each track in terms of track lengths and the number of

CDDBID: eb117b10
[22164FD]
artist=Céline DION
title=Let's Talk About Love
numtracks=16
compilationdisc=no
genre=Pop
year=1997
comment=
0=The Reason
1=Immortality
2=Treat Her Like A Lady
3=Why, Oh Why ?
4=Love Is On The Way
5=Tell Him (Avec Barbra Streisand)
6=Amar Haciendo El Amor
7=When I Need You
8=Miles To Go (Before I Sleep)
9=Us
10=Just A Little Bit Of Love
11=My Heart Will Go On (Chanson D'amour Du Film Titanic)
12=Where Is The Love ?
13=Be The Man
14=I Hate You Then I Love You (Avec Luciano Pavarotti)
15=Let's Talk About Love
frames=0,22580,41415,59812,81662,101655,123540,142347,161295,182290,208287,226792,247817
270010,290987,312245,335675
order=0 1 2 3 4 5 6 7 8 9 10 11 12 13 14 15

Fig. 2. Sample CDDB database record.

preceding tracks. The most basic information required to calculate these values is the CD table of contents (the CD track offsets, in "MSF" [Minutes, Seconds, Frames]). That is why tracks are often addressed on audio CDs using "MSF" offsets. The combination determines the exact disc frame where a song starts.

3.1 Available Tools for CDDB Information Retrieval

The process of CDDB database querying begins with submitting the content of the "frames" field to a database searching engine. It is assumed that this numerical data string is unique for each CD, because it is improbable that the numerical combination could be repeated for different albums. Sending this numerical string to a remote CDDB database results with transmitting back all data related to the album stored in the database, namely artist, title,..., genre, etc. This feature is exploited by a huge number of clients world-wide. However, as results from above, such a query can be made, provided users possess a copy of the CD record which metadata are searched for. If so, their computers can automatically get data from the CDDB database and display these data. Consequently, local catalogs of records (phonotecs) can be built up fast and very efficiently with the use of this system.

A possible benefit from the universal and unrestricted access to CDDB could be, however, much higher than just obtaining the textual information while having a copy of a record at a disposal. Namely, provided an adequate searching engine is employed, CDDB users could submit various kinds of queries in this largest set of data on recorded sound, without the necessity to gain an access to any CD record in advance. The purpose of such a data search mode could be different and much broader than building up catalogs of available records – it has various research, historic and cultural applications and connotations. The currently available searching engines are able to scan CDDB content for keywords or keyword strings. Usually, if the query sentence consists of several words, the logical operator AND is utilized (Boolean searching model), bringing in many occasions very poor results, because too many matches occur in the case famous artists' musical pieces are searched for or none of them occur, if a misspelled or mixed-up terms were provided by the operator. An important feature of the searching engine is also the ability to order matches according to users' expectations and adapt properly if attributes are close, but are not necessarily exact match. The latter assumption is easy to illustrate on the typical example presented in Fig. 3 showing that the CDDB database contains many records related to the same CD. That is because all CDDB users possessing records ale allowed to send and store remotely metadata utilizing various simplistic software tools (Fig. 4). Consequently, textual information related to the same CD records can be spelled quite much differently.

Artist	Album
Celine Dion	Let's Talk About Love
Céline Dion	Let's talk about love
Celine Dion	Let's Talk About Love
CELINE DION	CELINE DION LET'S TALK ABOU...

Fig. 3. Sample result of information retrieval from CDDB database obtained after sending content of "frames" field (4 different records were retrieved concerning the same disk).

Freedb Submit Wizard

This is the function for uploading discs to freedb's database. Fill in the appropriate information and press 'Next'. Please be careful with your spellings.

Artist:	Céline DION	
Album:	Let's Talk About Love	
Freedb Category:	Rock	
Genre:	Pop	Year: 1997

Additional information for this disc:

Her first and best world-wide known album

Fig. 4. Panel of typical software tool allowing users to transmit & store metadata in CDDB database.

3.2 Simple Search Questionnaires

An important factor in building a query questionnaire for searching in a music database is the knowledge on user's way to access music information. The question arises
whether all data contained in the album and track data fields are used by a person
searching for a particular recording or CD. A survey on this problem was performed
by Bainbridge *et al.* [1]. Some of these results are recalled in Tab. 2. It is most common that search starts with the name of the album or performer, the next most often
searched categories are title and date of the recording. However, another problem
concerns the fact that users rarely are able to fill in such a questionnaire with data,
which are exact to ones stored in the music database. Users experience difficulty in
coming up with crisp descriptions for several of the categories, indicating a need to
support imprecise metadata values for searches. Apart from not being able to give
crisply defined categories users often make mistakes, or are not certain when refer-
encing to the year of a recording. More typically, they would like to give the decade
or define time in other terms than singular year. Also giving information on lyrics,
may cause some problems such as how to transcribe non existing words. Another
category, which is difficult for users is deciding as to the genre of recording. All these
factors should be taken into account when building a questionnaire for searching a
database.

Table 2. Survey on user's queries on music [1].

Category	Description	Count	[%]
Performer	Performer or group who created a particular recording	240	58.8
Title	Name (or approximation) of work(s)	176	43.1
Date	Date that a recording was produced, or that a song was composed	160	39.2
Orchestra-tion	Name of instrument(s) and/or vocal range(s) and/or genders (male/female)	68	16.7
Album title	Name of album	61	15.0
Composer	Name of composer	36	8.8
Label	Name of organization which produced recording(s)	27	6.6
Link	URL providing a link to further bibliographic data	12	2.9
Language	Language (other than English) for lyrics	10	2.5
Other	Data outside from the above categories	36	8.8

Meanwhile, despite really expressive information that can be found on some Web-
pages [9], available searchers are very simplistic, basing just on text scanning with an
of application of simple logical operators in the case of multi-term queries (Boolean
search). The panel of such an electronic questionnaire was reproduced in Fig. 5.

3.3 Advanced Metadata Searching Attempts

As was said, recently, it is a rapid growth of interest observed in the so-called "semantic Web" concepts [13]. By the definition the Semantic Web is the representation
of data on the World Wide Web. The effort behind the semantic web is to develop a
vocabulary for a specific domain according to the recent ISO/IEC 13250 standard

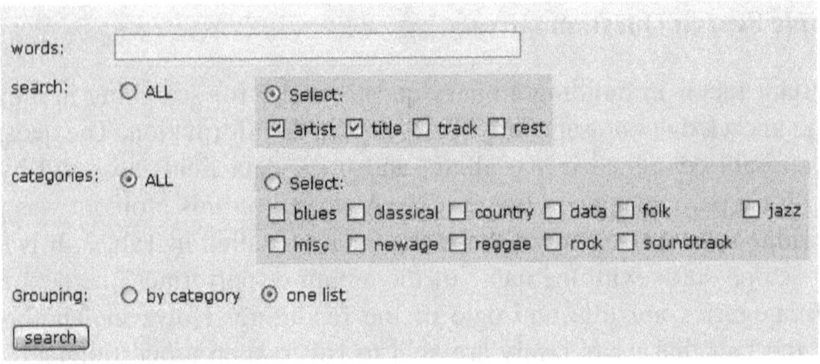

Fig. 5. Freedb.org simple searcher interface.

aiming at capturing semantics by providing a terminology and link to resources. It is based on the Resource Description Framework (RDF), which integrates a variety of applications using XML for syntax [14]. In other words, these are sets of organizing guidelines for publishing data in such a way as it can be easily processed by anyone. Within work done on semantic web automated reasoning, processing, and updating distributed content descriptions are also discussed. A so-called problem solving method (PSM) approach is taken in which algorithms from Knowledge Based Systems field are considered to perform inferences within expert systems. However, PSM are task specific. At the moment there seems to be still little consensus about characteristics of the Semantic Web, but many domains are already covered by some applications and test databases, or at least guidelines have been created.

Recent developments in MPEG-7 and MPEG-21 standards allow for multimedia content description, which includes audio and video features at lower abstraction level as well as at the higher conceptual level [7], [12], [20]. It can be used to describe the low-level features of the content. For example, low-level descriptors for musical data, are organized in groups of parameters such as: Timbral Temporal, Basic Spectral, Basic, Timbral Spectral, Spectral Basis, Signal Parameters [12]. The so-called Audio Framework that contains all these parameter groups includes 17 vector and scalar quantities. On the other hand, higher level descriptors can be used to provide conceptual information of the real-world being captured by the content. Intermediate levels of description can provide models that link low-level features with semantic concepts. Because of the importance of the temporal nature of multimedia, dynamic aspects of content description need to be considered. Therefore, the semantic multimedia web should be dedicated toward the conceptual as well as dynamical aspects of content description.

Even though a lot of research was focused on low-level audio features providing query-by-humming or music retrieval systems [2], [5], [11], [17–19], most web sites containing music only support category or text-based searching. Few reports can be found on attempts devoted to building advanced tools for mining the content of CDDB database, i.e. study by Bainbridge [1], Downie [3], and Pachet *et al.* [21]. One of reports on this subject was published by Pachet *et al.* [21]. They propose a method of classification based on musical data mining techniques that uses co-occurrence and correlation analysis for classification. The essence of the experiments performed by the authors of the cited work concerns processing extracted titles and establishing similarity measurements between them. The co-occurrence techniques were applied to two data sources: radio program, and CD compilation database.

The co-occurrence analysis is a well-known technique used to statistical linguistics in order to extract clusters of semantically related words. In the case of the study by Pachet *et al.* the analysis consisted in building a matrix with all titles in row and in column. The value at (i, j) corresponds to the number of times that titles i and j appeared together, either on the same sampler, on the same web page, or as neighbors in a given radio program.

Given a corpus of titles $S = (T_1,..., T_N)$, the co-occurrence may be computed between all pairs of titles T_i and T_j, [21]. The co-occurrence of T_i with itself is simply the number of occurrences of T_i in the considered corpus. Each title is thus represented as a vector, with the components of the vector being the co-occurrence counts with the other titles. To eliminate frequency effects of the titles, components of each vector are normalized according to:

$$Cooc_{norm}(T^1, T^2) = \left(\frac{Cooc(T^1, T^2)}{Cooc(T^1, T^1)} + \frac{Cooc(T^2, T^1)}{Cooc(T^2, T^2)} \right) / 2 \qquad (1)$$

The normalized co-occurrence values can directly be used to define a distance between titles. The distance will be expressed as:

$$Dist_1(T^1, T^2) = 1 - Cooc_{norm}(T^1, T^2) \qquad (2)$$

Given that the vectors are normalized, one can compute the correlation between two titles T^1 and T^2 as:

$$Sim(T^1, T^2) = \frac{Cov_{1,2}}{\sqrt{Cov_{1,1} \times Cov_{2,2}}} \qquad (3)$$

where: $Cov_{1,2}$ is the covariance between T^1 and T^2 and:

$$Cov(T^1, T^2) = E((T^1 - \mu_1) \times (T^2 - \mu_2)) \qquad (4)$$

E is the mathematical expectation and $\mu_j = E(T^i)$
The distance between T^1 and T^2 is defined as:

$$Dist_2(T^1, T^2) = 1 - (1 + Sim(T^1, T^2)) / 2 \qquad (5)$$

Experiments performed by Pachet *et al.* on a database of 5000 titles show that artist consistency may be not well enforced in the data sources. The correlation clustering generally indicates that items in a larger cluster tend to be classified according to their specific music genres, whereas the co-occurrence clustering is better suited for small clusters, indicating similarities between two titles only.

Unfortunately, the presented study by Pachet *et al.* was applied in a very fragmentary way, thus there is a need to perform more thorough analysis on music data in the context of searching for co-occurrence and similarity between data. Nevertheless, similar findings are applicable to address problems related to extracting clusters of semantically related words, and can be found in other references, i.e. [4], [15].

More universal and broader approach to searching for mutual dependencies among metadata (not only textual but also numerical) proposed by the authors is presented in the following paragraphs, basing on Pawlak's flow graph concept [24].

4 Data Mining in CDDB Database

The weakness of typical data searching techniques lays in lacking or not using any a priori knowledge concerning the queried dataset. More advanced methods assume stochastic search algorithms using randomized decisions while searching for solutions to a given problem, basing on a partial representation of data dependency expressed in terms of Bayesian networks, for example proposals formulated in papers of Ghazfan [4] or Turtle *et al.* [26]. The process of searching is divided into 2 phases: learning and query expansion. The learning phase stands for constructing a Bayesian network representing some of relationships among the terms appearing in a given document collection. The query expansion phase starts from giving a query by the user which are treated as terms that could be replaced in the process of propagating this information through the Bayesian network prepared in advance. The new terms are selected whose posterior probability is high, so that they could be added to the original query.

The abundant literature on techniques for searching data in databases describes many methods and algorithms for probabilistic search and data mining techniques, including decision trees application. There are no reports, however, on successful application of any of them to representing knowledge contained in the CDDB database. Meanwhile, this problem is vital and important, because this database contains almost complete knowledge on the sound recorded on digital disks. This knowledge, if extracted from the database can serve for various purposes, including an efficient support for data query made by millions of users.

As a method of data mining in the CDDB database we propose a system application which uses logic as mathematical foundations of probability for the deterministic flow analysis in flow networks. The flow graphs are then employed as a source of decision rules providing a tool for knowledge representation contained in the CDDB database. The new mathematical model of flow networks underlying the decision algorithm in question was proposed recently by Zdzislaw Pawlak [23], [24]. The decision algorithm allowed us to build an efficient searching engine for the CDDB database. The proposed application is described in subsequent paragraphs.

4.1 Probabilistic and Logical Flow Networks

In the kind of flow graphs proposed by Pawlak, flow is determined by Bayesian probabilistic inference rules, but also by the corresponding logical calculus which was originally proposed by Lukasiewicz. In the second case the dependencies governing flow have deterministic meaning [23]. The directed, acyclic, finite graphs are used in this context, defined as:

$$G = (N, B, \sigma) \tag{6}$$

$$B \subseteq N \times N \; ; \sigma : B \to \langle 0,1 \rangle \tag{7}$$

where:

 N – set of nodes
 B – set of directed branches
 σ - flow function

Input of $x \in N$ is the set:

$$I(x) = \{y \in N : (y, x) \in B\} \tag{8}$$

output of $x \in N$ is the set:

$$O(x) = \{y \in N : (x, y) \in B\} \tag{9}$$

Other quantities can be defined with each flow graph node, namely the inflow:

$$\sigma_+(x) = \sum_{y \in I(x)} \sigma(y, x) \tag{10}$$

and outflow:

$$\sigma_-(x) = \sum_{y \in O(x)} \sigma(x, y) \tag{11}$$

Considering the flow conservation rules, one can define the troughflow for every internal node as:

$$\sigma(x) = \sigma_+(x) = \sigma_-(x) \tag{12}$$

Consequently, for the whole flow graph:

$$\sigma(G) = \sigma_+(G) = \sigma_-(G), \qquad \sigma(G) = 1 \tag{13}$$

The factor $\sigma(x, y)$ is called **strength** of (x, y).

Above simple dependencies known from flow network theory were extended by Pawlak with some new factors and a new interpretation. The definitions of these factors are following:

certainty:

$$cer(x, y) = \frac{\sigma(x, y)}{\sigma(x)} \tag{14}$$

The certainty of the path $[x_1, ..., x_n]$ is defined by:

$$cer[x_1 ... x_n] = \prod_{i=1}^{n-1} cer(x_i, x_{i+1}) \tag{15}$$

coverage:

$$cov(x, y) = \frac{\sigma(x, y)}{\sigma(y)} \tag{16}$$

The coverage of the $[x_1, ..., x_n]$ is defined by:

$$cov[x_1 ... x_n] = \prod_{i=1}^{n-1} cov(x_i, x_{i+1}) \tag{17}$$

The strength of a path [x...y] is as follows:

$$\sigma[x...y] = \sigma(x)cer[x...y] = \sigma(y)cov[x...y] \tag{18}$$

The above factors have also clear logical interpretation Provided <x,y> – represents the set of all paths leading from x to y, **certainty** has the following property:

$$cer\langle x,y \rangle = \sum_{[x...y]\in\langle x,y\rangle} cer[x...y] \tag{19}$$

coverage:

$$cov\langle x,y \rangle = \sum_{[x...y]\in\langle x,y\rangle} cov[x...y] \tag{20}$$

and **strength** fulfills the condition:

$$\sigma\langle x,y \rangle = \sum_{[x...y]\in\langle x,y\rangle} \sigma[x...y] \tag{21}$$

It is important in this context, that the flow graph branches can be also interpreted as decision rules, similar to rough set algorithm decision rules [24]. That is because with every branch (x, y) a decision rule $x \to y$ can be associated. Consequently, a path $[x_1, x_2]$ can be associated with a rule string: $x_1 \to x_2, x_2 \to x_3,....x_{n-1} \to x_n$ or can be represented by a single rule of the form: $x^* \to x_n$, where x^* replaces the string: x_1, $x_2,...,x_{n-1}$.

This important finding was proved in Pawlak's papers and was completed with derivation of practically useful properties, for example:

$$cer\left(x^*, x_n\right) = cer[x_1...x_n] \tag{22}$$

$$cov\left(x^*, x_n\right) = cov[x_1...x_n] \tag{23}$$

$$\sigma\left(x^*, x_n\right) = \sigma(x_1)\cdot cer[x_1...x_n] = \sigma(x_n)\cdot cov[x_1...x_n] \tag{24}$$

Consequently, with every decision rule corresponding to graph branch the aforementioned coefficients are associated: flow, strength, certainty, and coverage factor. As was proved by the cited author of this applicable theory, these factors are mutually related as follows:

$$\sigma(y) = \frac{\sigma(x)\cdot cer\langle x,y \rangle}{cov\langle x,y \rangle} = \frac{\sigma(x,y)}{cov\langle x,y \rangle} \tag{25}$$

$$\sigma(x) = \frac{\sigma(y)\cdot cov\langle x,y \rangle}{cer\langle x,y \rangle} = \frac{\sigma(x,y)}{cer\langle x,y \rangle} \tag{26}$$

If x_1 is an input and x_n graph G output, then the path $[x_1...x_n]$ is complete, and the set of all decision rules associated with the complete set of the flow graph connections provides the decision algorithm determined by the flow graph.

Thus, in computer science applications of the flow graphs the concepts of probability can be replaced by factors related to flows, the latter representing data flows between nodes containing data. Knowledge on these flows and related dependencies can be stored in the form of a rule set from which the knowledge can be extracted. In contrast to many data mining algorithms described in literature, the described method is characterized by reasonably low computational load. That is why it provides very useful means for extracting the complete knowledge on mutual relations in large data sets, thus it can be applied also as a knowledge base of intelligent searching engine for the CDDB database.

4.2 Extracting Inference Rules from CDDB Database

Two databases prepared in the CDDB format were selected as objects of our experiments: a local database containing metadata related to approximately 500 CD disks and the original CDDB imported from *freedb.org* website (*rev. 20031008*). At first the much smaller local database was used in order to allow experiments without engaging too much computing power for flow graph modeling. Moreover, only 5 most frequently used terms were selected as labels of node columns. These are:

- **Album title** (optional *ASCII* string not exceeding 256 letters)
- **Album artist** (up to 5 words separated by spaces)
- **Year** of record issuing (4 decimal digits)
- **Genre** (type of music that can be according to CDDB standard: Blues,...,Classical,...,Country,..., Folk,..., Jazz,..., Rock,...,Vocal). It is together 148 kinds of musical genres
- **Track title** (optional ASCII string not exceeding 256 letters)
- The term **Number** is considered a decision attribute – in the CDDB database it is represented by unique digit/letter combination of the length equal to 8 (for example: *0a0fe010, 6b0a4b08, etc.*).

Once the number of a record is determined, which is associated with a concrete CD, it allows to retrieve all necessary metadata from the database (as presented in Fig. 2) and render them by automatic filling/replacing the fields of an electronic questionnaire. A questionnaire was prepared for searching CDDB databases employing knowledge extracted from flow graphs. The graph designed to represent data relations between chosen terms is illustrated in Fig. 6.

The process of knowledge acquisition was initiated for the smaller CDDB database with analyzing first letters of terms "Album Title", "Album Artist" and "Track Titles". This solution was adopted because of the small size of the experimental database. Otherwise the number of paths between nodes would be too small and the problem of CD records searching will be hard-defined in practice for most objects. Above restriction does not concern the full CDDB database which contains many records of selected performers as well as many records metadata of which contain the same words in the fields related to album or track titles.

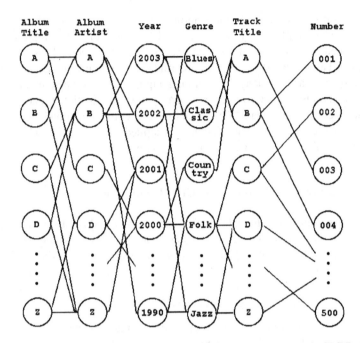

Fig. 6. Flow graph used to represent knowledge relevant to frequently made CDDB queries.

The layers "Album Title", "Album Artist" and "Track Title" contain nodes denoted with letters A, B, C,...,Z constituting disjoint subsets, the sum of which produces the full set of all elements contained in the database. The subsets are defined as a result of analyzing of all entries gathered in the database. For example it is possible to define individual nodes of the layer "Album Title" as follows: A – "sun", B – "love", ... It means that A is the subset of database elements containing the expression "sun" in the field "Album Title" and B is the subset of elements containing the expression "love" and so on. The subsets can be defined also by a string of expressions linked with the "OR" operator. In the latter case the subsets have to be disjoint, thus in order to satisfy that demand an additional algorithmic analysis is necessary at the stage of the graph building, however the analysis does not influence the further process of data querying. As results from above assumptions, two phases can be discerned in the process of building the graph:

a) initial phase (statistical), basing on the analysis of all expressions contained in the database

b) intelligent phase, basing on the knowledge related to querying history

In the phase (a) statistical similarity of expression is assessed resulting in distributing expressions among separate clusters containing mutually similar expressions. In our experiments simple statistical metrics were applied to that end. As it is described later on, the intelligent phase (b) is based on flow graphs. In Tab. 3 sample values of *certainty*, *strength* and *coverage* are gathered for branches linking layers: Album Artist - Genre for a small database containing 432 entries.

Table 3. Results of calculating data flow parameters between nodes: *Album Artist* and *Genre*.

	certainty	strength	coverage
A – DANCE	0.146	0.016	0.159
A – POP	0.146	0.016	0.206
A – ROCK	0.104	0.012	0.068
A – SOUNDTRACK	0.062	0.007	0.094
B – POP	0.097	0.007	0.088
B – ROCK	0.419	0.030	0.176
B – SOUNDTRACK	0.097	0.007	0.094
C – DANCE	0.120	0.007	0.068
C – POP	0.040	0.002	0.029
C – ROCK	0.080	0.005	0.027
C – SOUNDTRACK	0.200	0.012	0.156
D – DANCE	0.250	0.009	0.091
D – ROCK	0.125	0.005	0.027
D – SOUNDTRACK	0.125	0.005	0.062
E – DANCE	0.161	0.012	0.114
E – POP	0.097	0.007	0.088
E – ROCK	0.161	0.012	0.068
F – DANCE	0.069	0.005	0.045
F – POP	0.103	0.007	0.088
F – ROCK	0.241	0.016	0.095
F – SOUNDTRACK	0.069	0.005	0.062
VARIOUS – DANCE	0.100	0.019	0.182
VARIOUS – POP	0.113	0.021	0.265
VARIOUS – ROCK	0.175	0.032	0.189

A software implementation of the algorithm based on theoretical assumptions described in Par. 4.1 including also above discussed initial phase subalgorithm was prepared and implemented to a server having the following features: 2 *Athlon MP* 2,2 GHz processors, *Windows 2000*™ OS *MySQL* database server, *Apache*™ WWW server. The result of branch-related factors calculation is illustrated in Fig. 7.

The process of knowledge acquisition does not finish with determining the values of *certainty*, *coverage* and *strength* for each branch. The knowledge base should be prepared for servicing queries with any reduced term set. Correspondingly, the graph should be simplified in advance in order to determine data dependencies applicable to such cases. The knowledge base should be prepared in advance to serve such queries rather than assuming calculating new values of factors related to shorter paths each time a term is dropped (field left empty by the user). That is why in order to shorten the time needed for calculations made in response to a query, all terms are left-out consecutively, one of them at a time while the values of branch factors are calculated each time and stored. This solution lets users to get ready answer for each question almost immediately, independently of the amount of knowledge they possess on the CD record which is searched for. An example of a simplified flow graph is illustrated in Fig. 8. The dropping of the term "Album Artist" node layer entails among others the following calculations:

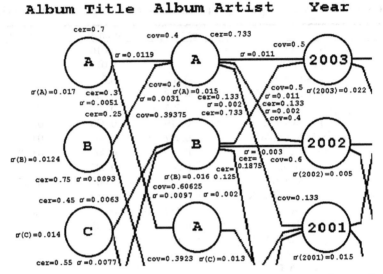

Fig. 7. Fragment of flow graph with marked values of **certainty, coverage** and **strength** calculated for branches.

A -> A -> 2003 ==> B -> 2003
0.0087=0.733*0.0119

C -> B -> 2002 ==> C -> 2002
0.0012=0.1875*0.0063

As was said in Par. 4.1, decision rules can be derived from flow graphs. Correspondingly, the following sample inference rules can be obtained from the graph showed in Fig.6, whose fragment is depicted in Fig. 7:

If *Album Title=B* **and** *Album Artist=A* **and** *Year=2003* **and** *Genre=genre_value* **and** *Track Title=track_title_value* **then** *Number=number_value*

If *Album Title=C* **and** *Album Artist=B* **and** *Year=2002* **and** *Genre=genre_value* **and** *Track Title=track_title_value* **then** *Number=number_value*

The values of: *genre_value, track_title_value and number_value* can be determined from the parts of the graph that are not covered by the figure (for caption resolution limitations). If the user did not provide *Album Artist* value, the direct data flows from the nodes *Album Title* to nodes *Year* can be analyzed as in Fig. 8. The inference rules are shorter in this case and adequate values of *certainty, coverage* and *strength* are adopted. For example the value of rule strength associated with the paths determined by node values: *Album Title=B -> Album Artist=A* (as in Fig. 7) equal to $\sigma=0.0031$ and $\sigma=0.0011$ are replaced by the new value of $\sigma=0.0023$ associated with the path: *Album Title=B -> Year=2003*. The shortened rules corresponding to the previous examples given above are as follows:

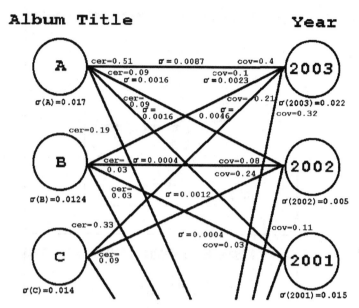

Fig. 8. Simplified flow graph (from Fig. 7) after leaving-out the term: "Album artist".

If *Album Title=B* and *Year=2003* and *Genre=genre_value* and *Track Title=track_title_value* then *Number=nember_value*

If *Album Title=C* and *Year=2002* and *Genre=genre_value* and *Track Title=track_title_value* then *Number=number_value*

The latter inference rules may adopt the same decision attribute (the number of the same CD record), however the rule strength (σ value) can be different in this case. The rule strength is a decisive factor for ordering searching results in the database. The principle of ordering matches is simple: the bigger the rule strength value is, the higher is the position of the CD record determined by the rule in the ordered rank of matches. This principle allows for descendant ordering of queried CD's basing on the rules derived from the analysis of optimal data flow in the graphs which represent the knowledge on CD records.

It is interesting in this context that the attributes in the decision rules can be reordered as long as a route between same nodes consisting of same branches is covered by the rule. The rule can be also reversed or the decision attribute can be swapped with any former conditional attribute. This feature of the decision system results from the principles of *modus ponens* and *modus tollens* valid for logical reasoning made on the basis of flow graphs. Correspondingly, the user of the searching engine based on this decision system, knowing the number of a CD can find information about the remaining metadata. What is also interesting in this context, is that the detailed (and practically impossible to memorize by the user) numbers of beginning packets of tracks might not be necessary, thus the CD album metadata can be searched for effectively without gaining an access to its physical copy. The panels of the constructed searching engine utilized for presenting search results are gathered in Fig. 9.

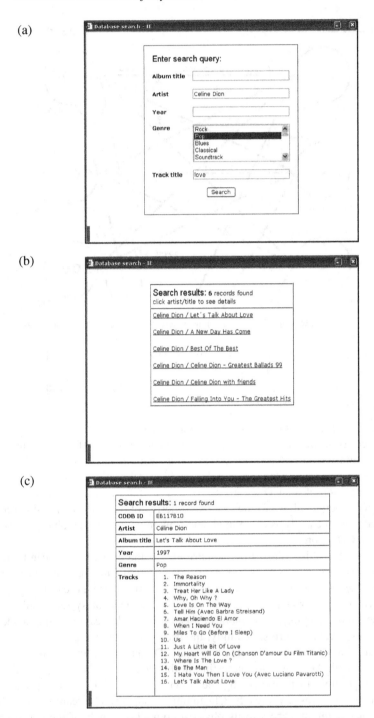

Fig. 9. Panels of the intelligent CDDB database searcher containing the results of a query (from Fig. 7): (a) query panel; (b) result of matches rendering; (b) retrieved metadata.

5   Conclusions

The knowledge base building process in the case of full CDDB database is computationally demanding. The practical limitation of time results from the update time limit of CDDB database. Consequently, the knowledge extraction algorithm application is practically justified if it is possible to complete all computing tasks on a typical server (full set of inference rule derivation) in time shorter that 1 day. As it results form our experiments, this demand is entirely fulfilled in the case of flow graphs application.

A serious problem worth further studying is dealing effectively with multiple key query strings. One of possible solutions is introducing more advanced algorithm allowing suppression of frequency effects of the performer/title, like the algorithm described in Par. 3.3 which uses co-occurrence and correlation analysis for classification of words in sentences. In contrast to previous proposals, however, the application of co-occurrence and correlation analysis may be restricted to information preprocessing stage, because the main part of knowledge can be still stored in Pawlak's flow graphs or in the rule base derived from these graphs.

Another direction of development of the application described in the paper is the currently introduced tagging system ID3v2, covering extended set of multimedia-related metadata. However, even at the present stage of its development, the application makes a practical step towards effective retrieval of knowledge on speech and music recorded in the compact disk format.

Acknowledgements

The research is sponsored by the Committee for Scientific Research, Warsaw, Grant No. 4T11D 014 22, and by the Foundation for Polish Science, Poland.

References

1. Bainbridge, D., Cunningham, S.J., Downie, J.S.: How People Describe Their Music Information Needs: A grounded Theory Analysis of Music Queries. Fourth International Conference on Music Information Retrieval (ISMIR), Baltimore, USA (2003)
2. Downie, J. S.: Music information retrieval. Annual Review of Information Science and Technology 37, Cronin, B., Medford, N. J.: Information Today, ch. 7, (2003) 295-340. Available from http://music-ir.org/downie_mir_arist37.pdf
3. Downie, J. S.: Toward the Scientific Evaluation of Music Information Retrieval Systems. Fourth International Conference on Music Information Retrieval (ISMIR), Baltimore, USA (2003)
4. Ghazfan, D., Indrawan, M., Srinivasan B.: Towards meaningful Bayesian belief networks, IPMU (1996) 841-846
5. Herrera, X., Amatriain, Battle, E., and Serra, X.: Towards Instrument Segmentation for Music Content Description: a Critical Review of Instrument Classification Techniques. Proc. Intern. Symposium on Music Information Retrieval, ISMIR 2000, (2000) http://ismir2000.indiana.edu/ 2000.

6. http://www.audiograbber.com-us.net/

7. http://www.darmstadt.gmd.de/mobile/MPEG7/

8. http://www.freedb.org

9. http://www.gracenote.com

10. http://www.id3.org

11. http://www.ismir.net/

12. http://www.meta-labs.com/mpeg-7-aud

13. http://www.semanticweb.org/

14. http://www.w3.org/2001/sw/

15. Klopotek, M. A., Intelligent Search Engines (in Polish). Akademicka Oficyna Wydawnicza EXIT Warsaw (2001) 304pp.

16. Komorowski, J., Pawlak, Z., Polkowski, L., Skowron, A.: Rough Sets: A Tutorial, Rough Fuzzy Hybridization: A New Trend in Decision-Making. Pal, S. K., Skowron, A. (eds.), Springer-Verlag (1998) 3-98

17. Kostek, B.: Soft Computing in Acoustics, Applications of Neural Networks, Fuzzy Logic and Rough Sets to Musical Acoustics, Studies in Fuzziness and Soft Computing, Physica Verlag, Heidelberg, New York (1999)

18. Kostek, B., and Czyzewski, A.: Representing Musical Instrument Sounds for their Automatic Classification, J. Audio Eng. Soc., vol. 49, No. 9, (2001) 768 – 785

19. Kostek, B.: Musical Instrument Classification and Duet Analysis Employing Music Information Retrieval Techniques. Proc. IEEE Special Issue on Engineering and Music, to be published in (2004)

20. Lindsay, A. T., and Herre, J.: MPEG-7 and MPEG-7 Audio – An Overview. J. Audio Eng. Soc., vol. 49, 7/8, (2001) 589-594

21. Pachet, F., Westermann, G., and Laigre, D.: Musical Data Mining for Electronic Music Distribution, WedelMusic, Firenze (2002)

22. Pal, S. K., Polkowski, L., Skowron, A.: Rough-Neural Computing. Techniques for Computing with Words. Springer Verlag, Berlin, Heidelberg, New York (2004)

23. Pawlak, Z.: Probability, Truth and Flow Graph. Electronic Notes in Theoretical Computer Science, International Workshop on Rough Sets in Knowledge Discovery and Soft Computing, Satellite event of ETAPS 2003, Warsaw, Poland, April 12-13, (2003) Elsevier, Vol. 82 (4) (2003)

24. Pawlak, Z.: Elementary Rough Set Granules: Towards a Rough Set Processor. In: Rough-Neural Computing. Techniques for Computing with Words. Pal, S.K., Polkowski L., Skowron A. (eds.). Springer Verlag, Berlin, Heidelberg, New York (2004) 5-13

25. Pohlman, K.: The Compact Disk Handbook, A-R editions (1992)

26. Turtle, H. R., Croft W. B.: Evaluation of an inference network-based retrieval model. ACM Trans. on Information Systems, Vol. 9 (3) (1991) 187-222

Data Decomposition and Decision Rule Joining for Classification of Data with Missing Values

Rafał Latkowski[1] and Michał Mikołajczyk[2]

[1] Warsaw University, Institute of Computer Science
ul. Banacha 2, 02-097 Warszawa, Poland
R.Latkowski@mimuw.edu.pl
[2] Warsaw University, Institute of Mathematics
ul. Banacha 2, 02-097 Warszawa, Poland
M.Mikolajczyk@mimuw.edu.pl

Abstract. In this paper we present a new approach to handling incomplete information and classifier complexity reduction. We describe a method, called D^3RJ, that performs data decomposition and decision rule joining to avoid the necessity of reasoning with missing attribute values. In the consequence more complex reasoning process is needed than in the case of known algorithms for induction of decision rules. The original incomplete data table is decomposed into sub-tables without missing values. Next, methods for induction of decision rules are applied to these sets. Finally, an algorithm for decision rule joining is used to obtain the final rule set from partial rule sets. Using D^3RJ method it is possible to obtain smaller set of rules and next better classification accuracy than classic decision rule induction methods. We provide an empirical evaluation of the D^3RJ method accuracy and model size on data with missing values of natural origin.

1 Introduction

Rough Set theory, proposed by Pawlak in 1982, creates a framework for handling the imprecise and incomplete data in information systems. However, in classic formalization it is not addressed to the problem of missing attribute values. Some methods for reasoning with missing attribute values were proposed by Grzymała-Busse, Stefanowski, Skowron, Słowiński, Kryszkiewicz and many others. Current findings on Granular Computing, Approximated Reasoning Schemes and Rough-Mereology (see, e.g., [41]) inspired research on new methods for handling incomplete information as well as better understanding of classifier and knowledge description complexity. In this paper we describe two of issues: reasoning under missing attribute values and reduction of induced concept description. A concatenation of solutions for problems related to these issues results in high quality classifier induction method, called D^3RJ.

The D^3RJ method is based on data decomposition and decision rule joining. The aim of this method is to avoid the necessity of reasoning with missing attribute values and to achieve better classification accuracy at the reduced

J.F. Peters et al. (Eds.): Transactions on Rough Sets I, LNCS 3100, pp. 299–320, 2004.

classification time. The D^3RJ method is based on more complex reasoning process, comparing the case of typical algorithms for induction of decision rules. The original incomplete data table is decomposed into data sub-tables without missing values. This is done using total templates that represent information granules describing the resulting data subset. Next, methods for induction of decision rules are applied to these sets. The classic decision rule induction methods are used here. In this way the knowledge hidden in data is extracted and synthesized in form of decision rules, that can also be perceived as information granules. Finally, an algorithm for decision rule joining is used to obtain classifier consisting of generalized rules built from previously induced decision rules. This final phase realizes an additional step of knowledge synthesization and can be perceived as transformation of simpler granules into the more complex ones. The D^3RJ method makes is possible to obtain smaller set of rules and to achieve similar or even better classification accuracy than standard decision rule induction methods known from literature.

In the following section the related work on missing values handling and decision rule joining is presented. In Section 3 we introduce some necessary formal concepts. In Section 4 overview of the D^3RJ method is provided. Section 5 describes the data decomposition phase. Next, the description of rule induction is provided. Section 7 describes the decision rule joining. In Section 8 contain empirical evaluation of the D^3RJ method. The final section presents some conclusions and remarks. This paper is an extended version of [30] where several issues related to decision rule joining were improved.

2   Related Work

2.1   Missing Attribute Values

The problem of reasoning with missing attribute values is known in machine learning and a lot of work has been already done for interpretation of the issues related to this problem as well as methods for reasoning with missing attribute values. However, there is no one satisfactory solution to the problems related to reasoning over incomplete data in the considered sense. In relational databases the nature of missing values was established and for more than a decade also the industrial standards fulfill the proposed logical framework and semantical meaning of the null values. Such an approach is not yet available in data mining at all and particularly, in the rough set theory and practice. Furthermore, it seems to be almost infeasible to discover one theoretical framework for dealing with missing attribute values and their role in induction learning that will fit in all aspects of Machine Learning. The findings in area of missing attribute values are rather loosely connected or even exclusive and do not form any coherent guidelines that would be applicable to a wide range of data mining problems.

The problem of missing values in inductive learning received its attention very early. In late '70 and early '80 there were proposed some findings of Friedman in [9], Kononenko et al. in [24] and Breiman et al. in [7] in this area. The proposed methods are addressed to induction of decision trees. The main idea is based on partitioning and replicating data objects and test nodes. In 1989 Quinlan published experimental evaluation of some proposed approaches (see [45]). The experimental evaluation proved, that the Kononenko's method that partition objects with missing values across all nodes is in most cases the best choice. His work influenced a lot the researchers and many later implementations of decision tree induction follow Kononenko's method used also in *C4.5*. This approach became widely applied due to its high performance and simple interpretation. Recent research made on the complexity of this method showed the great complexity breakdown that occurs when data contain many missing values (cf. [28]).

The methods presented above, for decision trees induced by recursive partitioning, build rather an isolated case that is thoroughly investigated. It is a consequence of popularity of decision trees in research and industry, as well as relative simpleness of decision tree induction algorithms. The other approaches for inducing classifiers directly from data with missing attribute values are usually loosely related to each other, but they perform quite well and are based on interesting ideas for dealing with missing values. Two recent examples of such a methods are *LRI* and *LazyDT*. Weiss and Indurkhya in [57] presented the Lightweight Rule Induction method that is able to induce decision rules over data with missing values. This method is trying to induce decision rules by ignoring cases with missing values on estimated test (descriptor). The proper functioning is obtained by redundancy of descriptors in decision rules as well as by normalization of the test evaluation. Friedman's Lazy Decision Tree method (see [10]) presents a completely different approach to classification process, called lazy learning. The decision tree is constructed on the basis of an object that is currently a subject to classification. Missing values are omitted in classified case and ignored in heuristical evaluation of tests.

Besides the methods that can work directly on data with missing attribute values, also the methods for missing values imputation or replacement were proposed. The simplest method — replacing the missing values with an unused domain value — is known from the beginning of the machine learning. However, this yields in significant decrease of classification accuracy. The applied imputation methods can be roughly categorized into simple ones, that do not build any special model of data or such a model is relatively simple, and more complex ones, that impute the missing values with respect to a determined model for a particular data. The most commonly used simple imputation methods are: imputation with mean or median value, imputation with most common value or imputing with mean, median or most common value, where the mean, median or most common value is calculated only over the objects from the same decision class (see, e.g., [18, 19]). There were proposed also some modifications of these methods, such as using the most correlated attribute instead of the decision class (e.g., [11]). The model based imputation methods are usually used with statistical learning methods and are not widely used in other machine learning algorithms like, e.g., decision rule induction. One of the best methods is the *EM* imputation (see, e.g., [13, 58]), where the Expectation-Maximization model is builded for the data and missing values are replaced by randomizing values with probability taken from the model. The EM imputation can be used together with the *Multiply Imputation* (see [46]), that is applied to improve the accuracy of calculating aggregates and other statistical methods. The imputation methods are inevitable in some applications, e.g., in data warehousing. However, in machine learning the imputation methods not always are competitive and their application is not justified or cannot be properly interpreted.

The problem of missing attribute values was investigated also within the rough set framework. We can mainly distinguish two kinds of approaches to this problem with respect to the modifications of the rough set theory they introduce. In the first group of approaches it is assumed that the missing value handling should be an immanent but special part of rough set theory. As the consequence approaches from this group consist in modification of the indiscernibility relation. In the second group of approaches we include all others that do not assume or do not require such a modification.

The practice of modifying of the indiscernibility relation is rather old and originates not directly from the rough set theory but rather from other mathematics areas like, e.g., universal algebra. The adaptation of concept "partiality" from universal algebra leaded to the *tolerance* or *symmetrical similarity* relation as a replacement for the indiscernibility relation. The successful application of symmetrical similarity relations were investigated among others by Skowron, Słowiński, Stefanowski, Polkowski, Grzymała-Busse and Kryszkiewicz (see, e.g., [19, 25, 44, 52]).

To overcome some difficulties in provided semantics of missing values (see, e.g., [52]) also the other types of the indiscernibility relation replacements were proposed. The one of them is the *nonsymmetric similarity* relation which was investigated in [14, 16, 49, 51–53]. To achieve yet more flexibility also the parametric relations were proposed, sometimes also with the fuzzy extension to the rough set concepts (see, e.g., [15, 51, 53]). All of this modifications enforce a certain semantic of the missing values. Such a sematic applies to all data sets and their attributes (i.e., properties of objects) identically and produce a bias in form of model assumptions. One should state, however, that this approach can be very successful in some applications and definitely produces superior results over the standard indiscernibility relation.

There are some other methods proposed within rough set framework that do not assume modification of the indiscernibility relation. The approach proposed by Grzymała-Busse in *LEM2* algorithm for decision rule induction is to modify the induction process itself (see [19, 20]). The special version of LEM2 algorithm omits the examples with unknown attribute values when building the block for that attribute. Than, a set of rules is induced by using the original LEM2 method.

The completely different approach is proposed in the *Decomposition Method*, where neither the induction process nor the indiscernibility relation is modified (see [27, 29]). In the decomposition method data with missing attribute values is decomposed into subsets without missing values. Then, methods for classifier induction are applied to these sets. Finally, a conflict resolving method is used to obtain final classification from partial classifiers. This method can be applied to any algorithm of classifier induction, also these ones, that cannot directly induce classifiers over data with missing values.

The decomposition method performs very good on data, but introduce some difficulties in interpreting last step of reasoning related to conflict resolving. In this chapter, among the decision rule joining, the idea of data decomposition is investigated. The most important improvement of the data decomposition in comparison to the previous research is avoiding the necessity of combining several different classifiers. The decision rules from resulting classifiers are subject to joining similarly as it is described in [33].

2.2   Decision Rule Induction

The decision rule induction problem has been extensively investigated not only within the rough set framework, but also in other fields. In machine learning several efficient algorithms have been proposed, like, e.g., Michalski's AQ algorithms or $CN2$ algorithms from Clark and Niblett. Rough sets can be used on different stages of rule induction and data processing. The most commonly used approaches are induction of certain and approximate decision rules by generating exhaustive, minimal or satisfactory set of decision rules (see [23, 50]). Such algorithms for decision rule induction were extensively investigated and are implemented in many software systems (see, e.g., [6, 17]).

There were proposed also methods for decision rule induction related to the local properties of data objects (see, e.g., [4, 5]). This approach combines advantages of lazy learning, e.g., reduced computational complexity in the learning phase with advantages taken from induction of rough set based decision rules.

In recent years also a similar problem to the decision rule induction has been investigated — the searching for association rules (see, e.g., [2, 21, 35, 36]). It is possible to represent a set of all the possible descriptors as a set of items. Then the problem of calculating the decision rules corresponds to searching for the sets of items. Each item set corresponds with one decision rule.

Decision rules express the synthesized knowledge extracted from data set. In our research we use the decision rule induction using the indiscernibility matrix and boolean reasoning techniques described in, e.g., [23, 47, 48]. Such decision rules represent some level of redundancy that from one point of view can increase the classification accuracy, while from the second one can result in too many decision rules. There were proposed some approaches for redundancy elimination as well as for classification accuracy improvement in the case of noisy or inexact data. These approaches are mainly based on shortening of decision rules (see, e.g., [4, 34]). The shortening techniques are very useful and with careful parameter assignment can improve classification accuracy and decrease the number of decision rules.

2.3   Decision Rule Joining

Decision rule joining is one of the methods reducing number of decision rules. Although there is not much done in area of decision rule joining or clustering, the reducing number of rules has already been investigated. The most common approach to reduction of the number of decision rules is filtering. The filtering methods assume some heuristical measure on decision rule evaluation and drop unpromising rules with respect to this heuristical evaluation.

One can reduce the number of rules by selecting a subset of all rules using for example quality-based filtering (see, e.g., [1, 40]). In this way we get fewer rules at the cost of reduced classification quality mainly (see, e.g., [54]). With such a reduced set of decision rules some new objects cannot be recognized because they are not matched by rules. Hence, with fewer rules it is more probable that an object will not be recognized. The essential problem is how to rate the quality of decision rules and how to calculate the weights for voting (see, e.g., [12]).

The quality-based filtering methods give low classification quality, but make fewer mistakes than many other decision systems. This is a consequence of smaller set of rules taking part in the voting, which results in lack of classi-fication for weakly recognized objects. This shows that dropping decision rules decrease important information about the explored data.

Recently some methods for decision rule joining and clustering were proposed. The System of Representatives, described in [33], is the method that offers a rule joining. This method achieves very good classification accuracy and model complexity reduction, but it is very time consuming. Therefore we utilize here a simplified method for rule joining that is less time consuming, called Linear Rule Joining (LRJ). This method also achieves good results and has been designed to cooperate with data decomposition method.

3   Preliminaries

3.1   Decision Tables

For the classification and the concept approximation problems we consider data represented in *information systems* called also *information tables* due to its nat-ural tabular representation (see, e.g., [23, 42]). A *decision system* (*decision table*) is an information system with a distinguished attribute called decision (see, e.g., [23, 42]).

Definition 1. *A decision table* $\mathbb{A} = (U, A, \{d\})$ *is a triple, where U is a non-empty finite set of objects called the universe and A is a non-empty set of at-tributes such that $a_i \in A$, $a_i : U \to V_i$ are conditional attributes and $d : U \to V_d$ is a special attribute called decision.*

This definition assumes that all objects have complete description. However, in a real world data frequently not all attribute values are known. Such attribute values that are not available are called *missing attribute values*. The above defini-tion of decision table does not allow an object to have an incomplete description. As a consequence of this fact the missing attribute values are rarely considered and they are not present in theoretical foundations of proposed methods. To be able to deal with missing attribute values we have to extend the definition of a decision table. There are two main concepts on how to give consideration to missing values. The first and more popular is to extend the attribute domains with a special element that denote absence of a regular attribute value. The other approach, taken from universal algebra, is to assume that attributes are *partial* functions in contrast to attributes without missing values assumed to be *total* functions. Both approaches are equivalent, but the first one is easier to implement in computer programs.

Definition 2. *A decision table with missing attribute values* $\mathbb{A} = (U, A, \{d\})$ *is a triple, where U is a non-empty finite set of objects called the universe and A is a non-empty set of attributes such that $a_i \in A$, $a_i : U \to V_i^*$, where $V_i^* = V_i \cup \{*\}$ and $* \notin V_i$, are conditional attributes and $d : U \to V_d$ is a special attribute called decision.*

The special symbol "$*$" denotes absence of the regular attribute value and if $a_i(x) = *$ we say that a_i is not defined on x. Such an approach is frequently used in all domains of computer science. For example in the relational databases a similar notion — "NULL" is used for representing missing attribute values in database record.

If all attribute values are known, the definition of the decision table with missing attribute values is equivalent to the definition of the decision table. From now on we will call decision tables with missing attribute values just decision tables, for short.

3.2   Total Templates

To discover knowledge hidden in data we should search for patterns of regularities in decision tables. We would like to focus here on searching for regularities that are based on the presence of missing attribute values. A standard tool for describing data regularities are *templates* (see, e.g., [37, 38]). The concept of template requires some modifications to be applicable in the incomplete decision table decomposition.

Definition 3. *Let $\mathbb{A} = (U, A, \{d\})$ be a decision table and let $a_i \neq *$ be a total descriptor. An object $u \in U$ satisfies a total descriptor $a_i \neq *$, if the value of the attribute $a_i \in A$ on this object u is not missing in $\mathbb{A}$, otherwise the object u does not satisfy total descriptor.*

Definition 4. *Let $\mathbb{A} = (U, A, \{d\})$ be a decision table. Any conjunction of total descriptors $(a_{k_1} \neq *) \wedge \ldots \wedge (a_{k_n} \neq *)$ is called a total template. An object $u \in U$ satisfies total template $(a_{k_1} \neq *) \wedge \ldots \wedge (a_{k_n} \neq *)$ if the values of attributes $a_{k_1}, \ldots, a_{k_n} \in A$ on the object u are not missing in $\mathbb{A}$.*

Total templates are used to discover regular areas in data without missing values. On the basis of the total templates we can create a granule system in following way. We consider decision sub-tables $\mathbb{B} = (U_{\mathbb{B}}, B, \{d\})$ of the decision table $\mathbb{A}$, where $U_{\mathbb{B}} \subseteq U$ and $B \subseteq A$. A template t uniquely determines a granule

$\mathcal{G}_t = \{\mathbb{B} = (U_\mathbb{B}, B, \{d\})\}$ consisting of such data tables $\mathbb{B}$ that all objects from $U_\mathbb{B}$ satisfies template t and all attributes $b \in B$ occur in descriptors of template t. In granule $\mathcal{G}_t$ exists the maximal decision table $\mathbb{B}_t = (U_{\mathbb{B}_t}, B_t, \{d\})$, such that for all $\mathbb{B}' = (U_{\mathbb{B}'}, B', \{d\}) \in \mathcal{G}_t$ the condition $U_{\mathbb{B}'} \subseteq U_{\mathbb{B}_t} \wedge B' \subseteq B_t$ is satisfied. Such maximal decision table has all attributes that occur in descriptors of template t and all objects from U that satisfy template t.

Once we have a total template t, we can identify it with the sub-table $\mathbb{B}_t$ of original decision table. Such a sub-table consists of the decision attribute, all attributes that are elements of total template and it contains all objects that satisfy template t. Obviously, the decision table $\mathbb{B}_t$ does not contain missing attribute values. We will use this fact later to present the data decomposition process in a formal and easy to implement way.

3.3   Decision Rules

Decision rules and methods for decision rule induction from decision data table without missing attribute values are well known in rough sets (see, e.g., [23, 42]).

Definition 5. *Let $\mathbb{A} = (U, A, \{d\})$ be a decision table. The decision rule is a function $\mathbb{R}: U \to V_d \cup \{?\}$, where $? \notin V_d$. The decision rule consist of condition φ and value of decision $d^\mathbb{R} \in V_d$ and can be also denoted in form of logical formula $\varphi \Rightarrow d^\mathbb{R}$. If the condition φ is satisfied for an object $x \in U$, then the rule classifies x to the decision class $d^\mathbb{R}$ ($\mathbb{R}(x) = d^\mathbb{R}$). Otherwise, rule $\mathbb{R}$ for x is not applicable, which is expressed by the answer $? \notin V_d$ ($\mathbb{R}(x) =?$).*

In above definition one decision rule describes a part of exactly one decision class (in mereological sense [41]). If several rules are satisfied for a given object, than voting methods have to be used to solve potential conflicts. The simplest approach assigns each rule exactly one vote. In more advanced approach the weights are assigned to decision rules to measure their strength in voting (e.g., using their support or quality).

Decision rule induction algorithms produce rules with conjunction of descriptors in the rule predecessor:

$$(a_{k_1}(x) = r^{k_1} \wedge \ldots \wedge a_{k_n}(x) = r^{k_n}) \Rightarrow d^\mathbb{R},$$

where $x \in U$, $a_{k_1}, \ldots, a_{k_n} \in A$, $r^{k_i} \in V_{k_i}$. For example:

$$\mathbb{R}: (a_1(x) = 1 \wedge a_3(x) = 4 \wedge a_7(x) = 2) \Rightarrow d^\mathbb{R}.$$

The D^{3}RJ method produces more general rules, where each descriptor can enclose subset of values. We call such rules the *generalized decision rules* (cf. [37, 56]). The generalized rules have the form:

$$(a_{k_1}(x) \in R^{k_1} \wedge \ldots \wedge a_{k_n}(x) \in R^{k_n}) \Rightarrow d^\mathbb{R},$$

where $R^{k_i} \subseteq V_{k_i}$. It is easy to notice, that any classic decision rule is also a generalized decision rule, where $R^i = \{r^i\}$. From now on we will assume that all decision rules are generalized.

The conditional part of a decision rule can be represented by ordered sequence of attribute value subsets $\left\{R^i\right\}_{a_i \in A}$ for any chosen liner order on A. For example, the decision rule $\mathbb{R}_1$, can be represented as:

$$\mathbb{R}_1 : (\{1\}, \emptyset, \{4\}, \emptyset, \emptyset, \emptyset, \{2\}) \Rightarrow d^{\mathbb{R}}.$$

The empty set denotes absence of condition for that attribute.

4   D³RJ

The D³RJ method is developed in the frameworks of Granular Computing and Rough-Mereology [41]. The processing consists of four phases called the data decomposition, decision rule induction, decision rule shortening and decision rule joining.

In the first phase the data that describes the whole investigated phenomenon is decomposed — partitioned into a number of subsets that describe, in a sense, parts of investigated phenomenon. Such a procedure creates an overlapped, but non-exhaustive covering that consist of elements similar to the covered data. These elements are data subsets and parts in the mereological sense of the whole, i.e., the original data. The data decomposition phase is aiming to avoid the problem of reasoning from data with incomplete object descriptions.

In the second phase information contained in parts, i.e., data subsets is transformed using inductive learning, to a set of decision rules. Each decision rule can be perceived as an information granule that correspond to knowledge induced from the set of objects that satisfy the conditional part of decision rule. The set of decision rules can be perceived as a higher level granule that represents knowledge extracted from the data subset. As it is explained later, we can apply any method of decision rule induction, including such ones that cannot deal with missing values. Often methods that make it possible to properly induce decision rules from data with missing values lead to inefficient algorithms or algorithms with low quality of classification. With help of a data decomposition decision rules are induced from data without missing values to take an advantage of lower computational complexity and more precise decision rules.

Third phase is the rule shortening. It is very useful because it reduces complexity of rule set and improves classifier resistance to noise and data disturbances.

In the fourth phase the set of classic decision rules is converted to the smaller and simplified set of more powerful representation of decision rules. In this phase decision rules are clustered and joined to a coherent classifier. The constructed generalized rules can be treated as the higher level granules that represent knowledge extracted from several decision rules — lower level granules. The main objectives of the decision rule joining are reduction of classifier complexity and simplification of knowledge representation.

The D³RJ method returns a classifier that can be applied to a data with missing attribute values in both, learning and classifying.

5 Data Decomposition

The data decomposition should be done in accordance to regularities in a real-world interest domain. We expect the decomposition to reveal patterns of missing attribute values with a similar meaning for the investigated real-world problem. Ideally, the complete sub-tables that are result of the decomposition should correspond to natural subproblems of the whole problem domain.

The result of data decomposition is a family of subsets of original data. Subsets of original decision table must meet some requirements in order to achieve good quality of inductive reasoning as well as to be applicable in case of methods that cannot deal with missing attribute values. We expect the decision sub-tables to exhaustively cover the input table, at least in the terms of objects, to minimize the possibility of loosing useful information. They should contain no missing values. It is also obvious that the quality of inductive reasoning depends on a particular partition and some partitions are better then others.

With the help of introduced concept of total template it is possible to express the goal of the data decomposition phase in terms of total templates. The maximal decision sub-table $\mathbb{B}_t \in \mathcal{G}_t$ is uniquely determined by template t. With such an assignment we can consider the data decomposition as a problem of covering data table with templates. The finite set of templates $S = \{t_1, t_2, \ldots, t_n\}$ determines uniquely a finite decomposition $D = \{B_{t_1}, B_{t_2}, \ldots, B_{t_n}\}$ of the decision table $\mathbb{A}$, where $B_{t_i} \in \mathcal{G}_{t_i}$ is a maximal decision sub-table related to template t_i. With such a unique assignment the decomposition process can be formally described in terms of total templates. The preference over particular decompositions can be translated to preference of particular set of templates.

We illustrate the data decomposition with an example. Let consider the following decision table:

	a	b	c	d
x_1	1	0	*	1
x_2	0	1	1	0
x_3	*	0	1	1
x_4	*	1	0	1

In above decision table 92 out of 127 nonempty combinations of seven possible total templates create proper data decompositions, i.e. that exhaustively cover all objects. For example, the total template $(a \neq *) \wedge (b \neq *)$, which covers objects x_1 and x_2, with the total template $(b \neq *) \wedge (c \neq *)$, which covers objects x_2, x_3 and x_4, create a proper data decomposition.

The problem of covering decision table with templates is frequently investigated (see, e.g., [37, 38]) and we can make an advantage of broad experience in this area. In our case the templates cover the original decision table in following sense. We say that an object x is *covered* by a template t if object x satisfies a template t. The decision table is covered (almost) completely, when (almost) all objects from U are covered by at least one template from a set of templates. The preferences or constraints on the set of templates are translated partially to preferences or constraints on templates and partially to constraints on an algorithm that generates the set.

The standard approach to the problem of covering decision table with templates is to apply a greedy algorithm. Such an approach is justified, because it is known that greedy algorithm is close to best approximate polynomial algorithms for this problem (see [8, 22, 32, 39]). The greedy algorithm generates the best template for a decision table with respect to a defined criterion and removes all objects that are covered by generated template. In subsequent iterations the decision table is reduced in size by objects that are already covered and the generation of the next best template is repeated. The algorithm continues until a defined stop criterion is satisfied. The most popular stop criterion is to have all objects covered by generated set of templates. Such a criterion is also suitable for our purpose. One should notice that a template generated in further iteration can cover objects already removed from decision table. This property allows, if it is necessary, to include a particular object in two or more data sub-tables related to a specific pattern of data.

5.1 Decomposition Criteria

Following the guidelines on covering decision table with templates we have to choose a preference measure that define the concept of best template. The template evaluation criterion should prefer decision sub-tables relevant to the data decomposition and to the approximated concept. This nontrivial problem was investigated in [26, 27, 29]. It is very difficult to define the proper criterion for an individual template for generating decompositions of high quality. This problem could be possibly solved with help of ontology knowledge base for investigated phenomenon, but for now such ontologies are not commonly available. We have to relay only on some morphological and data-related properties of decision table $\mathbb{B}_t$ in order to evaluate template t.

The frequently applied template evaluation function measures the amount of covered data with help of *template height* and *template width*. The template height, usually denoted as $h(t)$ is the number of covered objects, while the template width, denoted as $w(t)$ is the number of descriptors. To obtain a template evaluation function, also called template quality function, we have to combine these two factors to get one value. The usual formula is to multiply these two factors and get the number of covered attribute-value pairs.

$$q_1(t) = w(t) \cdot h(t) \tag{1}$$

The importance of with and height in q_1 can be easily controlled by manipulating the importance factor.

$$q_2(t) = w(t)^\beta \cdot h(t) \tag{2}$$

For such preference measures finding the best template is NP-hard (see, e.g., [37]), so usually also here approximated algorithms instead of exact one are used. It was also proved that there exist measures for which problem of searching for maximal template is PTIME. For example it is enough to replace the multiplication of width and height with addition and the resulting problem can be solved by a polynomial algorithm. Unfortunately, for the decomposition and generally for knowledge discovery problems measures that lead to polynomial complexity are inaccurate and unattractive.

The relation of template evaluation functions q_1 and q_2 with expected properties of decision tables relevant for inductive learning can be easily justified. From one point of view the quality of learning depends on the number of examples. It is proven that inductive construction of concept hypothesis is only feasible, when we can provide enough number of concept examples. A strict approach to this problem can be found in [55] where Vapnik-Chervonenkis dimension is presented as a tool for evaluating required number of examples. From the second point of view using inductive learning we try to discover relationships between decision attribute and conditional attributes. A precise description of concepts in terms of conditional attributes values is required to achieve good quality of classification. Without an attribute that values are important to concept description it is impossible accurately approximate a concept.

Methods that determine the best template with respect to the quality functions q_1 and q_2 are frequently investigated and well documented (see, e.g., [37, 38]). Unfortunately, in our case such quality functions do not sufficiently prefer the templates that are useful for data decomposition over the others. The experimental evaluation in [26] showed that a lot of templates with similar size (i.e. width and height) have very different properties for classifier induction and data decomposition.

There were proposed some other template evaluation functions (cf. [27, 29]) that perform much better than simple q_1 and q_2 functions presented above. These function have some similar properties to the feature selection criteria because the data decomposition itself depends on proper feature selection. The most important issue in selecting such measures is to solve the trade-off between computational complexity of function evaluation and the quality of resulting decomposition.

In rough sets some useful concepts to measure the information-related properties of data set are known, e.g., size of positive region or conflict measure. Based on these and similar concepts a number of template evaluation function were proposed and examined. One of the most promising heuristical template evaluation counts the average purity in each indiscernibility class:

$$G(t) = \sum_{i=1}^{K} \frac{\max_{c \in V_d} \mathrm{card}(\{y \in [x^i]_{\mathrm{IND}_t} : d(y) = c\})}{\mathrm{card}([x^i]_{\mathrm{IND}_t})}. \qquad (3)$$

In above formula K is the number of indiscernibility classes (classes of abstraction of the indiscernibility relation IND_t) and $[x^i]_{\mathrm{IND}_t}$ denotes the i-th indiscernibility class. The above formula is calculated for the maximal decision sub-table $\mathbb{B}_t$ related to the template t, in particular the indiscernibility relation IND_t is based on the attributes from the template t and the indiscernibility classes are constructed only from objects that do not have missing values on these attributes.

To ensure the expected properties of decomposition the heuristical template evaluation can be combined with size properties. Similarly to the q_2 function, one can incorporate an exponent to control the importance of each component of the formula.

$$q_3(t) = w(t)^\beta \cdot h(t) \cdot G(t)^\gamma \tag{4}$$

There is a number of possible heuristical evaluations functions that can be apply here. One can also use an approach known in feature selection as *wrapper* method, where the classifier induction algorithm is used to evaluate properties of investigated feature subset (see, e.g., [29]). The q_3 template evaluation function combining the heuristical function G with size properties showed in experiments to be reasonable good with respect to quality at the minimal computational cost, while, e.g., the functions based on classifier trials improve quality not so much at the enormous computational cost.

6   Decision Rule Induction

The data decomposition phase delivers a number of data tables free from missing values. Such data tables enable us to apply any classifier induction method. In particular, the methods for inducing decision rules, that frequently suffer from lack of possibility to induce rules from data with missing values can be used. On each data table returned from the decomposition phase we apply an algorithm for decision rule induction.

In D^3RJ we use a method inducing all possible consistent decision rules, called also optimal decision rules. This method induces decision rules based on indiscernibility matrix (see, e.g., [23, 47, 48]). The indiscernibility matrix, related to the indiscernibility relation, indicates which attributes differentiates each two objects from different decision classes. Using this matrix and boolean reasoning we can calculate a set of reducts.

A reduct is a minimal (in inclusion sense) subset of attributes that is sufficient to separate every two objects with different decision. For each reduct decision rule induction algorithm can generate many decision rules. Different reducts usually yields to different rule sets. These sets of rules are subject to joining, clustering and reduction in the next, decision rule joining phase.

The treatment of decision rule sets in D^3RJ differs from usual role of these sets. The obtained sets of decision rules, each one from one decision sub-table, are merged into one set of decision rules. It gives highly redundant set of decision rules, where each object is covered by at least one decision rule using its non-missing attribute values. The simplest reduction of obtained set of rules is that duplicate rules are eliminated. The more advanced classifier complexity reduction employed in D^3RJ is decision rule clustering and joining.

6.1   Rule Shortening

The decision rule shortening is a frequently utilized approach for achieving shorter and more noise-redundant decision rules (see, e.g., [2, 34, 60]). In shortening process unnecessary or weak descriptors in the conditional part of a decision rule are eliminated. The method for decision rule shortening drops some descriptors from conjunction φ in the left part of the rule.

The shortened decision rules can possibly misclassify objects. To control this phenomenon the parameter α of decision rule shortening is utilized, which steers the minimal possible accuracy of decision rule. In other words decision rule after shortening cannot misclassify more than $1 - \alpha$ objects. The side effect of the decision rule shortening is possibility of multiplication of decision rules, i.e., the result of shortening of one decision rule can be several decision rules. This effect is balanced from the other side by that one shortened decision rule can be a result of shortening of several decision rules. For example, the decision rule $\mathbb{R}$ can be shortened to decision rules $\mathbb{R}_1$, $\mathbb{R}_2$ and $\mathbb{R}_3$:

$$\mathbb{R}: \quad (a_1(x) = 1 \wedge a_3(x) = 4 \wedge a_7(x) = 2) \Rightarrow d,$$

$$\begin{aligned} \mathbb{R}_1: &\quad (a_1(x) = 1 \wedge a_3(x) = 4) \Rightarrow d, \\ \mathbb{R}_2: &\quad (a_1(x) = 1 \wedge a_7(x) = 2) \Rightarrow d, \\ \mathbb{R}_3: &\quad (a_3(x) = 4 \wedge a_7(x) = 2) \Rightarrow d. \end{aligned}$$

Continuing the example, the decision rule $\mathbb{R}_1$ can be result of shortening of decision rules $\mathbb{R}$ and $\mathbb{S}$:

$$\mathbb{S}: \quad (a_1(x) = 1 \wedge a_3(x) = 4 \wedge a_5(x) = 3) \Rightarrow d.$$

In practice we never observe increase of decision rule set after shortening. The decision rule shortening always decrease number of rules almost linearly with respect to the factor α.

7   Decision Rule Joining

The decision rule joining is employed at the end of D^{3}RJ method to reduce complexity of classifier and improve the classification quality. In the decision rule joining we allow to join only rules from the same decision class. It is possible to join two rules that have different decisions, but it would make this method more complicated. By joining rules with different decisions we calculate rules dedicated not for one decision but for a subset of possible decisions. These rules could be used to build hierarchical rule systems.

The main idea of decision rule joining is clustering that depends on distance computed from comparison of logical structures of rules. Similar rules are easy to join and by joining them we get rules that have similar properties.

Definition 6. *Let* $\mathbb{A} = (U, A, \{d\})$ *be a decision table and let* $\mathbb{R}_1$, $\mathbb{R}_2$ *be generalized rules calculated from the decision table* $\mathbb{A}$. *We define the distance function:*

$$\mathrm{dist}(\mathbb{R}_1, \mathbb{R}_2) = \begin{cases} \mathrm{card}(A) & \text{when } d^{\mathbb{R}_1} \neq d^{\mathbb{R}_2} \\ \sum_{a_i \in A} d_i(R_1^i, R_2^i) & \text{otherwise} \end{cases}$$

where:

$$d_i(X, Y) = \frac{\mathrm{card}((X - Y) \cup (Y - X))}{\mathrm{card}(V_i)}.$$

The above distance function is used for comparison of decision rule logical structures and for estimation of their similarity. This function differs from the one presented in [30]. It gives better results and is easier to interpret. Let us consider an example of simple rule joining:

$$\mathbb{R}_1 : (\{1\}, \{3\}, \emptyset, \{1\}, \{2\}, \emptyset, \{2\}) \Rightarrow d,$$
$$\mathbb{R}_2 : (\{2\}, \{3\}, \emptyset, \{2\}, \{2\}, \emptyset, \{3\}) \Rightarrow d.$$

If we suppose that each attribute a_i has a domain V_i with ten values $card(V_i) = 10$, then distance between these two rules is $dist(\mathbb{R}_1, \mathbb{R}_2) = 0.6$. After joining decision rules $\mathbb{R}_1$ and $\mathbb{R}_2$ we obtain a generalized decision rule:

$$\mathbb{R} : (\{1, 2\}, \{3\}, \emptyset, \{1, 2\}, \{2\}, \emptyset, \{2, 3\}) \Rightarrow d.$$

To illustrate on example the further classification abilities of created generalized decision rule lets consider following objects:

x_1: 1 3 3 1 2 3 2 x_3: 6 3 7 1 2 3 2
x_2: 2 3 1 2 2 5 2 x_4: 1 3 4 1 5 3 3

The objects x_1 and x_2 are classified by the generalized rule $\mathbb{R}$ to the decision class d, while the objects x_3 and x_4 are not recognized and the rule $\mathbb{R}$ returns the answer "?".

Moreover, we can join the generalized rules exactly in the same way as the classic ones. Formally speaking a new rule obtained from $\mathbb{R}_m$ and $\mathbb{R}_n$ have a form $\{R^i_{\mathbb{R}_m + \mathbb{R}_n}\}_{a_i \in A} \Rightarrow d$, where $R^i_{\mathbb{R}_m + \mathbb{R}_n} := R^i_m \cup R^i_n$. The D^{3}RJ method utilizes a decision rule joining algorithm as described in following points.

1. Let $X^{\mathbb{R}}$ be a set of all induced rules. We can assume that it is a set of generalized rules, because every classic rule can be interpreted as a generalized rule.
2. Let $\mathbb{R}_m \in X^{\mathbb{R}}$ and $\mathbb{R}_n \in X^{\mathbb{R}}$ be such, that $d^{\mathbb{R}_m} = d^{\mathbb{R}_n}$ and

$$dist(\mathbb{R}_m, \mathbb{R}_n) = \min_{i,j}\{dist(\mathbb{R}_i, \mathbb{R}_j) : \mathbb{R}_i, \mathbb{R}_j \in X^{\mathbb{R}} \wedge d^{\mathbb{R}_i} = d^{\mathbb{R}_j}\}.$$

3. If there exist $\mathbb{R}_m$ and $\mathbb{R}_n$ in $X^{\mathbb{R}}$ such that $dist(\mathbb{R}_m, \mathbb{R}_n) < \varepsilon$ then the set of rules $X^{\mathbb{R}}$ is modified as follows:

$$X^{\mathbb{R}} := X^{\mathbb{R}} - \{\mathbb{R}_m, \mathbb{R}_n\},$$

$$X^{\mathbb{R}} := X^{\mathbb{R}} \cup \{\mathbb{R}_{\mathbb{R}_m + \mathbb{R}_n}\},$$

where $\mathbb{R}_{\mathbb{R}_m + \mathbb{R}_n}$ is a new rule obtained by joining $\mathbb{R}_m$ and $\mathbb{R}_n$.
4. If the set $X^{\mathbb{R}}$ has been changed then we go back to step 2, otherwise the algorithm is finished.

We can assume that, for example, $\varepsilon = 1$. The algorithm ends when in the set $X^{\mathbb{R}}$ are no two rules from the same decision class that are close enough.

Presented method called Linear Rule Joining (LRJ) is very simple and efficient in time.

Table 1. Classification accuracy of the classic exhaustive decision rule induction and the D^3RJ method using various decomposition criteria and decision rule shortening.

α	No decomposition	$w \cdot h$	$w \cdot h \cdot G$	$w \cdot h \cdot G^8$
1.0	70.15	70.86	71.57	70.65
0.9	71.64	71.02	71.80	71.18
0.8	73.30	72.41	73.11	72.69
0.7	71.87	71.71	72.11	72.21
0.6	69.72	69.37	70.06	69.80
0.5	67.93	70.40	71.13	71.86
0.4	66.81	70.98	71.06	71.11
0.3	68.28	71.23	71.41	71.33
0.2	66.47	71.60	71.54	71.55
0.1	66.14	71.73	71.61	71.60

8 Empirical Evaluation

There were carried out some experiments in order to evaluate the D^3RJ method. Results were obtained using the ten-fold Cross-Validation (CV10) evaluation. The experiments were performed with different decomposition approaches as well as without using decomposition method at all. All data sets used in evaluation of the D^3RJ method were taken from *Recursive-Partitioning.com* [31]. The selection of these data sets was based on amount of missing attribute values and their documented natural origin. We selected following 11 data tables:

- att — AT&T telemarketing data, 2 classes, 5 numerical attributes, 4 categorical attributes, 1000 observations, 24.4% incomplete cases, 4.1% missing values.
- ech — Echocardiogram data, 2 classes, 5 numerical attributes, 1 categorical attribute, 131 observations, 17.6% incomplete cases, 4.7% missing values.
- edu — Educational data, 4 classes, 9 numerical attributes, 3 categorical attributes, 1000 observations, 100.0% incomplete cases, 22.6% missing values.
- hco — Horse colic database, 2 classes, 5 numerical attributes, 14 categorical attributes, 368 observations, 89.4% incomplete cases, 19.9% missing values.
- hep — Hepatitis data, 2 classes, 6 numerical attributes, 13 categorical attributes, 155 observations, 48.4% incomplete cases, 5.7% missing values.
- hin — Head injury data, 3 classes, 6 categorical attributes, 1000 observations, 40.5% incomplete cases, 9.8% missing values.
- hur2 — Hurricanes data, 2 classes, 6 numerical attributes, 209 observations, 10.5% incomplete cases, 1.8% missing values.
- hyp — Hypothyroid data, 2 classes, 6 numerical attributes, 9 categorical attributes, 3163 observations, 36.8% incomplete cases, 5.1% missing values.
- inf2 — Infant congenital heart disease, 6 classes, 2 numerical attributes, 16 categorical attributes, 238 observations, 10.5% incomplete cases, 0.6% missing values.
- pid2 — Pima Indians diabetes , 2 classes, 8 numerical attributes, 768 observations, 48.8% incomplete cases, 10.4% missing values.

Table 2. Number of decision rules using the classic exhaustive decision rule induction and the D^3RJ method using various decomposition criteria and decision rule shortening.

α	No decomposition	$w \cdot h$	$w \cdot h \cdot G$	$w \cdot h \cdot G^8$
1.0	9970.54	1149.67	1031.33	872.90
0.9	8835.55	1050.29	941.30	807.19
0.8	6672.00	862.11	783.09	677.45
0.7	4945.65	685.23	626.16	545.29
0.6	3114.22	384.32	349.19	308.29
0.5	1682.63	203.57	193.37	176.61
0.4	1158.45	164.12	159.44	150.85
0.3	661.78	74.09	75.77	72.65
0.2	366.80	43.77	44.95	42.85
0.1	227.59	35.49	36.25	34.00

– smo2 — Attitudes towards workplace smoking restrictions, 3 classes, 4 numerical attributes, 4 categorical attributes, 2855 observations, 18.7% incomplete cases, 2.5% missing values.

In presented results the exhaustive rule induction method was used to induce classifiers from the decision subtables. This method is implemented in the *RSES-Lib* software (see [6]). The data decomposition was done with the help of a genetic algorithm for best template generation (see [29]).

Table 1 presents a general comparison of the classification accuracy using the classic exhaustive decision rule induction with the D^3RJ method using various decomposition criteria and shortening factor values α in range from 0.1 to 1.0. Table contains the classification accuracy averaged over eleven tested data sets and ten folds of cross-validation (CV10). In the Table 2 the similar comparison is presented with respect to the number of decision rules. The detailed results are presented in next tables. From averages presented in Table 1 one can see that in general the classification accuracy of the D^3RJ method is similar or slightly worse than classic decision rules at the top of the table, but slightly better at the bottom of it, where the shortening factor is lower. It suggest that if the decision rules are more general and shorter then they are easier to join and the D^3RJ method performs better. Table 2 that present number of decision rules, shows that the D^3RJ method requires averagely 8 times less decision rules than the classic exhaustive decision rules, called also optimal decision rules. Thus, the reduction of the classification abilities is not as high as the reduction of the model size.

Table 3 presents detailed experimental results of D^3RJ method with use of template evaluation function $q = w \cdot h \cdot G$ and shortening factor $\alpha = 0.8$. The results are presented for the decomposition method without decision rule joining as well as with the decision rule joining. The decomposition method without decision rule joining uses the standard voting over all decision rules induced from sub-tables. The compression ratio presented in this table is the ratio of the number of decision rules without the decision rule joining to the number of

Table 3. The detailed empirical evaluation of the D^3RJ method using the shortening factor $\alpha = 0.8$, and template evaluation function $q = w \cdot h \cdot G$.

| | Before joining | | After joining | | Profits | |
| | | | | | Com-pres-sion | Imp-rove-ment |
Table	Accuracy	# Rules	Accuracy	# Rules		
att	60.50 ±4.39	2459.3 ±586.24	59.48 ±4.80	673.0 ±169.03	3.65	-1.02
ech	69.19 ±8.65	201.0 ±39.41	68.00 ±7.90	93.9 ±12.57	2.14	-1.19
edu	50.91 ±3.20	3580.9 ±61.20	54.20 ±3.97	397.2 ±22.84	9.02	3.29
hco	82.58 ±7.97	1440.1 ±527.19	83.94 ±6.88	391.2 ±142.73	3.68	1.36
hep	79.42 ±1.52	1454.5 ±104.70	79.42 ±1.52	1253.5 ±85.62	1.16	0.00
hin	72.51 ±4.46	436.2 ±17.88	72.01 ±4.72	285.5 ±15.38	1.53	-0.50
hur2	82.93 ±7.49	197.9 ±38.73	82.84 ±7.58	94.6 ±21.87	2.09	-0.09
hyp	95.23 ±0.09	420.2 ±39.26	95.29 ±0.16	150.0 ±8.16	2.80	0.06
inf2	70.22 ±9.67	4298.1 ±206.92	69.43 ±9.48	3866.3 ±192.67	1.11	-0.79
pid2	72.52 ±4.10	2606.3 ±121.75	70.84 ±3.92	222.9 ±16.86	11.69	-1.68
smo2	64.90 ±2.69	6108.8 ±64.77	68.72 ±0.85	1185.9 ±29.38	5.15	3.82
avg	72.81 ±4.93	2109.39 ±164.37	73.11 ±4.71	783.1 ±65.19	2.69	0.30

decision rules with the decision rule joining. The improvement is the difference of the classification accuracy between classification without and with decision rule joining. As we can see the decision rule joining not only reduces the number of decision rules, but also improves the classification accuracy. However, the improvement of the classification accuracy is not significant (Wilcoxon signed rank test p-value is 0.17) as well as the worsening in comparison to classic decision rule induction is not significant (Wilcoxon signed rank test p-value is 0.47). Reducing shortening factor gives the D^3RJ method advantage over both other approaches.

Table 4 presents detailed experimental results of D^3RJ method with use of template evaluation function $q = w \cdot h \cdot G^8$ and shortening factor $\alpha = 0.9$. Similarly to the previous table the results are presented for the decomposition method without decision rule joining as well as with the decision rule joining. The compression and improvement factors are also provided. The D^3RJ method using the $w \cdot h \cdot G^8$ criterion in the decomposition phase achieves similar results to the D^3RJ method using the $w \cdot h \cdot G$. As we can see, the decision rule joining significantly reduces model complexity and improves its predictive abilities. In this case the classification accuracy improvement is significant (Wilcoxon signed rank test p-value is 0.001), but the worsening in comparison to classic decision rule induction is not significant (Wilcoxon signed rank test p-value is 0.39). The D^3RJ method performs quite well requiring almost three times less decision rules then the decomposition method without rule joining and almost eleven times less then classic decision rule induction.

Table 4. The detailed empirical evaluation of the D^3RJ method using the shortening factor $\alpha = 0.9$, and template evaluation function $q = w \cdot h \cdot G^8$.

Table	Before joining		After joining		Profits	
	Accuracy	# Rules	Accuracy	# Rules	Com-pres-sion	Imp-rove-ment
att	55.20 ±2.21	2275.8 ±32.88	57.48 ±5.34	612.2 ±8.61	3.72	2.28
ech	67.34 ±9.94	157.3 ±33.67	65.58 ±7.09	58.1 ±9.51	2.71	-1.76
edu	47.72 ±5.09	4080.2 ±56.81	53.10 ±2.76	427.0 ±23.32	9.56	5.38
hco	81.79 ±6.26	2126.9 ±120.15	82.60 ±6.05	593.4 ±71.54	3.58	0.81
hep	79.46 ±4.93	758.4 ±177.42	81.41 ±5.99	611.1 ±162.98	1.24	1.95
hin	68.30 ±3.35	589.9 ±19.25	67.90 ±3.76	358.8 ±13.98	1.64	-0.40
hur2	76.10 ±7.86	77.3 ±12.02	74.69 ±11.72	31.7 ±7.58	2.44	-1.41
hyp	95.23 ±0.09	562.5 ±46.01	95.26 ±0.13	169.3 ±11.19	3.32	0.03
inf2	64.75 ±8.20	4854.0 ±488.92	66.08 ±8.96	4412.6 ±389.21	1.10	1.33
pid2	72.53 ±5.27	1953.7 ±136.74	71.09 ±5.16	149.9 ±13.81	13.03	-1.44
smo2	55.97 ±2.38	7897.4 ±57.71	67.81 ±0.99	1455.0 ±21.06	5.43	11.84
avg	69.49 ±5.05	2303.0 ±107.42	71.18 ±5.27	807.2 ±66.62	2.85	1.69

9   Conclusions

The presented method consists of two main steps. The first one, called the decomposition step, makes it possible to split decision table with missing attribute values into more tables without missing values. In the second step one classifier (decision system) is induced from decision tables returned from the first step by joining some smaller subsystems of decision rules.

   In the consequence we obtain a simple strategy for building decision systems for data tables with missing attribute values. Although the obtained decision rules are generated only for complete data, they are able to classify data with missing attribute values. It is done without using the missing values explicitly in the decision rule formula. For bigger decision tables the proposed approach works faster than one-pass classic decision rule induction. Moreover, we can use in this task a parallel computing because created subsystems are independent. It seems that in this way it is possible to solve many hard classification problems in relatively short time. The further advantage from the decision rule set reduction is reduction of time necessary for classification of test objects. The obtained results showed that the presented method is very promising for classification problems with missing attribute values in data sets.

Acknowledgments

The authors would like to thank professor Andrzej Skowron for his support while writing this paper. The research has been supported by the grant 3T11C00226 from Ministry of Scientific Research and Information Technology of the Republic of Poland.

References

1. Ågotnes, T.: Filtering large propositional rule sets while retaining classifier performance. Master's thesis, Department of Computer and Information Science, Norwegian University of Science (1999)
2. Agrawal, R., Srikant, R.: Fast algorithms for mining association rules in large databases. In Bocca, J.B., Jarke, M., Zaniolo, C., eds.: VLDB'94, Morgan Kaufmann (1994) 487–499
3. Alpigini, J.J., Peters, J.F., Skowron, A., Zhong, N., eds.: Rough Sets and Current Trends in Computing, Third International Conference, RSCTC 2002, Malvern, PA, USA, October 14-16, 2002, Proceedings. LNCS 2475, Springer (2002)
4. Bazan, J.G.: A comparison of dynamic and non-dynamic rough set methods for extracting laws from decision table. [43] 321–365
5. Bazan, J.G.: Discovery of decision rules by matching new objects against data tables. In Polkowski, L., Skowron, A., eds.: Rough Sets and Current Trends in Computing, RSCTC'98. LNCS 1424, Springer (1998) 521–528
6. Bazan, J.G., Szczuka, M.S., Wróblewski, J.: A new version of rough set exploration system. [3] 397–404
7. Breiman, L., Friedman, J.H., Olshen, R.A., Stone, P.J.: Classification and Regression Trees. Wadsworth International Group (1984)
8. Feige, U.: A threshold of $\ln n$ for approximating set cover (preliminary version). In: Proceedings of the 28th ACM Symposium on the Theory of Computing, ACM (1996) 314–318
9. Friedman, J.H.: A recursive partitioning decision rule for non-parametric classification. IEEE Trasactions on Computer Science **26** (1977) 404–408
10. Friedman, J.H., Kohavi, R., Yun, Y.: Lazy decision trees. In Shrobe, H., Senator, T., eds.: Proceedings of the AAAI96 and IAAI96. Volume 1., AAAI Press / The MIT Press (1996) 717–724
11. Fujikawa, Y., Ho, T.B.: Cluster-based algorithms for filling missing values. In: Proceedings of PAKDD-2002. LNCS 2336, Springer (2002) 549–554
12. Gago, P., Bento, C.: A metric for selection of the most promising rules. [62] 19–27
13. Ghahramani, Z., Jordan, M.I.: Supervised learning from incomplete data via an EM approach. In Cowan, J.D., Tesauro, G., Alspector, J., eds.: Advances in Neural Information Processing Systems. Volume 6., Morgan Kaufmann (1994) 120–127
14. Greco, S., Matarazzo, B., Słowiński, R.: Handling missing values in rough set analysis of multi-attribute and multi-criteria decision problems. [59] 146–157
15. Greco, S., Matarazzo, B., Słowiński, R.: Rough sets processing of vague information using fuzzy similarity relations. In Caldue, C.S., Paun, G., eds.: Finite vs. infinite: contribution to an eternal dilemma, Berlin, Springer (2000) 149–173
16. Greco, S., Matarazzo, B., Słowiński, R., Zanakis, S.: Rough set analysis of information tables with missing values. In: Proceedings of 5th International Conference Decision Sciences Institute. Volume 2. (1999) 1359–1362
17. Grzymała-Busse, J.W.: Lers–a system for learning from examples based on rough sets. In Słowinski, R., ed.: Intelligent Decision Support. Handbook of Applications and Advances of the Rough Sets Theory, Kluwer (1992) 3–18
18. Grzymała-Busse, J.W., Grzymała-Busse, W.J., Goodwin, L.K.: A closest fit approach to missing attribute values in preterm birth data. [59] 405–413
19. Grzymała-Busse, J.W., Hu, M.: A comparison of several approaches to missing attribute values in data mining. [61] 378–385

20. Grzymała-Busse, J.W., Wang, A.Y.: Modified algorithms LEM1 and LEM2 for rule induction from data with missing attribute values. In: Proceedings of RSSC'97 at the 3rd Joint Conference on Information Sciences. (1997) 69–72
21. Hipp, J., Myka, A., Wirth, R., Güntzer, U.: A new algorithm for faster mining of generalized association rules. [62] 74–82
22. Johnson, D.S.: Approximation algorithms for combinatorial problems. Journal of Computer and System Sciences **9** (1974) 256–278
23. Komorowski, J., Pawlak, Z., Polkowski, L., Skowron, A.: Rough sets: A tutorial. In Pal, S.K., Skowron, A., eds.: Rough Fuzzy Hybridization. A New Trend in Decision Making, Singapore, Springer (1999) 3–98
24. Kononenko, I., Bratko, I., Roškar, E.: Experiments in automatic learning of medical diagnostic rules. Technical report, Jozef Stefan Institute, Ljubljana (1984)
25. Kryszkiewicz, M.: Properties of incomplete information systems in the framework of rough sets. [43] 422–450
26. Latkowski, R.: Application of data decomposition to incomplete information systems. In Kłopotek, M.A., Wierzchoń, S.T., eds.: Proceedings of the International Symposium "Intelligent Information Systems XI", Physica-Verlag (2002)
27. Latkowski, R.: Incomplete data decomposition for classification. [3] 413–420
28. Latkowski, R.: High computational complexity of the decision tree induction with many missing attribute values. In Czaja, L., ed.: Proceedings of CS&P'2003, Czarna, September 25-27, Volume 2., Zakłady Graficzne UW (2003) 318–325
29. Latkowski, R.: On decomposition for incomplete data. Fundamenta Informaticae **54** (2003) 1–16
30. Latkowski, R., Mikołajczyk, M.: Data Decomposition and Decision Rule Joining for Classification of Data with Missing Values In Tsumoto, S., Komorowski, J., Grzymała-Busse, J.W., Słowiński, R., eds.: Rough Sets and Current Trends in Computing, RSCTC'2004, Springer (2004)
31. Lim, T.: Missing covariate values and classification trees. http://www.recursive-partitioning.com/mv.shtml, Recursive-Partitioning.com (2000)
32. Lovasz, L.: On the ratio of optimal integral and fractional covers. Discrete Mathematics **13** (1975) 383–390
33. Mikołajczyk, M.: Reducing number of decision rules by joining. [3] 425–432
34. Møllestad, T., Skowron, A.: A rough set framework for data mining of propositional default rules. In Raś, Z.W., Michalewicz, M., eds.: Foundations of Intelligent Systems — ISMIS 1996. LNCS 1079, Springer (1996) 448–457
35. Nguyen, H.S., Nguyen, S.H.: Rough sets and association rule generation. Fundamenta Informaticae **40** (1999) 383–405
36. Nguyen, H.S., Ślęzak, D.: Approximate reducts and association rules — correspondence and complexity results. [59] 137–145
37. Nguyen, S.H.: Regularity Analysis and its Application in Data Mining. PhD thesis, Warsaw University, Institute of Computer Science (1999)
38. Nguyen, S.H., Skowron, A., Synak, P.: Discovery of data patterns with applications to decomposition and classification problems. In Polkowski, L., Skowron, A., eds.: Rough Sets in Knowledge Discovery 2: Applications, Case Studies and Software Systems, Physica-Verlag (1998) 55–97
39. Nigmatullin, R.G.: Method of steepest descent in problems on cover. In: Memoirs of Symposium Problems of Precision and Efficiency of Computer Algorithms. Volume 5., Kiev (1969) 116–126
40. Øhrn, A., Ohno-Machado, L., Rowland, T.: Building manageable rough set classifiers. In Chute, C.G., ed.: Proceedings of the 1998 AMIA Annual Symposium. (1998) 543–547

41. Pal, S.K., Polkowski, L., Skowron, A., eds.: Rough-Neural Computing: Techniques for Computing with Words. Springer (2004)

42. Pawlak, Z.: Rough sets: Theoretical aspects of reasoning about data. Kluwer, Dordrecht (1991)

43. Polkowski, L., Skowron, A., eds.: Rough Sets in Knowledge Discovery 1: Methodology and Applications. Physica-Verlag (1998)

44. Polkowski, L., Skowron, A., Żytkow, J.M.: Tolerance based rough sets. In Lin, T.Y., Wildberger, A.M., eds.: Soft Computing, San Diego Simulation Councils Inc. (1995) 55–58

45. Quinlan, J.R.: Unknown attribute values in induction. In Segre, A.M., ed.: Proceedings of the Sixth International Machine Learning Workshop, Morgan Kaufmann (1989) 31–37

46. Rubin, D.B.: Multiple Imputation for Nonresponse in Surveys. John Wiley & Sons, New York (1987)

47. Skowron, A.: Boolean reasoning for decision rules generation. In Komorowski, H.J., Raś, Z.W., eds.: ISMIS 1993. LNCS 689, Springer (1993) 295–305

48. Skowron, A., Rauszer, C.: The discernibility matrices and functions in information systems. In Słowiński, R., ed.: Intelligent Decision Support. Handbook of Applications and Advances in Rough Sets Theory, Dordrecht, Kluwer (1992) 331–362

49. Słowiński, R., Vanderpooten, D.: A generalized definition of rough approximations based on similarity. IEEE Transactions on Data and Knowledge Engineering **12** (2000) 331–336

50. Stefanowski, J.: On rough set based approaches to induction of decision rules. [43] 500–529

51. Stefanowski, J., Tsoukiàs, A.: On the extension of rough sets under incomplete information. [59] 73–81

52. Stefanowski, J., Tsoukiàs, A.: Incomplete information tables and rough classification. International Journal of Computational Intelligence **17** (2001) 545–566

53. Stefanowski, J., Tsoukiàs, A.: Valued tolerance and decision rules. [61] 212–219

54. Swets, J.A.: Measuring the accuracy of diagnostic systems. Science **240** (1988) 1285–1293

55. Vapnik, V.N.: The Nature of Statistical Learning Theory. Springer, New York (1995)

56. Wang, H., Düntsh, I., Gediga, G., Skowron, A.: Hyperrelations in version space. Journal of Approximate Reasoning **36** (2004)

57. Weiss, S.M., Indurkhya, N.: Lightweight rule induction. In: Proceedings of the International Conference on Machine Learning ICML'2000. (2000)

58. Witten, I.H., Frank, E.: Data Mining: Practical Mashine Learning Tools and Techniques with Java Implementations. Morgan Kaufmann (2000)

59. Zhong, N., Skowron, A., Ohsuga, S., eds.: New Directions in Rough Sets, Data Mining, and Granular-Soft Computing, RSFDGrC '99. LNCS 1711, Springer (1999)

60. Ziarko, W.: Variable precision rough sets model. Journal of Computer and System Sciences **46** (1993) 39–59

61. Ziarko, W., Yao, Y.Y., eds.: Rough Sets and Current Trends in Computing, Second International Conference, RSCTC 2000 Banff, Canada, October 16-19, 2000, Revised Papers. LNCS 2000, Springer (2001)

62. Żytkow, J.M., Quafafou, M., eds.: Principles of Data Mining and Knowledge Discovery, Second European Symposium, PKDD '98, Nantes, France, September 23-26, 1998, Proceedings. LNCS 1510, Springer (1998)

Rough Sets and Relational Learning

R.S. Milton[1], V. Uma Maheswari[2], and Arul Siromoney[2]

[1] Department of Computer Science
Madras Christian College, Chennai – 600 059, India
[2] Department of Computer Science and Engineering
College of Engineering Guindy
Anna University, Chennai – 600 025, India
asiro@vsnl.com

Abstract. Rough Set Theory is a mathematical tool to deal with vagueness and uncertainty. Rough Set Theory uses a single information table. Relational Learning is the learning from multiple relations or tables. Recent research in Rough Set Theory includes the extension of Rough Set Theory to Relational Learning. A brief overview of the work in Rough Sets and Relational Learning is presented.

The authors' work in this area is then presented. Inductive Logic Programming (ILP) is one of the main approaches to Relational Learning. The generic Rough Set Inductive Logic Programming model introduces a rough setting in ILP. The Variable Precision Rough Set Inductive Logic Programming model (VPRSILP model) extends the Variable Precision Rough Set model to ILP.

In the cVPRSILP approach based on the VPRSILP model, elementary sets are defined using attributes that are based on a finite number of clauses of interest. However, this results in the number of elementary sets being very large. So, only significant elementary sets are used, and test cases are classified based on their proximity to the significant elementary sets.

The utility of this approach is shown in classification experiments in predictive toxicology.

1 Introduction

Rough set theory [1–4], introduced by Zdzislaw Pawlak in the early 1980s, is a mathematical tool to deal with vagueness and uncertainty. Rough set theory defines an indiscernibility relation that partitions the universe of examples into elementary sets. In other words, examples in an elementary set are indistinguishable. A concept is rough when it contains at least one elementary set that contains both positive and negative examples. The indiscernibility relation is defined based on a single table.

Relational Learning is based on multiple relations or tables. Inductive Logic Programming (ILP) [5, 6] is one of the approaches to Relational Learning. A brief survey of research in Rough Sets and Relational Learning is presented later in this paper.

J.F. Peters et al. (Eds.): Transactions on Rough Sets I, LNCS 3100, pp. 321–337, 2004.
© Springer-Verlag Berlin Heidelberg 2004

The authors' work is in the intersection of Rough Sets and ILP. The gRS–ILP model [7, 8] introduces a rough setting in Inductive Logic Programming. It describes the situation where any induced logic program cannot distinguish between certain positive and negative examples. Any induced logic program will either cover both the positive and the negative examples in the group, or not cover the group at all, with both the positive and the negative examples in this group being left out.

The Variable Precision Rough Set (VPRS) model [9] allows for a controlled degree of misclassification. The Variable Precision Rough Set Inductive Logic Programming (VPRSILP) model [10] is an extension of the gRS–ILP model using features of the VPRS model. The cVPRSILP approach [11] uses clauses as the attributes. Test cases are classified based on their proximity to significant elementary sets. An illustrative experiment in toxicology is presented.

2   Relational Learning

A *relation* is represented as a table consisting of rows and columns. Each *row* has the description of one object. A *column* corresponds to an *attribute* of the objects in the table. Each attribute has a set of possible values. An entry in the table has the value of the attribute (corresponding to that column) for that object (corresponding to that row).

Typical learning approaches assume that the data is stored in a single relation or table. However, in many applications, data is organized in multiple relations. *Relational learning* is learning from multiple relations. One of the goals of relational learning is to build models to predict the value of some target attribute (or attributes) of a target relation (or relations) from the other attributes of the objects of the target relation and the objects to which they are related. A training data set, with the values of the target attributes filled in, is presented to the learner, and the learner must produce a model that accurately predicts the values of target attributes on some other data set, where they are unknown [12].

Typical data mining approaches look for patterns in a single relation of a database. For many applications, squeezing data from multiple relations into a single table (known also as *propositionalisation*) requires much thought and effort and can lead to loss of information. An alternative for these applications is to use Multi-Relational Data Mining (MRDM). Multi-relational data mining is the multi-disciplinary field dealing with knowledge discovery from relational databases consisting of multiple tables. Multi-relational data mining can analyze data from a multi-relation database directly, without the need to transfer the data into a single table first. Present MRDM approaches consider all of the main data mining tasks, including association analysis, classification, clustering, learning probabilistic models and regression [13].

Attribute-value representations have the expressive power of propositional logic. Using first-order logic has advantages over propositional logic. First-order logic provides the richer expressive power required to represent objects having an internal structure and relations among them. Unlike propositional learning systems, the first-order approaches do not require that the data be composed into a single relation. They can take into account data organized in several database relations with various connections existing among them. First-order logic can also handle recurrent relations [14].

The most prominent approach to relational learning is that of Inductive Logic Programming (ILP). ILP is the research area formed at the intersection of logic programming and machine learning. ILP uses background knowledge, and positive and negative examples to induce a logic program that describes the examples. The induced logic program consists of the original background knowledge along with an induced hypothesis.

Researchers from a variety of backgrounds (including machine learning, statistics, inductive logic programming, databases, and reasoning under uncertainty) are beginning to develop techniques to learn statistical models from relational data. This area of research is known as Statistical Relational Learning (SRL) (http://robotics.stanford.edu/srl/). There are multiple paths toward the common goal of SRL. One path begins with machine learning and statistical methods for 'flat' or attribute-value representations, and expands these approaches to incorporate relational structure A second path extends techniques for relational learning (originally in nonprobabilistic domains) to incorporate stochastic models. ILP models have been capable of representing deterministic dependencies among instances for years. However, only recently statistical models have been developed to exploit the dependencies in relational data [15].

3   Rough Sets and Relational Learning

Relational Learning is broadly the learning from multiple relations or tables. This section is divided into two subsections. The first subsection presents a survey of research that extends Rough Sets to multiple relations or tables, but does not appear to consider the Learning aspects. However, these approaches can potentially be extended to Relational Learning. The second subsection presents a survey of research in Rough Sets and Relational Learning.

3.1   Rough Sets and Relational Data

The research discussed in this section focuses primarily only on the data representation, that is, on extending Rough Sets to be able to handle data that is from multiple relations or tables.

The notion of Rough Relations is discussed in [16]. This paper presents the formal basis for Rough Sets and multiple approximation spaces, where an approximation space corresponds to a table. This paper contains the extensions of the ideas referred in Professor Pawlak's work in 1981 [17]. This work on Rough Relations has been extended by several researchers. A good survey of this work is available in [18].

There are also several extensions of Rough Sets to multiple relations in Relational Databases. The standard relational database model [19] is extended using the features of Rough Set Theory in [20]. Other work in Rough Sets and Relational Databases include [21]. A Relational Information System is defined in [22] and used to analyse Relational Databases.

A framework is proposed in [23] for defining and reasoning about rough sets based on definite extended logic programs. A rough-set-specific query language is also introduced. A language is presented in [24] based on the rough set formalism. It caters for implicit definition of rough relations by combining different regions of other rough relations. The semantics of the language is obtained by translating it to the language of extended logic programs.

There are also extensions of Rough Set Theory to First–Order Logic [25–27]. Knowledge acquisition and processing with the use of rough set theory and the use of Prolog as a tool is discussed in [28]. There has also been some work relating rough sets to Computational Learning Theory [29].

3.2   Rough Sets and Relational Learning

Some of the research in Rough Sets and Relational Learning uses Rough Sets as a preprocessing stage to help or improve the learning algorithm. Other work integrates Rough Sets into the learning algorithm.

Some of the early work includes [30], where a Rough Set based approach to inducing schema design rules from examples is presented.

Concepts from rough set theory are used in an ILP algorithm for determining a minimal set of arguments for a newly invented predicate in [31, 32].

The EAGLE system [33] is a fuzzy–set rough–set ILP system. The use of fuzzy set theory to handle quantitative numeric values makes the Eagle a powerful tool. The grouping into granules takes the label of the example (positive or negative) into consideration, and so a granule consists of only positive examples or only negative examples.

A preliminary definition that integrates the basic notions of Rough Sets with ILP is found in [7]. More formal definitions are found in [34, 8]. This work is extended in [10] where the ILP Learning Algorithm is modified.

Rough Set theory is applied to ILP to deal with imperfect data [35, 36]. Rough problem settings are proposed for incomplete background knowledge (where essential predicates / clauses are missing) and indiscernible data (where some examples belong to both sets of positive and negative training examples). The rough settings relax the strict requirements in the standard normal problem setting for ILP, so that rough but useful hypotheses can be induced from imperfect data.

Another Rough Set approach to treating imperfect data in ILP is found in [37]. The approach is based on neighborhood systems, due to the generality of the language. A first-order decision system is introduced and a greedy algorithm for finding a set of rules (or clauses) is given.

A method is proposed in [38] that removes irrelevant clauses and reduces extensional background knowledge (i.e., consisting of only ground atoms) in first order learning. This work is further extended in [39] which has an excellent survey of work in Rough Sets and Relational Learning. A learning algorithm for learning first order logic rules using rough set theory is presented in [14].

4   The VPRSILP Model

4.1   Inductive Logic Programming

The basics of ILP are presented in several books such as [40–42]. The semantics of ILP systems are discussed in [43]. In ILP systems, background (prior) knowledge B and evidence E (consisting of positive evidence E^+ and negative evidence E^-) are given, and the aim is then to find a hypothesis H such that certain conditions are fulfilled.

In the *normal semantics*, the background knowledge, evidence and hypothesis can be any well-formed logical formula. The conditions that are to be fulfilled by an ILP system in the normal semantics are

Prior Satisfiability: $B \wedge E^- \not\models \square$
Posterior Satisfiability: $B \wedge H \wedge E^- \not\models \square$
Prior Necessity: $B \not\models E^+$
Posterior Sufficiency: $B \wedge H \models E^+$

However, the *definite semantics*, which can be considered as a special case of the normal semantics, restricts the background knowledge and hypothesis to being definite clauses. This is simpler than the general setting of normal semantics, since a definite clause theory T has a unique minimal Herbrand model $\mathcal{M}^+(T)$, and any logical formula is either true or false in the minimal model. The conditions that are to be fulfilled by an ILP system in the definite semantics are

Prior Satisfiability: all $e \in E^-$ are false in $\mathcal{M}^+(B)$
Posterior Satisfiability: all $e \in E^-$ are false in $\mathcal{M}^+(B \wedge H)$
Prior Necessity: some $e \in E^+$ are false in $\mathcal{M}^+(B)$
Posterior Sufficiency: all $e \in E^+$ are true in $\mathcal{M}^+(B \wedge H)$

The Sufficiency criterion is also known as *completeness* with respect to positive evidence and the Posterior Satisfiability criterion is also known as *consistency* with the negative evidence.

The special case of definite semantics, where evidence is restricted to true and false ground facts (examples), is called the *example* setting. The example setting is thus the normal semantics with B and H as definite clauses and E as a set of ground unit clauses. The example setting is the main setting of ILP employed by the large majority of ILP systems.

4.2   Formal Definitions of the gRS–ILP Model

The generic Rough Set Inductive Logic Programming (gRS–ILP) model introduces the basic definition of elementary sets and a rough setting in ILP [7, 8]. The essential feature of an elementary set is that it consists of examples that cannot be distinguished from each other by any induced logic program in that ILP system. The essential feature of a rough setting is that it is inherently not possible for certain positive and negative examples to be distinguished, since both these positive and negative examples are in the same elementary set. The basic definitions formalised in [34] follow.

The ILP system in the example setting of [43] is formally defined as follows.

Definition 1. *An* ILP system in the example setting *is a tuple* $S_{es} = (E_{es}, B)$, *where*
(1) $E_{es} = E_{es}^+ \cup E_{es}^-$ *is the* universe, *where* E_{es}^+ *is the set of positive examples (true ground facts), and* E_{es}^- *is the set of negative examples (false ground facts), and*
(2) B *is a background knowledge given as definite clauses such that (i) for all* $e^- \in E_{es}^-$, $B \nvdash e^-$, *and (ii) for some* $e^+ \in E_{es}^+$, $B \nvdash e^+$.

Let $S_{es} = (E_{es}, B)$ be an ILP system in the example setting. Then let $\mathcal{H}(S_{es})$ (also written as $\mathcal{H}(E_{es}, B)$) denote the set of all possible definite clause hypotheses that can be induced from E_{es} and B, and be called the *hypothesis space* induced from S_{es} (or from E_{es} and B). Further, let $\mathcal{P}(S_{es})$ (also written as $\mathcal{P}(E_{es}, B) = \{P = B \wedge H \mid H \in \mathcal{H}(E_{es}, B)\}$) denote the set of all the programs induced from E_{es} and B, and be called the *program space* induced from S_{es} (or from E_{es} and B).

The aim is to find a program $P \in \mathcal{P}(S_{es})$ such that the next two conditions hold: (iii) for all $e^- \in E_{es}^-$, $P \nvdash e^-$, (iv) for all $e^+ \in E_{es}^+$, $P \vdash e^+$.

The following definitions of Rough Set ILP systems in the gRS–ILP model (abbreviated as *RSILP systems*) use the terminology of [43].

Definition 2. *An* RSILP system in the example setting *(abbreviated as RSILP–E system) is an ILP system in the example setting,* $S_{es} = (E_{es}, B)$, *such that there does not exist a program* $P \in \mathcal{P}(S_{es})$ *satisfying both the conditions (iii) and (iv) above.*

Definition 3. *An* RSILP–E system in the single–predicate learning context *(abbreviated as RSILP–ES system) is an RSILP–E system, whose* universe E *is such that all examples (ground facts) in* E *use only one predicate, also known as the* target predicate.

A *declarative bias* [43] restricts the set of acceptable hypotheses, and is of two kinds: *syntactic bias* (also called *language bias*) that imposes restrictions on the form (syntax) of clauses allowed in the hypothesis, and *semantic bias* that imposes restrictions on the meaning, or the behaviour of hypotheses.

Definition 4. *An* RSILP–ES system with declarative bias *(abbreviated as RSILP–ESD system) is a tuple* $S = (S', L)$, *where*
(i) $S' = (E, B)$ *is an RSILP–ES system, and*
(ii) L *is a declarative bias, which is any restriction imposed on the hypothesis space* $\mathcal{H}(E, B)$.
We also write $S = (E, B, L)$ *instead of* $S = (S', L)$.

For any RSILP–ESD system $S = (E, B, L)$, let
$\mathcal{H}(S) = \{H \in \mathcal{H}(E, B) \mid H$ is allowed by $L\}$, and
$\mathcal{P}(S) = \{P = B \wedge H \mid H \in \mathcal{H}(S)\}$.
$\mathcal{H}(S)$ (also written as $\mathcal{H}(E, B, L)$) is called the *hypothesis space* induced from S (or from E, B, and L). $\mathcal{P}(S)$ (also written as $\mathcal{P}(E, B, L)$) denotes the set of all the programs induced by S, and is called the *program space* induced from S (or from E, B, and L).

An equivalence relation on the universe of an RSILP–ESD system is now defined.

Definition 5. *Let $S = (E, B, L)$ be an RSILP–ESD system. An indiscernibility relation of S, denoted by $R(S)$, is a relation on E defined as follows: $\forall x, y \in E$, $(x, y) \in R(S)$ iff $(P \vdash x \Leftrightarrow P \vdash y)$ for any $P \in \mathcal{P}(S)$ (i.e. iff x and y are inherently indistinguishable by any induced logic program P in $\mathcal{P}(S)$).*

It follows directly from the definition of $R(S)$, that for any RSILP–ESD system S, $R(S)$ is an equivalence relation.

Definition 6. *Let $S = (E, B, L)$ be an RSILP–ESD system. An elementary set of $R(S)$ is an equivalence class of the relation $R(S)$. For each $x \in E$, let $[x]_{R(S)}$ denote the elementary set of $R(S)$ containing x. Formally,*
$[x]_{R(S)} = \{y \in E \mid (x, y) \in R(S)\}$.
A composed set of $R(S)$ is any finite union of elementary sets of $R(S)$.

Definition 7. *An RSILP–ESD system $S = (E, B, L)$ is said to be in a rough setting iff*
$\exists e^+ \in E^+ \; \exists e^- \in E^- \; ((e^+, e^-) \in R(S))$.

4.3   Formal Definitions of the VPRSILP Model

The formal definitions of the VPRSILP model are defined in [10].

A parameter β, a real number in the range $(0.5, 1]$, is used in the VPRS model as a threshold in elementary sets that have both positive and negative examples. This threshold is used to decide if that elementary set can be classified as positive or negative, depending on the statistical occurrence of positive and negative examples in it.

Definition 8. *A Variable Precision RSILP–ESD system   (abbreviated as VPRSILP–ESD system) is a tuple $S = (S', \beta)$, where*
(i) $S' = (E, B, L)$ is an RSILP–ESD system, and
(ii) β is a real number in the range $(0.5, 1]$.
It is also written $S = (E, B, L, \beta)$ instead of $S = (S', \beta)$.

The definitions of hypothesis space, program space, equivalence relation, elementary sets, composed sets and rough setting defined above for RSILP–ESD systems hold for the VPRSILP–ESD system.

The following definitions use the VPRS terminology from [44].

Definition 9. *The conditional probability* $P(E^+ \mid [x]_{R(S)})$ *is defined as*

$$P(E^+ \mid [x]_{R(S)}) = \frac{P(E^+ \cap [x]_{R(S)})}{P([x]_{R(S)})} = \frac{\mid E^+ \cap [x]_{R(S)} \mid}{\mid [x]_{R(S)} \mid}$$

where $P(E^+ \mid [x]_{R(S)})$ *is the probability of occurrence of event* E^+ *conditioned on event* $[x]_{R(S)}$.

It is noted that $P(E^+ \mid [x]_{R(S)}) = 1$ if and only if $[x]_{R(S)} \subseteq E^+$; $P(E^+ \mid [x]_{R(S)}) > 0$ if and only if $[x]_{R(S)} \cap E^+ \neq \emptyset$; and $P(E^+ \mid [x]_{R(S)}) = 0$ if and only if $[x]_{R(S)} \cap E^+ = \emptyset$.

Definition 10. *The* β*–positive region of* S, $Pos_\beta(S)$, *is defined as*
$Pos_\beta(S) = \bigcup_{P(E^+ \mid [x]_{R(S)}) \,>=\, \beta, \text{ for all } [x]_{R(S)} \text{ in } R(S)} \{[x]_{R(S)}\}$
The β*–negative region of* S, $Neg_\beta(S)$, *is defined as*
$Neg_\beta(S) = \bigcup_{P(E^+ \mid [x]_{R(S)}) \,<\, \beta, \text{ for all } [x]_{R(S)} \text{ in } R(S)} \{[x]_{R(S)}\}$

Definition 11. *The* β*–restricted program space of* S, $\mathcal{P}_\beta(S)$ *(also written as* $\mathcal{P}_\beta(E, B, L, \beta)$*), is defined as*
$\mathcal{P}_\beta(S) = \{P \in \mathcal{P}(S) \mid P \vdash x \Rightarrow x \in Pos_\beta(S)\}$.
Any $P \in \mathcal{P}_\beta(S)$ *is called a* β*–restricted program of* S.

Our aim is to find a hypothesis H such that $P = B \wedge H \in \mathcal{P}_\beta(\mathcal{S})$.

The VPRSILP model has been applied in illustrative experiments to determine the transmembrane domains in amino acid sequences [10] and to analyse and classify web log data [45, 46].

5   The cVPRSILP Approach

In this section, the cVPRSILP approach [11] based on the VPRSILP model is outlined. In the VPRSILP model, the equivalence relation is defined using all possible induced logic programs in the program space. In the cVPRSILP approach, the equivalence relation is defined using attributes that are based on a finite number of clauses of interest. A more formal presentation follows.

In [47], two finite, nonempty sets U and A are considered, where U is the universe of objects, and A is a set of attributes. With every attribute $a \in A$ is associated a set V_a of its values, called the domain of a. The set of attributes A determines a binary relation R on U. R is an indiscernibility relation, defined as follows: xRy if and only if $a(x) = a(y)$ for every $a \in A$; where $a(x) \in V_a$ denotes the value of attribute a for object x. Obviously R is an equivalence relation. Equivalence classes of the relation R are referred to as elementary sets.

In the cVPRSILP approach, let $A = \{A_1, \ldots, A_{i_{max}}\}$ be the set of attributes, with $V_a = \{\texttt{true}, \texttt{false}\}$ for every $a \in A$. Every $A_i \in A$ is associated with the clauses of interest $C'_i, i = 1, ..., i_{max}$, such that $A_i = \texttt{true}$ if the example can be derived from $C'_i \wedge B$, and $A_i = \texttt{false}$ otherwise. In this context, it is seen that the relation R using these attributes forms an equivalence relation. This equivalence relation is the same as that in definition 5, when all possible programs $P \in \mathcal{P}$ are of the form $C'_i \wedge B$.

An example falls into a particular elementary set based on which programs $C'_i \wedge B$ cover the example. The training examples are used to populate the elementary sets. The elementary sets fall into either the β–positive or the β–negative region, depending on the value of β and the number of positive and negative training examples in the elementary set.

A test example is predicted as being positive or negative, depending on whether its elementary set is in the β–positive or the β–negative region.

6   Proximity to Significant Elementary Sets

6.1   Significant Elementary Sets

In the cVPRSILP approach clauses are used as attributes. This results in the number of elementary sets being very large. Many of these elementary sets are too small to be considered as representative of either positive or negative trend. However, some of the elementary sets are *significant elementary sets*, having sufficiently large number of examples with a significant difference between the number of positive and negative examples. An elementary set $[x]_R$ is said to be a significant elementary set if and only if

$$(|[x]_R| > \alpha_{size}) \wedge (\, (|[x]_R \cap E^+| \sim |[x]_R \cap E^-|) > \alpha_{diff} \,)$$

where α_{size} and α_{diff} are two user-defined parameters.

Unfortunately, several of the test examples in the universe do not fall in these few significant elementary sets. It is proposed in this paper to classify a test case based on its *proximity* to the significant elementary sets.

6.2   Proximity Based on Common Attributes

It is possible to define the proximity of an example to an elementary set in several ways. It is defined here based on the *common attributes*, the number of attributes that have the value `true` in both the example and the elementary set. The proximity between an example x and an elementary set $[y]_R$ based on common attributes is

$$Proximity_{ca}(x, [y]_R) = | \, \{ a \in A \mid a(x) = \texttt{true} \wedge a(y) = \texttt{true} \} \, |$$

A test example is predicted as positive or negative in the following manner. The proximity of the test example to each of the significant elementary sets is calculated. The significant elementary set with which the test example has the highest proximity is chosen. The test example is predicted as positive if this elementary set is in the β–positive region, and as negative if this elementary set is in the β–negative region.

6.3   Proximity Based on Weighted Common Attributes

The proximity of an example to an elementary set can also be defined based on the *weighted common attributes*, by assigning weights to the attributes. The weight associated with an attribute is the difference between the number of positive and negative training examples that have the value of this attribute as true.

Let P_a be the number of positive training examples that have the value of the attribute a as true, and similarly N_a for negative training examples.

$$P_a = |\{x \in E^+ \mid a(x) = \text{true}\}|$$

$$N_a = |\{x \in E^- \mid a(x) = \text{true}\}|$$

The weight w_a of an attribute $a \in A$ is

$$w_a = \frac{P_a \sim N_a}{P_a + N_a}.$$

The proximity of an example to an elementary set is the sum of the weights of the attributes that have the value true in both the example and the elementary set. The proximity between an example x and an elementary set $[y]_R$ based on weighted common attributes is defined formally as

$$Proximity_{wca}(x, [y]_R) = \sum_{\forall a \in A \mid (a(x)=\text{true}) \wedge (a(y)=\text{true})} w_a$$

6.4   Average Proximity

The next two measures take the overall effect of all the significant elementary sets.

The proximity of the example to each of the significant elementary sets is determined. The average proximity of the example to all the significant positive elementary sets is the sum of the proximity of the example to each of the significant positive elementary sets divided by the number of significant positive elementary sets. The average proximity of the example to all the significant negative elementary sets is similar. The example is predicted as positive or negative depending on whether the average proximity of the example to all the significant positive elementary sets is more, or the average proximity of the example to all the significant negative elementary sets is more.

The sum of the proximities of an example to all the significant positive elementary sets is

$$\sum_{[y]_R \in Pos_\beta} Proximity_{ca}(x, [y]_R)$$

The average proximity of an example with all the positive elementary sets (average positive proximity) is

$$AvgPosProximity_{ca} = \frac{\sum_{[y]_R \in Pos_\beta} Proximity_{ca}(x, [y]_R)}{|\{[y]_R \in Pos_\beta\}|}$$

Similarly, the average proximity of an example with all the significant negative elementary sets (average negative proximity) is

$$AvgNegProximity_{ca} = \frac{\sum_{[y]_R \in Neg_\beta} Proximity_{ca}(x, [y]_R)}{\mid \{[y]_R \in Neg_\beta\} \mid}$$

A test example is predicted positive if $AvgPosProximity_{ca} > AvgNegProximity_{ca}$, and as negative otherwise.

The sum of the weighted proximities of an example with all the significant positive elementary sets is

$$\sum_{[y]_R \in Pos_\beta} Proximity_{wca}(x, [y]_R)$$

The average weighted proximity of an example with all the significant positive elementary sets (average positive weighted proximity) is

$$AvgPosProximity_{wca} = \frac{\sum_{[y]_R \in Pos_\beta} Proximity_{wca}(x, [y]_R)}{\mid \{[y]_R \in Pos_\beta\} \mid}$$

Similarly, the average weighted proximity of an example with all the significant negative elementary sets (average negative weighted proximity) is

$$AvgNegProximity_{wca} = \frac{\sum_{[y]_R \in Neg_\beta} Proximity_{wca}(x, [y]_R)}{\mid \{[y]_R \in Neg_\beta\} \mid}$$

A test example is predicted positive if $AvgPosProximity_{wca} > AvgNegProximity_{wca}$, and as negative otherwise.

7   The cVPRSILP Model and Application to Predictive Toxicology

The rodent carcinogenicity tests conducted within the US National Toxicology Program by the National Institute of Environmental Health Sciences (NIEHS). has resulted in a large database of compounds classified as carcinogens or otherwise. The Predictive Toxicology Evaluation project of the NIEHS provided the opportunity to compare carcinogenicity predictions on previously untested chemicals. This presented a formidable challenge for programs concerned with knowledge discovery. The ILP system Progol [48] has been used in this Predictive Toxicology Evaluation Challenge [49, 50].

An illustrative experiment is performed using the cVPRSILP model. The dataset used is the Predictive Toxicology Evaluation Challenge dataset found at http://web.comlab.ox.ac.uk/oucl/research/areas/machlearn/cancer.html.

In this experimental illustration, two predicates has_property and atm with four properties and three atom types are considered. These have been heuristically chosen based on visual inspection of clauses induced by Progol. Further studies are in progress to arrive at a more systematic choice.

The maximum number of predicates in a clause is taken as 2 and a finite set of clauses of interest is generated. Approximately 100 rules (clauses) are generated.

Each of the clauses of interest is treated as an attribute, and every example is placed in the appropriate elementary set, based on the subset of clauses which cover that example. Each elementary set falls in the β–positive or the β–negative region, depending on the chosen value of β. (In this illustration, we use the value of 0.5, but place elementary sets in the β–positive region only if the value of the conditional probability is *greater* than β. In the VPRS model, the value of β is greater than 0.5, but elementary sets are place in the β–positive region if the value of the conditional probability is greater than or equal to β.)

An example is predicted positive if its elementary set falls in the β–positive region, and is predicted negative if the elementary set falls in the β–negative region.

The following tables present the results of the ten-fold cross-validation and the summary of the results. The percentage accuracy is calculated as the percentage of test cases predicted correctly (actual positive test cases predicted positive and actual negative test cases predicted negative) to the total number of test cases (including the unclassified test cases). The results of fold 8 are not available due to an error in the experiment.

Fold	Positive			Negative			% Acc
	Pred +	Pred -	Unclass	Pred +	Pred -	Unclass	
0	7	3	6	8	1	5	50
1	11	2	5	5	5	5	48
2	10	3	5	5	5	5	45
3	8	3	7	3	5	7	33
4	10	2	6	2	4	8	37
5	5	2	10	8	4	3	40
6	8	4	6	4	6	5	36
7	7	3	8	7	2	6	42
9	10	4	5	6	3	4	50
	76	26	58	48	35	48	

	Actual Positive	Actual Negative	
Predicted Positive	76	48	124
Predicted Negative	26	35	61
Unclassified	58	48	106
	160	131	291

The following tables present the results of the experiment using proximity based on common attributes.

Fold	Positive		Negative		% Acc
	Pred +	Pred -	Pred +	Pred -	
0	8	8	9	5	56
1	12	6	12	3	72
2	13	5	7	8	60
3	9	9	4	11	39
4	12	6	8	6	62
5	10	7	9	6	59
6	11	7	11	4	66
7	12	6	9	6	63
9	12	7	9	4	62
	99	61	78	53	

	Actual Positive	Actual Negative	
Predicted Positive	99	78	177
Predicted Negative	61	53	114
	160	131	291

The following tables present the results of the experiment using proximity based on weighted common attributes.

Fold	Positive		Negative		% Acc
	Pred +	Pred -	Pred +	Pred -	
0	8	8	10	4	60
1	12	6	12	3	72
2	14	4	7	8	63
3	10	8	5	10	45
4	13	5	8	6	65
5	10	7	11	4	65
6	12	6	9	6	63
7	16	2	9	6	75
9	11	8	9	4	62
	106	54	80	51	

	Actual Positive	Actual Negative	
Predicted Positive	106	80	186
Predicted Negative	54	51	105
	160	131	291

The following tables present the results of the ten-fold cross-validation and the summary of the results, using average proximity.

Fold	Positive		Negative		% Acc
	Pred +	Pred -	Pred +	Pred -	
0	8	8	2	12	66
1	11	7	0	15	78
2	13	5	8	7	60
3	7	11	6	9	48
4	11	7	6	8	59
5	6	11	2	13	59
6	7	11	4	11	54
7	12	6	6	9	63
9	8	11	2	11	59
	83	77	36	95	61.21

	Actual Positive	Actual Negative	
Predicted Positive	83	36	119
Predicted Negative	77	95	172
	160	131	291

The following tables present the results using average weighted proximity.

Fold	Positive		Negative		% Acc
	Pred +	Pred -	Pred +	Pred -	
0	8	8	5	9	56
1	11	7	2	13	72
2	14	4	8	7	63
3	9	9	9	6	45
4	13	5	7	7	62
5	9	8	3	12	65
6	9	9	5	10	57
7	14	4	6	9	69
9	11	8	3	10	65
	98	62	48	83	62.17

	Actual Positive	Actual Negative	
Predicted Positive	98	48	146
Predicted Negative	62	83	145
	160	131	291

The average prediction accuracy of the cVPRSILP approach without using any proximity measure is 42.33%, using the proximity based on common attributes is 59.89%, and using proximity based on weighted common attributes is 63.33%. Similar results are achieved using the average proximity measures. The average prediction accuracy using the average proximity based on common attributes is 61.21%, and using the average proximity based on weighted common attributes is 62.17%.

8   Conclusions

This paper presents an overview of research in the intersection of Rough Set Theory and Relational Learning. It then presents the cVPRSILP approach, where clauses are used as the attributes of the VPRSILP model. The results of an illustrative example in toxicology are presented.

The VPRS model uses statistical trend information in the data. It appears likely that VPRSILP can be extended to SRL. Work is currently in progress to extend the existing work on the VPRSILP model to SRL.

References

1. Pawlak, Z.: Rough sets. International Journal of Computer and Information Sciences **11** (1982) 341–356
2. Pawlak, Z.: Rough Sets — Theoretical Aspects of Reasoning about Data. Kluwer Academic Publishers, Dordrecht, The Netherlands (1991)
3. Pawlak, Z., Grzymala-Busse, J., Slowinski, R., Ziarko, W.: Rough sets. Communications of ACM **38** (1995) 89–95
4. Komorowski, J., Pawlak, Z., Polkowski, L., Skowron, A.: Rough sets: A tutorial. In Pal, S.K., Skowron, A., eds.: Rough Fuzzy Hybridization: A New Trend in Decision-Making. Springer-Verlag (1999) 3–98
5. Muggleton, S.: Inductive logic programming. New Generation Computing **8** (1991) 295–318
6. Muggleton, S.: Scientific knowledge discovery through inductive logic programming. Communications of the ACM **42** (1999) 43–46
7. Siromoney, A.: A rough set perspective of Inductive Logic Programming. In Raedt, L.D., Muggleton, S., eds.: Proceedings of the IJCAI-97 Workshop on Frontiers of Inductive Logic Programming, Nagoya, Japan (1997) 111–113
8. Siromoney, A., Inoue, K.: The generic Rough Set Inductive Logic Programming (gRS–ILP) model. In Lin, T.Y., Yao, Y.Y., Zadeh, L.A., eds.: Data Mining, Rough Sets and Granular Computing. Volume 95., Physica–Verlag (2002) 499–517
9. Ziarko, W.: Variable precision rough set model. Journal of Computer and System Sciences **46** (1993) 39–59
10. Uma Maheswari, V., Siromoney, A., Mehata, K.M., Inoue, K.: The Variable Precision Rough Set Inductive Logic Programming Model and Strings. Computational Intelligence **17** (2001) 460–471
11. Milton, R.S., Uma Maheswari, V., Siromoney, A.: The Variable Precision Rough Set Inductive Logic Programming model — a Statistical Relational Learning perspective. In: Workshop on Learning Statistical Models from Relational Data (SRL 2003), IJCAI-2003. (2003)
12. Hulten, G., Domingos, P., Abe, Y.: Mining massive relational databases. In: Workshop on Learning Statistical Models from Relational Data (SRL 2003), IJCAI-2003. (2003)
13. Dzeroski, S., Raedt, L.D., Wrobel, S.: Multirelational data mining 2003: Workshop report. In: 2nd Workshop on Multi-Relational Data Mining (MRDM 2003), ICKDDM-2003. (2003)
14. Stepaniuk, J., Honko, P.: Learning first-order rules: A rough set approach. Fundamenta Informaticae **XXI** (2003) 1001–1019

15. Getoor, L., Jensen, D., eds.: Workshop on Learning Statistical Models from Relational Data (SRL 2003), IJCAI-2003. (2003)
16. Pawlak, Z.: On rough relations. Bulletin of the Polish Academy of Sciences Technical Sciences **34** (1981) 587–590
17. Pawlak, Z.: Rough relations. ICS PAS Reports **435** (1981)
18. Stepaniuk, J.: Rough Relations and Logics. In: Rough Sets in Knowledge Discovery 1. Methodology and Applications. Physica–Verlag (1998)
19. Codd, E.F.: A relational model of data for large shared data banks. Communications of the ACM **13** (1970) 377–387
20. Beaubouef, T., Petry, F.E.: A Rough Set model for relational databases. In: Rough Sets, Fuzzy Sets and Knowledge Discovery, Springer–Verlag (1994) 100–107
21. Machuca, F., Millán, M.: Enhancing the exploitation of data mining in relational database systems via the rough sets theory including precision variables. In: Proceedings of the 1998 ACM Symposium on Applied Computing. (1998) 70–73
22. Wróblewski, J.: Analyzing relational databases using rough set based methods. In: Proceedings of IPMU 2000. Volume 1. (2000) 256–262
23. Vitoria, A., Maluszynski, J.: A logic programming framework for rough sets. In Alpigini, J.J., Peters, J.F., Skowron, A., Zhong, N., eds.: Rough Sets and Current Trends in Computing — Third International Conference, RSCTC 2002. Lecture Notes in Artificial Intelligence 2475, Pennsylvania, USA, Springer (2002) 205–212
24. Vitória, A., Damasio, C.V., Małuszyński, J.: From rough sets to rough knowledge bases. Fundamenta Informaticae (2003) Accepted for publication.
25. Lin, T., Liu, Q.: First order rough logic I: Approximate reasoning via rough sets. Fundamenta Informaticae **27** (1996) 137–153
26. Lin, T., Liu, Q., Zuo, X.: Models for first order rough logic applications to data mining. In: Proc. 1996 Asian Fuzzy Systems Symposium, Special Session Rough Sets and Data Mining, Taiwan (1996) 152–157
27. Parsons, S., Kubat, M.: A first–order logic for reasoning under uncertainty using rough sets. Journal of Intelligent Manufacturing **5** (1994) 211–223
28. Mrózek, A., Skabek, K.: Rough rules in Prolog. In Polkowski, L., Skowron, A., eds.: Rough Sets and Current Trends in Computing — First International Conference, RSCTC 1998. Lecture Notes in Artificial Intelligence 1424, Warsaw, Poland, Springer (1998) 458–466
29. Yokomori, T., Kobayashi, S.: Inductive learning of regular sets from examples: a rough set approach. In: Proc. of 3rd Intl. Workshop on Rough Sets and Soft Computing. (1994) 570–577
30. Yasdi, R., Ziarko, W.: An expert system for conceptual schema design: a machine learning approach. Machine Learning and Uncertain Reasoning **3** (1987)
31. Stahl, I., Weber, I.: The arguments of newly invented predicates in ILP. In Wrobel, S., ed.: Proceedings of the 4th International Workshop on Inductive Logic Programming — ILP94. Volume 237 of GMD-Studien., Gesellschaft für Mathematik und Datenverarbeitung MBH (1994) 233–246
32. Stahl, I.: The efficiency of bias shift operations in ILP. In Raedt, L.D., ed.: Proceedings of the 5th International Workshop on Inductive Logic Programming — ILP95, Dept. of Computer Science, K.U.Leuven, Belgium (1995) 231–246
33. Martienne, E., Quafafou, M.: Learning logical descriptions for document understanding: A Rough Sets-based approach. In Polkowski, L., Skowron, A., eds.: Rough Sets and Current Trends in Computing — First International Conference, RSCTC 1998. Lecture Notes in Artificial Intelligence 1424, Warsaw, Poland, Springer (1998) 202–209

34. Siromoney, A., Inoue, K.: Elementary sets and declarative biases in a restricted gRS–ILP model. Informatica **24** (2000) 125–135
35. Liu, C., Zhong, N.: Rough problem settings for Inductive Logic Programming. In N.Zhong, ad S.Ohsuga, A., eds.: New Directions in Rough Sets, Data Mining, and Granular–Soft Computing — 7th International Workshop, RSFDGrC'99. Lecture Notes in Artificial Intelligence 1711, Yamaguchi, Japan, Springer (1999) 168–177
36. Liu, C., Zhong, N.: Rough problem settings for ILP dealing with imperfect data. Computational Intelligence **17** (2001) 446–459
37. Midelfart, H., Komorowski, J.: A Rough Set approach to Inductive Logic Programming. In Ziarko, W., Yao, Y., eds.: Rough Sets and Current Trends in Computing — Second International Conference, RSCTC 2000. Lecture Notes in Artificial Intelligence 2005, Banff, Canada, Springer (2000) 190–198
38. Stepaniuk, J.: Rough sets and relational learning. In Zimmermann, H.J., ed.: Proceedings of EUFIT. (1999)
39. Stepaniuk, J.: Knowledge discovery by application of rough set models. In: Rough Set Methods and Applications — New Developments in Knowledge Discovery in Information Systems. Physica–Verlag (2000) 137–233
40. Muggleton, S., ed.: Inductive Logic Programming. Academic Press (1992)
41. Lavrač, N., Džeroski, S.: Inductive Logic Programming: Techniques and Applications. Ellis Horwood (1994)
42. Bergadano, F., Gunetti, D.: Inductive Logic Programming: From Machine Learning to Software Engineering. The MIT Press (1995)
43. Muggleton, S., Raedt, L.D.: Inductive logic programming: Theory and methods. Journal of Logic Programming **19** (1994) 629–679
44. An, A., Chan, C., Shan, N., Cercone, N., Ziarko, W.: Applying knowledge discovery to predict water-supply consumption. IEEE Expert **12** (1997) 72–78
45. Uma Maheswari, V., Siromoney, A., Mehata, K.M.: The Variable Precision Rough Set Inductive Logic Programming model and web usage graphs. In: New Frontiers in Artificial Intelligence — Joint JSAI 2001 Workshop Post–Proceedings. Volume 2253., Lecture Notes in Computer Science, Springer (2001) 339–343
46. Uma Maheswari, V., Siromoney, A., Mehata, K.M.: The Variable Precision Rough Set Inductive Logic Programming model and future test cases in web usage mining. In Inuiguchi, M., Tsumoto, S., Hirano, S., eds.: Rough Set Theory and Granular Computing, Physica–Verlag (2003)
47. Pawlak, Z., Skowron, A.: Rough set rudiments. In: Bulletin of International Rough Set Society. Volume 3. (1999)
48. Muggleton, S.: Inverse entailment and Progol. New Generation Computing **13** (1995) 245–286
49. Srinivasan, A., King, R., Muggleton, S., Sternberg, M.: The predictive toxicology evaluation challenge. In: Proceedings of the Fifteenth International Joint Conference Artificial Intelligence (IJCAI-97). Morgan-Kaufmann (1997) 1–6
50. Srinivasan, A., King, R., Muggleton, S., Sternberg, M.: Carcinogenesis predictions using ILP. In Lavrač, N., Džeroski, S., eds.: Proceedings of the Seventh International Workshop on Inductive Logic Programming. Springer-Verlag, Berlin (1997) 273–287 LNAI 1297.

Approximation Space for Software Models

James F. Peters[1] and Sheela Ramanna[2]

[1] Department of Electrical and Computer Engineering, University of Manitoba
Winnipeg, Manitoba R3T 5V6 Canada
jfpeters@ee.umanitoba.ca

[2] Department of Applied Computer Science, University of Winnipeg
Winnipeg, Manitoba R3B 2E9 Canada
s.ramanna@uwinnipeg.ca

> *...There is deep and important underlying structural correspondence between the pattern of a problem and the process of designing a physical form which answers that problem.*
> –Christopher Alexander, 1971

Abstract. This article introduces an approximation space for graded acceptance of proposed software models for system design relative to design patterns that conform to a system design standard. A fundamental problem in system design is that feature values extracted from experimental design models tend not to match exactly patterns associated with standard design models. It is not generally known how to measure the extent that a particular system design conforms to a standard design pattern. The rough set approach introduced by Zdzisław Pawlak provides a ground for concluding to what degree a particular model for a system design is a part of a set of a set of models representing a standard. To some extent, this research takes into account observations made by Christopher Alexander about the idea of form in Plato's philosophy, which is helpful in arriving at an understanding about design patterns used to classify similar models for system designs. The basic assumption made in this research is that every system design can be approximated relative to a standard, and it is possible to prescribe conditions for the construction of a set of acceptable software models. The template method and memento behavioral design patterns are briefly considered by way of illustration. An approximation space for software models is introduced. In addition, an approach to the satisfaction-based classification of software models is also presented.

Keywords: Approximation space, classification, design pattern, rough sets, software model.

1 Introduction

This article introduces an approach to classifying software models for system design in the context of an approximation space defined in the context of rough sets [9-14]. Considerable work has been done on approximation spaces in the context of rough sets [26-28, 18] as well as generalized approximation spaces [21, 8]. It is well-known

J.F. Peters et al. (Eds.): Transactions on Rough Sets I, LNCS 3100, pp. 338–355, 2004.
© Springer-Verlag Berlin Heidelberg 2004

that models for system design seldom exactly match what might be considered a standard. This is to be expected, since system designs tend to have an unbounded number of variations relative to an accepted design pattern. Consider, for example, the variations in the implementation of design patterns in architecture made possible by pattern languages [1-3]. It is this variation in actual system designs that is a source of a difficult classification problem. This problem is acute in reverse engineering a legacy system. It is not generally known how to measure the extent that a particular system design conforms to a standard design pattern. It is usually the case that the feature values of a particular system design approximately rather than exactly match a standard pattern. An approach to a solution of the system design classification problem is proposed in this article in the context of rough sets and a particular form of approximation space.

This research has been influenced by recent work on classification [4], information systems and infomorphisms [29], design patterns in general [3, 5], design patterns for intelligent systems [15-16], pattern recognition [31,34], the ideas concerning knowledge, classification and machine learning in rough set theory [10] and work on an approximation space approach to intelligent systems [18, 23], and [29]. In general, a model for a system design is represented by a set of interacting objects where each object is an instance of a class (a description of a set of objects that share the same attributes, operations, and semantics). A pattern is a conjunction of feature values that are associated with a decision rule. In particular, a system design pattern is a conjunction of feature values relative to the structure and functionality of a set of classes used in designing components of a system. The inclusion of structural as well as functional features in design patterns has been used in solving object recognitions problems and 3D object representation [31], and is implicit in the description of architectural design patterns [3], and in software design [5]. Patterns commonly found in models for system designs can be gleaned from class, interaction, and state diagrams from the Unified Modeling Language (UML) [7], and, especially, [6]. In this article, only class and collaboration diagrams are considered. In addition, it is possible to distinguish creational (creation of objects), structural (class structures) and behavioral (interaction between objects) design patterns. This article is limited to a consideration of behavioral design patterns.

This paper has the following organization. Basic concepts from rough sets and UML underlying the proposed approach to classifying system design models are briefly presented in Section 2. Design features and construction of software model decision tables are briefly considered in Sections 3 and 4, respectively. Design patterns in the context of rough sets are briefly covered in Section 5. Approximation of sets of software model decisions is considered in Section 6. An approximation space for software models is considered in Section 7. A framework for classification of software models within a satisfaction-based, parameterized approximation space is given in Section 8.

2 Basic Concepts

This section covers some fundamental concepts in rough sets and UML that underlie the presentation in the later sections of this paper.

2.1   Rough Sets

The rough set approach introduced by Zdzisław Pawlak [14] provides a ground for concluding to what degree a set of design models representing a standard are a part of a set of candidate design models. In this section, we briefly consider set approximation in rough set theory, and some of the rudiments of UML. For computational reasons, a syntactic representation of knowledge is provided by rough sets in the form of data tables. Informally, a data table is represented as a collection of rows each labeled with some form of input, and each column is labeled with the name of an attribute that computes a value using the row input. Formally, a data (information) table IS is represented by a pair (U, A), where U is a non-empty, finite set of objects and A is a non-empty, finite set of attributes, where $a:U \rightarrow V_a$ for every $a \in A$. For each $B \subseteq A$, there is associated an equivalence relation $Ind_{IS}(B)$ such that $Ind_{IS}(B) = \{(x, x') \in U^2 \mid \forall a \in B. a(x) = a(x')\}$.

If $(x, x') \in IndIS(B)$, we say that objects x and x' are indiscernible from each other relative to attributes from B. This is a fundamental concept in rough sets. The notation B(x) denotes a block of B-indiscernible objects in the partition of U containing X. For $X \subseteq U$, the set X can be approximated using B by constructing a B-lower and B-upper approximations denoted by $B_*(X)$ and $B^*(X)$, where $B_*(X) = \{x \mid B(x) \subseteq X\}$ and $B^*(X) = \{x \mid B(x) \cap X \neq \varnothing\}$. A lower approximation $B_*(X)$ of a set X is a collection of objects that can be classified with full certainty as members of X using the knowledge represented by attributes in B. By contrast, an upper approximation $B^*(X)$ of a set X is a collection of objects representing both certain and possible uncertain knowledge about X. Whenever $B_*(X) = B^*(X)$, the collection of objects can be classified perfectly, and forms what is known as a crisp set. In the case $B_*(X$ is a proper subset of $B^*(X)$, then the set X is considered rough (inexact) relative to B.

2.2   Unified Modeling Language (UML): Collaboration Diagrams

One type of UML diagrams is briefly considered in this section, namely, collaboration diagram (dynamic, message oriented view of interacting objects).

A collaboration diagram reveals the behavior of objects interacting with each other, which is helpful in visualizing how objects work together. Objects in such a diagram are represented by boxes. An object label is of the form "object name : Class name", where the object name or the :Class name can appear by themselves. An association between two objects is denoted by a straight line. An interaction between two objects is denoted by a label of the form "<numeral> : [condition] message →". The numeral indicates the position of a message in a sequence of interactions that define a behavior. The "[condition]" denotes an optional boolean condition that must be satisfied before an interaction between objects can occur. A message such as "createMemento( )" in Fig. 1 denotes a form of stimulus for some action that an object in an interaction must perform. The "→" indicates the direction of an interaction.

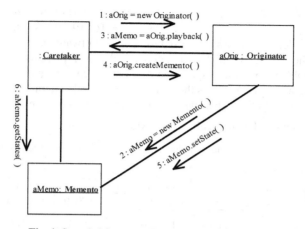

Fig. 1. Sample Memento Collaboration Diagram.

Example 2.2.1. Collaboration Diagram for a Memento Pattern.

A collaboration diagram for the Memento pattern is given in Fig. 1. In this diagram, there is a sequence of six interactions between three objects, where the Caretaker stimulates the Originator to save its internal state, which it transmits to the Mememto object.

3 Design Features

In this section, some of the structural and functional features extractable from a typical software model are briefly described (see Table 1).

Table 1. Sample Features.

A	Explanation
a1	Number of objects
a2	Number of 1-way messages
a3	Number of 2-way messages
a4	Number of 3-way messages
a5	Number of abstract classes
a6	Number of interfaces
a7	**Context** ∈ {x | x = algorithm skeleton = **as** ∈ [70.0, 70.9] *or* snapshot of internal state = **sis** ∈ [71.0, 71.9] }
a8	**Consequence** ∈ {x | x = factor out common code for reuse = **fcc** ∈ [80.0, 80.9] *or* internal state retrievable = **isr** ∈ [81.0,81.9] }
a9	**Problem solved** ∈ {x | x = algorithm operations realized with subclasses = **aos** ∈ [90.0, 90.9] *or* internal state storage = **iss** ∈ [91.0, 91.9] }
a10	Coupling a10(X) = 1 − ((# of loosely coupled classes in X)/card(X)), where high values indicate tight coupling, and low values indicate loose coupling.
Design	Design ∈ {null (0), template method (1), memento (2)}

Such features are described briefly in [31], and are part of the more general rough set approach to pattern recognition (see, e.g., [15-16, 19-20, 23, 26-28, 30, 34]). A design pattern identifies a form [34] used to classify software models. Features a7, a8, a9 measure the extent to which a model for a design represents a context, consequence, or problem solution, respectively, in a standard software model. The decisions represent a sample decisions that classify behavioral patterns relative to standard behavioral patterns. In this paper, we have restricted it to three decisions.

Example 3.1. Sample Feature Values for a Memento Design.

Let X_{Fig1} denote the set of objects in Fig. 1. The features of memento model in Fig. 1 are summarized as follows.

X\A	a1	a2	A3	a4	a5	a6	a7	a8	a9	a10	D
$X_{Fig.1}$	3	2	0	1	0	0	71.9	81.9	91.9	1	2

The model in Fig. 1 has 3 classes (a1), 2 instances of 1-way message passing (a2), zero 2-way message passing (a3), and 1 instance of 3-way message passing (a4). There are zero abstract classes (a5) and zero interfaces (a6). The values of a7 (context), a8 (problem solved), and a9 (consequence) are high because the model is considered very close to the standard model for a memento pattern. That is, the values of a7, a8, and a9 are set high because it is assumed that the model in Fig. 1 flawlessly satisfies the context, problem, and consequence requirements for the memento design pattern.

Example 3.2. Objects with Loose Coupling.

The coupling feature (a10) has a value of 1, since none of the classes in the model in Fig. 1 is loosely coupled. The objects in the Collaboration diagram in Fig. 1 illustrate what is known as tight coupling (each object in the model interacts directly with the other objects in the model).

This model can be improved by decoupling the Caretaker from the Memento object. A new model that accomplishes this is given in Fig. 2. In the new collaboration

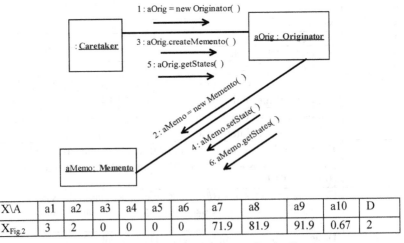

X\A	a1	a2	a3	a4	a5	a6	a7	a8	a9	a10	D
$X_{Fig.2}$	3	2	0	0	0	0	71.9	81.9	91.9	0.67	2

Fig. 2. Loose Coupling in the Memento Design Pattern.

diagram in Fig. 2, the Client requests a retrieval of saved internal states in step 6 by sending its request to the Originator, which responds by forwarding the Caretaker's request to Memento. In effect, this induces loose coupling in the model, since the Caretaker interacts with Memento only indirectly (with the help of the Originator). The features of the new Memento model in Fig. 2 are summarized in the Legend for that Figure. The coupling feature $a10(X_{Fig.2}) = 1 - 1/3 = 0.67$ stems from the fact that 1 out the 3 classes in Fig. 2 is decoupled from the Memento class, namely, the Caretaker class.

Example 3.3. Template Method Design Pattern.

The template method design pattern is the second pattern considered in the information tables constructed in this article. The context of the template method (tm) design pattern is to define the skeleton for an algorithm in an abstract class, and defer to subclasses some of the steps in the algorithm (see, e.g., Fig. 3). The problem to solve with the tm pattern is discover ways to redefine one or more steps in an algorithm without changing the basic structure of the algorithm given a templateMethod( ). A consequence of the tm pattern is that reusable parts of an algorithm are factored out as expressed in the templateMethod( ) of the abstract class (represented by the lollipop – o symbol in Fig. 3), where parts of an algorithm are specialized. The feature values for the sample tm model in Fig. 3 are summarized in the Legend.

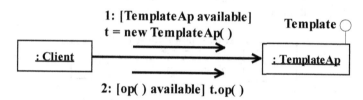

Fig. 3. Basic Template Method Model.

Legend

X\A	a1	a2	a3	a4	a5	a6	a7	a8	a9	a10	D
$X_{Fig.3}$	2	0	0	0	1	0	71.9	81.9	91.9	0	1

For example, in an application requiring forecasting, more than one method may be used to perform forecasts (see, e.g., Fig. 4). This is the case in forecasting traffic volume where one method depends on radar used to record traffic movements, and another method depends on a network of infrared sensors to monitor traffic. The Client object in Fig. 4 represents a traffic control station, which switches between forecasting methods depending on the situation. The feature values for the weather forecaster model in Fig. 4 are summarized in the Legend for Fig. 4.

The Client in Fig. 4 is tightly coupled to the ForecaseAp1 and ForecastAp2 classes. Hence, $a10 (X_{Fig.4}) = 1 - 0/3 = 1$.

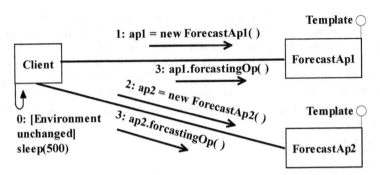

Fig. 4. Forecasting Collaboration Diagram.

Legend

X\A	a1	a2	a3	a4	a5	a6	a7	a8	a9	a10	D
$X_{Fig.4}$	3	2	0	0	1	0	71.9	81.5	91.5	1	1

4 Construction of Software Model Decision Table

In this section, a new decision table reflecting design patterns and corresponding Design decisions from earlier examples as well as new examples is given. Consider first the following null pattern.

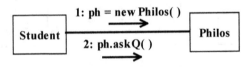

Fig. 5. Sample Model Not Matching A Standard Model.

Legend

X\A	a1	a2	a3	a4	a5	a6	a7	a8	a9	a10	D
$X_{Fig.5}$	2	1	0	0	0	0	70.0	80.0	90.0	1	0

Since the a7 (Context), a8 (Consequence), and a9 (Problem Solved) indicate that the model in Fig. 5 does not match the pattern for either template method or memento, the Design decision is zero. In an expanded view of design patterns, the collaboration diagram corresponds to a design pattern that satisfies what is known as the Law of Demeter (in time of need, ask a neighbor for help). The law of Demeter applies in the model in Fig. 5, where Student calls the askQ( ) method in the Philos class).

In Fig. 6, the object named Student plays a dual role as suggested by its actions, namely, acquiring knowledge by calling the askQ( ) method in Philos and instructing to archive its knowledge by calling saveKD( ) in Philos. The Philos object responds by calling saveAns( ) in the Memento object to save its answer to a question asked by the student. From this we can infer that Philos captures part of its internal state each time it asks Memento to save one of its answers, but this is only a fraction of its internal state. Hence, the values a7, a8, and a9 are low.

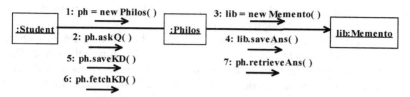

Fig. 6. Sample Model Matching Memento.

Legend

X\A	a1	a2	a3	a4	a5	a6	a7	a8	a9	a10	D
$X_{Fig.6}$	3	2	0	0	0	0	71.2	81.2	91.2	0.67	2

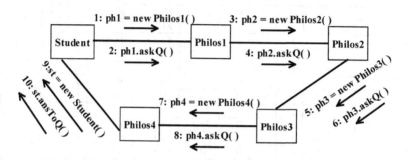

Fig. 7. Sample Model Not Matching A Standard Model.

Legend

X\A	a1	a2	a3	a4	a5	a6	a7	a8	a9	a10	D
$X_{Fig.7}$	5	5	0	0	2	1	70.0	80.5	90.5	0.4	0

The model in Fig. 7 is an example of what is known as Chain of Responsibility [5], where more than one object is designed to handle a request (when one philosopher does not know the answer to question, it asks another philosopher).

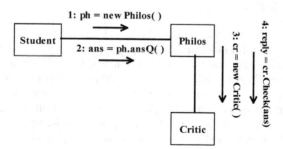

Fig. 8. Sample Model Not Matching A Standard Model.

Legend

X\A	a1	a2	a3	a4	a5	a6	a7	a8	a9	a10	D
$X_{Fig.8}$	3	2	0	0	0	0	70.9	80.9	90.9	0.67	0

However, because we limited the consideration of design patterns to the template method and memento, the Design decision is zero.

The model in Fig. 8 has the structure of what is known as a Proxy pattern [5], where the Philos object acts as a surrogate for another object, namely, Critic. This is a very common design pattern in system models. Again, because the model does not match either of the standard models being considered, the Design decision is zero. Table 2 presents a software model decision table derived from Figures 1 to 8 and other software models.

Table 2. Software Model Decision Table.

X\A	a1	a2	a3	a4	a5	a6	a7	a8	a9	a10	Design
X1	3	3	0	1	0	0	71.5	81.5	91.5	0.67	2
X2	2	1	0	0	1	0	70.1	80.1	90.1	1	1
X3	5	6	0	2	1	0	71.7	81.9	91.9	0.6	2
X4	3	3	1	1	1	0	71.5	81.55	91.45	1	2
X5	3	3	1	1	1	0	71.5	81.6	91.5	1	1
X6	4	4	1	1	1	1	71.5	81.5	91.5	0.5	2
X7	3	3	0	0	1	0	70.5	80.5	90.5	1	1
X8	2	0	2	0	0	0	70.5	80.5	90.5	1	1
X9	8	6	0	2	1	0	71.7	81.9	91.9	0.6	2
X10	3	3	1	1	1	0	71.5	81.55	91.45	1	2
X11	5	5	1	1	1	1	71.5	81.5	91.5	0.5	2
X12	4	1	0	0	1	0	70.5	80.5	90.5	1	1
X13	3	2	0	1	0	0	71.9	81.9	91.9	1	2
X14	3	2	0	0	0	0	71.9	81.9	91.9	0.67	2
X15	2	0	0	0	1	0	71.9	81.9	91.9	0	1
X16	3	2	0	0	1	0	71.9	81.5	91.5	1	1
X17	2	1	0	0	0	0	71.0	81.0	91.0	1	0
X18	3	2	0	0	0	0	71.2	81.2	91.2	0.67	2
X19	5	5	0	0	2	1	70.0	80.5	90.5	0.4	0
X20	3	2	0	0	0	0	70.9	80.9	90.9	0.67	0

5   Design Patterns

To some extent, this research takes into account observations by Christopher Alexander about the idea of form in Plato's philosophy in arriving at an understanding about pattern classification [2]. The basic assumption made in this research is that every design has a formal (ideal) model with distinguishing features that constitute a design pattern, and it is possible to prescribe conditions for the construction of a set of equivalent design models. In general, a pattern is a conjunction of feature values that are associated with a decision rule. Let a1, ..., ak, X, d be features, set of objects (instances of classes in a model for a design), and design decision, respectively. A pattern rule has the form $a1(X) \wedge \ldots \wedge an(X) \Rightarrow d$, where d is a decision about the design represented by a pattern. In particular, a design pattern is a conjunction of feature values relative to the structure and functionality of a model represented by set of objects (instances of classes) used in designing components of a system model. In this article, the decision associated with a pattern identifies a particular type of design.

6 Approximation of Sets of Software Model Decisions

Software models (collections of interacting objects) populate the universe underlying Table 2. The term *model* denotes a description of something. In this work, a *software model* is a description is in the form of a diagram. This is an engineering use of the term model, which is an abstraction (see, e.g., [22]). Set approximation from rough sets can be used to advantage to reason about software design decisions about software models represented in Table 2.

Example 6.1. Approximating a Set of Decisions about Software Models.

From Table 2, for simplicity, let U = {X1, X2, X3, X4, X5, X6, X7, X8, X9, X10, X11, X12}, and let D = {X | Design(X) = 2}. Let B = {a7 (Context), a10 (Coupling)}. Then consider the approximation of set D relative to the set of attributes B from Table 2.

Decision Value=2
Equivalence Classes for Attributes: {a7, a10}
[attr. values]:{equivalence classes}
[70.1, 1] : {X2}
[70.5, 1] : {X12, X7, X8}
[71.5, 0.5] : {X11, X6}
[71.5, 0.67] : {X1}
[71.5, 1] : {X10, X4, X5}
[71.7, 0.6] : {X3, X9}

Decision Class Stimuli: {X1, X10, X11, X3, X4, X6, X9}
$B_*(D)$ = {X1, X11, X3, X6, X9}
$B^*(D)$ = {X1, X10, X11, X3, X4, X5, X6, X9}
$BN_A(D)$ = {X4, X5, X10}   [B-boundary region of X]
U - B*(D) = {X2, X7, X8, X12}   [B-outside region of X]

$$\alpha_B(D) = \frac{|B_*(D)|}{|B^*(D)|} = \frac{5}{8} = 0.625$$

Consequently, the set D is considered rough or inexact, since the accuracy of approximation of D is less than 1. A visualization of approximating the set of design decisions relative to our knowledge in B is shown in Fig. 9.

The non-empty boundary region in Fig. 9 is an indication that the set D cannot be defined in a crisp manner.

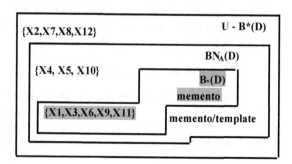

Fig. 9. Approximation of Set of Design Decisions.

7 Approximation Space for Software Models

In some cases, a set of objects in a software design will represent the basic model for a design pattern only partially, especially in cases where a set of objects in a software model represents overlapping, incomplete, or combinations of design patterns. In other words, there is the possibility that a particular set of objects does not match exactly the model for a particular pattern. In fact, we can organize our approximate knowledge about a set of software models by considering the designs in the context of an approximation space 21, 26-28]. A generalized approximation space is a system GAS = (U, I, v) where

- U is a universe, which is a non-empty set of objects.
- I : U → $\wp$(U) is an uncertainty function where $\wp$(U) denotes the   powerset of U.
- v : $\wp$(U) × $\wp$(U) → [0, 1] denotes a rough inclusion function.

The uncertainty function defines for every object x in U a set of similarly described objects. Values of an uncertainty function I usually define a covering of U. A set X ⊆ U is definable on GAS if it is a union of objects computed by the uncertainty function. An uncertainty function defined with rough set theory usually induces tolerance relations [32]. The rough inclusion function v defines the degree of inclusion between two subsets of U. An example of a rough inclusion function is a generalized rough membership function v : $\wp$(U) × $\wp$(U) → [0, 1], where

$$v(X,Y) = \begin{cases} \dfrac{|X \cap Y|}{|X|}, & \text{if } X \neq \varnothing, X \subset Y \\ 1, & \text{otherwise} \end{cases}$$

A complete presentation of generalized approximation spaces is given in [21]. In this work, we use a specialized form of approximation space that is parameterized [26]. A parameterized approximation space $PAS_{\#,\$} = (U, I_\#, v_\$)$ where #, $ denote vectors of parameters, where

- U is a universe, which is a non-empty set of objects.
- $I_\#$: U → $\wp$(U) is an uncertainty function where $\wp$(U) denotes the   powerset of U.
- $v_\$$: $\wp$(U) x $\wp$(U) → [0, 1] denotes rough inclusion

A common parameter for a PAS is gleaned from a subset of the attributes of an information system (U, A), where U is a universe, A is a set of attributes, and # = $ = B for B ⊆ A. Other forms of parameters for a PAS have been considered in approximate reasoning (see, e.g., [18]) and rough neural computing (see, e.g., [8]).

Example 7.1. Rough Inclusion Relative to a Standard.

Consider the case where the universe U consists of objects used in modeling system designs where X, Y ⊆$\wp$(U), the powerset of U. Let X denote a set of objects in a model for a system design, and let B(Y) denote sets of equivalent models of system designs relative to knowledge in the feature set B. Then v_B : $\wp$(U) × $\wp$(U) → [0, 1], where

$$v_B(X,Y) = \frac{|X \cap B(Y)|}{|B(Y)|}$$

The value of $v_B(X, Y)$ provides a measure of the degree to which the set X representing some standard is a subset of the block of equivalent models represented by B(Y). In this example, rough inclusion is limited to the parameter B. Notice also that in contrast to the usual rough membership function, rough inclusion has two independent variables, namely, X and Y (only parameter B is fixed). For some problems, it is convenient to define an uncertainty set function of the form $I_{\#}: \wp(U) \to \wp(U)$. This is the case in setting up an approximation space for design patterns.

Example 7.2. Design Pattern Approximation Space.

Consider the approximation space (U, I_B, v_B) where

- $I_B : \wp(U) \to \wp(U)$, where $I_B(X)$ is an uncertainty function
- $v_B : \wp(U) \times \wp(U) \to [0, 1]$, where $v_B(X, Y)$ is rough inclusion

The parameter used in this example is a feature set B. Let X, $Y \subseteq \wp(U)$. The uncertainty function $I_B(X)$ defines for every object X in U a set of similarly described objects relative to our knowledge expressed by the feature set B. In particular, B(X) denotes a block of B-indiscernible design patterns in a partition of the universe U using the indiscernibility relation. In this example, let the uncertainty function $I_B(X)$ compute $B^*(X)$. Then rough inclusion is defined as follows.

$$v_B(X,Y) = \frac{|I_B(X) \cap B(Y)|}{|B(Y)|} = \frac{|B^*(X) \cap B(Y)|}{|B(Y)|}$$

In effect, rough inclusion $v_B(X, Y)$ measures the extent that an upper approximation is included in a partition of the universe containing B-equivalent sets representing a design standard. Let B = {a7, a9} for the first 12 rows in Table 2, where a7, a9 denote Context and Problem Solved by a model for a design, respectively. Relative to the knowledge represented by the feature set B and the set D = {X | Design(X) = 2, memento design pattern}, the following upper approximation can be computed.

Equivalence Classes for Attributes: {a7, a9}
[attr. values]:{equivalence classes}
[70.1, 90.1] : {X2}
[70.5, 90.5] : {X12, X7, X8}
[71.5, 91.45] : {X10, X4}
[71.5, 91.5] : {X1, X11, X5, X6}
[71.7, 91.9] : {X3, X9}
Decision Class Stimuli: {X1, X10, X11, X3, X4, X6, X9}
Upper Approximation: $B^*(D)$ = {X1, X3, X4, X5, X6, X9, X10, X11}

In cases where design decisions are both in the boundary and outside the boundary, v_B will be less than 1. This is shown in Table 3.

It is also possible to set up an approximation space for design decisions relative to lower approximations. This is done because it makes sense to consider the degree of

Table 3. Rough Inclusion Table.

B(X)	I_B	v_B
B(X1)={X1,X5,X6,X11}	$B^*(D)$	$\|B(X1) \cap B^*(D)\| / \|B(X1)\| = 1$
B(X2) = {X2}	$B^*(D)$	$\|B(X2) \cap B^*(D)\| / \|B(X2)\| = 0$
B(X3) = {X3, X9}	$B^*(D)$	$\|B(X3) \cap B^*(D)\| / \|B(X3)\| = 1$
B(X4) = {X4, X10}	$B^*(D)$	$\|B(X4) \cap B^*(D)\| / \|B(X4)\| = 1$
B(X7) = {X7, X8, X12}	$B^*(D)$	$\|B(X7) \cap B^*(D)\| / \|B(X7)\| = 0$

inclusion of a lower approximation reflecting certain knowledge in various blocks of B-equivalent objects in a universe. This is shown in the next example.

Example 7.3. Lower Approximation as a Standard.

Consider an approximation space (U, I_B, v_B) with B equal to a feature set, and where

$$I_B : \wp(U) \rightarrow \wp(U), \text{ where } I_B(X) = B_*(X)$$

$$v_B: \wp(U) \times \wp(U) \rightarrow [0,1], \text{ where } v_B(X,Y) = \frac{|B(X) \cap B_*(Y)|}{|B_*(Y)|}$$

Again, let B = {a7, a9} in Table 2, where a7, a9 denotes Context and Problem Solved by a model for a design, respectively. Relative to the knowledge represented by the feature set B and the set D = {X | Design(X) = 2, memento design pattern}, the following lower approximation can be computed.

[70.1, 90.1] : {X2}
[70.5, 90.5] : {X12, X7, X8}
[71.5, 91.45] : {X10, X4}
[71.5, 91.5] : {X1, X11, X5, X6}
[71.7, 91.9] : {X3, X9}
$B_*(D) = \{X3, X4, X9, X10\}$

In this case, assume that $I_B(D)$ computes $B_*(D)$, and compute $v_{\{a7,a9\}}(X, D)$. Then construct Table 4.

Table 4. Rough Inclusion Table.

B(X)	I_B	v_B
B(X1)={X1,X5,X6,X11}	$B_*(D)$	$\|B(X1) \cap B_*(D)\| / \|B(X1)\| = 0$
B(X2) = {X2}	$B_*(D)$	$\|B(X2) \cap B_*(D)\| / \|B(X2)\| = 0$
B(X3) = {X3, X9}	$B_*(D)$	$\|B(X3) \cap B_*(D)\| / \|B(X3)\| = 0.5$
B(X4) = {X4, X10}	$B_*(D)$	$\|B(X4) \cap B_*(D)\| / \|B(X4)\| = 0.5$
B(X7) = {X7, X8, X12}	$B_*(D)$	$\|B(X7) \cap B_*(D)\| / \|B(X7)\| = 0$

The lower approximation $B_*(D)$ represents certain knowledge about a particular set of models for designs relative to our knowledge reflected in B. That is, every set of classes in the lower approximation has been classified as a memento design. A comparison of the lower approximation with each of the equivalence classes ranging over the software models provides an indication of where mismatches occur. The degree of mismatch between the lower approximation and a block of equivalent designs is measured using v_B. In two cases in Table 2, $v_{\{a7,a9\}}(X3, D) = v_{\{a7,a9\}}(X4, D) = 0.5$, there is an overlap between the models for designs in B(X3), B(X4) and $B_*(D)$. This indicates that the models in $B_*(D)$ have some relation to the models in blocks B(X3), B(X4) containing X3 and X4. At this point, these results are inclusive. The issue now is to devise a scheme provided by an approximation space relative to standards for classifying models for system models and a measure of when a model for a design is acceptable.

8 Framework for Classification of Software Models

In this section, the notion of a classification is considered in the context of satisfaction relations. A formal definition of a classification was given by Barwise [4], and elaborated in the context of information systems and rough set theory in [29].

Definition 8.1. *Classification.* A Classification Λ is defined by the tuple.

$$\Lambda = \left(\Sigma_\Lambda, C, \models_\Lambda \right)$$

where Σ_Λ, C are set of types used to classify objects and a set of objects (called tokens) to be classified, respectively, and a binary relation such that $\models_\Lambda \subseteq \Sigma_\Lambda \times C$. A type is a set. Let $x \in C$ and $\alpha \in \Sigma_\Lambda$. The notation $x \models_\Lambda \alpha$ reads "x satisfies α relative to classification Λ".

Example 8.1. Classification of Design Models.

Consider the classification $\Lambda = \left(\Sigma_\Lambda, C, \models_\Lambda \right)$ where Σ_Λ is a set of features used in a design pattern and C is a set of design models where each model $X \in C$ is a set of objects (instances of classes) that represent the structure and functionality of a system. The relation $X \models_\Lambda \alpha$ asserts that design model X in some sense satisfies α.

The satisfaction relation for a classification of design models can be defined relative to rough inclusion. Recall that a pattern is a conjunction $\wedge(a1, \ldots, ak)(X)$, of k feature values. When it is clear from the context what is meant, $B \subseteq \Sigma_\Lambda$ is called a design pattern. Then $X \models_\Lambda B$ asserts that design model X in some sense satisfies every $\alpha \in B$. Let $X, Y \in \wp(C)$ (powerset of design models). A partition of C is defined using the indiscernibility relation Ind(B) as follows:

X Ind(B) Y if, and only if, a(X) = a(Y) for every $\alpha \in B$

That is, two models X and Y are indiscernible from each other in the partition C/Ind(α) in the case where a(X) = a(Y) for every $\alpha \in B$. The block of the partition C/Ind(B) containing X is denoted by B(X). Then rough inclusion $v_B : \wp(C) \times \wp(C) \rightarrow [0,1]$ is defined as follows.

$$v_B(X,Y) = \frac{\left| B(X) \cap Y \right|}{\left| B(X) \right|}$$

The notation $X \models_{\Lambda,Y} B$ reads "X satisfies α relative to classification Λ and standard Y for X, Y $\in \wp(C)$, and for every $\alpha \in B$".

Definition 8.2. *Rough Inclusion Satisfaction Relation.* Let the threshold $th \in [0, 1)$. In addition, let X, Y $\in$ $\wp(C)$, B $\subseteq \Sigma_\Lambda$. Then $X \models_{\Lambda,Y} B$ iff $v_B(X,Y) \geq th$. That is, X satisfies pattern B if, and only if, the rough inclusion $v_\alpha(X,Y)$ value is greater than some preset threshold th.

The threshold *th* serves as a filter for candidate models for system designs relative to a standard represented by a set of acceptable models. In the case where the overlap between a set of models X for a system design and a standard Y is greater than threshold *th*, then X has been satisfactorily classified relative to the standard.

Example 8.2. Rough Inclusion Relative to a Standard.

Let B $\subseteq \Sigma_\Lambda$, X, $Y_{Std} \in \wp(C)$ denote a subset of the set of types, a subset of the power-erset of C, and a standard for a system design, respectively. We want to measure the overlap of partition B(X) with Y_{Std} using rough inclusion. Then define rough inclusion $v_B : \wp(C) \times \wp(C) \to [0,1]$ as follows:

$$v_B(X, Y_{Std}) = \frac{|B(X) \cap Y_{Std}|}{|Y_{Std}|}$$

Table 5. Classification Table.

B (X)	v_B	Result
B(X), where X = X19, X2, X12, X20, X17, X5, X16, X13	$\|B(X) \cap Y_{Std}\| / \|Y_{Std}\| = 0$	reject
B(X1) ={X1,X11,X6}	$\|B(X2) \cap Y_{Std}\| / \|Y_{Std}\| = 0.38$	**accept**
B(X3) = {X3, X9}	$\|B(X3) \cap Y_{Std}\| / \|Y_{Std}\| = 0.25$	reject
B(X4) = {X4, X10}	$\|B(X4) \cap Y_{Std}\| / \|Y_{Std}\| = 0.25$	reject
B(X18) = {X18}	$\|B(X7) \cap Y_{Std}\| / \|Y_{Std}\| = 0.13$	reject

Example 8.3. Classification Standard as a Lower Approximation.

The lower approximation of a set of design decisions provides a basis for classifying each object in the universe with certainty (see Table 5).

```
[attr. values]:{equivalence class}
[70.0, 80.5, 90.5] : {X19}
[70.1, 80.1, 90.1] : {X2}
[70.5, 80.5, 90.5] : {X12, X7, X8}
[70.9, 80.9, 90.9] : {X20}
[71.0, 81.0, 91.0] : {X17}
[71.2, 81.2, 91.2] : {X18}
[71.5, 81.5, 91.5] : {X1, X11, X6}
[71.5, 81.55, 91.45] : {X10, X4}
[71.5, 81.6, 91.5] : {X5}
[71.7, 81.9, 91.9] : {X3, X9}
[71.9, 81.5, 91.5] : {X16}
[71.9, 81.9, 91.9] : {X13, X14, X15}
```
$Y_{Std} = B_*(D) = \{X1, X3, X4, X6, X9, X10, X11, X18\}$

In effect, the lower approximation of a set of design decisions makes an ideal candidate for a standard in classifying the models contained in each partition of the universe. Let $Y_{Std} = B_*(X)$ be a lower approximation of X that defines a standard relative to pattern B = {a7, a8, a9} in Table 2. Also, let D = {X | Design(X) = 2, memento design pattern}, the following lower approximation can be computed. Assume that the threshold th = 0.3 in the rough inclusion satisfaction relation. In what follows, assume that Y_{Std} presents a collection of similar models for a system Sys, where each Y $\in$ Y_{Std} is a model for the design of a subsystem of Sys. Further, assume that each partition of the universe represented by B(X) contains candidate models for the design of Sys. The outcome of a classification is to cull from partitions of C those models that satisfy the standard to some degree. Next construct a classification table (see Table 5). From Table 5, B(X1) satisfies the standard. The choice of the threshold is somewhat arbitrary and will depend on the requirements for a particular project. Returning now to the notion of an approximation, it is possible to construct a new form of approximation space that is satisfaction based. This can be achieved by noting that the set C can be viewed as a universe of objects. The set Σ_Λ can be viewed a set of attributes. Recall that an attribute a is a mapping of the form a: X $\rightarrow$ Va and noting that an attribute can also be defined by a set, namely,

$$a = \left\{ v_{a_1}, ..., v_{a_{|Va|}} \right\}$$

In effect, an attribute is a type. Next, let X, Y $\in$ $\wp(C)$ and let $B \subseteq \Sigma_\Lambda$. Then, for example, define an uncertainty function I_B as follows:

$$I_B(X) = B_*(X)$$

The assumption here is that the lower approximation $B_*(Y)$ represents a standard. Then a block B(X) in the partition of C satisfies the standard provided that the overlap between B(X) and $B_*(Y)$ is greater than a preset threshold th. That is,

$$B(X) \models_{\Lambda,Y} B \text{ iff } \mu_B\left(X, I_B(Y)\right) \geq th$$

In effect, we obtain the following result.

Proposition 8.1. PAS_{sat} = (C, I_B, v_B, $\models_{PAS}$) is a satisfaction-based, parameterized approximation space.

9 Conclusion

This paper has presented a framework for classifying models for system design in the context of rough set-based approximation spaces and a schema for classification introduced by Barwise and Seligman in 1997. The basic goal in this research has been to establish an approach to measuring to what extent a collection of similar design models approximate a set of standard models for a system design. The construction of sets of similar design models is based on knowledge gained from design patterns, especially those patterns constructed from reducts. It should be noted that the approach to approximate knowledge about design models is pertinent not only to software systems, but also to classification of models of embedded systems and ordinary physical systems. One of the challenges in this work is to develop a methodology for reverse engineering existing system designs. This is very difficult to accomplish because the features of existing systems designs belong to composite design patterns. To make sense of an existing system, it is necessary to unravel the composite patterns in the context of an approximation space, to decompose composite patterns into separate patterns that correspond to well-understood, standard design patterns. This is a bit like trying to assemble a replica of the brain of a dinosaur from knowledge gained from fragments available to palaeontologist (extracted, e.g., from partially accessible, partially incomplete fossil remains) so that the replica approximates the structure and functionality of the dinosaur brain. Since it common for models for subsystem designs to overlap, a subsystem model extracted from a complete system model of a legacy system has the appearance of a fragment, something incomplete when compared with a standard. Hence, it is appropriate to apply approximation methods described in this article relative to design patterns for fragmentary subsystem design models.

Acknowledgements

The research by James Peters and Sheela Ramanna has been supported grants from the Natural Sciences and Engineering Research Council of Canada (NSERC) grant 185986 and 194376, respectively. In addition, the research by James Peters has also been supported by grants from Manitoba Hydro.

References

1. Alexander, C.: The Timeless Way of Building. Oxford University Press, UK (1979)
2. Alexander, C.: Notes on the Synthesis of Form. Harvard University Press, Cambridge, MA (1964)
3. Alexander, C., Ishikawa, S., Silverstein, M., Jacobson, M., Fiksdahl-King, S. Angel, I.: A Pattern Language. Oxford University Press, UK (1977)
4. Barwise, J., Seligman, J.: Information Flow. The Logic of Distributed Systems. Cambridge University Press, UK (1997)
5. Gamma, E., Helm, R., Johnson, R., Vlissides, J.: Design Patterns: Elements of Reusable Object-Oriented Software. Addison-Wesley, Toronto (1995)
6. Holt, J.: UML for Systems Engineering. Watching the Wheels. The Institute of Electrical Engineers, Herts, UK (2001)
7. OMG Unified Modeling Language Specification. Object Management Group, http://www.omg.org.
8. Pal, S.K., Polkowski, L., Skowron, A. (eds.): Rough-Neural Computing. Techniques for Computing with Words. Springer-Verlag, Heidelberg (2004)
9. Pawlak, Z.: Rough sets. International J. Comp. Inform. Science. 11 (1982) 341-356
10. Pawlak, Z.: Rough sets and decision tables. Lecture Notes in Computer Science, Vol. 208, Springer Verlag, Berlin (1985) 186-196
11. Pawlak, Z.: On rough dependency of attributes in information systems. Bulletin Polish Acad. Sci. Tech., 33(1985) 551-599
12. Pawlak, Z.: On decision tables. Bulletin Polish Acad. Sci. Tech., 34 (1986) 553-572
13. Pawlak, Z.: Decision tables – a rough set approach, Bulletin ETACS, 33 (1987) 85-96

14. Pawlak, Z.: Rough Sets. Theoretical Reasoning about Data. Kluwer, Dordrecht (1991)
15. Peters, J.F.: Design patterns in intelligent systems. Lecture Notes in Artificial Intelligence, Vol. 2871, Springer-Verlag, Berlin (2003) 262-269
16. Peters, J.F., Ramanna, S.: Intelligent systems design and architectural patterns. In: Proceedings IEEE Pacific Rim Conference on Communication, Computers and Signal Processing (PACRIM'03) (2003) 808-811
17. Peters, J.F., Ramanna, S. : Towards a software change classification system: A rough set approach. Software Quality Journal, 11, (2003) 121-147
18. Peters, J.F., Skowron, A., Stepaniuk, J., Ramanna, S.: Towards an ontology of approximate reason. Fundamenta Informaticae, Vol. 51, Nos. 1, 2 (2002) 157-173
19. Polkowski, L. and Skowron, A. (eds.): Rough Sets in Knowledge Discovery. Vol. 1, Physica-Verlag, Heidelberg (1998a)
20. Polkowski, L. and Skowron, A. (eds.): Rough Sets in Knowledge Discovery. Vol. 2, Physica-Verlag, Heidelberg (1998b)
21. Polkowski, L.: Rough Sets. Mathematical Foundations. Physica-Verlag, Heidelberg (2002)
22. Sandewall, E.: Features and Fluents. The Representation of Knowledge about Dynamical Systems. Vol. 1. Clarendon Press, Oxford (1994)
23. Skowron, A.: Toward intelligent systems: Calculi of information granules. In: Hirano, S. Inuiguchi, M., Tsumoto S. (eds.), Bulletin of the International Rough Set Society, Vol. 5, No. 1 / 2 (2001) 9-30
24. Skowron, A., Polkowski, L: Synthesis of decision systems from data tables. In: Lin, T.Y., Cercone, N. (eds.): Rough Sets and Data Mining: Analysis for Imprecise Data, Kluwer Academic Publishers, Boston (1997) 259-300
25. Skowron, A. and Rauszer, C.: The Discernibility Matrices and Functions in Information Systems. In: Slowinski, R. (ed.): Intelligent Decision Support: Handbook of Applications and Advances of the Rough Sets Theory, Kluwer Academic Publishers, Dordrecht (1992) 331-362
26. Skowron, A., Stepaniuk, J.: Tolerance approximation spaces. Fundamenta Informaticae, 27 (1996) 245-253
27. Skowron, A., Stepaniuk, J.,: Information granules and approximation spaces. In: Proc. of the 7th Int. Conf. on Information Processing and Management of Uncertainty in Knowledge-based Systems (IPMU'98), Paris (1998) 1354-1361
28. Skowron, A. Stepaniuk, J.: Information granules: Towards Foundations of Granular Computing. Int. Journal of Intelligent Systems, 16, (2001) 57-85
29. Skowron, A., Stepaniuk, J. Peters, J.F.: Rough sets and infomorphisms: Towards approximation of relations in distributed environments. Fundamenta Informaticae, Vol. 54, Nos 2, 3 (2003) 263-277
30. Skowron, A. and Swiniarski, R.W. Information granulation and pattern recognition. In Pal, S.K, Polkowski, L., Skowron, A. (eds.): Rough-Neural Computing, Springer-Verlag, Berlin, (2002) 599-636
31. Stark, L., Bowyer, K.: Achieving generalized object recognition through reasoning about association of function to structure. IEEE Trans. on Pattern Analysis and Machine Intelligence, Vol. 13, No. 10 (1991) 1097-1104
32. Stepaniuk, J.: Knowledge discovery by application of rough set model. In: Polkowski, L., Tsumoto, S,. Lin T.Y. (eds.): Rough Set Methods and Applications. New Developments in Knowledge Discovery in Information Systems, Studies in Fuzziness and Soft Computing, Vol. 56. Physica-Verlag, Heidelberg (2000) 137-234
33. Stepaniuk, J. Approximation spaces, reducts and representatives. In: [21] (1998) 295-306
34. Watanabe, S.: Pattern Recognition: Human and Mechanical. Wiley, London (1985)

Application of Rough Sets
to Environmental Engineering Models

Robert H. Warren[1], Julia A. Johnson[2], and Gordon H. Huang[3]

[1] University of Waterloo, Waterloo, ON, N2L 3G1 Canada
rhwarren@uwaterloo.ca
[2] Laurentian University, Sudbury, ON, P3E 2C6 Canada
julia@cs.laurentian.ca
[3] University of Regina, Regina, SK, S4S 0A2 Canada
huangg@uregina.ca

Abstract. Rough Sets is a method of dealing with domains characterised by inconsistent and incomplete information. We apply this method to the problem of Environmental Engineering modelling. The solid waste management problem, in particular, is the subject of our analysis.

Modelling large engineering problems is difficult because of the volume of information processed and the number of modelling decisions being made. In many cases, a chicken and the egg problem presents itself when modelling new or one-of-a-kind systems: The model is needed to gain the knowledge necessary for constructing the model.

The generally accepted solution is to iteratively verify the importance of a parameter or model change until a concise model is created which appropriately supports the decision making process. We improve on this process by using Rough Sets to actively search for simplifying assumptions of the model and validate the process using a municipal solid waste management case.

1 Introduction

Since Pawlak introduced Rough Sets as a machine learning method, a number of applications have been explored. In this paper we apply the Rough Sets method to the problem of modelling Environmental Engineering systems as a method of obtaining domain knowledge.

The process is reviewed in detail and results contrasted with current methods. We apply it to the solid waste management case of the Hamilton-Wentworth region, where municipal consumer waste is collected, transformed and disposed of under cost and environmental constraints. Finally, we close with a brief literature review of the application of Rough Sets to other problems.

2 Rough Sets Basics

The Rough Sets method is reviewed within the context of environmental engineering modelling. The method used for the conversion of linear data sets to discrete ones is also discussed[1].

---

[1] This section is reproduced from [1] with minor modifications.

J.F. Peters et al. (Eds.): Transactions on Rough Sets I, LNCS 3100, pp. 356–374, 2004.

Table 1. The relationship between the decision to use a particular pickup route and the road's condition.

	Conditional Attributes			Decision Attribute
Row	Weight Capacity	Alternate Available	Road Surface	Use Route
e_1	Low	No	Good	Yes
e_2	Low	Yes	Good	No
e_3	Low	Yes	Bad	No
e_4	High	No	Good	Yes
e_5	Med.	Yes	Very Good	Yes
e_6	Med.	Yes	Good	No
e_7	High	No	Very Good	Yes
e_8	High	No	Very Good	Yes

Rough Sets [2] is a non-deterministic, machine learning method which generates rules based on examples contained within information tables such as Table 1. This table contains a series of examples that relate the decision to use a particular garbage pickup route based on road conditions.

Each row within the information table is considered an element, which we represent using the notation $(e_1, e_2, e_3, \ldots)$, within an universe (U) defined by the information table.

2.1 Key Definition

In this approach we distinguish between two types of attributes: conditional attributes and decision attributes. Conditional attributes are those which we define as being the causal factors in the relationship (**Weight Capacity, Alternate Available, Road Surface**), and the decision attribute is the attributes which defines the consequence, or the class of the element (**Use Route**).

We define a specific value instance of an attribute a concept. For example, a **Good** value for the conditional attribute **Alternate Available** represents the concept that an alternative to the choice of this route exist.

Likewise, the value of the decision attribute **Use Route** as **Yes** represents the concept of the route being used to transport waste. As a convention, we refer to a specific value instance of the decision attribute as a class.

2.2 Operating Principles

There are two main concepts to the area of Rough Sets: the concept of indiscernible relations and the concept of boundary approximated, non-deterministic rules.

Indiscernibility Relations. An indiscernibility relation occurs between two elements, when they cannot be differentiated with respect to a certain attribute. For example, with respect to column **Use Route**, the elements of set $(e_1, e_4, e_5, e_7, e_8)$ and (e_2, e_3, e_6) are indiscernible because they all have the same column values.

We can also generate indiscernibility classes to identify redundant or dispensable columns within the information table. By definition, if a column is contained within a superset of attributes, but not within the subset of indiscernible attributes, it is redundant.

Table 2. The indiscernibility sets for all three conditional attributes, and the decision attribute.

Column	indiscernibility set
Weight Capacity	$(e_1, e_2, e_3), (e_4, e_7, e_8),(e_5, e_6)$
Alternative	$(e_1, e_4, e_7, e_8), (e_2, e_3, e_5, e_6)$
Road	$(e_1, e_2, e_4, e_6), (e_3), (e_5, e_7, e_8)$
Use Route	$(e_1, e_4, e_5, e_7, e_8), (e_2, e_3, e_6)$

Table 3. The indiscernibility classes of two attributes combinations are listed.

Column	indiscernibility set
Weight Capacity + Alternative	$(e_1),(e_2,e_3),(e_4,e_7,e_8),(e_5,e_6)$
Weight Capacity + Road	$(e_1,e_2),(e_3),(e_4),(e_5),(e_6),(e_7,e_8)$
Alternative + Road	$(e_1,e_4),(e_2,e_6),(e_3),(e_5),(e_7,e_8)$
Use Route	$(e_1, e_4, e_5, e_7, e_8), (e_2, e_3, e_6)$

Table 2 lists the indiscernibility class for each individual column and by inspection it is obvious that no indiscernibility class is a subset of the Use Route column. Thus, we can conclude that no single column is alone able to act as an indiscernibility set of the conditional attribute.

Trying combinations of two attributes at a time, we see from Table 3 that the combination of columns Alternative and Road produces an indiscernibility set which defines the same relation as the decision attribute Use Route. Thus, we can state that while both columns Alternative and Road are necessary to the relation, column Route Volume can be considered redundant.

A redundant attribute does not necessarily mean that the attribute is useless, and as is in most cases of reduction, some form of loss occurs. While in this case we have noted that the Weight Capacity attribute is not part of the indiscernibility class, the attribute still holds potentially useful information: all high capacity routes are used, a potentially useful fact that is ignored by our indiscernible attribute set.

Boundary Areas Applied to Rule Generation. The concept of indiscernibility relation is the primary means of generating rules in the Rough Sets approach. For example, by the contents of Table 3, specifically the indiscernibility set (e_1,e_4) for attributes Alternative and Road, we can generate the deterministic Rule 1.

Rule 1 $(Alternative, No) \land (Road, Good) \Rightarrow (UseRoute, Yes)$

Or, we can also use the relation to generate non-deterministic rules, as in the case of the indiscernibility set (e_2, e_3, e_5, e_6) in Table 2 for attribute Alternative.

Rule 2 $(Alternative, No) \Rightarrow (UseRoute, Yes)$

We know that Rule 2 holds true for elements (e_2, e_3, e_6), but not for element (e_5). Hence, using the indiscernibility relations we can obtain rules which while non-deterministic, are not totally false either, and can provide partial information when deterministic rules are not available.

Defining Rule Sets According to Boundary Lengths. The use of indiscernibility relations can be accounted for mathematically, using set theory which we augment with the notion of a bounded rule set. We use the notion of boundaries as a means of referring to concepts with a mix of deterministic, and non-deterministic rules.

The **lower approximation**, or lower bound, contains the indiscernibility relations of one, or multiple, attribute(s) which are subsets of the class indiscernibility relation. Hence, the lower bound is the collection of all deterministic rules which determine the conditions to a concept. The indiscernibility set of a conditional attribute for a certain value forms part of the lower approximation only if it is a subset of the indiscernibility set of the decision attribute value we are looking for.

By Section 2.2, the indiscernibility set of the column **Weight Capacity** with respect to value **High** is (e_4, e_7, e_8) and the indiscernibility set of the decision attribute for class **Yes** is $(e_1, e_4, e_5, e_7, e_8)$. Thus, since $(e_4, e_7, e_8) \subseteq (e_1, e_4, e_5, e_7, e_8)$ we can state that the indiscernibility set of the column **Weight Capacity** with respect to value **High** forms part of the lower bound approximation.

The mathematical generalisation is:

$$Set_{Attribute=Value} \subseteq \underline{Y} \iff (Set_{Attribute=Value} \subseteq U_{Class}) \tag{1}$$

The **upper approximation**, or upper bound, contains indiscernibility sets which are not subsets of the class indiscernibility set, but have a non-empty intersection with the class indiscernibility relation. Thus, the indiscernibility set of the column **Weight Capacity** with respect to value **Low** is (e_1, e_2, e_3) and is not a subset of class **Yes**. However, it's intersection with the class set is non-empty, (e_1) which signifies that while the rule is non-deterministic, it does provide some representation of the desired concept. The mathematical generalisation for the upper bound is:

$$Set_{Attribute=Value} \subseteq \overline{Y} \iff ((Set_{Attribute=Value} \cup U_{Class}) \neq \emptyset) \tag{2}$$

The use of the upper and lower bound concept allows us to provide for instances where both deterministic and non-deterministic information is available. Thus, by using the lower bound rules only we can generate certain rules which cover a limited number of possibilities. By adding the upper approximation, a complete coverage of all possibilities s achieved at a price in certainty.

2.3 Rule Generation and Interpretation

The RS1 algorithm [3, 4] is used to generate lower and upper bound rough set rules from information tables. It functions by selecting a sequence of columns from which indiscernibility sets are generated incrementally, and from these upper and lower bound rules are generated.

Rule 3

$(WeightCapacity, Med) \wedge (RoadSurface, VeryGood) \Rightarrow (UseRoute, Yes)$

Rule 4

$(WeightCapacity, Low) \wedge (RoadSurface, Good) \Rightarrow (UseRoute, Yes)$

By this time, the only remaining elements within the universe are (e_1, e_2), and since only one column remains, we can simply state the remaining class element through the deterministic Rule 5.

Rule 5

$(WeightCapacity, Low) \wedge$
$(RoadSurface, Good) \wedge$
$(AlternateAvailable, No) \Rightarrow (UseRoute, Yes)$

It should be noted that the interpretation of rules can sometimes add insight into the system under study, beyond the traditional cause and effect interpretation.

In the case of rules created using RS1, the order of the columns within the rules is important. Since a quality factor is used to select the columns, these are pre-sorted in their order of importance; and thus we can gain insight on which factor is the most influential in the relation.

Furthermore, by inspecting the differences between rules of different classes, we can gain insight into the system. For example, deterministic Rule 6 states that Very Good roads are always used; while non-deterministic Rule 7 states that Good roads are sometimes used.

Rule 6

$(Road, VeryGood) \Rightarrow (UseRoute, Yes)$

Rule 7

$(Road, Good) \Rightarrow (UseRoute, Yes)$

Rule 8

$(Road, Bad) \Rightarrow (UseRoute, No)$

Contrasted with Rules 8, where all bad roads are never used, we can state that having a Good road is a pre-requisite to the route being used, while a Very Good road will always be used.

2.4   Evaluation Metrics for Rough Set Rules

In the course of the Rough Sets method, both deterministic and non-deterministic rules are generated. In both cases, the rule is a generalisation of the information contained within the table. But because we use concepts such as indiscernibility relations to reduce the information set, the rules may have different levels of importance depending on how often they appear in the information table.

Thus, a number of performance metrics are available for us to use which qualify not only how often the rule may occur, but how well it predicts the occurrence of certain classes. Figure 1 is a graphical representation of the need for such metrics: within the universe (U) a rule will select a set of elements (B), within which a limited number are relevant (A). Thus, we need measures which quantify these different proportions.

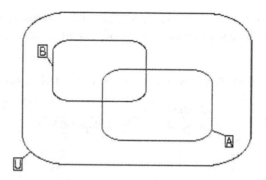

Fig. 1. Graphical representation of precision and recall.

Coverage. The Coverage of a rule refers to the total number of set elements which satisfy the requirements of the Conditional attributes. Because the size of the universe changes over the course of rule generation, we normalise the measure using the total number of elements within the universe. The measure (3) expresses the likelihood of occurrence of this condition, and by extension is an indicator of how generalised the rule is according to the data set.

$$Coverage = \frac{A \cap B}{B} \tag{3}$$

Thus, a rule which has a coverage of 0.5 will cover half of the example set, and a gross generalisation of the data set behavior. A rule with a coverage of 0.0001 applies to a very narrow segment of the data set and is likely to be more of an exception than a generalisation.

Support. Support is an indicator which measures the credibility of the rule according to the data set, or the normalised measure of positive examples covered by this rule.

$$Support = \frac{A \cap B}{A} \tag{4}$$

While the literature differs on an exact definition of support, we use here a measure absolute (4) to the data set. The use of an absolute result allows us to track the performance of a group of rules as more columns are used in the same rule group.

Furthermore, the use of an absolute reference enables us to track the positive coverage of the rule by comparing against the total number of positive examples in the set.

2.5   Example: Continuous Data in Information Table

A separation between two distinct data types is evident: continuous data and discrete data. The two are inherently different because while discrete values are by definition limited, continuous data is infinite in the number of data points that it provides. This causes a problem for Rough Sets because RS deals with attributes on the basis of indiscernibility sets, which require a limited number of possible values, or symbols, for each attribute.

Table 4. The Weight Capacity column has been replaced with continuous data.

| ROW_ID | Conditional Attributes | | | Decision Attribute |
	Weight Capacity (tons)	Alternate Available	Road Surface	Use Route
e_1	50	No	Good	Yes
e_2	42	Yes	Good	No
e_3	38	Yes	Bad	No
e_4	100	No	Good	Yes
e_5	84	Yes	Very Good	Yes
e_6	79	Yes	Good	No
e_7	150	No	Very Good	Yes
e_8	160	No	Very Good	Yes

As an example, Table 4 is a variation of Table 1 where the Weight Capacity column has been replaced with continuous data sets. Hence, while the attribute originally had several indiscernibility sets (e_1, e_2, e_3), (e_4, e_7, e_8) and (e_5, e_6), our continuous version cannot form any, and the attribute is ignored as part of the rule generation method.

Thus, in order for continuous information to be used, preprocessing of the data is needed to pre-generalise the continuous data set. This can be done either by using quantisation or manually pre-labelling the continuous information. A large amount of literature [5] does exist on the topic of quantisation which provides a simple way of dealing with continuous data. This comes at a cost of having the discretization done in a way which maybe inappropriate: *(Weight Capacity, RANGE_BIN_6)* does not necessarily provide any clear information.

The alternative of manually discretizing the continuous data provides a tailor-made solution which is appropriate to the situation, for example *(Weight Capacity, High)*. This comes at the cost of additional work and does require expert grasp of the domain at hand.

Thus, the Rough Set method can be used to generate deterministic and non-deterministic rules from information tables. However, its use within the Environmental Engineering context is problematic when continuous data sets are used. Quantisation and user driven methods are needed to convert continuous data set to discrete sets that can be used by the Rough Set methods. Though in some cases quantisation is a choice, the use of user driven methods not only allows conversion of the data, but does so in a way that ensures that the correct information is extracted from the data set.

3   Environmental Engineering Modelling Primer

Environmental Engineering deals with the relationships between human beings and their environment. As with most engineering disciplines, a mathematical model of the system is created and used to support decision making.

The model is an attempt to reduce the number of elements that need to be taken into consideration while sufficiently mimicking the system's behavior. A "perfect" model would require a complete copy of the system itself. Thus we need to create a model which only takes into account the specific elements which influence and interact with the functions we wish to study.

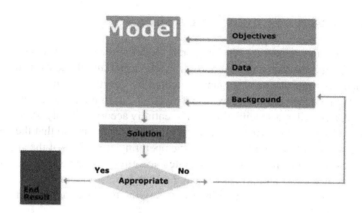

Fig. 2. The modelling decision cycle.

To balance the goal of a simple, manageable model with the requirement for one which we may consider as "correct", we need to understand how the system works. In many cases this knowledge is either not available or is our reason for creating a model in the first place.

During the creation of the model, there exists a learning process where the elements relevant to the model objective are identified. While a large amount of the modelling decisions can be made through inspection, other decisions need to be made through analysis. This can include sensitivity analysis, the rate of change of parameters is related to the rate of change of the output variable, or it can entail creating another independent model for this purpose only.

Figure 2[2] is a representation of the process flow which is necessary to arrive at a "correct" model. Not only must the modelling activity support the decision making process, it also serves to discover the underlying behavior of the system.

What compounds the problem is that the information required to create an instance of the model may not be available. For example, if we wish to model transportation in a city, traffic flow data may only be available as a daily average whereas an hourly average would have been preferable. This requires some additional experimentation to adapt the model to the information available without compromising the correctness of the model.

Finally, environmental engineering is a difficult field in that its objectives tend to be uncertain and open to debate. For a number of political and policy reasons, environmental engineering is required to provide alternatives and suggestions, as opposed to other engineering fields which might seek the "one true answer". Where the question might have been in another field "How can we reduce the number of garbage collection

---

[2] The authors would like to thank Tudor Whiteley for his help with the diagrams.

vehicles?", the query in an environmental model could read as "How can we lower disposal costs?". This further complicates the process: not only must the modeler create a model based on uncertain data, but it must also be done without a clear objective.

4   Rough Sets as a Learning Mechanism

When modelling an engineering system, a draft model of the system is constructed and through sensitivity analysis and trial and error, the important elements of the system are identified. We improve on this manual method by generating a list of draft model instances which are then processed by Rough Sets. The draft model does not need to be sophisticated, targeted to a specific decision or entirely accurate: it only serves to represent the overall system behavior. However, it must be constructed so that the instances it does generate represent a wide variety of the operating conditions of the system. The data generated from this model is then tabulated according to the conditional and decision attributes of interest and processed using a Rough Sets rule prediction algorithm. The rules thus generated can then be interpreted by the modeler for appropriate background knowledge.

We review our process in step-by-step detail:

1) **Generic Model Creation:** Initially we create a very general model of the system which is not intended for decision making and may not be completely accurate. The intent is to capture as many elements as possible of the system within a single unified model. Where information is missing or knowledge lacking, a new parameter is inserted into the model to allow for the testing of the effects of the specific unknown condition. In this manner, we are able not only to deal with unknown conditions but can also avoid the limitations of certain modelling techniques, such as non-linearities in linear programming. For example, this could be used to verify the impact of economies of scale within a linear program.

2) **Creating Target Data Set:** From this complex model we generate a series of model instances whose values are chosen to cover all of the possible operating ranges of the system. This can be done as an enumeration or as a combinatorial problem driven by the model constraints. When all of the instances have been created, they are solved to determine the state of the system under those particular conditions. Unsolvable models are especially important in that they determine the hard limits under which the system can operate. Some care must be taken in choosing the ranges of parameter values that should be used and the number of data points to use. In cases such as decision variables (eg: facility present or not) this is straightforward, but in the case of continuous parameters some inspection is necessary.

3) **Preprocessing Continuous Data into Domain Objects:** At this stage we label the numerical data of the model instances. This is done to generate indiscernibility relations from what is essentially continuous data and to ensure that the latter rule generation is "aesthetic" with respect to the model domain. In the case of design decisions and/or specific model conditions, it is easy to assign a descriptive label to the data value [6]. Hence, a label for a resource constraint set to 0 could be

labelled as Facility_not_available. A more complex problem occurs in the case of continuous numerical data, since recognising the significance of the data values is non trivial. A straight-forward solution to this problem is to draw on the significant body of literature in quantisation [7], at the cost of potentially meaningless labels such as RANGE_BIN_1. A certain amount of inspection is necessary on the users part to choose a method appropriate to the problem; a simple solution may involve trivial quantisation based on data set tiers.

4) Tabulating Domain Objects: Given a set of domain objects, we tabulate them in groupings of conditional and decisional attributes in order to evaluate relationship pairs that are of interest. This step can be treated as a combinatorial problem, relating any possible system parameter or decision to a model decision variable. However, it is likely that the user will decide himself on the tables to be processed based on his area of interest. These intermediate tables also provide an early glimpse into the system's behavior through simple inspection of the relations.

5) Applying Rough Sets: The tables are processed using a modified version (See Section 5.3) of the RS1 algorithm by Wong and Ziarko [4, 3] and sets of rules, along with precision and recall metrics, are stored in a report for each table. The precision and recall information are added along with the rules to enable the user to judge the validity of the rules being generated.

6) Rule Interpretation: Rules for each of the tables are reviewed and interpreted by the modeler to gain domain knowledge about the system. It is here that adding domain-specific labels to the numerical model information gets most of its value by enabling Rough Sets to generate human-readable rules specific to the problem. Another important element of this step is that the interpretation of the precision and recall figures must be done with the initial data set in mind. While some rules may seem to have a high or low performance, it is only by normalising these scores against the table population that an accurate judgement can be made.

4.1 Synopsis

The above process enables the modeler to test multiple hypotheses about the system in a very short time without committing to specific design decisions. Rough Sets is particularly well suited to this type of discovery because of its inherent support for uncertainty in information systems. To test the methodology, it was applied to a solid waste management case and the conclusion compared to previous work on that case.

5   The Hamilton-Wentworth Solid Waste Management Case

As a proof of concept we used the solid waste management case of the Hamilton-Wentworth region in Southwestern Ontario, Canada. This case is particularly applicable to Rough Sets methodology because it is a large case with a number of different alternatives which must be explored. Furthermore it is a typical environmental engineering problem in that several conflicting objectives are being pursued both in terms of optimising day-to-day operations and observing capital cost restrictions. This solid waste management case study was originally presented in [8–10] where grey linear programming was used to deal with the cases' imprecise data values. Inexact non-linear programming was later [11] used in an attempt to reduce the level on uncertainty within the system.

5.1 Case Description

The case deals with the collection, transportation and elimination of the municipal waste of the cities surrounding Hamilton. It includes waste routing, transfer station use, incineration, composting, recycling and landfill disposal as well as capacity constraints and capital investment costs.

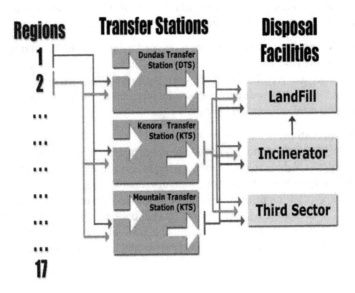

Fig. 3. Hamilton-Wentworth waste flow.

The Hamilton-Wentworth region totals an area of about 1,100 square KM. The six cities within this area have a population of slightly less than half a million people and are sub-divided into 17 waste collection regions. While no data on seasonal variations are available for the regions, the waste volume generated is known to be within certain value ranges[3].

The case makes allowances for the use of three transfer stations (Kenora, Mountain and Dundas) which sort, compact and route waste traffic from each of the regions to specific disposal plants. Each transfer station has an unique construction cost and waste processing capacity which must be taken into account by planners. While it is possible to avoid using transfer stations, their use is preferred to reduce the final volume routed

---

[3] The raw data tables are not reproduced here for space considerations, but can be found in [1].

to the disposal facilities. Each waste route has an individual cost per ton and is routed through a waste transfer station and not directly to a disposal facility.

The disposal facilities include a landfill, incinerator, composting and recycling facilities. While the operating costs of the landfill and incinerators are relatively low, preference is given to composting and recycling (Third Sector) to extend the landfill lifetime. Incineration is a disposal option within the model which has the benefit of generating electrical power. However, potential emissions legislation makes this only a short term solution. The original intent of the case was to study the optimal allocation of the solid waste flows to routes and the changes in the allocations with various facility configurations.

5.2 Application of Rough Sets Methodology to SWM Case

The objective of the Hamilton-Wentworth case was to study changes in optimal (lowest cost) waste distributions based on the available waste processing and disposal facilities. We applied the method to the case to generate rules to relate the availability of facilities and disposal methods to operating costs.

1) **Generic Model Creation:** The approach to modelling the system was to use a Linear Programming (LP) model to optimise the flow of the waste to the disposal sites at lowest cost. We could easily have used another type of modelling method, but we chose a simple modelling solution to highlight the capacity of our approach to learn from simple, generic models.

2) **Data Set Generation:** The scenarios generated concentrated on the availability of waste transfer stations, the forced use of higher cost disposal alternatives such as recycling and incineration and the effects of uncertain waste flow on overall system planning. Because each scenario is independent, we implemented missing facilities and disposal alternatives by changing the constraint's parameters on each model instance. In all, 18,432 different instances of the model were created and solved using an openly available LP solver [12]. Some model instances had no solutions and were marked as such within the table.

3) **Domain Objects:** Labels were assigned to parameters and operating cost. In the case of transfer stations and disposal facility, a simple Yes/No labelling system was used to indicate facility availability. For operating cost several sets of labels were generated: a tier label based on a Low, Medium or High cost, a cost minimisation label Optimal, Non-optimal (based on a cost within 10% of the lowest observed cost) and lastly, a label based on whether the data set had a cost that was Preferable or Non-preferable. In all cases, a simple label Bad was used to mark data sets with no possible waste flow solution. Finally, waste streams were assigned labels High or Low based on their waste volume assignment.

4) **Information Tables and Rule Generation:** Tables were generated to explore the relationship between the different elements of the system. For our purposes we attempted to relate the presence of any available facility to the operating cost, in a manner similar to Table 5. Other tables related the uncertainty in waste flow to

Table 5. Facilities related to the system cost.

Conditional Attributes					Decision Attributes
KTS_ AVAILABLE	MTS_ AVAILABLE	DTS_ AVAILABLE	incinerator_ AVAILABLE	ThirdSector_ AVAILABLE	COST_ RANGE
Values					
YES	YES	YES	YES	YES	OPTIMAL
NO	NO	NO	NO	NO	NON-OPTIMAL
-	-	-	-	-	BAD

variations in operating costs and a combined objective table was created where the presence of transfer facilities was related to low landfill use with a medium to low operating cost.

5) Application of Rough Sets: The data tables were processed by our version of the RS1 algorithm and the rules, along with precision and recall metrics, were printed in reports. The modifications made to the RS1 algorithm proved to be adequate for the algorithm to produce relevant rules even in situations where no deterministic rules where initially present. Table 6 is a typical listing of the results that were obtained from the algorithm with Table 5.

Table 6. Selected rules returned by the Rough Sets algorithm.

	Condition Attributes						Rule Performance	
#	KTS AVAIL.	DTS AVAIL.	MTS AVAIL.	3rdS. AVAIL.	incinerator AVAIL.	Class	Coverage	Support
1	NO	-	-	-	-	BAD	0.5	0.416
2	YES	-	-	-	-	BAD	0.5	0.00173
3	NO	NO	-	-	-	BAD	0.25	0.25
4	NO	YES	-	-	-	BAD	0.25	0.166
5	YES	NO	-	-	-	BAD	0.25	0.001
6	YES	YES	-	-	-	BAD	0.25	0.001
7	NO	NO	NO	-	-	BAD	0.125	0.125
8	NO	NO	YES	-	-	BAD	0.125	0.125
9	NO	YES	NO	-	-	BAD	0.125	0.125
10	-	-	-	-	YES	HIGH	0.125	0.125
11	NO	-	-	-	-	LOW	0.5	0.0555
12	YES	-	-	-	-	LOW	0.5	0.321
13	NO	-	-	-	NO	LOW	0.25	0.0417
14	NO	-	-	-	YES	LOW	0.25	0.0139
15	YES	-	-	-	NO	LOW	0.25	0.240
16	YES	-	-	-	YES	LOW	0.25	0.0802
17	YES	YES	-	-	NO	LOW	0.125	0.125

6) Rule Interpretation: We inspect here the rules contained within Table 6, which relate the presence of transfer stations and certain disposal methods to the operating cost. From the first rule, which has strong coverage and support, we can see that absence of the Kenora Transfer Station (KTS) causes the system to be infeasible, in which cases the system is incapable of processing the waste and the model fails. Similarly, both rules 3, 7 and 8 support the notion that the KTS is generally necessary to a working system. Rules 4 points out that the Dundas Transfer Station (DTS) can make the system work without the KTS, but in all cases at least one of them must be available for the system to be feasible. This knowledge is valuable in that through a detailed inspection of the model capacity data, we would discover that the KTS has the largest waste handling capacity with low waste routing costs. Normally, this knowledge would be acquired through a detailed inspection of the model data. With our Rough Sets based method we are able to inspect a short list of rules and extract the relevant information directly and quickly.

Another rule of interest is rule 10, which states that the usage of the incinerator will always lead to high operating costs. Interestingly, rule 14 and 16 point out that the system is heavily influenced by the presence of the KTS even when the incinerator facility is in use. Again, this knowledge could be extracted from the model information, but this method reduces the amount of work required.

5.3   Rough Sets and Case Algorithm Implementation

Rough Sets algorithm RS1 was implemented in Java based on the description given in [4, 3] using the Postgresql Database Management System (DBMS) [13] to store and manipulate the indiscernibility relations. The use of a DBMS to store and manipulate the model information permitted the evaluation of a large number of model instances without the normal overhead of flat files. These were generated by a series of configurable java programs coupled with a freely-available java LP solver [12].

One of the issues that was encountered during the implementation of the RS1 algorithm was a problem in its analysis of information tables where no single-column deterministic rules were present. The algorithm depends on a quality factor α, which determines the order in which columns will generate rules. When no deterministic rules are present, there is no difference between either of the decision attribute sets and the value α remains set at 1 and in Equation (5).

$$\alpha = 1.0 - \frac{\overline{Y} - \underline{Y}}{U} \tag{5a}$$

$$\alpha = 1.0 - \frac{\overline{Y} - \overline{Y}}{U} \tag{5b}$$

$$\alpha = 1.0 - \frac{0}{U} \tag{5c}$$

$$\alpha = 1.0 \tag{5d}$$

In our modified algorithm, the Shannon disorder formula [14] is used as an alternate selection measure as represented in Equation 6 when the α value remains stuck at 1.

$$Disorder = \sum_{d}(\frac{N_d}{N_t}) * (\sum_{c} -\frac{N_{dc}}{N_d} \log_2(\frac{N_{dc}}{N_d})) \tag{6}$$

Where N_d is the number of items in branch d.
Where N_t is the total number of items.
Where N_{dc} is the number of items in branch d with class c.

Furthermore, the original algorithm would periodically trim the set of examples from which it would generate rules. Under the $\overline{Y} = \underline{Y}$ condition, this would result in the immediate trimming of the entire universe of examples as in Equation (7).

$$U = U - [(U - \overline{Y}) \cup \underline{Y}] \tag{7a}$$

$$U = U - [(U - \overline{Y}) \cup \overline{Y}] \tag{7b}$$

$$U = U - U \tag{7c}$$

$$U = \emptyset \tag{7d}$$

To prevent a premature null universe, we elected not to trim the universe of examples whenever the α value remained at 1. A side effect is that our implementation of the algorithm will occasionally generate redundant rules.

For convenience the rules returned by the Rough Sets implementation were inserted into other DMBS tables which enabled them to be queried at will. Rule generation was extremely quick, with model and rules generation taking only a few minutes.

6 Discussion

Overall the use of Rough Sets to acquire domain knowledge is valuable in that it dramatically speeds up a tedious process. While there exists a certain overhead in determining which model instances to generate and in interpreting rules, it becomes most reasonable as the complexity of the models being used grows.

The results of the analysis proved to be similar to other results [15] which reported operating costs ranging from $263,461 to $430,769, while ours ranged from $215,930 to $325,526. While differences in modelling and scenario account for our lower cost figures, the results are sufficiently close to one another to be acceptable.

Several conclusions were extracted from our inspection of returned rules. Firstly, the KTS facility is one of the most important in terms of the ability of the system to dispose of the waste at a reasonable cost. However, this generalisation changes when we force the use of incineration as an alternative to the landfill.

Under these conditions, the use of incineration tends to favour the use of the Mountain Transfer Station (MTS) facility over others, even through the use of the KTS facility is still required for low system operating costs. These generalisations agree with a manual analysis of the linear programming model. The low costs and high capacity of the

KTS facility make it an important element of the waste disposal system. Similarly, the MTS facility has a lower transportation cost to the incinerator than most other facilities even when the high inbound costs are considered. This explains the relationship between the MTS facility and the waste incineration scenarios.

This knowledge is useful in that in both cases we now know the simplifying assumptions which allow us to create a model targeted to objectives while ensuring that the assumptions made are correct. Through detailed analysis of several iterative models, it is possible to acquire this same knowledge. However, our Rough Sets based process allows us to extract this information from a coarse model quickly while retaining a certain amount of discretion over the area which we wish to study. There is a certain amount of data preparation required to create the tables and model instances. However, much of this work needs to be done as part of the final model analysis and thus does not represent excessive pre-requisite.

7   Related Work

The Rough Sets method has been applied to many problems with inconsistent and incomplete information including : web based learning [16], word sense disambiguation [17], deadlock detection in Petri nets [18] and software specification [19]. A few of the works most relevant to this project are examined below:

One domain of application worked on by [20] was product scheduling for a tile manufacturing company. The work in [21, 22] differs from optimisation techniques [23, ?] and constraint programming [24, 25] in that reasoning under uncertainty based on rough set methods were used. The objective was to provide close to optimal schedules that satisfied the requirements "to acceptable degree" [26, 27]. Similarly, agent based technologies were applied to the production scheduling problem as in [28].

[29] used Rough Sets theory to analyse diagnosis decision tables of diesel engines, with the objective of diagnosing valve faults. Their results showed that the attributes from specific cylinder heads were more important than other attributes in making a decision. The authors stated that this finding agreed with expert opinion because certain cylinders were observed to be the most sensitive to changing vibrations.

A genetic encoding for rough set evolutionary computing was proposed by [30] for the classification of highway segments. Using data from traffic counters distributed over the highway system, the authors used Rough Set to classifying highways into classes such as: commuter, business, long distance and recreational. These classes were then used to develop guidelines for the construction, maintenance and evolutionary upgrade of highway sections from one class to another.

Other authors focused on the application areas of business failure prediction, database marketing and financial investment [31]. Their conclusions were that Rough Sets methods appeared to be an effective tool for the analysis of financial information tables relating firms to multi-valued attributes. The example used focused on assessing the risk involved in investing in a firm. The rules that were generated were useful in revealing the financial policy applied in the selection of viable firms and also could be used to evaluate another sample of firms which seek first time financing from the ETEVA bank.

[32] applied Rough Sets to the data-mining of telecommunications network monitoring data of a cable Internet provider, with the objective of supporting network management as an objective. Their results included several network management rules which proved novel to the managers of the network, relating line signal levels to long-term quality of service issues.

Finally, [33] examined a cement production control example where Rough Sets was used for the control mechanism during the production process. The two control parameters (rotation speed of kiln and burn speed of coal) were related to the conditions of the process (temperature, burn area, etc...) and the rules generated by Rough Sets paradigm were found to be a match for expert determined control rules.

8   Conclusion and Future Work

Our method is novel in that it uses Rough Sets to generate relevant domain knowledge to support environmental system modelling. While the method requires a certain amount of user guidance, it is capable of reducing the time spent experimentally determining the necessary model elements. We applied our method to a solid waste management case and we verified our method by obtaining much the same conclusions as other studies. Finally, during the course of the implementation an issue with the RS1 algorithm was identified and corrected to enable it to handle larger datasets than it could handle previously. The value of our work lies in showing the use of Rough Sets in a model solving practical problem. Presented methods can find many applications in real life problems.

References

1. R. H. Warren, "Domain knowledge acquisition from solid waste management models using rough sets," Master's thesis, University of Regina, Faculty of Graduate Studies, Regina, Saskachewan, Canada, 2000.
2. Z. Pawlak, J. Grzymala-Busse, R. Slowinski, and W. Ziarko, "Rough sets," *Communications of the ACM*, vol. 38, no. 11, pp. 89–95, 1995.
3. S. Wong and W. Ziarko, "Algorithm for inductive learning," *Bulletin of Polish Academy of Sciences*, vol. 34, no. 5, pp. 271–276, 1986.
4. S. K. M. Wong, W. Ziarko, and R. L. Ye, "Comparison of rough-set and statistical methods in inductive learning," *International Journal of Man-Machine Studies*, vol. 24, pp. 53–72, 1986.
5. A. Wakulicz-Deja, M. Boryczka, and P. Paszak, "Discretization of continuous attributes on decision system in mitochondrial encephalomyopathies," in *Rough Sets and Current Trends in Computing RSCTC-98*, pp. 481–490, 1998.
6. J. R. Dunphy, J. J. Salcedo, and K. S. Murphy, "Intelligent generation of candidate sets for genetic algorithms in very large search spaces," in *Rough Sets, Fuzzy Sets, Data Mining, and Granular Computing: 9th International Conference, RSFDGrC 2003*, vol. 2639, pp. 453–457, Springer-Verlag Heidelberg, January 2003.
7. J. Dougherty and Al, "Supervised and unsupervised discretization of continuous features," in *Proceedings of the Twelfth International Conference on Machine Learning*, pp. 194–202, 1995.

8. G. H. Huang, *Grey Mathematical Programming and its application to Waste Management Planning under Uncertainty*. PhD thesis, McMaster University, Department of Civil Engineering, Hamilton, Ontario, Canada., 1994.

9. G. H. Huang, B. W. Baetz, and G. G. Patry, "Grey fuzzy dynamic programming: Application to municipal solid waste management planning problems," *Civil Engineering Systems*, vol. 11, pp. 43–73, 1994.

10. G. H. Huang, B. W. Baetz, and G. G. Patry, "Grey fuzzy integer programming: an application to regional waste management planning uncer uncertainty.," *Socio-Economic Planning Science*, vol. 29, no. 1, p. 17, 1995.

11. X. Wu, "Inexact nonlinear programming and its application to solid waste management and planning," Master's thesis, University of Regina, Faculty of Engineering, Regina, Saskachewan, Canada, 1997.

12. T. J. Wisniewski and Al., *Simplex Java Applet*. Optimization Technology Center, Argonne National Labratory and Northwestern University, http://www-fp.mcs.anl.gov/otc/Guide/CaseStudies/simplex/applet/SimplexTool.html, January 1997.

13. T. P. development team, *PostgreSQL*. Open Source Software, 2003.

14. C. E. Shannon, *The Mathematical Theory of Communication*. Urbana, Illinois: The University of Illinois Press, 1949.

15. G. H. Huang, B. W. Baetz, and G. G. Patry, "Grey fuzzy dynamic programming: application to municipal solid waste management planning problems," *Civil Engineering Systems*, vol. 11, pp. 43–74, 1994.

16. A. Liang, M. Maguire, and J. Johnson, "Rough set based webct learning," in *Lecture Notes in Computer Science (Web-Age Information Management)*, pp. 425–436, 2000.

17. J. Johnson, X. D. Yang, and Q. Hu, "Word sense disambiguation in the rough," in *Proc. Fourth Symposium on Natural Language Processing SNLP*, pp. 206–223, 2000.

18. J. Johnson and H. Li, "Rough set approach for deadlock detection in petri nets," in *Proc. International Conference on Artificial Intelligence IC-AI*, pp. 1435–1439, 2000.

19. J. Johnson, "Specification from examples," in *Proc. International Conference on Advances in Infrastructure for Electronic Business, Science, and Education on the Internet*, 2000. L'Aquila, Italy, July 31-August 6.

20. J. A. Johnson, "Rough scheduling," in *Proceedings of the 5th Joint Conference in Information Science JCIS 2000*, pp. 162–165, 2000.

21. J. Johnson, "Rough mereology for industrial design," in *Rough Sets and Current Trends in Computing* (L. Polkowski and A. Skowron, eds.), pp. 553–556, Springer, 1998.

22. J. Johnson, "Rough scheduling," in *Proc. 5th Joint Conference in Information Science JCIS 2000*, vol. 1, pp. 162–165, 2000.

23. S. Abraham, V. Kathail, and B. Beitrich, "Meld scheduling: A technique for relaxing scheduling constraints," *International Journal of Parallel Programming*, vol. 26, no. 4, pp. 349–381, 1998.

24. R. Barták, "Theory and practice of constraint propagation," in *3rd Workshop on Constraint Programming for Decision and Control CPDC2001*, (Gliwice, Poland), pp. 7–14, 2001.

25. D. Lesaint, "Inferring constraint types in constraint programming," in *Eighth International Conference on Principles and Practice of Constraint Programming*, (Ithaca, NY), 2002.

26. J. Komorowski, L. Polkowski, and A. Skowron, "Rough sets: A tutorial." http://www.folli.uva.nl/CD/1999/library/coursematerial/Skowron/Skowron.pdf, 2002.

27. A. Skowron, "Approximate reasoning by agents in distributed environments," in *IAT'01*, (Maebashi, Japan), 2001. http://kis.maebashi-it.ac.jp/iat01/pdf/iat_invited/IAT01_SKO_SLE.pdf.

28. W. Shen and H. Ghenniwa, eds., *Journal of Intelligent Manufacturing: Special Issue on Agent-Based Manufacturing Process Planning and Scheduling*. Kluwer Academic Publishers, to appear.

29. L. Shen, F. E. H. Tay, L. Qu, and Y. Shen, "Fault diagnosis using rough sets theory," *Computers in Industry*, vol. 43, no. 1, pp. 61–72, 2000.

30. P. Lingrass, "Unsupervised rough set classification using genetic algorithms," *Journal of Intelligent Information Systems*, vol. 16, no. 3, pp. 215–228, 2001.

31. F. E. H. Tay and L. Shen, "Economic and financial prediction using rough sets model," *European Journal of Operational Research*, vol. 141, no. 3, pp. 643–661, 2002.

32. M. Chartrand and J. Johnson, "Network mining for managing a broadband network," in *Data Mining 2003* (G. Rzevski, P. Adey, and D. Russell, eds.), p. to appear, Southampton, UK: Computational Mechanics, 2003.

33. Z. Hui, J. Johnson, W. Bin, and L. Z. Gui, "Control rules generating method based on rough sets," *Journal of Computer Engineering and Application*, vol. 39, no. 13, pp. 98–100, 2003. written in Chinese, English translation available from julia@cs.laurentian.ca.

Rough Set Theory and Decision Rules in Data Analysis of Breast Cancer Patients

Jerzy Załuski[1], Renata Szoszkiewicz[1],
Jerzy Krysiński[2], and Jerzy Stefanowski[3,*]

[1] Department of Chemotherapy, Wielkopolska Oncology Center
15 Garbary Str, 61-866 Poznań, Poland
[2] Department of Pharmacy, Medical University of Bydgoszcz
13 Jagiellońska Str, 85-067 Bydgoszcz, Poland
`jerzy.krysinski@wp.pl`
[3] Institute of Computing Science, Poznań University of Technology
Piotrowo 3A, 60-965 Poznań, Poland
`jerzy.stefanowski@cs.put.poznan.pl`

Abstract. In this paper an approach based on the rough set theory and induction of decision rules is applied to analyse relationships between condition attributes describing breast cancer patients and their treatment results. The data set contains 228 breast cancer patients described by 16 attributes and is divided into two classes: the 1st class - patients who had not experienced cancer recurrence; the 2nd class - patients who had cancer recurrence. In the first phase of the analysis, the rough sets based approach is applied to determine attribute importance for the patients' classification. The set of selected attributes, which ensured high quality of the classification, was obtained. Then, the decision rules were generated by means of the algorithm inducting the minimal cover of the learning examples. The usefulness of these rules for predicting therapy results was evaluated by means of the cross-validation technique. Moreover, the syntax of selected rules was interpreted by physicians. Proceeding in this way, they formulated some indications, which may be helpful in making decisions referring to the treatment of breast cancer patients. To sum up, this paper presents a case study of applying rough sets theory to analyse medical data.

Keywords: rough sets, decision rules, attribute selection, classification performance, medical data analysis, breast cancer.

1   Introduction

Breast cancer is the most frequently occurring malignant tumour in the case of Polish women. Approximately 11 thousands new incidences of disease are reported annually. Breast cancer morbidity rate increases at the age of 35 – 39 years, and it reaches its maximum value at the age of 60.

---

* Corresponding author.

J.F. Peters et al. (Eds.): Transactions on Rough Sets I, LNCS 3100, pp. 375–391, 2004.
© Springer-Verlag Berlin Heidelberg 2004

It is not possible to identify reasons for the majority of breast cancer incidences. Risk factors of breast cancer occurrence are: female sex, senior age, family record of breast cancer incidence at young age, menopause occurring at old age, long lasting supplementary hormone therapy, indolent proliferative breast diseases as well as BRCA1 and BRCA2 genes carrying.

An associated treatment is a general rule of breast cancer treatment: in cases of early stage of the disease it is a primary surgery treatment and/or radiotherapy; and the primary chemotherapy and/or hormonetherapy in cases, where disease becomes general. At relevant stages of this disease the selection of local or systemic treatment methods is based on an evaluation of the following elements: histological degree of cancer malignancy, clinical and pathological characteristic of primary tumour, as well as, axilla lymph glands, lack or presence of distant metastasis, time span between the date of primary treatment and the date of cancer recurrence, type of the applied primary treatment and clinical characteristics of recurrence, manifestation of ER/PgR receptors and HER2 receptor, menopause status and patients age.

A decision on a supplementary treatment should be based on an estimation of particular recurrence risk (prognostic determination based on known prognostic factors) and potential benefits resulting from supplementary treatment application.

In this paper, we will apply an approach based on the rough set theory to analyse the influence of prognostic factors on the results of the breast cancer treatment. An identification of the most important factors / attributes describing patients and discovering the relationships between attribute values and results of patients' treatment is a crucial issue from the medical point of view. It could help the physicians to adopt adequate treatment strategies for the breast cancer patients. Let us notice that although the rough set theory has been quite often applied to analyse medical data (for some reviews see e.g. [11, 16, 24, 27, 37]), it has not been used for the breast cancer problems yet. We will present an analysis of the medical data which were collected in the Wielkopolska Oncology Center.

The paper is organized as follows. In section 2, the breast cancer problem is more precisely discussed from the medical point of view. Moreover, a short discussion of related works is given. Then, section 3 contains a brief description of the analysed medical data. In section 4, basic information about the rule induction and rough sets approach are summarized. Section 5 includes a description of the performed analysis of the attribute importance and rule induction. The clinical discussion of the obtained results is presented in section 6. Final remarks and conclusions are grouped in section 7.

2   Background and Related Work

Breast cancer still confronts us with many unsolved problems and open questions regarding its early detection, optimal treatment, prediction of outcome and response to therapy. The proper determining a prognosis of a patient's treatment is extremely important, because the general therapeutic strategy strongly depends on an accurate disease evaluation as well as the known patient survival studies have shown that breast cancer appears to take two distinct forms. The first form is a fatal one within a short time, usually two years after first diagnosis, while the other is characterized by a long-term survival. Many node negative patients, with the first form of the breast tumour, have a poor prognosis and therefore need aggressive therapy. Such patients should be distinguished from those node negative patients who have a good prognosis and could go without further treatment after surgery [15, 7]. The identification and better understanding of factors predicting response to chemotherapy might also help toward understanding the basic mechanisms behind response to cytotoxic agents in malignant disease. It would also assist the clinician in selecting the right candidates for chemotherapy and spare the others from unnecessary toxicity. Therefore, it is very important to identify factors (attributes, features) that can predict the natural history of disease, allow determination of the best treatment strategy and predict the outcome following such interventions. Single predictive factors or a combination of factors should indicate those patients liable to profit by or respond to an adjuvant systemic chemo- or hormone- therapy as well.

While establishing prognostic factors in neoplastic diseases, different statistical methods were traditionally applied, such as: univariate-, multivariate- and multiple regression analysis [15, 7, 5, 13, 21, 14, 2] or Cox's proportional hazard model [35, 39].

Other methods could come from Artificial Intelligence [17]. Besides identification of the most important factors (subsets of attributes), these methods present regularities discovered from data in a form of symbolic representation, called *knowledge* (derived from data), which could be inspected by humans. For instance, an induction of decision trees by means of Assistant Professional learning system was considered for breast cancer data coming from University of Lubliana [3]. An interesting discussion on the use of symbolic machine learning methods in the medical data analysis is presented in [17]. Yet another non-statistical approach, which has been already successfully used in the analysis of medical and biological data, are *rough sets* [28, 19, 20, 27, 24, 30, 29, 10]. According to the discussion presented in [27], the rough set theory and rule induction approach is particularly well suited:

— to identify the most significant attributes for the classification of patients,
— to discover the dependencies between values of the attributes and the classification of patients.

This paper presents the use of rough sets in analysing data of breast cancer patients. According to our best knowledge this approach has not been applied to this problem yet. Let us shortly comment related literature results on the analysis of the influence of prognostic factors on the therapy results.

The outcome of breast cancer therapy depends on several known and unknown factors which may also interact with one and other. Prognostic factors should apply to the biological characteristics of the tumour, such as its aggressiveness, tumour cell proliferation and metastasizing potential. The determination of such factors should be feasible, reproducible and easily interpretable by the clinicians, so that he can draw therapeutic consequences from these factors.

Unfortunately, we must admit that no single prognostic parameter or combination of parameters exist that would fulfill all the criteria mentioned, however, several factors have been recommended for clinical use and other are currently tested [15, 7]. The prognostic factors in the treatment of breast cancer have been widely described and statistical methods were employed to establish them [15, 7, 2, 1, 6, 8, 22, 12]. Published papers concern mainly the influence of single factors of epidemiological, anatomo-morphological, hormono-cellular, genetic- molecular groups on the effects of breast cancer treatment. In the hitherto available statistical studies there has been no information on interrelations among co-occurring various attributes. Such interrelations are presented in this paper with the use of the rough set method.

3   The Breast Cancer Data

The retrospective analysis covered the treatment results of 228 breast cancer patients (all women after mastectomy) divided into a set of 162 patients who had no recurrence of cancer (called class 1) and a set of 66 patients who had had cancer recurrence (called class 2). All the data concern patients treated in the Wielkopolska Oncology Center. The physicians chose a sample of patients from the years 1990-1991 for whom the long term therapeutic results were known. The reader can notice that this data set is unbalanced considering cardinalities of decision classes. The class 2 is a minority one (contains 29.39% patients). However from the medical point of view it is more important than class 1, because it includes patients with breast cancer recurrence (so, more risky / dangerous ones). This situation of unbalanced classes often occurs in medical data and requires particular attention during the analysis (see discussion and specific techniques presented in [27, 30, 10, 34])

For the analysis of patients the information system was defined, where a set of 228 objects (patients) was described by a set of attributes. The set of attributes includes sixteen condition attributes and one decision attribute. The condition attributes characterise the patients, while the decision attribute defines the patients' classification according to the cancer recurrence, i.e. whether the patients did not experience cancer recurrence or they had the one. The condition attributes, portraying patients, have a good clinical quality. They are easy to measure, analyse and available for doctors. Moreover, they can be adopted and applied at every oncological ward. These condition attributes were tailored so that a doctor, without subjecting a patient to expensive and long lasting tests, is quickly able to learn prognosis features, which help to make decision on an application of chemo-, hormone- and/or radiotherapy in case of patients who underwent surgery treatment.

The list of chosen sixteen condition attributes with characteristics of their domains is presented in Table 1. One should notice that some of them were originally defined on numerical scales (e.g. age, hormonal activity, tumour size). However, their exact measurement values were transformed (discertized) into qualitative intervals according to existing medical standards. For simplicity, these

Table 1. Characteristics of condition attributes

No.	Attribute	Domain
1.	Age (years)	<30; 31-50; 51-60; >60
2.	Menopause	before; during; after
3.	Hormonal activity (years)	<10; 11-20; 21-30; 31-40; >40
4.	Number of parturition	0; 1; 2; 3; >3
5.	Breast cancer in the 1st and 2nd generation	yes; no
6.	Tumor size (mm)	no data; <40; >40
7.	Positive lymph nodes	no affected; 1-3; 4-8; >8
8.	Infiltration of node capsule	yes; no
9.	Arrest in microvessels	yes; no
10.	Bloom malignancy	I; II; III
11.	Type of surgery	tumorectomy; Halsted; Patey; simple mastectomy
12.	Type of adjuvant therapy	none; radio-; chemo-; hormono-; radio+chemo-; radio+hormono-; chemo+hormono-; radio+chemo+hormono-
13.	Target of adjuvant radiotherapy	no; cicatrix+lymph nodes; lymph nodes
14.	Neoadjuvant therapy	yes; no
15.	Type of adjuvant chemotherapy	no; CMF; anthracyclines
16.	Type of adjuvant hormonotherapy	no; tamoxifen

intervals where coded by numbers. So, a conventional code number corresponds to quantitative or qualitative value of an attribute. For example, in Table 1 the category "menopause" originally takes three values: "before", "during" and "after". They correspond to the following codes "1", "2' and "3", respectively.

4   Methodology

The breast cancer information system is analysed using the rough set theory. Similarly to previous medical applications [19, 20, 27, 24, 30, 28, 29] the following elements of the rough set theory are used:

- creating classes of indiscernibility relations (atoms) and building approximations of the objects' classification,
- evaluating the ability of attributes to approximate the objects' classification; the measure of the quality of approximation of the classification (or level of dependency), defined as the ratio of the number of objects in the lower approximations to the total number of objects, is used to for this aim,
- examining the significance of attributes by observing changes in the quality of approximation of the classification caused by removing or adding given attributes,
- discovering cores and reducts of attributes (a reduct is the minimal subset of attributes ensuring the same quality of the classification as the entire set of attributes; a core is an intersection of all reducts in the information system).

We skip formal definitions, the reader can find them, e.g. in [23, 16, 24].

The first elements are useful to check whether analysed data are consistent and available attributes can sufficiently approximate the patients' classification. In general, high values of the quality of classification and accuracies (ideally equal to 1.0 or close to this value) show that classes of patients' classification can be well described employing considered attributes. Low values may indicate the necessity for changing the input representation of the data table (e.g. looking for better attributes or performing some transformations of existing ones, see [30]). As medical data sets are quite often inbalanced (i.e. numbers of objects in decision classes are significantly different and the minority class usually requires particular attention from the medical point of view), it is useful to analyse rather an accuracy for every classes (in particular, for minority but more important one) than the total accuracy. On the other hand, the highest quality and precise approximation do not automatically guarantee good results of next steps of the analysis, e.g. looking for reducts or rules.

Then, the analysis of an attribute importance based on rough sets can be performed by using the concept of the significance of attributes [23]. The higher this value, the more important attribute. Apart from removal of single attributes, we can also consider sequences of elimination one attribute by one and receiving a kind of stepwise hierarchy of attribute importance (see e.g. a case study of the highly selective vagotomy [34]). However, these concepts may not be efficient in some data sets (see e.g. discussion in [29], so other more complicated approaches are necessary. In case of rough sets we can use more heuristic approaches, e.g. connected with dynamic reducts [16] or dividing the set of attributes into some disjoint subsystems and analysing the significance of attributes inside these subsystems (see [18, 28]) or an idea of using, so-called, data templates [16].

The attribute selection and identification of 'the best' subsets is also connected with rough set approaches to looking for reducts and a core of attributes.

In general, the concept of reducts seems to be so attractive as it may lead to reducing the data size by removing redundant attributes. In other words, we can keep only those attributes that preserve approximation of classification and reject such attributes the removal of which cannot worsen the classification. There are, however, two obstacles which one should be aware of. First of all, complexity of computing all reducts in the information system is exponential in time. Moreover, the information system usually contains more reducts than one. Sometimes, this number is very high, in particular if: the number of attributes is too large comparing to the number of patients; there are too many self-dependent attribute subsets; observations describing patients have "individual" character and are difficult to be generalized. Let us also remark that the interpretation of too many possible reducts by practitioners is often impossible. In medical applications, it is not however necessary to look for all reducts and one is often satisfied by computing a limited subset of reducts and/or choosing the most satisfactory subset according to some criteria (see, e.g., discussions in [19, 20, 28, 24].

In this study we used a strategy of adding to the core the attributes having the highest discriminatory power [24]. The core of attributes is chosen as a starting reduced subset of attributes. It usually ensures lower quality of approximation of the object classification than all attributes. A single remaining attribute is temporarily added to the core and the influence of this adding on the change of the quality is examined. Such an examination is repeated for all remaining attributes. The attribute with the highest increase of the quality of classification is chosen to be added to the reduced subset of attributes. Then, the procedure is repeated for remaining attributes. It is finished when an acceptable quality of the classification is obtained. If there are ties (as we can choose few attributes with exactly the same increase of the quality), several possible ways of adding are checked.

Examining dependency between values of condition attributes and the patients' classification is available by means of decision rules. A decision table can be seen as a set of learning examples which enable induction of decision rules. If the decision table is consistent, rules are induced from examples belonging to decision classes. Otherwise, decision rules are generated from approximations of decision classes. As a consequence of using the rough approximations, induced decision rules are categorized into certain (exact) and approximate (possible) ones, depending on the used lower and upper approximations (or boundaries), respectively [9, 33]. Decision rules are represented in the following form:

IF (a_1, v_1) and (a_2, v_2) and ... and (a_n, v_n) THEN $Class_j$

where a_i is the i-th attribute, v_i its value and $Class_j$ is j-th decision class.

Induction of decision rules from decision tables is a complex task and a number of various algorithms have been already proposed (see e.g. [9, 33, 34, 31]). In this study we used our implementation [25] of the LEM2 algorithm introduced by Grzymala [9]. This algorithm induces a set of discriminating rules from learning examples belonging to approximations of decision classes, i.e. so called, minimal cover of learning examples. These rules distinguish positive examples, i.e. objects belonging to the lower approximation of the decision class, from other objects (negative examples). The rule induction scheme consists of creating a first rule by choosing sequentially the 'best' elementary conditions according to some heuristic criteria. Then, learning examples that match this rule are removed from consideration. The process is repeated iteratively while some significant examples remain still uncovered.

To interpret each single discovered rule we use the measure of their strength [34]. It is a number of objects in the information system whose description satisfies the condition part of the rule. Generally, one is interested in discovering the strongest rules [33, 34].

The main evaluation of the set of induced rules is done on the basis of it classification performance. The overall evaluation is measured by a coefficient called classification accuracy, which is a ratio of the number of correctly classified testing examples to the total number of testing examples.

The estimation of the classification accuracy is performed by means of the random train-and-test approach, i.e. a standard 10-fold cross validation technique [38]. While performing classification of the testing examples, possible ambiguity in matching their description to condition parts of decision rules was solved using partially matched rules (according to idea of looking for the closest rules introduced in [32]).

5   The Analysis of the Breast Cancer Data

5.1   Rough Approximations and Selection of Attributes

The analysis of the breast cancer data was performed by means of the ROSE software developed at Poznan Univeristy of Technology [25], available at http://www-idss.cs.put.poznan.pl/rose.

According to rough set theory, classes of indiscernible objects, i.e. atoms, were found. Their number was equal to 204. The majority of them are single element atoms. We can interpret this observation as an evidence of "difficulty" of the breast cancer data. Single cases are distinguished from others and their descriptions are quite "individual". The number of inconsistent examples is equal to 3 patients. In the next step, approximations of decision classes were created and their accuracies were calculated. The results are presented in Table 2.

Table 2. Rough approximations of the patients' classification

class	number of patients	cardinality of lower approx.	cardinality of upper approx.	accuracy of class
1	162	160	163	0.9876
2	66	65	68	0.9846

Then, considering all condition attributes, the initial quality of the classification is equal to 0.9868. We can comment that available attributes can sufficiently approximate patients' classification.

Calculating a core resulted in the following subset of attributes {attribute 1 - age of patient; attribute 3 - period of hormonal activity; attribute 4 - number of parturition; attribute 6 - tumour size, attribute 7 - positive lymph nodes; attribute 10 - Bloom malignancy}. The quality of classification of the core is equal to 0.8772. This value is quite high comparing to the value of the total quality of the classification. We have also analysed the significance of attributes belonging to the core. The results are given in Table 3. This analysis has focused our attention on the attributes: 4 - number of parturition, 7 - positive lymph nodes and 10 - Bloom malignancy .

Table 3. Significance of attributes of approximating the patients' classification

Attribute	1	3	4	6	7	10
Significance	0.018	0.004	0.075	0.013	0.053	0.039

Following the procedure of looking for all reducts 7 subsets of attributes were obtained:

1. {Age, Menopause, Hormonal activity, Number of parturition, Tumor size, Positive lymph nodes, Arrest in microvessels, Bloom malignancy, Target of adjuvant radiotherapy }
2. {Age, Menopause, Hormonal activity, Number of parturition, Tumor size, Positive lymph nodes, Bloom malignancy, Type of surgery, Target of adjuvant radiotherapy}
3. {Age, Hormonal activity, Number of parturition, Tumor size, Positive lymph nodes, Bloom malignancy, Type of adjuvant therapy, Target of adjuvant radiotherapy}
4. {Age, Hormonal activity, Number of parturition, Tumor size, Positive lymph nodes, Infiltration of node capsule, Bloom malignancy, Type of adjuvant therapy}
5. {Age, Hormonal activity, Number of parturition, Tumor size, Positive lymph nodes, Bloom malignancy, Target of adjuvant radiotherapy, Neoadjuvant therapy}
6. {Age, Hormonal activity, Number of parturition, Tumor size, Positive lymph nodes, Bloom malignancy, Type of adjuvant therapy, Neoadjuvant therapy}
7. {Age, Hormonal activity, Number of parturition, Tumor size ,Positive lymph nodes, Bloom malignancy, Target of adjuvant radiotherapy, Type of adjuvant chemotherapy}

Let us remind that the condition attributes pertaining to the reduct have different effects on the quality of classification and the class accuracy. However, this number of reducts is too high for physicians and the further analysis. Therefore, the choice of one subset of attributes was required. We used strategy of adding the most discriminating attributes to the core.

A computational experiment was performed based on addition of respective attributes to the core and observation of the increase of the classification quality. An addition of attribute 12 (i.e., type of adjuvant therapy) still gave a high quality level equal to 0.97 and this quality of classification was the highest for the smallest number of condition attributes.

Therefore, the following set of attributes was finally selected: the attribute 1 - age of patient; the attribute 3 - period of hormonal activity; the attribute 4 - number of parturition; the attribute 6 - tumour size, the attribute 7 - positive lymph nodes; the attribute 10 - Bloom malignancy; the attribute 12 - type of adjuvant therapy.

5.2   Rule Induction

First, we induced a minimum set of rules using the algorithm LEM2 (considering all 16 attributes). The obtained set contained 60 rules (59 certain and 1 approximate), 32 rules described patients from class 1, and 27 rules correspond to class 2.

Then, we used the same algorithm for the reduced information table (i.e. for in the previously selected 7 attributes - see the previous subsection). We generated 72 decision rules, including 3 nondeterministic ones (i.e. 35 rules for class 1 and 34 rules for class 2).

Ten-fold cross-validation evaluation was performed for the reduced and non-reduced information system. The mean accuracy of classification and the standard deviation determined in the case of the non-reduced information system were equal to 70.63% and 8.4 respectively. For reduced information system the mean accuracy of classification was equal to 71.5%, and standard deviation – 6.06. This indicates that classification accuracies in both information systems are similar to each other, i.e. correctness of decisions in the reduced information system was still adequate. The comparison of these results with classification results coming from other machine learning systems is given in Section 5.4.

The interesting results include classification accuracies (sensitivities) calculated for single decision class – see Table 4.

Table 4. Classification performance of decision rules

Rule set	Overall accuracy	Class 2 classification
all attributes	$70.63 \pm 8.4\%$	24.77%
selected attributes	$71.55 \pm 6.1\%$	39.69%

5.3   Analysis of the Syntax of Induced Decision Rules

Although the number of induced rules is quite (even too) high, the physicians were interested in analysing the syntax of rules, which could lead to identify interesting relationships between the most characteristic values of some attributes for patients belonging to decision classes. This could help physicians in discussing, the so called, model of patients in the given class. To analyse the limited number of the most "important" patterns, we decided to focus attention only on the strongest rules, i.e. rule supported by sufficient number of facts / observations in the analysis information systems. Therefore, the measure of rule strength was used to filter the set of induced rules. Tables 5 and 6 present the strongest decision rules obtained for the complete, unreduced set of sixteen condition attributes, where the rules should be supported by at least 10 objects (class 1) or by at least 5 objects (class 2).

In class 1 (referring to no cancer recurrence), the strongest rules are formed by the objects: without infiltration of node capsule, age: 31÷50 years, with tumor size < 40 mm, without neoadjuvant therapy (rule 1); age 51÷60 years, hormonal activity 31÷40 years, without infiltration of node capsule, no arrest in microvessels, without adjuvant radiotherapy, without neoadjuvant therapy (rule 2); with Bloom malignancy I ° (rule 3); without infiltration of node capsule and without adjuvant therapy (rule 4); no parturition, no records of breast cancer in the 1-st and 2-nd generation, tumor size < 40 mm, without neoadjuvant therapy (rule 5).

In the class 2 (cancer recurrence), the strongest rules are formed by the objects: positive lymph nodes > 8, neoadjuvant therapy, no arrest in microvessels (rule 1); positive lymph nodes 4÷8, no records of breast cancer in the 1st and 2nd generation, Patey mastectomy, without adjuvant chemotherapy (rule 2); before menopause, hormonal activity 21÷30 years, positive lymph nodes > 8 (rule 3); tumor size > 40 mm, with arrest in microvessels, (rule 4).

The similar analysis was performed for the reduced information system. In the class of "recurrence free patients", an example of such rules is the rule 1: patients aged 31-50 years, with tumour <40 mm and no involvement of lymph nodes. Moreover, this class would include all patients with Bloom I (rule 2); patients aged 51-60 years, with hormonal activity time of 31-40 years, without involvement of lymph nodes, receiving adjuvant hormonal therapy (rule 3); patients without involvement of lymph nodes, receiving adjuvant therapy (rule 4); nullipara with tumour size < 40 mm and no involvement of lymph nodes (rule 5); with Bloom II, tumour size < 40 mm, women who had 3 parturition and with hormonal activity time of 31-40 years (rule 6) and women with history of 1 parturition, receiving adjuvant hormonal therapy (rule 7).

The decision rules in the class 2 (patients with cancer recurrence) are weaker; they are supported by a lower number of objects. This results from a lower number of patients who belong to this class (67 objects). Three rules supported by 4 objects are as follows: patients who had 3 parturition, aged > 60 years, with tumour size < 40 mm, Bloom III ° ; patients who had 3 parturition, tumour size > 40 mm, Bloom II ° , and patients aged > 60 years with hormonal activity time 31-40 years, tumour size < 40 mm and 4-8 lymph nodes involved.

Table 5. Strong decision rules for the non-reduced information system and the class 1

No.	Condition attribute	Rule Strength
1.	Age: 31-50, tumor size: <40 mm, no infiltration of node capsule, no adjuvant chemotherapy;	31
2.	Age: 51-60, hormonal activity: 31-40 years, no infiltration of node capsule;	20
3.	Bloom malignancy: I;	19
4.	No infiltration of node capsule, no adjuvant therapy;	17
5.	Without parturition, no record in breast cancer in the 1st and 2nd generation, tumor size: <40mm, no neoadjuvant therapy;	15
6.	No affected lymph nodal, no neoadjuvant therapy, adjuvant chemotherapy: CMF;	13
7.	Before menopause, hormonal activity: 31-40 years, no infiltration of node capsule, no arrest in microvessels;	13
8.	Number of parturition: >3, Bloom malignancy: II, no affected lymph nodes, no neoadjuvant therapy;	12
9.	Hormonal activity: 31-40 years, 3 parturition, tumor size: <40 mm, Bloom malignancy: III;	12
10.	Age: 31-50 years, 2 parturition, Bloom malignancy: III, no infiltration of node capsule;	11
11.	1 parturition, adjuvant hormonotherapy;	10

Table 6. Strong decision rules for the reduced information system and the class 2

No.	Condition attribute	Rule Strength
1.	Positive lymph nodes: >8, with neoadjuvant therapy, no arrest in microvessels;	9
2.	Positive lymph nodes: 4-8, no record of breast cancer in the 1st and 2nd generation, no arrest in microvessels, Patey's mastectomy, without adjuvant chemotherapy;	6
3.	Before menopause, hormonal activity: 21-30 years, positive lymph nodes: > 8	5
4.	Tumor size: >40, with arrest in microvessels;	5

Considering the above decision rules one can try to consider a generalised "model of a patient" from the recurrence free class and a patient from the class with recurrence.

- In the class 1, such a patient will demonstrate: Bloom I $^\circ$, axillary lymph nodes not involved or less than three involved lymph nodes, tumour size < 40 mm.
- In the class 2, the model patient will have: Bloom III $^\circ$ and more than four involved axillary lymph nodes.

The decision rules obtained for the reduced and non-reduced sets of condition attributes allow considering some hypothesis concerning prognosis for breast cancer patients.

The group of patients without recurrence will consist of patients having: Bloom I malignancy grade, tumour size < 40 mm, no involvement of axillary lymph nodes and no infiltration of the node capsule.

The factors associated with poor prognosis (the group of patients with recurrence of the disease) will include Bloom III $^\circ$ malignancy and more than 4 involved axillary lymph nodes.

Our data analysis has not shown unequivocal influence of age, menopause status, number of parturition and adjuvant therapy on prognosis.

5.4   Comparison with Other Methods

It is interesting to compare results of our approach against other methods. As our approach discovers knowledge represented in a symbolic way, we consider such other non-rough sets discovery methods that induce such or a similar kind of knowledge representation. Therefore, we decided to use implementations of CN2 rule induction algorithm [4], C4.5 system inducing decision trees [26] (also with option transforming a tree into rules) and another implementation of decision tree induction based on the ideas coming from Assistant 86 system [3]. Summary of classification results is given in Table 7.

As one can see the results obtained by our approach are comparable taking into account overall classification accuracy.

6   Clinical Discussion

The hitherto search for prognostic factors in the treatment of breast cancer has been based on statistical methods [7]. Influence of respective factors on curing patients with breast cancer is not unequivocal. No analysis has been performed on the relationships between various factors and on how they can affect patients' prognoses. The described study is an attempt to make a non-statistical analysis of the relationships between the attributes describing breast cancer patients and the effect of their treatment (cancer recurrence or no recurrence).

In the present discussion it is hard to compare the obtained analysis results with publications on prognostic factors in breast cancer, because, as it has already been mentioned, these papers mainly refer to influence of single factors on

Table 7. Classification performance of different systems

Set of attributes	Algorithm	Classification accuracy
All Attributes	C4.5 tree	71.1 ± 8.1 %
	C4.5 rules	72.9 ± 8.4 %
	Assistant	71.05 ± 6.3 %
	CN2	71.05 ± 2.8%
Selected	C4.5	72.18 ± 9.5%
7 attributes	Assistant	72.18 ± 7.5 %
	CN2	68.42 ± 3.9 %

the treatment result. Whereas, here the rough sets method analyses the cause-effect relationships by studying all attributes' influence on the treatment result, and the decision rules present the most significant relationships.

The condition attributes which are the most significant for the quality of classification and also prophetically the strongest, are understandable and self-evident from the clinical standpoint. These are: tumour size, number of involved axillary lymph nodes, Bloom malignancy grade. The condition attributes insignificant for the quality of classification are less important from the prognostic viewpoint. These are: menopause status, arrest in microvessels, type of surgery, record of breast cancer in the 1st and 2nd generation.

The condition attribute – "infiltration of node capsule" seems to be meaningful in the decision rules obtained from the unreduced information system. However, in the course of experiment in reducing the number of condition attributes, it turned out that it is insignificant for the quality of classification.

The number of patients who experienced cancer recurrence is low (class 2), and this influences strength of the obtained decision rules. In order to evaluate actual influence of condition attributes on cancer recurrence, it is necessary to analyse a larger group of patients who experienced cancer recurrence or to use more sophisticated methods for handling unbalanced data (attempts to propose such new classification strategies of using rules have already been described in [10, 34]).

7   Final Remarks

This paper presents a case study of applying rough set theory and rule induction to analyse medical data on the treatment of breast cancer patients. From the methodological point of view, the rough set theory was used to evaluate the ability attributes (related to the prognostic factors for the results of breast cancer treatment) to approximate the patients' classification. Using the measures quality of classification and the attribute significance helped us to identify the important attributes. This allowed us to select the subset of attributes. The methodology used in this part of the analysis followed the general methodological schema presented in [20, 18, 24, 27].

Although statistical methods have been widely applied to analyse the prognostic factors in the treatment of breast cancer, the standard approaches concern mainly the influence of single factors only. We do not discuss why the more advanced multi-dimensional statistical methods are not popular but would like to stress that the rough sets allowed to examine interrelations among subsets of attributes.

Then, the LEM2 algorithm was chosen to generate the sets of rules. This kind of knowledge representation is easy for human inspection and interaction, so the clinical interpretation was discussed. The classification performance of induced rules (both for all and selected attributes) is comparable with results of using other learning systems.

On the other hand, one can notice the difficulty of this data set. There are too many decision rules, some of them are supported by too small number of examples. Another disadvantage is a weak recognition of patients from class 2 (see the classification results in Table 4). This is typical for medical, unbalanced data sets. Therefore, some research on using more sophisticated classification strategies should be undertaken.

Acknowledgment

The first three authors want to acknowledge support from State Committee for Scientific Research, research grant no. 8T11E 042 26.

References

1. Adami H.O, Maker B, Holmberg B, Personn I, Stone B, The relationship between survival and age at diagnosis in breast cancer. New Engl J Med 1986; 315, pp. 559-563.
2. Bonnier P, Romain S, Charpin C, et al., Age as a prognostic factor in breast cancer: relationships to pathologic and biological features. Int J Cancer 1995; 62, pp. 138-144.
3. Cestnik B., Kononenko I., Bratko I., Assistant 86. A knowledge elicitation tool for sophisticated users, In: Bratko I., Lavrac N. (eds.), Progress in Machine Learning, Sigma Press, Wilmshow, 1987, pp. 31-45.
4. Clark P., Niblett T., The CN2 induction algorithm. Machine Learning, 3, 1989, pp. 261–283.
5. Collin F, Chassevent A, Bonichon F, Bertrand G, Terrier P, Coindre JM., Flow cytometric DNA content analysis of 185 soft tissue neoplasms indicates that s-phase fraction is a prognostic factor for sarcoma. Cancer 1997; 79, pp. 2371-2379.
6. Corle D, Sears M, Olson K., Relationship of quantitative estrogen-receptor level and clinical response to cytotoxic chemotherapy in advanced breast cancer. Cancer 1984; 54, pp. 1554-1561.
7. Dhingra K, Hortobagyi GN., Critical evaluation of prognostic factors. Sem Oncol 1996; 23, pp. 436-445.
8. Gitsch G, Sevelede P. Microvessel density and vessel invasion in lymph-node negative breast cancer: effect on recurrence free survival. Int J. Cancer 1995; 62, pp. 126-131.

9. Grzymala-Busse JW., LERS - a system for learning from examples based on rough sets. In: Slowinski R, editor. Intelligent decision support. Handbook of application and advances of the rough sets theory. Dordrecht, Boston, London: Kluwer Academic Publishers; 1992. pp. 3-18.

10. Grzymala-Busse J.W., Grzymala-Busse W.J., Zhang X.: Increasing sensitivity of preterm birth predication by changing rule strength, in: Proceedings of the VIIIth Intelligent Information Systems, Ustron 14-18 June 1999, IPI PAN Press, pp. 127-136.

11. Grzymala-Busse J.W., Hippe Z., A Search for the Best Data Mining Method to Predict Melanoma. In: Proc. of the Rough Sets and Current Trends in Computing Conference RSCTC'2002, Springer Verlag LNCS, 2002, 538-545

12. Harris AL, Nicholson S, Sainsbury JRC, Wright C, Farndon JR., Epidermal growth factor receptor and other oncogenes as prognostic markers. National Cancer Inst Monogr 1992; 11, pp. 181-187.

13. Heslin MJ, Lewis JJ, Nadler E, et al., Prognostic factors associated with long-term survival for retroperitoneal sarcoma: implication for management. J Clin Oncol 1997; 15, pp.2832-2839.

14. International Germ Cell Cancer Collaborative Group. International germ cell consensus classification: a prognostic factor – based staging system for metastatic germ cell cancers. J Clin Oncol 1997; 15, pp.594-603.

15. Kleist von S., Prognostic factors in breast cancer: theoretical and clinical aspects (review). Anticancer Res 1996; 16, pp. 3907-3912.

16. Komorowski J., Pawlak Z., Polkowski L. Skowron A., Rough Sets: tutorial, In: Pal S.K., Skowron A. (eds.), Rough fuzzy hybridization. A new trend in decision-making. Honkong, Springer-Verlag, 1999, p. 3-98.

17. Kononenko I., Bratko I., Kukar M., Application of machine learning to medical diagnosis, In: Michalski R.S., Bratko I, Kubat M. (eds.), Machine learning and data mining, John Wiley & Sons, 1998, p. 389-408.

18. Krusinska E, Slowinski R, Stefanowski J., Discriminant versus rough sets approach to vague data analysis. Applied Stochastic Model Data Analysing, 1992, 8, pp. 43-56.

19. Krysiński J., Rough sets in the analysis of the structure-activity relationships of antifungal imidazolium compounds. J Pharm Sci 1995; 84, pp. 243-248.

20. Krysiński J., Rough sets approach to the analysis of the structure-activity relationship of quaternary imidazolium compounds. Arzneim-Forsch Drug Res 1990; 40, pp. 795-799.

21. Maestu I, Pastor M, Codina-Gomez J, et al., Pretreatment prognostic factors for survival in small-cell lung cancer: a new prognostic index and validation of three known prognostic indices on 341 patients. Ann Oncol 1997; 8, pp.547-553.

22. McGuire WL, Clark GM., Prognostic factors and treatment decisions in axillary-node-negative breast cancer. New Engl J Med 1992, 326, pp. 1756-1761.

23. Pawlak Z., Rough sets. Theoretical aspects of reasoning about Data. Dordrecht, Boston, London: Kluwer Academic Publishers; 1991.

24. Pawlak Z., Slowinski K., Stefanowski J., Rough set theory in the analysis of medical data (In Polish: Teoria zbiorów przyblizonych w analizie danych medycznych). In: E.Kacki, J.L.Kulikowski, A.Nowakowski, E.Waniewski (eds.), Systemy komputerowe i teleinformatyczne w suzbie zdrowie, Tom 7 w serii Biocybernetyka i in+ynieria biomedyczna, Akademicka Oficyna Wydawnicza EXIT, Warszawa, 2002, pp. 253-268.

25. Predki B, Slowinski R, Stefanowski J., Susmaga R, Wilk S., Rough set data explorer. Bulletin of the International Rough Sets Society, 1998; 2, pp. 31-34.

26. Quinlan J. R., C4.5: Programs for Machine Learning. Morgan Kaufmann, San Mateo CA, 1993

27. Slowinski K., Stefanowski J., Siwinski D., Application of rule induction and rough sets to verification of magnetic resonance diagnosis, Fundamenta Informaticae, vol 53, no 3/4, 2002, 345-363.

28. Slowinski K, Stefanowski J., Multistage rough set analysis of therapeutic experience with acute pancreatitis, rough sets in knowledge discovery. In: Polkowski L, Skowron A, (eds.) Studies in Fuzziness and Soft Computing. Heidelberg, New York: Physica-Verlag, 1998, pp. 272-294.

29. Slowinski K., Stefanowski J., Medical information systems - problems with analysis and way of solution, In: Pal S.K., Skowron A. (eds.), Rough fuzzy hybridization. A new trend in decision-making. Honkong, Springer-Verlag, 1999, pp. 301-315.

30. Slowinski K., Stefanowski J., Twardosz W., Rough set theory and rule induction techniques for discovery of attribute dependencies in experience with multiple injured patients. Bulletin of the Polish Academy of Sciences, Technical Sciences, 1998, vol. 46, no. 2, pp. 247-263.

31. Skowron A., Boolean reasoning for decision rules generation, In: Komorowski J., Ras Z. (eds.) Methodologies for Intelligent Systems, LNAI 689, Springer-Verlag, Berlin, 1993, pp. 295-305.

32. Stefanowski J., Classification support based on the rough sets, Foundations of Computing and Decision Sciences, 1993, 18 (3-4), pp. 371-380.

33. Stefanowski J., On rough set based approaches to induction of decision rules. In: Rough Sets in Data Mining and Knowledge Discovery, Vol 1, Polkowski L., Skowron A., (eds.), Physica Verlag, Heidelberg, 1998, 500–529.

34. Stefanowski J., Algorithms of rule induction for knowledge discovery. (In Polish), Habilitation Thesis published as Series Rozprawy no. 361, Poznan Univeristy of Technology Press, Poznan, 2001.

35. Sutton L, Chastang C, Ribaud P, et al., Factors influencing outcome in de novo myelodysplastic syndromes treated by allogenic bone marrow transplantation: a long-term study of 71 patients. Blood 1996; 88, pp.358-365.

36. Tsumoto, S. Ziarko, W. Shan. N. Tanaka, H. Knowledge discovery in clinical databases based on variable precision rough sets model. In: Proc. of the Nineteenth Annual Symposium on Computer Applications in Medical Care, New Orleans, 1995. Journal of American Medical Informatics Association Supplement, 270-274.

37. Tsumoto S., Discovery of rules about complications. Proc. 7th Workshop New Directions in Rough Sets, Data Mining, and Granular-Soft Computing., In: (Zhong N., et al., eds.), Springer Verlag LNAI 1711, 1999, 29 - 37.

38. Weiss S.M., Kulikowski C.A., Computer Systems That Learn: Classification and Prediction Methods from Statistics, Neural Nets, Machine Learning and Expert Systems, Morgan Kaufmann, San Francisco, 1991.

39. Wigren T, Oksanen H, Kellokumpu-Lehtinen P., A practical prognostic index for inoperable non-small-cell lung cancer. J Cancer Res Clin Oncol 1997; 123, pp. 259-266.

Independent Component Analysis, Principal Component Analysis and Rough Sets in Face Recognition

Roman W. Świniarski[1,2] and Andrzej Skowron[3]

[1] Department of Mathematical and Computer Sciences
San Diego State University
5500 Campanile Drive San Diego, CA 92182, USA
rswiniar@sciences.sdsu.edu
[2] Institute of Computer Science, Polish Academy of Sciences
Ordona 21, 01-237 Warsaw, Poland
[3] Institute of Mathematics, Warsaw University
Banacha 2, 02-097 Warsaw, Poland
skowron@mimuw.edu.pl

Abstract. The paper contains description of hybrid methods of face recognition which are based on independent component analysis, principal component analysis and rough set theory. The feature extraction and pattern forming from face images have been provided using Independent Component Analysis and Principal Component Analysis. The feature selection/reduction has been realized using the rough set technique. The face recognition system was designed as rough-sets rule based classifier.

1 Introduction

Face recognition is one of the most classical pattern recognition projects. Many techniques have been developed for face recognition based on facial images [4,9,11,13–15,17–19]. One of the prominent methods is based on principal component analysis [11,17,18]. Despite of the fact that many face recognition systems have been designed, new robust face recognition methods are still expected. Independent Component Analysis (ICA) [2,3,5,8,16] is one of new powerful methods of discovery latent variables of data that are statistically independent.

We have tried to explore a potential of ICA methods combined with rough set methods (hybrid method) in face recognition. For comparison, we have also studied an application of classic Principal Component Analysis (PCA) nn-supervised technique combined with rough set method for extraction and selection of facial features as well as for classification. For feature extraction and reduction techniques from face images we have applied the following sequence of processing operations:

J.F. Peters et al. (Eds.): Transactions on Rough Sets I, LNCS 3100, pp. 392–404, 2004.

1. ICA for feature extraction and reduction, and pattern forming of facial images.
2. Rough set method for feature selection and data reduction.
3. Rough set based method for rule based classifier design.

Similar processing sequence has been applied for PCA for feature extraction and reduction, and pattern forming of facial images.

ICA based patterns, with feature selected by rough sets have shown significant predisposition for robust face recognition.

The paper is organized as follows. First, we present a brief introduction to rough set theory and its application to feature selection, data reduction and rule based classifier design. Then, we provide a short introduction to ICA and its application to feature extraction and data reduction. The following chapter briefly describes PCA and its application to feature extraction and reduction. The paper concludes with description of numerical experiments in recognition of 40 classes of faces represented by gray scale images [11] using proposed methods.

2   Rough Set Theory and Its Application of Rough Sets to Feature Selection

The rough sets theory has been developed by Professor Pawlak [10] for knowledge discovery in databases and experimental data sets. This theory has been successfully applied to features selection and rule based classifier design [4,6,7,10,12,13, 14]. We will briefly describe a rough set method.

Let us consider an *information system* given in the form of the decision table

$$DT =< U,\ C \cup D,\ V,\ f >,\tag{1}$$

where U is the *universe*, a finite set of N objects $\{x_1, x_2, ..., x_N\}$, $Q = C \cup D$ is a finite set of *attributes*, C is a set of *condition* attributes, D is a set of *decision* attributes, $V = \bigcup_{q \in C \cup D} V_q$, where V_q is the set of *domain* (*value*) of attribute $q \in Q$, $f : U \times (C \cup D) \to V$ - is a total *decision function* (information function, decision rule in DT) such that $f(x,\ q) \in V_q$ for every $q \in Q$ and $x \in U$.

For a given subset of attributes $A \subseteq Q$ a relation

$$IND(A) = \{(x,\ y) \in U :\ for\ all\ a \in A,\ f(x,\ a) = f(y,\ a)\},\tag{2}$$

is an *equivalence relation* on universe U, called the *indiscernibility relation*. By A^* we denote $U/IND(A)$, i.e., the partition of U defined by $IND(A)$.

For a given information system S a given subset of attributes $A \subseteq Q$ determines the approximation space $AS = (U, IND(A))$ of S. For a given $A \subseteq Q$ and $X \subseteq U$ (a concept X), the A-*lower approximation* $\underline{A}X$ of set X in AS and the A-*upper approximation* $\bar{A}X$ of set X in AS are defined as follows:

$$\underline{A}X = \{x \in U : [x]_A \subseteq X\} = \bigcup\{Y \in A^* : Y \subseteq X\},\tag{3}$$

$$\bar{A}X = \{x \in U : [x]_A \cap X \neq \emptyset\} = \bigcup\{Y \in A^* : Y \cap X \neq \emptyset\}.\tag{4}$$

A *reduct* is the essential part of an information system (related to a subset of attributes) which can discern all objects discernible by the original information system. We have applied of rough sets reduct in the technique of feature selection-reduction of face images.

From a decision table DT, decision rules can be derived. Let $C^* = \{X_1, X_2, \cdots, X_r\}$ be C-definable classification of U and $D^* = \{Y_i, Y_2, \cdots, Y_l\}$ be D-definable classification of U. A class Y_i from a classification D^* can be identified with the decision i $(i = 1, 2, \cdots, l)$. Then *ith decision rule* is defined by

$$Des_C(X_i) \Longrightarrow Des_D(Y_j) \text{ for } X_i \in A^* \text{ and } Y_j \in D^*. \tag{5}$$

These decision rules are logically described as follows:

$$\textit{if} \text{ (a set of conditions) } \textit{then} \text{ (a set of decisions)}.$$

The set of decision rules for all classes $Y_j \in D^*$ is denoted by

$$\{\tau_{ij}\} = \{Des_C(X_i) \Longrightarrow Dec_D(Y_j) : \text{ where } X_i \cap Y_j \neq \emptyset \text{ for } X_i \in A^*, Y_j \in D^*\}.$$

The set of decision rules for all classes $X_i \in D^*$, where $i = 1, 2, \ldots, r$, generated by the set of decision attributes D (D-definable classes in S) is called the *decision algorithm* resulting from the decision table DT.

In our approach C is a relevant relative reduct [1] of the original set of condition attributes in a given decision table.

2.1   Rough Sets for Feature Reduction/Selection

Many feature extraction methods for different types of row data (like images or time-series) do not guarantee that attributes (elements) of extracted feature patterns will be the most relevant for classification tasks. One of possibilities for selecting features from feature patterns is to apply rough sets theory [3,5,7,8]. Specifically, defined in rough sets computation of a reduct can be used for selection of some of extracted features constituting a reduct [6,7,10,12] as reduced pattern attributes. These attributes will describe all concepts in a training data set. We have used rough set method to find of reducts from the discretized feature patterns and to select features forming the reduced pattern based on chosen reduct. Choosing of one specific reduct from the set of reducts is yet another searching problem [4,6,7,12,23,14]. We have selected reduct based on dynamic reduct concept introduced in [1,3].

3   ICA – An Introduction

ICA is the unsupervised computational and statistical method for discovering intrinsic hidden factors in the data [2,3,5,8]. ICA exploits higher-order statistical dependencies among data and discovers a generative model for the observed multidimesional data. In the ICA model, observed data variables are assumed to be linear mixtures of some unknown independent sources (independent components). A mixing system is also assumed to be unknown. Independent components are assumed to be nongaussian and mutually statistically independent. ICA can be applied to feature extraction from data patterns representing time series, images or other media [2,5,8,16].

The ICA model assumes that the observed sensory signals x_i are given as the pattern vectors $\mathbf{x} = [x_1, x_2, \cdots, x_n]^T \in \mathbf{R}^n$. The sample of observed patterns are given as a set of N pattern vectors $T = \{\mathbf{x}_1, \mathbf{x}_2, \cdots, \mathbf{x}_N\}$, that can be represented as a $n \times N$ data set matrix $\mathbf{X} = [\mathbf{x}_1, \mathbf{x}_2, \cdots, \mathbf{x}_N] \in \mathbf{R}^{n \times N}$ which contains patterns as its columns. The ICA model for the element x_i is given as linear mixtures of m source independent variables s_j

$$x_i = \sum_{j=1}^{m} h_{i,j} s_j, \quad i = 1, 2, \cdots, n, \tag{6}$$

where x_i is observed variable, s_j is the independent component (source signals) and $h_{i,j}$ are mixing coefficients. The independent source variables constitute the source vector (source pattern) $\mathbf{s} = [s_1, s_2, \cdots, s_m]^T \in \mathbf{R}^m$. Hence, the ICA model can be presented in the matrix form

$$\mathbf{x} = \mathbf{H s}, \tag{7}$$

where $\mathbf{H} \in \mathbf{R}^{n \times m}$ is $n \times m$ unknown mixing matrix where row vector $\mathbf{h}_i = [h_{i,1}, h_{i,2}, \cdots, h_{i,m}]$ represents mixing coefficients for observed signal x_i. Denoting by $\mathbf{h}_{c,i}$ columns of matrix $\mathbf{H}$ we can write

$$\mathbf{x} = \sum_{i=1}^{m} \mathbf{h}_{c,i} s_i. \tag{8}$$

The purpose of ICA is to estimate both the mixing matrix $\mathbf{H}$ and the sources (independent components) $\mathbf{s}$ using sets of observed vectors $\mathbf{x}$.

The ICA model for the set of N patterns $\mathbf{x}$, represented as columns in matrix $\mathbf{X}$, can be given as

$$\mathbf{X} = \mathbf{H S}, \tag{9}$$

where $\mathbf{S} = [\mathbf{s}_1, \mathbf{s}_2, \cdots, \mathbf{s}_N]$ is the $m \times N$ matrix which columns correspond to independent component vectors $\mathbf{s}_i = [s_{i,1}, s_{i,2}, \cdots, s_{i,m}]^T$ discovered from the observation vector $\mathbf{x}_i$. Once the mixing matrix $\mathbf{H}$ has been estimated, we can compute its inverse $\mathbf{B} = \mathbf{H}^{-1}$, and then the independent component for the observation vector $\mathbf{x}$ can be computed by

$$\mathbf{s} = \mathbf{B x}. \tag{10}$$

The extracted independent components s_i are as independent as possible, evaluated by an information-theoretic cost criterion such as minimum Kulback-Leibler divergence kurtosis, negenropy [2,3,8].

3.1   Preprocessing

Usually ICA is preceded by preprocessing, including centering and whitening.
   Centering

Centering of $\mathbf{x}$ is the process of subtracting its mean vector $\boldsymbol{\mu} = E\{\mathbf{x}\}$ from $\mathbf{x}$:

$$\mathbf{x} = \mathbf{x} - E\{x\}. \tag{11}$$

Whitening (sphering)
   The second frequent preprocessing step in ICA is decorrelating (and possibly dimensionality reducing), called *whitening* [16]. In whitening the sensor signal vector $\mathbf{x}$ is transformed using formula

$$\mathbf{y} = \mathbf{W}\mathbf{x}, \quad so \quad E\{\mathbf{y}\mathbf{y}^T\} = \mathbf{I}_l, \tag{12}$$

where $\mathbf{y} \in \mathbf{R}^l$ is the $l - dimensional$ ($l \leq n$) whitened vector, and $\mathbf{W}$ is $l \times n$ whitening matrix. The purpose of whitening is to transform the observed vector $\mathbf{x}$ linearly so that we obtain a new vector $\mathbf{y}$ (which is white) which elements are uncorrelated and their variances are equal to unity. Whitening allows also dimensionality reduction, by projecting of $\mathbf{x}$ onto first l eigenvectors of the covariance matrix of $\mathbf{x}$.
   Whitening is usually realized using the eigen-value decomposition (EVD) of the covariance matrix $E\{\mathbf{x}\mathbf{x}^T\} \in \mathbf{R}^{n \times n}$ of observed vector $\mathbf{x}$

$$\mathbf{R}_{\mathbf{xx}} = E\{\mathbf{x}\mathbf{x}^T\} = \mathbf{E}_{\mathbf{x}}\boldsymbol{\Lambda}_{\mathbf{x}}^{\frac{1}{2}}\boldsymbol{\Lambda}_{\mathbf{x}}^{\frac{1}{2}}\mathbf{E}_{\mathbf{x}}^T. \tag{13}$$

Here, $\mathbf{E}_{\mathbf{x}} \in \mathbf{R}^{n \times n}$ is the orthogonal matrix of eigenvectors of $\mathbf{R}_{\mathbf{xx}} = E\{\mathbf{x}\mathbf{x}^T\}$ and $\boldsymbol{\Lambda}$ is the diagonal matrix of its eigenvalues

$$\boldsymbol{\Lambda}_{\mathbf{x}} = \mathrm{diag}(\lambda_1, \lambda_2, \cdots, \lambda_n), \tag{14}$$

with positive eigenvalues $\lambda_1 \geq \lambda_2 \cdots \geq \lambda_n \geq 0$. The whitening matrix can be computed as

$$\mathbf{W} = \boldsymbol{\Lambda}_{\mathbf{x}}^{-1/2}\mathbf{E}_{\mathbf{x}}^T, \tag{15}$$

and consequently the whitening operation can be realized using formula

$$\mathbf{y} = \boldsymbol{\Lambda}_{\mathbf{x}}^{-1/2}\mathbf{E}_{\mathbf{x}}^T\,\mathbf{x} = \mathbf{W}\mathbf{x}. \tag{16}$$

Recalling that $\mathbf{x} = \mathbf{H}\,\mathbf{s}$, we can find from the above equation that

$$\mathbf{y} = \boldsymbol{\Lambda}_{\mathbf{x}}^{-1/2}\mathbf{E}_{\mathbf{x}}^T\,\mathbf{H}\,\mathbf{s} = \mathbf{H}_w\mathbf{s}. \tag{17}$$

We can see that whitening transforms the original mixing matrix $\mathbf{H}$ into a new one, $\mathbf{H}_w$

$$\mathbf{H}_w = \boldsymbol{\Lambda}_{\mathbf{x}}^{-1/2}\mathbf{E}_{\mathbf{x}}\mathbf{H}. \tag{18}$$

Whitening makes it possible to reduce the dimensionality of the whitened vector, by projecting observed vector into first l ($l \leq n$) eigenvectors corresponding to first l eigenvalues $\lambda_1, \lambda_2, \cdots, \lambda_l$ of the covariance matrix $\mathbf{E_x}$. Then, the resulting dimension of the matrix $\mathbf{W}$ is $l \times n$, and there is reduction of the size of observed transformed vector $\mathbf{y}$ from n to l.

Output vector of whitening process can be considered as an input to ICA algorithm. The whitened observation vector $\mathbf{y}$ is an input to unmixing (separation) operation

$$s = \mathbf{B}y, \tag{19}$$

where $\mathbf{B}$ is an original unmixing matrix.

An approximation (reconstruction) of the original observed vector $\mathbf{x}$ can be computed as

$$\tilde{\mathbf{x}} = \mathbf{B}s, \tag{20}$$

where $\mathbf{B} = \mathbf{W}_w^{-1}$.

For the set of N patterns $\mathbf{x}$ forming as columns the matrix $\mathbf{X}$ we can provide the following ICA model

$$\mathbf{X} = \mathbf{B}\,\mathbf{S}, \tag{21}$$

where $\mathbf{S} = [s_1, s_2, \cdots, s_N]$ is the $m \times N$ matrix which columns correspond to independent component vectors $s_i = [s_{i,1}, s_{i,2}, \cdots, s_{i,m}]^T$ discovered from the observation vector $\mathbf{x}_i$. Consequently we can find the set $\mathbf{S}$ of corresponding independent component vectors as

$$\mathbf{S} = \mathbf{B}^{-1}\mathbf{X}. \tag{22}$$

3.2   Fast ICA

The estimation of the mixing matrix and independent components has been realized using Karhunen and Oja FastIca algorithm [16]. In computational efficient Karhunen and Oja ICA algorithm the following maximization criterion has been exploited

$$J(\tilde{\mathbf{s}}) = \sum_{i=1}^{m} |E\{\tilde{s}_i^4\} - 3[E\{\tilde{s}_i^2\}]^2|. \tag{23}$$

This equation corresponds to 4-th order cumulant kurtosis. ICA assumes that independent components must be nongaussian. The base for estimating the ICA model is assumption about nongaussianity. One can observe that that kurtosis is a normalized version of the fourth moment $kurt(y) = E\{(y^4)\}$. Kurtosis has been frequently used as a measure of nongaussianity in ICA and related fields. In practical computationally, kurtosis can be estimated by using the fourth moment of the sample data.

A negentropy, as robust measure of nongaussianity can be used for approximations of a kurtosis.

$$J(\mathbf{y}) = H(\mathbf{y}_{gauss}) - H(\mathbf{y}), \tag{24}$$

where $\mathbf{y}_{gauss}$ is a Gaussian random variable with the same covariance matrix as $\mathbf{y}$. Since it requires estimation of the probability density of y, hence the approximations of negentropy can be written as

$$J(y) \approx \frac{1}{12}E\{y^3\}^2 + \frac{1}{48}\,\text{kurt}(y)^2 \approx \tag{25}$$

$$\approx \sum_{i=1}^{p} k_i[E\{G_i(y)\} - E\{G_i(\nu)\}]^2, \tag{26}$$

where k_i are some positive constants, and ν is a standardized Gaussian variable of zero mean and unit variance. The variable $y = \mathbf{w}^T\mathbf{x}$ is assumed to be of zero mean and unit variance, and the functions G_i are some nonquadratic functions. The above measure can be used to test nongaussianity. It is always non-negative, and equal to zero if y has a Gaussian distribution.

If one uses only one nonquadratic function G, the approximation results in

$$J(y) \approx [E\{G(y)\} - E\{G(v)\}]^2. \tag{27}$$

Karhunen and Oja have introduced a very efficient method of maximization suited for this task. It is here assumed that the data is preprocessed by centering and whitening. The FastICA learning rule finds a direction, i.e. a unit vector $\mathbf{w}$ such that the projection $\mathbf{w}^T\mathbf{x}$ maximizes nongaussianity. Nongaussianity is here measured by the the criterion being an approximation of negentropy given in [2,5,8].

3.3   Feature Extraction Using ICA

In feature extraction which is based on independent component analysis [2,3,5,8] one can consider an independent component s_i as the i-th feature of the recognized object represented by the observed pattern vector $\mathbf{x}$. The feature pattern can be formed from m independent components of the observed data pattern. The use of ICA for feature extraction is partly motivated by results in neurosciences, revealing that the similar principle of pattern dimensionality can be found in the early processing of sensory data by the brain.

In order to form the ICA patterns we propose the following procedure:

1. Extraction of n_f element feature patterns $\mathbf{x}_f$ from the recognition objects. Composing the original data set T_f containing N cases $\{\mathbf{x}_{f,i}^T, c_i\}$. The feature patterns are represented by matrix $\mathbf{X}_f$ and corresponding categorical classes are represented by column $\mathbf{c}$.
2. Heuristic reduction of feature patterns from the matrix $\mathbf{X}_f$ into n_{fr} element reduced feature patterns $\mathbf{x}_{fr}$ (with resulting patterns $\mathbf{X}_{fr}$). This step could be directly possible for example for features computed as singular values of image matrices.
3. Pattern forming through ICA of reduced feature patterns $\mathbf{x}_{fr}$ from the data set $\mathbf{X}_{fr}$.
   (a) Whitening of the data set X_{fr} including reduced feature patterns of dimensionality n_{fr} into n_{frw} element whitened patterns $\mathbf{x}_{rfw}$ (projected reduced feature patterns into n_{frw} principal directions).

(b) Reduction of the whitened patterns $\mathbf{x}_{frw}$ into first n_{frwr} element reduced whitened patterns $\mathbf{x}_{frwr}$ through projection of reduced feature patterns into first principal directions of data.

4. Computing the unmixing matrix $\mathbf{W}$ and computing reduced number n_{icar} of independent components for each pattern $\mathbf{x}_{frwr}$ obtained from whitening using ICA (projection patterns $\mathbf{x}_{frwr}$ into independent component space).

5. Forming n_{icar} element reduced ICA patterns $\mathbf{x}_{icar}$ from corresponding independent components of whitened patterns, with the resulting data set $\mathbf{X}_{icar}$. Forming a data set T_{icar} containing pattern matrix $\mathbf{X}_{icar}$ and original class column $\mathbf{c}$

6. Providing rough sets based processing of the set T_{icar} containing ICA patterns $\mathbf{x}_{icar}$. Discretizing pattern elements and finding relative reducts from set T_{icar}. Choosing one relevant relative reduct. Selecting the elements of patterns $\mathbf{x}_{icar}$ corresponding to chosen reduct and forming the final pattern $\mathbf{x}_{fin}$. Composing the final data set $T_{final,d}$ containing discrete final patterns $\mathbf{x}_{fin,d}$ and class column. Composing the real valued data set T_{fin} from the set T_{icar} choosing elements of real-valued pattern using selected relative reduct.

3.4   ICA and Rough Sets

The ICA does not guarantee that selected first independent components, as a feature vector, will be the most relevant for classification. As opposed to PCA, ICA does not provide an intrinsic order for the representation features of a recognized object (for example an image). Thus, one cannot reduce a ICA pattern just by removing its trailing elements (which is possible for PCA patterns). Selecting features from independent components is possible through application of rough sets theory [3,5,7,8]. Specifically, defined in rough sets computation of a reduct can be used for selection some of independent components based attributes constituting a reduct. These reduct-based independent component based features will describe all concepts in a data set. The rough set method is used for finding of reducts from the discretized reduced ICA patterns. The final pattern is formed from reduced ICA patterns based on the selected reduct.

The results of discussed method of feature extraction/selection depend on a data set type and designer decisions: a) selection of dimension of the independent component space, b) discretization method applied, and (c) the selection of a reduct, etc.

4   ICA-Faces. Rough-Set-ICA-Faces

ICA tries to find a linear transformation of patterns using basis as statistically independent as possible, and to project feature patterns into independent component space. Figure 1 depicts an example of ICA-face. By applying rough sets processing one can find the final pattern composed with independent component patterns corresponding to elements of selected reduct. We can call this pattern a rough-set-ICA-face.

Fig. 1. ICA-face

5   Face Recognition Using ICA and Rough Sets

We have applied independent component analysis and rough sets for the face recognition. We have considered data set of images with 40 classes of 112×92 pixel gray scale images with 10 instances per class [11]. Figure 2 presents examples of face classes. We have applied the following method of feature extraction and pattern forming from 112×92 pixels gray scale face images [11]. The original feature patterns have been extracted from the lower 92×92 pixels sub-image of an original image. First, we have sampled that subimage taking its every second pixel and we have formed the sampled 46×46 pixels subimage. This subimage has been considered as a recognized object. The feature pattern has been created as a $n_{or} \times n_{or}$ ($n_{or} = 46 \times 46 = 2116$) element vector $\mathbf{x}_{orig} = [\mathbf{x}_{row}^{1}, \cdots, \mathbf{x}_{row}^{2116}]^{T}$ containing concatenated rows of a sampled subimage. As a result of that phase of the pattern forming it was obtained the $N \times (n_{or} + 1)$ feature data set T_f with N ($N = 400$) cases containing $n_{or} = 2116$ element feature patterns and associated classes $(\{\mathbf{x}_{or}^{i}\}^{T}, class\}$ ($class = 1, 2, \cdots, 40$).

Then we have applied ICA (including the whitening phase) to the pattern part $\mathbf{X}$ of the data set T_f in order to transform feature patterns into independent component space and to reduce pattern dimensionality.

Additionally the rough set technique has been applied for feature selection and data set reduction.

The original feature patterns have been projected into reduced 100 element independent component space (including 120 element whitening phase). Then we have found 14 element reduct from the ica patterns using rough set method and we have formed 14-element final patterns. The rough sets rule based classifier has provided 88.75% of classification accuracy for the test set.

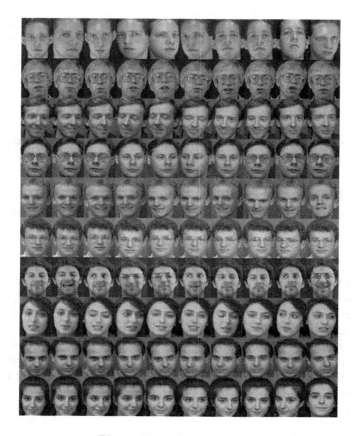

Fig. 2. Example of face classes

6   PCA for Feature Extraction and Reduction from Facial Images

Reduction of pattern dimensionality may improve the recognition process by considering only the most important data representation, possibly with uncorrelated elements retaining maximum information about the original data and with possible better generalization abilities. We have applied PCA, with the resulting Karhunen-Loéve transformation (KLT) [4], for the orthonormal projection (and reduction) of feature patterns $\mathbf{x}_f$ composed with concatenated rows of sampled face images.

6.1   PCA

We generally assume that our knowledge about a domain is represented as a limited size sample of N random n-*dimensional patterns* $\mathbf{x} \in \mathbf{R}^n$ representing extracted feature patterns. We assume that the pattern part of data set X of feature patterns can be represented by a $N \times n$ data pattern matrix

$\mathbf{X} = [\mathbf{x}^1, \mathbf{x}^2, \ldots, \mathbf{x}^N]$. The training data set can be characterized by the square $n \times n$ dimensional *covariance* matrix $\mathbf{R}_x$. Let the eigenvalues of the covariance matrix $\mathbf{R}_x$ are arranged in the decreasing order $\lambda_1 \geq \lambda_2 \geq \cdots \lambda_n \geq 0$ (with $\lambda_1 = \lambda_{max}$), with the corresponding orthonormal eigenvectors $\mathbf{e}^1, \mathbf{e}^2, \cdots, \mathbf{e}^n$. Then the optimal linear transformation

$$\mathbf{y} = \hat{\mathbf{W}}\mathbf{x}, \tag{28}$$

is provided using the $m \times n$ optimal Karhunen-Loéve transformation matrix $\hat{\mathbf{W}}$

$$\hat{\mathbf{W}} = \left[\mathbf{e}^1, \mathbf{e}^2, \ldots, \mathbf{e}^m\right]^T, \tag{29}$$

composed with m rows being the first m orthonormal eigenvectors of the original data covariance matrix $\mathbf{R}_x$. The optimal matrix $\hat{\mathbf{W}}$ transforms the original n-*dimensional* patterns $\mathbf{x}$ into m-*dimensional* $(m \leq n)$ feature patterns $\mathbf{y}$

$$\mathbf{Y} = (\hat{\mathbf{W}}\mathbf{X}^T)^T = \mathbf{X}\hat{\mathbf{W}}^T, \tag{30}$$

minimizing the mean least square reconstruction error. The PCA can be effectively used for the feature extraction and the dimensionality reduction by forming the m-dimensional $(m \leq n)$ feature vector $\mathbf{y}$ containing only the first m most dominant principal components of $\mathbf{x}$. The open question remains, which principal components to select as the best for a given processing goal [3]. One of possible methods (criteria) for selection of a dimension of a reduced feature vector $\mathbf{y}$ is to choose a minimal number of the first m most dominant principal components $y_1, y_2, \cdots, y_m$ of $\mathbf{x}$ for which the mean square reconstruction error is less than heuristically set the error threshold ϵ.

6.2   Numerical Experiment

The original face feature patterns have been projected into reduced 100 element principal component space.

Then we have found 19 element reduct from the PCA patterns using rough set method and we have formed 19-element final patterns. The rough sets based rule based classifier has provided 86.25% of classification accuracy for the test set.

6.3   Eigen-faces. Rough-Set-eigen-Faces

PCA of face feature patterns (obtained by concatenation of image rows) makes it possible to find the principal components of data set and to project feature patterns into principal component space. PCA intends to find the representation of patterns based on uncorrelated basis variables. The optimal Karhunen-Loeve transform matrix is composed with eigenvectors of data covariance matrix. Rows of optimal transform matrix represent orthogonal bases. Figure 3 depicts an example of eigen-face. By applying rough sets processing one can find the final pattern composed with reduced principal component patterns corresponding to elements of selected reduct. We call this pattern a rough-set-eigenface.

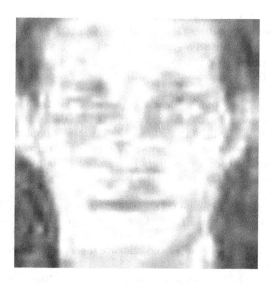

Fig. 3. Eigen-face

7 Conclusion

A proposed hybrid processing technique including ICA and rough sets has demonstrated significant predispositions for face recognition. The proposed method outperforms a hybrid method which comprises PCA and rough sets.

Acknowledgements

The research has been supported by the grant 3T11C00226 from Ministry of Scientific Research and Information Technology of the Republic of Poland.

References

1. Bazan, J., Skowron, A., and Synak, P.: Dynamic reducts as a tool for extracting laws from decision tables. In *Proc. of the Symp. on Methodologies for Intelligent Systems*, Charlotte, NC, October 16-19, 1994, Lecture Notes in Artificial Intelligence, 869 Springer-Verlag (1994) 346-355
2. Bell, A.J., Sejnowski, T.J.: An information-maximization approach to blind separation and blind deconvolution. *Neural Computation*, 7 (1995)1129-1159
3. A. Cichocki, R.E. Bogner, L. Moszczynski, Pope, A.: Modified Herault-Jutten algorithms for blind separation of sources. *Digital Signal Processing*, 7, (1997) 80 - 93
4. Cios, K., Pedrycz, W., Swiniarski, R.: Data Mining Methods for Knowledge Discovery. Kluwer Acad. Publ., Boston (1998)
5. Comon, P.: Independent component analysis - a new concept? *Signal Processing*, 36 (1994) 287-314

6. Grzymała-Busse, J.W.: Knowledge acquisition under uncertainty - A rough set approach. *Journal of Intelligent & Robotic Systems*, 1(1) 3–16

7. Grzymała-Busse, J.W.: LERS-a system for learning from examples based on rough sets. In: *Intelligent Decision Support. Handbook of Applications and Advances of the Rough Set Theory*, Słowiński, R. (ed.), Kluwer Academic Publishers, (1992) 3–18

8. Hyvrinen, A., Oja, E.: Independent component analysis by general nonlinear Hebbian-like learning rules. *Signal Processing*, 64(3) (1998) 301-313

9. Jonsson, J. , Kittler, J., Li, J.P. , Matas, J.: Learning Support Vectors for Face Verification and Recognition, Proceedings of the Fourth IEEE International Conference on Automatic Face and Gesture Recognition (2000) 26-30

10. Pawlak, Z.: Rough sets, Theoretical aspects of reasoning about data, Kluwer, Dordrecht (1991)

11. Samaria, F., Harter, A.: Parametrization of stochastic model for human face idntification. Proceedings of IEEE Workshop on Application of Computer Vision, 1994. ORL database is available at www.cam-orl.co.uk/facedatabase.html.

12. Skowron, A.: The rough sets theory and evidence theory. *Fundamenta Informaticae*, 13 (1990) 245-262

13. Swiniarski, R., Hargis, L.: Rough Sets as a Front end of Neural Networks Texture Classifiers. *Neuralcomputing Journal*. A special issue on Rough-Neuro Computing, 36 2001) 85-102

14. Swiniarski, R.: An Application of Rough Sets and Haar Wavelets to Face Recognition. In Proceedings of RSCTC'2000 In the Proceedings of The International Workshop on Rough Sets and Current Trends in Computing. Banff Canada, October 16-19 (2000) 523-530

15. Swiniarski, R. Skowron, A.: Rough Sets Methods in Feature Selection and Recognition. *Pattern Recognition Letters*. 24(6). (2003) 833-849

16. Swets, D.L, Weng. J.J.: Using discriminant eigenfeatures for image retrieval. *IEEE Trans. on Pattern Recognition and Machine Intelligence.* 10(9)(1996) 831-836

17. The FastICA MATLAB package.
Available at http://www.cis.hut.fi/projects/ica/fastica/

18. Turk, M.A. Pentland, A.P.: Face Recognition Using Eigenspaces. Proc. CVPR'91, June (1991) 586-591

19. Turk, M., Pentland, A.: Face recognition using eigenfaces. In: Proc. IEEE Conf. on Computer Vision and Pattern Recognition. (1991) 586-591

Author Index